Microcomputer Engineering

Second Edition

Microcomputer Engineering

Second Edition

Gene H. Miller

Kettering University
Flint, Michigan

PRENTICE HALL, Upper Saddle River, New Jersey 07458

Library of Congress Cataloging-in-Publication Data

Miller, Gene H.
    Microcomputer engineering/Gene H. Miller. —2nd ed.
      p. cm.
    Includes index.
    ISBN: 0-13-895368-6
    1. Microcomputers.   2. Assembler language (Computer program.
    language)   3. 68HC11 (Microprocessor) I. Title.
    TK7888.3.M548   1998
    004.165—dc21   98–24745
                   CIP

Publisher: **TOM ROBBINS**
Acquisitions editor: **ALICE DWORKIN**
Editorial Production: **BOOKMASTERS, INC.**
Editor-in-Chief: **MARCIA HORTON**
Assistant VP of production and manufacturing: **DAVID W. RICCARDI**
Managing editor: **EILEEN CLARK**
Full Service/Manufacturing Coordinator: **DONNA M. SULLIVAN**
Creative Director: **JAYNE CONTE**
Cover design: **BRUCE KENSELAAR**
Copy editor: **MARY GINN**

© 1999 by Prentice-Hall, Inc.
Upper Saddle River, New Jersey 07458

Printed in the United States of America

10 9 8 7 6 5 4 3

ISBN 0-13-895368-6

Prentice-Hall International (UK) Limited, *London*
Prentice-Hall of Australia Pty. Limited, *Sydney*
Prentice-Hall of Canada, Inc., *Toronto*
Prentice-Hall Hispanoamericana, S. A., *Mexico*
Prentice-Hall of India Private Limited, *New Delhi*
Prentice-Hall of Japan, Inc., *Tokyo*
Pearson Education Asia Pte. Ltd., *Singapore*
Editora Prentice-Hall do Brasil, Ltda., *Rio de Janeiro*

I dedicate this second edition to my wife *Margo*
whom I thank for all her love and encouragement

Contents

Chapter 5   Advanced Assembly Language Programming   187

Preface

To the Student

Microcomputer Engineering, Second Edition, was written for students studying engineering and related disciplines in a first microcomputers course. Knowledge of microprocessors and single-chip microcomputers is essential to the design of products, manufacturing equipment, and laboratory instrumentation. Many schools now require at least one microcomputers course that includes laboratory work with these devices. Practicing engineers with some digital systems background can learn this material without an instructor by using this book. The best background includes courses in electrical circuits, electronics, digital systems, and high-level language programming. Digital systems is the most important of these.

This textbook will help you learn the fundamentals of microcomputers; it is not a computer manual. Examples demonstrate conceptual topics, since most people learn examples before they can generalize. Small sections cover only a few details at a time. I introduce the computer instructions only when the topic requires them; a table of the instructions appears only in an appendix.

You should experience the computer by hands-on work with microcomputer training hardware. In using that hardware, you will see all its features simultaneously, and that is a problem. You cannot force parts of the hardware to stop operating until you learn about them; the hardware operates whether you know about it or not. You must learn to ignore those things that you are not yet prepared to comprehend. Similarly, in this book unnecessary details are ignored until you are ready for them. Implement all the small program examples on trainer hardware as you read for the best learning experience.

I recommend the Motorola M68HC11EVB or S68HC11EVBU microcomputer trainers for implementing the examples and exercises. Appendix C contains information about the M68HC11EVB trainer, most of which also applies to the S68HC11EVBU trainer. Additional reference material is available through the Internet on the World Wide Web (WWW) at www.prenhall.com/miller_microcomputer.

I sincerely hope that this book will be useful to students. If the opportunity arises, I will improve it based upon the recommendations and other feedback that I receive. I invite and urge you to write to me to comment on this book; student opinion about books is hardly ever heard, but always appreciated.

To the Instructor

Microcomputer Engineering is a complete course-teaching aid that encourages hands-on laboratory work. An associated WWW site at www.prenhall.com/miller_microcomputer supports the text material.

The Motorola 68HC11 single-chip microcomputer/microprocessor is the only hardware discussed. I recommend the Motorola M68HC11EVB or the S68HC11EVBU microcomputer trainer with an IBM PC compatible personal computer system running Windows NT for laboratory work. Considering the low cost of microcomputer trainers, I prefer them over software simulators. I also recommend the PFE freeware editor and the 2500AD assembler. Material related to this support equipment and software is provided on the Web site. Other equipment and software can be used, although with greater effort. The Web site also describes how to set up a complete laboratory.

The text makes teaching easy. The order of the chapters is the order I present the material. I want students working on a trainer as soon as possible, so I deliberately simplified the early topics. Reading assigned by section is possible since the sections are mostly independent of one another. Small laboratory/homework exercises are at the end of each chapter. Larger projects are on the Web site.

Chapter 1 reviews the digital systems and computer fundamentals required. The section on number systems is essential to understanding later chapters.

Chapter 2 introduces some of the microprocessor registers, the programming model, and the functions of these registers. Following this, some instructions and the addressing modes are introduced with straightforward examples. At the end of this chapter, a section discusses designing and writing a small machine language program.

I encourage laboratory/homework exercises on the trainer that demonstrate the instruction set, addressing modes, and a machine language program.

Chapter 3 changes direction and introduces assembly language. All examples thereafter use assembly language.

Motorola notation and terminology are used so you can reference Motorola manuals without confusion. Anything in program examples specifically related to the particular assembler program used for their development has been removed. Exercises in writing simple programs using an editor and assembler are appropriate, but only use the instructions from Chapter 2 at this point.

The student is now beginning to understand what a microcomputer is and how to program it. Because bad habits develop quickly, discussing program design before people have much programming experience is important. Chapter 4 concentrates on the established ideas for writing cost-effective and useful programs. Most microcomputer book authors avoid this topic apparently assuming that good software engineering is unnecessary; this author assumes exactly the contrary.

Chapter 5 introduces most of the remaining instructions using assembly language. Instructions are grouped according to their usual function such as bit manipulation, comparison, BCD arithmetic, and stack operations. Subroutines and several parameter-passing techniques are covered in detail.

Chapter 6 covers hardware, input/output concepts, and input/output programming. Microprocessor buses, memory and I/O chips, and input/output synchronization techniques are the main topics. Little electronics knowledge is required. The I/O section includes both polling and interrupt techniques. The interrupt material is quite extensive including discussion of concurrency and reentrancy problems and their solutions.

Chapter 7 covers the I/O hardware capabilities of the 68HC11 chip. Operation in both the single-chip microcomputer and expanded modes, chip versions, chip configuration, and pin connections are first. The next topics include hardware reset and details of the interrupt system. The following I/O hardware sections discuss the real-time clock, programmable timer, pulse accumulator, serial communications interface, and the analog-to-digital converter—all with programming examples. The next section on fail-safe operation discusses the COP timer, clock failure detection, and illegal instruction response. The hardware expansion section covers microprocessor bus extension, the SPI bus, and the port replacement unit. Finally, special hardware operations such as stopping the clock are covered.

Chapter 8 introduces the concept of an operating system and discusses the principles of real-time systems and multitasking systems.

Chapter 9 contains a major design example that encompasses much of the material in the first eight chapters. The goal is to develop a working product based on a relatively simple real-time multitasking operating system with preemptive priority scheduling. The product is a basic weather station that displays wind speed and direction on both digital displays and on a CRT. Design specifications for the weather station are stated first, then the complete operating system software is presented and discussed. Finally, all the tasks required to meet the weather station specifications are presented and discussed. The simplest working approach to the software is chosen in all cases. While reading the discussion of the software in this chapter, also "read the code."

Chapter 9 provides all the program code to make a working weather station. The required laboratory hardware is common to most student labs. The anemometer and wind vane are easily simulated with simple electronics if necessary.

Developing realistic student projects is often difficult because of the time required for students to get started. The design project in Chapter 9 provides a significant foundation for building additional projects. The projects can be modifications and improvements to the operating system, additions to the weather station, or development of entirely new applications of the operating system. In all these cases, the complexity is great enough to challenge students without requiring many hours of preliminary work. The modular nature of the multitasking operating system encourages student teamwork.

Appendix A documents the 68HC11 instruction set with an instruction set table, a reverse instruction set table, and op code maps. Appendix B tabulates the internal 68HC11A8 I/O hardware control registers. Appendix C explains the Motorola M68HC11EVB microcomputer trainer and includes a tutorial on its use. Much of this appendix should be useful even if other trainers are used. It's particularly useful to the student doing hands-on projects in a laboratory. Appendix D contains answers to selected exercises.

See the associated Web site for additional material such as documentation on other microcomputer trainers, the PFE editor program, and assembler programs; lab setup information; source code for the examples; and laboratory exercises. A glossary of terms used in the text is also available.

Acknowledgments

Special thanks are due to several people who directly helped me during the development of this book. First, I thank my wife, Margo, who encouraged me to produce a second edition even after enduring my efforts on the first edition. Next, I thank all the people at Prentice Hall, and especially Alice Dworkin, for supporting this effort. Thanks to Professor James McDonald at Kettering University, Harry Tyrer at University of Missouri at Columbia, and Eric Schwartz at University of Florida for their reviews. Finally, I thank my many students who have made my life both interesting and challenging.

I gratefully acknowledge the information in this book obtained from the manuals published by Motorola, Inc. The reader is encouraged to consult Motorola manuals for further information when needed.

Colophon

I created this entire book on a Dell XPS Pro200n personal computer containing a 200 MHz Intel Pentium Pro® microprocessor, 64-megabytes of memory, and a 9-gigabyte hard disk. A Nokia 21-inch monitor with a Matrox Millennium graphics board provided WYSIWYG page displays. The text was composed and spell-checked using the WordPerfect® word processor. I did the page layout and generated the table of contents and index with Ventura Publisher. Figures were drawn using GEM Draw or the Ventura Publisher table feature. Frequent backups of the hard disk were made to a Hewlett Packard DAT tape drive using the tape backup software provided with Microsoft's Windows NT® 4.0 Workstation operating system. The book was printed to disk as PostScript® files which were supplied to Prentice Hall.

Every effort was made to keep the size of the book the same as the first edition. To effect this, some material from the first edition has been moved to the Web site. The use of the Web also provides the opportunity to add some hardware-specific material that would have enlarged the book considerably.

Chapter 1

Computer Fundamentals

Microprocessors and microcomputers are so commonplace today that almost everybody, regardless of their education, is aware of them. Elementary school students in most developed countries routinely work with personal computers containing microprocessors. The microprocessor has become a fundamental device for engineering design. It is as common as the resistor, transistor, and circuit board have been in the past.

Studying microcomputers should be fun. Many people play with microprocessors as their hobby. These hobbyists generally are self-taught and have learned to program by trial and error rather than through an organized educational program.

Effort is required to learn the many fundamental concepts covered by this book; the fun comes when you apply them in the laboratory. Because the cost of quite powerful hardware is so low, you can probably experience the microcomputer yourself. Most of the material on microcomputers is very abstract, so working with actual hardware helps you learn the concepts; then the material is easy to learn and, to repeat, fun to learn.

Unfortunately, some people think of computers as trial-and-error devices. They don't think understanding is necessary. They believe that enough tries will result in a working system. Trial and error simply doesn't work when designing sophisticated, complex, and costly systems. Lack of understanding results in subtle bugs in software that testing may not find; some things happen only under obscure circumstances. Such bugs can be catastrophic. They lead to manufacturers recalling products to repair defects. Such recalls usually eliminate any possibility of a profit.

Your emphasis should be on a full understanding of the operation, design, and programming of the microcomputer. Use your hands-on experience to reinforce the concepts rather than to prove that you can make things work by trial and error. And by all means, have fun!

1.1 NUMBERS AND NUMBER SYSTEMS

The fundamentals of number systems are important to the use of computers. You will not need to calculate using binary and other numbers frequently—the machine will do that. The reason to study numbers is to better understand how the various kinds of numbers are formed and the limitations on their use. Binary numbers are especially difficult for humans to use, so a coding scheme is used to isolate us from the actual binary numbers.

Octal, Hexadecimal, and Binary Numbers

Let's begin with a review of decimal numbers, and then we can extrapolate to other number systems. Decimal, or base ten, numbers have digits that have ten different values called 0 through 9. The base is ten because there are ten different digit values.

Decimal numbers are built from the digits by placing them to the left or right of a decimal point. The position relative to the point is important, so the decimal number system is called a *weighted number system*. Each position as you move to the left has a weight or value that is the base ten raised to an integer exponent or power. The first position to the left of the point has weight of the base raised to the power zero, so its weight is always one regardless of the base. The next position to the left has weight of the base raised to the power one, and so on counting to the left. Fractional numbers have weights defined with the base to a negative exponent. For example, the decimal number 123.45 means

$$123.45_{10} = (1 \times 10^2) + (2 \times 10^1) + (3 \times 10^0) + (4 \times 10^{-1}) + (5 \times 10^{-2})$$

which can be written as

$$123.45_{10} = (1 \times 100) + (2 \times 10) + (3 \times 1) + (4 \times \frac{1}{10}) + (5 \times \frac{1}{100})$$

You will most frequently use integer numbers without a fractional part when programming microcomputers. Usually integer numbers are written without the point, which is assumed to be at the right end of the number.

Octal numbers

Next, extend the ideas illustrated with the decimal number system to the octal or base eight number system. Octal numbers use the digits 0 through 7.

Let's convert the number 123_8 to its decimal equivalent by using the definition of the number. Therefore,

$$123_8 = (1 \times 8^2) + (2 \times 8^1) + (3 \times 8^0) = 83_{10}$$

One serious difficulty with octal numbers is the lack of English words to describe them. No word is available for octal 100! Therefore, we simply use the words we have with a modifier. For octal 123, say *octal one hundred twenty-three*. Of course, you now must think about two number systems at once to understand the meaning. But decimal is second nature to us, so mixing two systems is easy after a little practice.

Hexadecimal numbers

Now consider the base 16, or hexadecimal, number system. Because our ordinary decimal digits do not have 16 values, we use letters of the alphabet. So the hexadecimal digits are 0 through 9 and A through F.

Hexadecimal numbers have the same language problem as octal numbers. So say the number 123_{16} as *hexadecimal one hundred twenty-three*, or just *hex one hundred twenty-three*. Hexadecimal numbers with letter digits usually are spoken by reciting the digits. For example, say *C three five F* for the number C35F.

Here is the conversion of 123_{16} to the equivalent decimal number:

$$123_{16} = (1 \times 16^2) + (2 \times 16^1) + (3 \times 16^0) = 291_{10}$$

Notice that 123_{16} is a much larger number than 123_{10} because the digits have much bigger weights.

Finally, the word is spelled "hex*a*decimal," not "hex*i*decimal." This common spelling error is often found in manuals and textbooks.

Binary numbers

Finally, let's look at the binary, or base 2, number system. The binary digits are very restricted in that they only have the values 0 and 1. Binary numbers also are weighted numbers, so the conversion to decimal is done as with other numbers. For example,

$$101_2 = (1 \times 2^2) + (0 \times 2^1) + (1 \times 2^0) = 5_{10}$$

A problem for humans using binary numbers is the communication of long yet practical numbers. If you try to read the number 1101010110010111 to someone, it would sound like *one one zero one zero one zero one one zero zero one zero one one one*. Reading the number as a decimal number is no better. Clearly, normal humans will not easily communicate large binary numbers in spoken, written, or keyboard form by these methods.

Consider an alternative representation of binary numbers. A straightforward approach is to convert the binary number to its equivalent number in the decimal number system. Then people could communicate using the decimal equivalents of the binary numbers. Unfortunately, conversion to decimal is quite difficult, especially with large numbers.

Converting binary numbers to other number systems

Consider both 3-bit and 4-bit binary numbers as Table 1-1 shows. The 3-bit numbers have eight different values ranging from 0 to 7 which match the octal digits. The 4-bit numbers have 16 different values ranging from decimal 0 to 15, or from hexadecimal 0 to F. The 4-bit numbers match the hexadecimal digits.

TABLE 1-1   NUMBER CONVERSIONS

Decimal	Octal	Binary 3-bit	Hexadecimal	Binary 4-bit
0	0	000	0	0000
1	1	001	1	0001
2	2	010	2	0010
3	3	011	3	0011
4	4	100	4	0100
5	5	101	5	0101
6	6	110	6	0110
7	7	111	7	0111
8	10	—	8	1000
9	11	—	9	1001
10	12	—	A	1010
11	13	—	B	1011
12	14	—	C	1100
13	15	—	D	1101
14	16	—	E	1110
15	17	—	F	1111

Let's convert a 16-bit binary number to octal by grouping the digits in bunches of three bits, and then writing the octal digits for the binary bunches.

$$1101010110010111 = 1\ 101\ 010\ 110\ 010\ 111 = 152627_8$$

Similarly, the same binary number can be written in hexadecimal by making bunches of four bits and writing the hexadecimal equivalents,

$$1101010110010111 = 1101\ 0101\ 1001\ 0111 = D597_{16}$$

You may wonder if the binary number was correctly converted to octal or hexadecimal. The conversion is correct, but only because the bases of all these number systems are a power of two.

Now we are mixing three number systems together. A person may refer to a 16-bit binary number by saying *hex one hundred twenty-three*. The pattern is decimal, the representation is hexadecimal, but the number is binary. An example of the conversion of the spoken version to binary is

$$hex\ one\ hundred\ twenty\text{-}three = 0000000100100011$$

The conversion from binary to octal or hex is so easy that you can do it in your head as you read the numbers! Therefore, binary numbers are commonly represented in octal or hexadecimal form. You will easily memorize the bit patterns for 3-bit and 4-bit numbers by using them frequently. When the numbers you work with are 8-bit and 16-bit numbers, the better choice is hexadecimal because the numbers fit better.

Hexadecimal number representation of binary numbers is used in the remainder of this book. Any exceptions are either obvious or the number base is noted by use of the appropriate subscript. Hexadecimal numbers are printed using the Helvetica typeface for further clarity. Also, hexadecimal representation is used regardless of the meaning of the number—think of hexadecimal as a pattern-matching scheme to make use of binary numbers easier.

Two's Complement Signed Numbers

You will need signed positive and negative numbers for many applications. Several techniques can represent signed numbers. However, the most used is the two's complement number system.

An important restriction

Because you will be writing two's complement numbers on paper, you must heed an important restriction on the use of these numbers. *All numbers used in a calculation must have the same number of bits.* That is, you must add an 8-bit number to another 8-bit number and get an 8-bit number. You cannot add an 8-bit number to a 6-bit number; instead, you must convert the 6-bit number to the equivalent 8-bit number and then add. Likewise, the sum of two 8-bit numbers is always an 8-bit number.

The size restriction is of no importance in the computer, and it is seldom mentioned in textbooks. Without exception, computer hardware will operate on numbers with the same number of bits. For example, 8-bit hardware can add two 8-bit numbers because eight adder circuits work at once. Adding other sizes of numbers with 8-bit hardware is not possible. Of course, you can write a program to use multiple 8-bit numbers to represent bigger numbers, but the hardware still inherently works on 8-bit numbers. So the restriction is only important when you write numbers on paper.

Working with two's complement numbers

Let's begin by looking at the construction of a two's complement number. All two's complement binary numbers have a sign bit and a numerical value. The left bit of the number is the sign bit.

Positive numbers.   Look first at the positive numbers only because they are easier to read. A positive sign is a binary zero in the left bit; the rest of the bits determine the numerical value. So the positive numbers appear to be unsigned binary numbers with at least one leading zero bit. Here are some examples:

$$00000101 = +5_{10} \qquad 01000100 = +68_{10}$$

Don't be confused by the similarity to unsigned numbers. The zero sign bit makes the number a positive number. When you write the decimal equivalent value of a positive two's complement number, you should always write it with a plus sign to remind yourself that it represents a positive number.

Negative numbers. The negative numbers in the two's complement number system are more difficult to read. They don't look like ordinary binary numbers. Let's approach the construction of the negative numbers in a strange way by first stating a rule to change the sign of a number.

The rule to change the sign of any two's complement number is a two-step algorithm as follows:

- *Step 1.* Complement ALL the bits of the number, including the sign bit. That is, change all the zeros to ones and the ones to zeros.

- *Step 2.* Add one to the result.

Now you can make a negative number by writing the positive number and changing its sign. Let's change the signs of the numbers $+5_{10}$ and $+68_{10}$. Beginning with the number $+5_{10}$,

- *Step 1.* Complementing 00000101 yields 11111010.

- *Step 2.* Adding one yields 11111011, or -5_{10}.

The second number is $+68_{10}$, so

- *Step 1.* Complementing 01000100 yields 10111011.

- *Step 2.* Adding one yields 10111100, or -68_{10}.

The resulting numbers do indeed have a one in the left bit and thus meet the definition of a negative number. Now look at how strange the rest of the number is. The negative two's complement numbers are difficult to read.

But you probably ask how you can tell if these numbers really are -5_{10} and -68_{10}. Let's see what happens if +5 is added to −5 and $+68_{10}$ is added to -68_{10}. Of course, you would like to get zero:

$$
\begin{array}{rr}
00000101 & +5 \\
+\ \ 11111011 & -5 \\
\hline
00000000 & 0
\end{array}
\qquad\qquad
\begin{array}{rr}
01000100 & +68 \\
+\ \ 10111100 & -68 \\
\hline
00000000 & 0
\end{array}
$$

As you see, the result of adding these numbers was indeed zero. Of course, getting zero depends upon the restriction that each number must have the same number of bits. Any carry outside the number is lost.

Of course, zero was expected! The rule to change the sign was derived to make this happen. The essence of the derivation is to ask: What number can be added to a positive number so the answer will be zero? When this problem is solved, you get the rule to change the sign.

The rule to change the sign applies to all two's complement signed numbers. Let's apply it to −5:

- *Step 1.* Complementing 11111011 yields 00000100.

- *Step 2.* Adding one yields 00000101, or +5.

Of course, the result is correct—changing the sign of −5 gave +5. Don't make the error of thinking that the rule only applies to positive numbers. It applies equally well to both positive and negative numbers.

Two's complement overflow.   Consider adding the following numbers. The first two are large positive numbers, and the second two are large negative numbers.

```
      01000101              10000100
   +  01000011           +  10111100
      10001000              01000000
```

Notice that the first result is negative and the second is positive! The only appropriate comment is that the answers are wrong. Adding two positive numbers cannot result in a negative sum, and adding two negative numbers cannot result in a positive sum. The problem is caused by answers too big to fit an 8-bit number. The name of this problem is *two's complement overflow error*. A two's complement overflow is really a sign error, although the rest of the answer is also incorrect. Checking only the sign of the answer will always detect this error regardless of what calculation, including changing the sign, you are doing.

The properties of zero.   Next let's apply the *change the sign* rule to the number zero, which you should recognize as a positive number.

- *Step 1.* Complementing 00000000 yields 11111111.

- *Step 2.* Adding one yields 00000000.

The result is still positive zero. This exercise illustrates an important property of the two's complement number system; that is, only one zero exists in the number system and it is positive. A negative zero does not exist. Because zero is an important number that is used frequently, a number system that has only one zero is convenient. Therefore, even though changing the sign of zero caused a two's complement overflow, we choose to ignore the error and define zero as a special number.

One other number is not affected by the change the sign rule; it is the most negative number, 10000000. This number, hex 80, does not change sign, because no positive number has a magnitude as large as its magnitude. Consequently, changing the sign of 80 truly causes a two's complement overflow error.

Subtraction.   You also can subtract two's complement numbers. A common error is thinking that subtraction must be done by changing the sign of the subtrahend and adding.

Of course, this solution is correct—but it is not the only solution. Let's subtract two numbers by both subtracting and adding the negative number:

$$
\begin{array}{rl} & 00000011 \quad +3 \\ - & 00000101 \quad +5 \\ \hline & 11111110 \quad -2 \end{array}
\qquad
\begin{array}{rl} & 00000011 \quad +3 \\ + & 11111011 \quad -5 \\ \hline & 11111110 \quad -2 \end{array}
$$

Both answers are correct. However, remember that a two's complement overflow can occur when two's complement numbers are either added or subtracted. Again, if the signs don't make sense, a two's complement overflow has occurred and the answer is wrong.

Sign extension. Converting an 8-bit two's complement number into an equivalent 16-bit number is often necessary. The conversion is necessary when you want to add or subtract mixed size numbers. The number with fewer bits must be extended to the same number of bits as the other number. The process of converting a number to the equivalent number with more bits is called *sign extension* because you need only reproduce the sign bit enough times to achieve the required number of bits. The result will always have the equivalent value. For example,

$$01001111_{8\text{-bit}} = 0000000001001111_{16\text{-bit}} = +79_{10}$$

$$10111000_{8\text{-bit}} = 1111111110111000_{16\text{-bit}} = -72_{10}$$

In each case, the sign bit was reproduced eight times at the left end of the number. Eight zeros were put to the left of the positive number and eight ones to the left of the negative number.

Modular number systems

Numbers are usually visualized as points along a line. The line begins at the origin and extends to infinity in both the positive and negative directions. The numbers in a modular number system are visualized as points along a circle that has a fixed size. If you count through these numbers far enough, you come back to your starting point. When using binary numbers with a fixed number of bits, the number of bits determines the size of the circle.

The numbers on the face of an analog clock illustrate a common modular number system. One hour past 12 o'clock is called one o'clock because 12 different hour numbers represent time—the clock is a modulo 12_{10} system. Similarly, many automobile odometers are modulo 100,000 systems because the odometer returns, or rolls over, to zero after the car travels 100,000 miles.

The two's complement number system is a modular number system. Figure 1-1(a) uses 8-bit numbers to show that half the circle is positive and the other half is negative.

The circular representation also helps you see the effects of changing the sign. An axis is drawn through the circle including the numbers 00 and 80 in Figure 1-1(b). Changing the sign of a number flips its location from a position on one side of the axis to the equivalent position on the other side. Notice that flipping the numbers 00 and 80 does not change their signs.

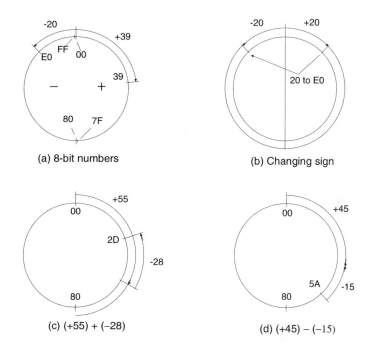

(a) 8-bit numbers

(b) Changing sign

(c) (+55) + (−28)

(d) (+45) − (−15)

Figure 1-1   Circular representation of 8-bit two's complement numbers.

Adding and subtracting can be visualized with line segments placed end to end. However, the line segments must fit on the same circle, which means the numbers must have the same number of bits. Figure 1-1(c) illustrates the addition of a positive and a negative number by placing the tail of one line segment at the head of the other.

Subtraction is more complicated. Draw the minuend beginning at zero, and then draw the subtrahend with its arrow head at the arrow head of the minuend. Figure 1-1(d) illustrates the subtraction of a negative number from a positive number.

The problem called two's complement overflow should now be apparent. For example, adding two positive numbers can result in a negative answer because the sum ends up on the wrong side of the circle. You should also see that adding a positive and a negative number can never cause a two's complement overflow.

The reason for two's complement numbers

It's easy to lose sight of the real reason for two's complement numbers as you do all the manipulations discussed above. The ultimate reason is that the computer hardware can treat them just like unsigned binary numbers. For example, adding and subtracting are no different for unsigned binary numbers and two's complement signed binary numbers. In the discussion above, the manipulations of the two's complement numbers treated the sign just like any other bit. In other words, we ADDED the signs! The hardware adds all the bits including the sign bits just as it would do with unsigned numbers. Different hardware is not required for each kind of number.

You should ask how unsigned and signed numbers are distinguished. The answer is that *you* must keep them straight! The computer hardware neither knows nor cares about the difference. The difference is in your interpretation of the numbers.

Don't think that you don't need to understand the differences between signed and unsigned numbers. You will find hardware in the computer specifically designed to work on only one type of number. For example, most computers have hardware to change the sign of a two's complement signed number. Trying to change the sign of an unsigned number results in garbage.

Binary-coded Decimal Numbers

Binary-coded decimal, or BCD, numbers are binary codes for decimal digits. Each digit of a decimal number is represented by four binary bits. For example,

$$73_{10} = 01110011_{BCD}$$

$$91_{10} = 10010001_{BCD}$$

To clarify the point, if each of these numbers is treated as an ordinary unsigned binary number and converted to decimal, the results will not be the same as the BCD value.

BCD coding is very convenient for certain input/output devices that work with decimal numbers. The device makes or uses a 4-bit code for each decimal digit.

The difficult part of using BCD numbers is making calculations with them. Consider the addition of two pairs of BCD numbers,

```
      01110011    73                01110011    73
   +  00000101     5             +  00011001    19
      01111000    78                10001100    8C
```

The first addition is straightforward, but the second one did not result in a proper BCD number. The least-significant digit is incorrectly greater than 9. The answer is converted to a correct BCD number by a simple rule. Add a number containing zero at each position with a correct BCD digit and a six at each position with an incorrect BCD digit. The following examples demonstrate the effect:

```
      01110011    73                01101001    69
   +  00011001    19             +  01000001    41
      10001100    8C                10101010    AA
   +  00000110    06             +  01100110    66
      10010010    92                00010000    10
```

The result of adding the correct number is that each digit is converted to a correct BCD form for 8-bit numbers. The conversion process works because the group of four bits has 16 values, while the decimal digit has only 10—adding a 6 adjusts for this difference. Of course, any carries that occur are included in the addition.

1.2   DIGITAL SYSTEM FUNDAMENTALS

You need some knowledge of digital systems design to gain full understanding of micro-computers. Later parts of this book refer to the fundamental concepts and other information reviewed in this section.

Signals, Functions, and Hardware

Digital electronic circuits, also called *logic circuits*, use binary signals. Proper notation to represent these circuits and signals is necessary to communicate ideas correctly. Also, the fundamental logic circuit devices must be understood. Making the logical connection between physical hardware and abstract notation is a very important task.

Binary signals

A *signal* is an electrical quantity that conveys information. Most electronic circuits use voltage as the signal. Usually signals require only very low power levels.

Analog circuits use the size of a voltage to represent a numeric quantity. Analog circuits have the practical problem of maintaining the accuracy of the signal under various circumstances. Changes in circuits due to temperature and aging effects cause inaccurate signal values. Electrical noise pollutes the signals and generally cannot be removed. Inaccurate components also cause incorrect signals.

Digital circuits use only two sizes of voltage to represent numeric quantities, so the signals are called *binary signals*. In digital circuits, close is good enough. If the actual signal voltage is close to the ideal voltage, the electronic circuit treats it as if it were ideal because only two ideal voltages exist. For the same reason, digital circuits have few problems due to inaccurate components and temperature and aging effects. Noise problems are dramatically reduced. Furthermore, digital circuits can be easily manufactured as integrated circuits. Integrated circuit technology is well suited to the manufacture of large quantities of simple circuits. Therefore, digital circuits can be made at very low cost, which makes them very practical.

In many applications, only two signal values cannot accurately represent the information. Then multiple signals are used to represent many more possible values. Of course, each additional signal requires an additional electronic circuit. The use of multiple circuits certainly increases the cost and reduces reliability of the system. Yet, the electronic hardware in integrated circuits is simple, is very reliable, does not require great accuracy in the components, and is available at low cost.

Binary signal notation

A binary signal has only two ideal voltages. The electronic circuit with a sensor shown in Figure 1-2 generates such a signal. The sensor is a simple toggle switch. The switch handle can be in either the up position or the down position. It is a binary switch because a mechanism prevents any other position. The mechanism operates switch contacts that operate an electronic circuit. The electronics generates the binary signal; when the handle is changed from one position to the other, the signal changes value.

Different digital circuits use different voltages. Some common DC voltage pairs are −12 and +12 volts, −5 and 0 volts, 0 and +5 volts, and 0 and +10 volts. Repeatedly stating these particular voltages makes discussing digital circuits difficult. To avoid this problem, a well-accepted naming convention distinguishes the two voltages without specifying the numerical voltage. The two voltages are called the *high level* and the *low level* and are defined as follows:

- *The high level is the more positive of the two voltages.*

- *The low level is the less positive of the two voltages.*

The words *more* and *less* are used in an algebraic sense—the sign of the voltage is significant. So −12 volts is less positive than 0 volts, though the magnitude is larger. Most people now describe the operation of digital electronic circuits using the words *high* and *low*.

Binary variables

A binary variable is an algebraic variable that can have only two values. Many names are used for the two values: *true and false*, *one and zero*, *operated and normal*, *asserted and deasserted*, or *active and inactive*. Although one pair of names may be better suited for a certain application, they all mean the same thing. The words *true*, *one*, *operated*, *asserted*, and *active* are equivalent. Also, *false*, *zero*, *normal*, *deasserted*, and *inactive* are equivalent.

Binary variable notation

A binary variable is usually named by a symbol that reflects its application. For example, the variables SW1, SW2, and LT1 may represent the conditions of two toggle switches and an indicator light. Each of these devices has two useful states. If the switch is a vertical

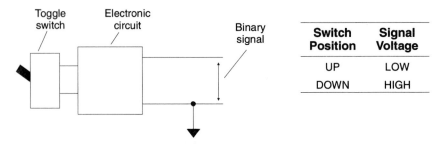

Figure 1-2   Binary signals.

toggle switch, the up position is the asserted condition and the down position is the deasserted condition. A push button switch is asserted when the button is pushed, and deasserted when the spring returns the device to its normal condition. A light is asserted when illuminated and deasserted when dark.

Tying variables to signals

To correctly use digital hardware, abstract binary variables must be related to physical hardware. Several types of notation are used to make this connection. Probably the best notation is *mixed-logic* notation. This notation encompasses all configurations of hardware and the other notations in one simple notation. More restrictive notations only apply correctly to certain hardware, so the logical meaning must be adjusted to fit the hardware.

Mixed-logic notation.   The purpose of logic notation is to describe the binary event that controls a binary signal generated by hardware. A binary variable represents the binary event. A logic level, high or low, represents the binary signal. Mixed-logic notation is a way of labeling signal wires with both the binary variable and the corresponding binary signal level. *The wire is labeled with a binary variable and the voltage level that corresponds to the true or asserted case of the variable.*

Figure 1-3 illustrates the notation. The wire from logic Device 1 carries a signal representing the variable RESET such that a true or asserted RESET variable makes the wire low as designated by the letter L in parentheses. Consequently, a false or deasserted RESET makes the wire high. The wire from logic Device 2 represents START where an asserted START signal makes the wire high as designated by the letter H. Therefore, a deasserted or false START signal makes the wire low. The notation always logically connects the true case of the labeled variable with the corresponding signal voltage.

The example shows that some devices use *high-asserted* signals while others use *low-asserted* signals. There are valid and useful reasons for a single circuit to use both high-asserted and low-asserted signals. Usually, designing circuits where the assertion level is chosen for convenience will be easier. However, the asserted signal level is sometimes chosen for safety or for compatibility with other equipment.

Motorola logic notation.   Motorola product documentation usually does not follow the mixed-logic notation. Sometimes Motorola labels hardware only with a name that describes the function performed. Often, you can interpret a Motorola label as a binary variable.

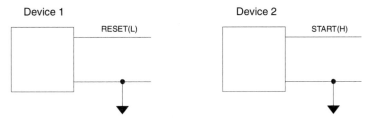

Figure 1-3   Mixed-logic notation.

However, further study of the documentation is often necessary to determine the associated voltage level.

Sometimes Motorola labels binary variables with a bar on top, such as $\overline{PB1}$. Interpret this notation to mean that the true case of the variable is associated with the low voltage. Then, the lack of a bar on a variable implies that the true case is associated with the high level. The notation $\overline{PB1}$ probably means that a signal wire is low if push button switch 1 is pushed. Someone can mistakenly omit the bar and change the assertion level of the signal with this notation very easily.

Logic functions

The fundamental logic functions are AND, OR, and COMPLEMENT. Using the word NOT when describing the complement function is convenient. Truth tables for the logic functions are shown in Table 1-2. The table also shows logic equations for the functions using the distinctive notation for the logical operators. Functions of two variables are shown, but these can be expanded to any number of variables. The derived function EXCLUSIVE OR (EOR) is shown because computer programs frequently use this function. If you compare the EOR function to the OR function, you see that EOR excludes the both case—the EOR function is zero when both variables are ones.

Be careful! The bar on top of a variable in a logic equation always means the complement of the variable. However, some companies use a bar on top to mean that a signal is low-asserted. Don't confuse a binary signal that is low-asserted and the complement of a binary variable.

Logic hardware

Although the fundamental logic hardware is very easy to understand, describing it and drawing circuit diagrams can be confusing because many different notations are used. Probably, the best notation available is one that clearly illustrates both the logic function and the hardware levels used. Such notation is compatible with the mixed-logic notation for binary signals. Only IEEE standard symbols are used in this book. However, Motorola manuals sometimes use a different notation, so that variation is also discussed.

TABLE 1-2   LOGIC FUNCTIONS

AND			OR			COMPLEMENT		EXCLUSIVE OR		
X	Y	F1	X	Y	F2	X	F3	X	Y	F4
0	0	0	0	0	0	0	1	0	0	0
0	1	0	0	1	1	1	0	0	1	1
1	0	0	1	0	1			1	0	1
1	1	1	1	1	1			1	1	0
$F1 = X \wedge Y$			$F2 = X \vee Y$			$F3 = \overline{X}$		$F4 = X \veebar Y$		

Device symbols.   Logic *gates* are electronic hardware devices that perform the fundamental logic functions on binary signals. Distinctive symbols that depend on shape denote the circuits that do the fundamental functions. A shape is used for AND, another for OR, and a third shape is used for an identity element. The identity element simply passes the signal through from the input to the output. Its logic function is to reproduce its input at its output. Its circuit function may be to reproduce the signal at a higher possible current level or at a different logic voltage level. Here, from left to right, are the most commonly used symbols for AND, OR, and identity function devices:

Note that no input or output signal wires are shown on these symbols. The reason is that the inputs and outputs require binary signals that are asserted or active at a certain voltage level. One device may require a high level to assert an input, while another device requires the low level for assertion. Additional standard symbols can designate which is the asserted level for each input and output wire.

Polarity indicators.   To designate high-asserted inputs or outputs, show a wire connected to the symbol with no additional symbol. To designate low-asserted inputs or outputs, show a wire with a *polarity indicator*. The standard symbol for a polarity indicator is a right triangle. *The presence of the polarity indicator means low-assertion, and the absence of the polarity indicator means high-assertion.* Here are examples of devices with high-asserted inputs and low-asserted outputs:

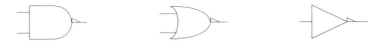

The standard polarity indicator is the triangular symbol. However, many people still use an older symbol described as a *bubble*. The following devices also have high-asserted inputs and low-asserted outputs, but use the bubble notation:

In this book, these two variations of polarity indicators are interchangeable. When circuit drawings are shown, the standard polarity indicator is used simply because it is the standard. However, Motorola manuals use the bubble symbol. Interpret bubbles in Motorola manuals as indicators of low-assertion.

To clarify the notation, Figure 1-4 describes two circuit devices by tables of voltages illustrating the input to output relationships. The figure also shows the corresponding circuit

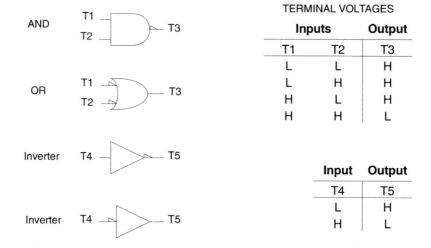

Figure 1-4   Two symbols describe each logic gate.

symbols. Two different symbols correctly describe each logic device, as you can see by examining the figure carefully. The first device can be described by these sentences:

- Terminal 3 is low if both terminal 1 AND terminal 2 are high.

- Terminal 3 is high if either terminal 1 is low OR terminal 2 is low OR if both terminals 1 and 2 are low.

Similarly, the identity element has two symbols. The particular device shown in the figure is called an *inverter* or *level changer* because the output voltage level is opposite the input voltage level.

When designing logic circuits, you must select the symbol that conveys the intent of the circuit. If the device is used for ANDing, draw an AND symbol. If it is used for ORing, then draw the OR symbol.

As a final note, if you have learned logic symbols where the bubble is described by the word NOT, carefully consider changing your point of view. You will find the polarity indicator approach to be much better as you work with practical logic circuits. Most companies are using and many new textbooks are written with this better notation.

Combinational Logic Networks

A *combinational logic network* (CLN) makes output signals now that are a function only of the input signals now. A CLN, which is built from a collection of AND, OR, and inverter devices, performs a more complicated logic operation than does a single device. The CLN outputs are determined solely by the combination or pattern of the current input signals. Time is not relevant to combinational logic networks—assume that the circuits respond instantaneously. The response is not instantaneous, but this assumption is practical for most applications. Combinational logic networks are simpler than the alternative, the synchronous sequential network, which has outputs now that depend on the inputs sometime in the past.

CLN example

The example circuit is a controller for the dome light in a two-door automobile. The purpose is to turn on the dome light if the left door is open or if the right door is open. Consequently, the light is on when both doors are open. A logic equation that describes this problem with binary variables is

$$\text{LIGHT} = \text{LDOPEN} \lor \text{RDOPEN}$$

where the definition of the binary variables should be apparent. This equation describes the logical operation of the circuit without any information about the physical circuit.

Figure 1-5 is a complete circuit drawing of the combinational logic circuit. The designer had a supply of the two hardware devices shown in Figure 1-4 for building the circuit. The hardware uses +5 V_{DC} as the high level and zero volts as the low level. The circuit drawing shows an OR symbol for the logic gate because the problem is described by an equation using OR.

Carefully consider how the door position-sensing switches work. A closed door presses on the push button and operates the switch. The switches are drawn in their normal positions with the doors closed. When a door is opened, the spring in the switch moves the mechanism so the switch terminals are electrically connected. The fact that the two switches are wired differently is strange and not very practical, but the mixed-logic notation clarifies the wiring difference easily. For example, when the left door is open, the switch connects the signal wire to the common ground making the wire low. When the left door is closed, the switch breaks the connection, so the resistor pulls the wire to the high level. The inverter that controls the light is designed to handle the current required for the light; typical integrated circuit inverters would not have the required current capability.

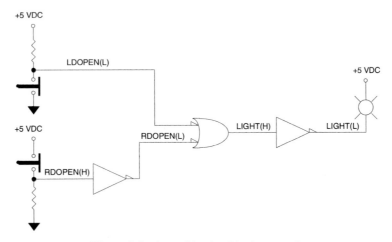

Figure 1-5   A combinational logic network.

Flip Flops and Registers

Complex digital circuits require the ability to remember or store binary information so it can be used later. Flip flops and registers are the basic devices for storing information. The addition of these devices to a circuit makes the circuit much more complex than a CLN because time must then be considered.

Flip flops

A *flip flop* is a 1-bit memory device. It remembers a single binary signal. The most common flip flop in computer systems is the D-flip flop. Figure 1-6 illustrates the symbols for two D-flip flops and the transition table that describes their operation. The input signal that is to be remembered is applied to the D input, and the remembered signal is the output at Y.

The D-flip flop is a *clocked* flip flop. A clocked flip flop can change its output only if a timing signal called the *clock* activates the flip flop. To activate the flip flop, the clock signal must be changed from one level to another; time cannot be indicated with a static signal. The transition table in Figure 1-6 shows how the D-flip flop acts when it changes from the current state to the next state at the clock change.

The output of a flip flop responds to its input signal or signals when the clock signal changes from the logic zero level to the logic one level, but it does not respond on the opposite change. The change that causes the flip flop to respond is called the *active transition* of the clock signal.

Dynamic indicators.   An input to a device, such as a clock signal, is designated as *transition sensitive* by a standard symbol called a *dynamic indicator*. The symbol is a small equilateral triangle as shown on the clock leads in Figure 1-6. The dynamic indicator says the active transition is a change from logic zero to logic one.

In Figure 1-6, the left flip flop clock has both a polarity indicator and a dynamic indicator, so its output responds when the clock lead is changed from high to low. The right flip flop does not have a polarity indicator, so it responds to a change from low to high.

Registers

A *register* is a collection of flip flops that are clocked as a unit. Together, the flip flops remember several binary signals that together represent a binary number. Figure 1-7 illus-

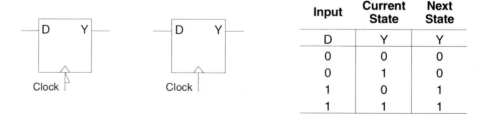

Input	Current State	Next State
D	Y	Y
0	0	0
0	1	0
1	0	1
1	1	1

Figure 1-6   Two D-flip flops and their transition table.

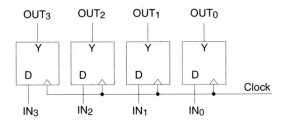

Figure 1-7 4-bit register using parallel transfers.

trates a 4-bit register. This register stores all the bits of the input number at once, and all the bits in the register can be read at once on the output wires. Storing or reading all the bits at once is called a *parallel transfer*.

Shift registers. Some registers are little more than a collection of flip flops with a common clock lead. Others contain control hardware to enable internal operations on the number stored in the register—a shift register is an example. A *shift register* can move the bits in the register one position at a time. For example, the register may shift all its bits left one position for each active clock transition. Usually, a new bit value is supplied to the hardware to replace the right bit, so a new number can be stored in the register one bit at a time. Similarly, the bits in the register may be read one bit at a time by reading the left bit as the register is shifted left. Loading or reading the register one bit at a time is called a *serial transfer*. Many varieties of shift registers are available, including some that can shift either left or right according to a control signal. Some can also be parallel loaded or read.

Load-controlled registers. The register shown in Figure 1-7 does a parallel load anytime that the register is clocked. Sometimes a register has extra control circuitry to allow the register to retain its information and ignore the input number even if it is clocked. This is called a *load-controlled register*, and it has an additional input signal called *load*. When the load input is asserted or active, the register loads from the input on an active clock transition; otherwise, the number in the register is unaffected.

Register symbols. The choice of a symbol to represent a register depends on the amount of detail needed. Sometimes every signal wire must be shown. But more frequently, the symbol need only show the number in the register or whether the data transfer is parallel or serial.

Figure 1-8 illustrates some ways of representing registers. The first drawing shows all the details. The second uses boxes to represent the flip flops and 1s and 0s in the boxes to show the states of the flip flops. The third drawing is simplified by showing a box for the entire register without showing the individual flip flops. And finally, the last drawing is the most common notation where the number stored in the register is labeled as a hexadecimal number with only the outline of the register shown.

The figure also shows a notation for parallel and serial transfers. If a register can do both serial and parallel transfers, both notations may be combined. For example, combining a serial load with a parallel read of a register is common practice. Such a register is described as *doing a serial-to-parallel conversion.*

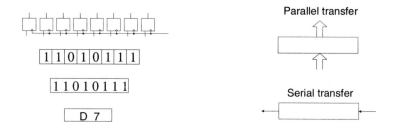

Figure 1-8 Register notation and serial and parallel transfers.

Buses and transfer gates

A *bus* is a group of binary signal wires that collectively carry a binary number. A bus may be used to parallel transfer a number from one register to another. Figure 1-9 shows a 3-bit bus bringing a number to three AND gates, and a 3-bit bus is connected to the outputs of the three AND gates.

A transfer gate is an AND gate used to pass a binary signal to another wire, or to block it. Consider the three AND gates in Figure 1-9. When the input labeled TRANSFER is asserted or true making the wire high, the output of each AND gate is the same as its other input—low in gives low out, and high in gives high out. When TRANSFER is not asserted making the wire low, the output of the AND gate is always low, in effect blocking the transfer of the input signal. In this example, use high for logic 1 and low for logic 0 so a deasserted transfer lead causes the outputs of the AND gates to be all 0s. The three transfer gates together either transfer the number from the input bus to the output bus, or they put the number zero on the output bus.

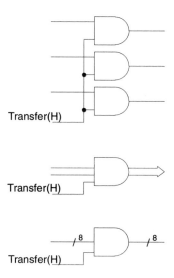

Figure 1-9 Buses and transfer gates.

Figure 1-9 also shows a symbol that represents any number of transfer gates with an input and output bus. The double line is commonly used to represent a bus. Sometimes the double line has a large arrow head to indicate the direction of signal flow if it is not apparent. An alternative bus symbol is a single line labeled with the number of wires in the bus.

Memory

A *memory* is a device that stores many binary numbers that can be accessed electronically. Many types of memories using a variety of technologies are available. The numbers may be stored by physical wiring, magnetic fields, electric fields, or the active operation of an electronic circuit. Later chapters discuss some of these technologies further. Therefore, a memory will be defined now as a collection of registers made from flip flops.

Memory signals

The registers within a memory need to be accessed only one at a time. Furthermore, a number may be loaded or written into the register or a number may be read from the register, but writing to the register and reading its contents at the same time is never necessary. Due to these restrictions, the memory hardware is very simple.

Because only one register is accessed at a time, each register is assigned a binary code number by the memory hardware. The number distinguishes each register from the others. The code number is called an *address*. The memory address is similar to the postal address on a mailbox.

A memory is a package that interacts with other devices through several groups of binary signals. The signal wires include those to carry a data number, those to select or address the register, those to determine whether read or write will happen, and a timing signal to tell the memory when to respond to the other signals. The wires that carry the data number to or from the memory are called the *data bus*. The wires that carry the address to the memory are called the *address bus*. Figure 1-10 shows the binary signals on a block diagram.

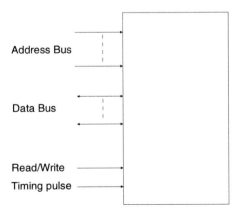

Address Bus

Data Bus

Read/Write
Timing pulse

Figure 1-10   Memory signals.

When another device makes signals to write a number into a memory register, we speak of that device as *storing* to the memory. The other device must supply an address on the address bus, the data number on the data bus, a write signal, and the timing pulse to make the memory write.

When the other device reads a number from a memory register, we speak of *loading* a number from the memory into the other device. The other device now supplies an address, a read signal, and a timing pulse. During a read operation, the memory unit puts the data number on the data bus. So numbers travel on the data bus in two directions—the read/write signal controls the direction.

Memory model

Initially you will not need to know all the signal details to use a memory. An abstract picture called a *model* is an adequate representation of a memory. Figure 1-11 shows a memory model. In this figure, each box represents one 8-bit register containing a hexadecimal number. The 16-bit address of each register is the hexadecimal number shown to the left of each register. The memory model shows none of the control signals. Usually a memory model will not show all the registers in a memory and therefore will have breaks.

Words and bytes

When a number is read from or written to the memory, a certain number of bits are transferred at once in parallel. This collection of bits is called a *word* of memory; the word size is the same as the size of a register. So the *word size* is the length of a register measured in bits. In effect, a word is a group of bits that are accessed together.

However, another word is used to denote a contiguous group of bits; the word is *byte*. Originally, the word *byte* meant any size group of bits. Usually, these bits are part of some larger number. The most common size group is eight bits. Through common usage, the word *byte* has evolved to mean a group of eight bits.

Because many memories are built with 8-bit registers with a word size of 8-bits or one byte, the word *byte* and the word *word* have come to have the same meaning to many people. Be careful! Not everyone uses these words as synonyms. Many memories have 16-bit or 32-bit words.

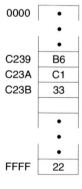

Figure 1-11   Memory model.

Memory length

Each register within a memory has a unique address code. The address is a binary number. The number of combinations of the bits within a number is 2 raised to the power N where N is the number of bits in the number. For example, an 8-bit number has 2^8, or 256_{10}, combinations. A 10-bit number has 1024_{10} combinations. A 16-bit number has $65,536_{10}$ combinations. Therefore, a memory with a 10-bit address cannot have more than 1024_{10} registers.

The number of registers in a memory device will almost always be a power of two. For example, if a memory is designed for a 10-bit address, it will have 1024_{10} registers. To include fewer registers is to waste some of the capability of the hardware.

The number 1024_{10} is close to the number 1000_{10}, which engineers and scientists describe by the prefix *kilo*, so people who work with digital systems have redefined kilo to mean 1024_{10}! The abbreviation for kilo is the capital letter K.

A memory with 1024_{10} 8-bit registers is described as a 1-kilobyte memory. Furthermore, a memory with $65,536_{10}$ 8-bit registers is called a 64-kilobyte memory. The memory shown in Figure 1-11 is a 64-kilobyte memory because the addresses have 16 bits.

Sometimes people also use kilo with bit in describing memories, so a memory device may have 256_{10} kilobits of memory. The abbreviation for kilobyte is KB, and the abbreviation for kilobit is Kb. It is easy to overlook whether the B is capitalized and get the wrong meaning from the abbreviation.

Synchronous Sequential Networks

A *synchronous sequential network* (SSN) is a circuit made from a combinational logic network and clocked flip flops. The output signals from a SSN depend on both current input signals and the input signals in the past. Control systems have the same requirements—they must have information about both the past and the present condition of the device being controlled. A mathematical point of view holds that this requirement is the essence of differentiation and integration. So, SSNs are essential to the design of digital control systems. Synchronous sequential networks are also interesting because a computer is a very large and sophisticated SSN. Most of the devices connected to computers also contain SSNs.

SSN block diagram

The block diagram shown in Figure 1-12 illustrates the configuration of a synchronous sequential circuit in the most general way. This block diagram represents any SSN including some trivial cases. The essence of it is the combination of a combinational logic circuit with memory devices or flip flops.

State of the network.   The *state* of a synchronous sequential network is the combined states of the flip flops. Together, the flip flops represent stored information. Individually, the outputs of the flip flops are called *state variables* or feedback variables.

External inputs.   The external inputs, or just *inputs*, are the signals used by a network in generating next state and output information. Similarly, the state variables are sometimes

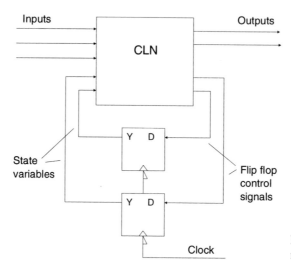

Figure 1-12 Synchronous sequential
network block diagram.

called *internal inputs* because they also affect the next state and outputs. Both are inputs to the combinational logic network.

External outputs.   The external outputs, or just *outputs*, are the control signals generated by the synchronous sequential network; they are the purpose for having the SSN. External outputs clearly are signals that go outside the SSN. The external outputs depend upon the current external inputs and the current state of the network—the inputs and the current condition of the flip flops.

Flip flop control signals.   The combinational logic network generates the *control signals* for the flip flops. The control signals determine the information to be stored at the time of the next active clock transition. Together, the external inputs and the current state of the network determine what information will be stored. Therefore, the next state of the SSN depends on the current inputs and the information stored in the flip flops in the past.

The clock.   The *clock* signal is usually a periodic or repeating pulse or square wave signal. The clock signal provides timing so the sequential network can operate properly.

When the active transition of the clock signal occurs, the outputs of the flip flops can change. Changes will propagate through the CLN to the input side of the flip flops at various speeds. The characteristics of the circuit components at the time of the discussion determine the speed. These characteristics can vary depending upon temperature, replaced parts, or even changed position of adjacent components. All signal changes must reach steady state before the next active clock transition. If the memory control signals were still changing at the clock transition, incorrect state changes would occur.

An upper limit on the clock speed is determined by the characteristics of the circuit components. If the clock is changed to a higher frequency, the network will be unreliable because the flip flop control signals may still be changing when the next active clock transition occurs. With an appropriate clock signal, the SSN will reliably change from state to state as determined by the inputs and thus generate the correct output signals.

Notice that the output signals can change at times other than an active clock transition because input changes may change the outputs.

State transition diagram

A graphical tool makes describing the complicated operation of a SSN easy. The *state transition diagram* is the most common graphical tool. It shows the input to output relationship of the SSN at each state. It also shows how the states will change at active clock transitions.

Figure 1-13 illustrates a state transition diagram. Circles containing a descriptor represent the states of the network. As the figure shows, the states of the flip flops, which together determine the state of the SSN, are written in the circle. The flip flop outputs are labeled with Ys.

At a given state, the SSN has a certain input to output relationship. The relationship is shown on an arrow that points outward from a state circle. The input condition X is written to the left of a slash. It corresponds to the output condition Z written to the right of the slash. Each input combination requires an arrow, but sometimes the combinations are grouped on one arrow if the output is the same for all combinations. The input/output relationship is associated with the state at the tail end of the arrow.

The arrows that connect the state circles illustrate the state changes. Which change of state takes place is determined by the inputs at the time of the active clock transition, so each arrow has an associated input condition even if the change is back to the same state. The diagram does not explicitly show the time at which a state transition occurs.

Example SSN

The state transition diagram in Figure 1-13 describes a two-speed light flasher for a light connected to the output Z. If the input X is zero, the output oscillates between zero and one at a slow speed synchronized to the clock. If the input is one, the output oscillates twice as fast, making a fast flasher.

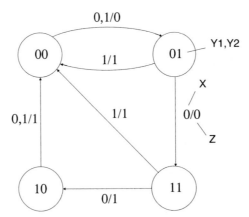

Figure 1-13   State transition diagram for two-speed light flasher.

You can follow the operation by tracing the state transition diagram while the input is kept at a constant value.

Figure 1-14 is a circuit diagram for the light flasher. You can trace the binary signals in the circuit as the circuit changes state on each active clock transition. You will find tracing the binary signals much more difficult than reading the state transition diagram.

1.3   DIGITAL SYSTEMS

A *digital system* is a network of registers and CLNs controlled by a synchronous sequential network. The SSN is therefore called the *control unit* because it controls other hardware in the system rather than directly generating the system output signals. The hardware controlled by the control unit may be very simple or very complex.

Digital circuits can be designed in ways other than using a control unit with other hardware. Instead, a larger SSN can always do the job. An advantage to the control unit approach is that the system is less complex. Similarly, building and repairing the hardware is easier. Changing the design when new algorithms are required is easier. On the contrary, the speed of the system may be slower.

Synchronous Serial Communications Example

In many applications of digital hardware, 8-bit numbers must be transmitted from one circuit device to another, or from one printed circuit board to another. Parallel transfers provide the fastest communication because all the bits of a number are sent at once over multiple paths.

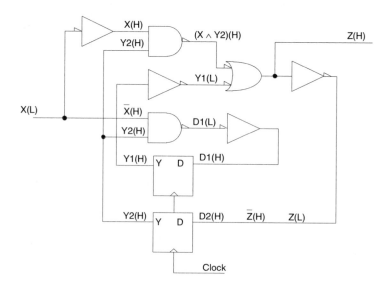

Figure 1-14   Light flasher circuit.

But parallel transfers require many pins and interconnecting wires—one wire for each bit of the number. Simpler serial hardware requiring only a single signal wire can send the number one bit at a time. The serial transfer is inherently slower than an equivalent parallel transfer. However, synchronous serial transfers over a path with little delay achieve high transfer rates, easily making serial transfers practical.

The building blocks

The digital system shown in Figure 1-15 will transmit 8-bit numbers in both directions simultaneously using three signal wires between the two devices. So this circuit does two-directional duplex serial communication. The control unit controls the shift registers, flag, and pulse counter in response to a start signal. Its complexity is similar to that of the light-flashing circuit. Only the master device has a control unit; the serial transfer in the slave device is controlled by the master device. The other hardware does all the significant work.

Here is a description of the building blocks of this digital system shown in Figure 1-15:

- *Shift registers.*   These shift-left registers move the contents of the register one bit to the left for each active shift or clock transition. The right bit is an input that comes from the other device, and the left bit is the output signal to the other device. The shift registers can also be parallel loaded and parallel read from within the system using the parallel register.

- *Flag.*   A *flag* is a flip flop that indicates the occurrence of something. When a flag is set, information is ready to be used. If a flag is cleared, it means wait.

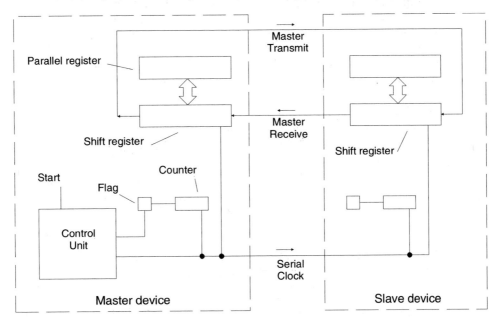

Figure 1-15   Synchronous serial communication system.

- *Pulse counter.* The counter circuits in the master and slave devices count the pulses to determine when an 8-bit transfer is complete. After eight pulses, the counter sets the flag to indicate that a transmission is complete.

System operation

The control algorithm is very simple, only requiring eight pulses on the serial clock line to do an 8-bit transfer. Before a transmission begins, other hardware parallel loads the shift registers in both the master and slave devices. Next, the master device is told to start a transmission. The control unit clears the flag and then makes a string of eight pulses. At the eighth pulse, the counters set the flags indicating the transmission is complete. When other devices respond to the flags, they will parallel read the shift registers and use the transmitted data.

Because the output of one 8-bit shift register is connected as the input of the other, together they effectively make one 16-bit circular shift register. Of course, one register is within the master device, and the other is in the slave device.

When the shift leads are pulsed together, all the bits in the two registers move to the left one place. The bit at the left of the master register is transmitted to the slave register and enters the right end of the slave register. The bit at the left end of the slave register is transmitted to the master register and enters the right end of the master register. After eight pulses, the numbers in the two registers have been exchanged. The effect is serial communication between the two systems containing the shift registers with only three signal wires. Only a few details have been omitted from this discussion.

The transmission system sends information in both directions at once, one bit at a time; therefore, it is called a *duplex serial communication system.*

Instruction-controlled Information Processor Example

An information processor is a digital system that manipulates binary numbers according to an algorithm. An instruction-controlled processor can perform more than one algorithm. An instruction code number tells the system which algorithm to perform. In many designs, some parts of the different algorithms will be common to several instructions.

The following example is an instruction-controlled digital system. It carries out one of two simple algorithms beginning when a start signal changes to the asserted condition. The instruction number tells the control unit which sequence of states to follow. Each sequence therefore carries out a different algorithm.

The building blocks

A collection of hardware devices can manipulate numbers. Manipulating numbers is information processing; therefore, calling this hardware the *processor* is appropriate. Figure 1-16 shows the block diagram of a processor with a control unit. The control unit is a synchronous sequential network that controls the processor hardware to carry out each possible algorithm.

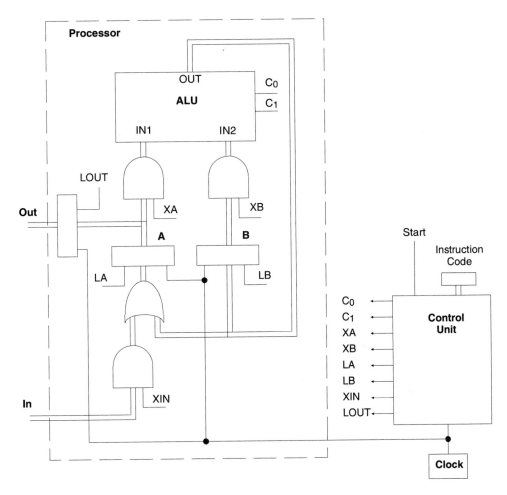

Figure 1-16   Instruction-controlled digital system.

Here is a description of the building blocks that form the processor hardware shown in Figure 1-16:

- *Arithmetic logic unit.*   The ALU is a combinational logic network that does the functions shown in Table 1-3. This ALU performs very simple operations. You can design this one easily if you have some digital design experience. More complex yet practical ALUs perform dozens of operations.

- *A and B registers.*   The A and B registers are load-controlled registers. They parallel load a number from the incoming bus on the active clock transition if the load lead is asserted. Otherwise, the register is unaffected, and the number in the register remains the same as it was.

TABLE 1-3   ALU OPERATION

Controls C_0C_1	Operation
00	IN1 + IN2 $\rightarrow$ OUT
01	IN1 + IN2 + 1 $\rightarrow$ OUT
10	IN1 + $\overline{\text{IN2}}$ $\rightarrow$ OUT
11	IN1 + $\overline{\text{IN2}}$ + 1 $\rightarrow$ OUT

- *Transfer gates.*   These gates are two-input AND gates that pass the input number to the output when the transfer lead is asserted. If transfer is not asserted, the output is the number zero.

- *Input port.*   The collection of OR gates passes one number or another number to its output. The unwanted number must be zero in order to not affect the output. The input port has transfer gates that can output the number zero; therefore, external hardware doesn't affect the output of the OR gates except during input.

- *Output port.*   The output register can capture a number from the bus at an appropriate time and hold the output signals for external devices to respond.

- *Control unit.*   The control unit is the synchronous sequential network that controls all the hardware in the digital system. The control unit interprets the instruction code and generates the proper sequence of control signals.

System operation

The processor hardware proposed here could perform many different operations, but let's consider only two of them. In this discussion, assume that all numbers are unsigned binary numbers. First, let's input a number and send it directly to the output port. Second, let's input a number, multiply it by two, and then send the product to the output port.

Here are the steps necessary to input a number and output it to the output port. Carefully follow Figure 1-16 as you read these steps.

- *Step 1.*   Pass the input number through the transfer gates and load into the A register.

- *Step 2.*   Load the output register from the bus carrying the contents of the A register.

Here are the steps necessary to input a number, multiply it by two, and send the result to the output port:

- *Step 1.*   Pass the input number through the transfer gates and load it into the A register.

- *Step 2.*   Add the number in the A register to zero and load the result into the B register. To do this step, pass the number in the A register through the transfer gates to the ALU; don't assert the B side transfer gates so that the number zero goes to the ALU; tell the ALU to add the two numbers; and load the B register.

- *Step 3.*   Add the two numbers and put the answer in A, effectively multiplying the number by two. To do this step, transfer the numbers to the ALU; tell the ALU to add; and load the sum into the A register.

- *Step 4.*   Load the output register from the bus carrying the contents of the A register.

Control unit operation

The control unit is a synchronous sequential network that makes all the signals that control the processor hardware. These signals are shown in Figure 1-16 and include C_0, C_1, XA, XB, LA, LB, XIN, and LOUT. The control unit inputs include a start signal and a 1-bit instruction code. The instruction code determines which of two operations the system will perform. An external device will make the start signal change from 0 to 1 to indicate a start of operation.

Instruction code is 0.   Figure 1-17 shows the state transition diagram for the control unit. Let's follow the signals that the control unit generates as it moves through its states. Assume that the system user has set the current instruction to 0. The operation will be the transfer of the input number to the output register. Refer to the figure for the inputs and outputs as you follow the states of the control unit. The d's on the state transition diagram are *don't cares*, meaning that the output could be either zero or one. Here are the details:

- *State 0.*   While the start signal is 0, the controller doesn't care what the processor does, provided it does not disturb the output. The processor can't change the output, because a number may be in the output register from a previous operation. When the start signal changes to 1, the outputs remain the same, and the SSN changes state at the next active clock transition. Note that the clock transition also clocks registers A and B, but they are not important at this point.

- *State 1.*   Because the instruction is 0, the SSN will go to state 2 next. While at state 1, the outputs must transfer the input number to the A register through the OR gates. Therefore, the other number applied to the OR gates must be zero. The control unit deasserts the transfer signals XA and XB to send zeros to the ALU; it sends 00 to the ALU control signals telling it to add; it asserts the A register load lead; and it asserts the XIN transfer gates to bring the input number to the A register. When the active clock transition comes, the A register loads the input number.

- *State 2.*   Regardless of the inputs, the SSN will next go to state 6. At state 2, it transfers the number in the A register to the output register. The control unit asserts the LOUT signal to enable the output register to load on the next active clock transition. Because no other useful numbers are in the system, the control unit doesn't care about the remaining control signals.

Instruction code is 1.   Now let's consider the operation when the instruction is 1, telling the control unit to send two times the input number to the output. Here are the details:

- *State 0.*   The operation is the same as state 0 described above.

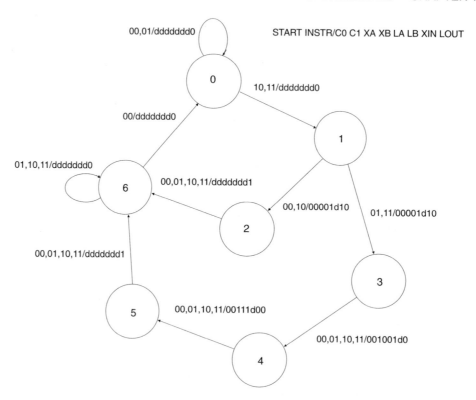

Figure 1-17 State transition diagram for information processor.

- *State 1.* Because the instruction is 1, the SSN will go to state 3 next. At state 1, the control unit brings the input number into the A register the same as state 1 above.

- *State 3.* Regardless of the inputs, the SSN will next go to state 4. At state 3, a copy of the number in A is loaded into the B register. To do this step, the control unit asserts XA to transfer the A register to the ALU; does not assert XB so that zero is sent to the ALU; tells the ALU to add, which makes a copy of the number in A; and asserts LB to load the number into the B register. Notice LA and LOUT are not asserted, so the numbers in these registers are not disturbed.

- *State 4.* Regardless of the inputs, the SSN will next go to state 5. At state 4, two times the input number is loaded into A. The control unit asserts both XA and XB to transfer two copies of the number to the ALU, tells the ALU to add, and asserts LA to load the output of the ALU into the A register. The control unit no longer needs the B register, so it doesn't care if B is loaded. However, it must not assert XIN because it must apply zero to the OR gates so that the correct number loads into A. The output register is not loaded, to prevent unwanted changes in the output number.

- *State 5.* The next state will always be state 6 because the operation will be complete. At state 5, the control unit loads the result into the output register by asserting LOUT. No other hardware is used, so all the other outputs are don't cares.

Observations

The information processor discussed here is very simple. Building it in a laboratory is not difficult. But consider making a more complex system; say, one that has eight different instruction codes. If the design of the processor hardware is adequate, only the control unit need be changed. Probably making the changes will be relatively easy compared to constructing a new system. The total complexity of the system will be only slightly greater than that of the example.

A complex system can be designed and built with very little extra hardware. The reason is the algorithmic approach used in the system design. The hardware does only small operations at a time and uses relatively simple hardware to do them. Of course, the time required to complete the total operation depends on the complexity of the operation. If the hardware is fast enough, this is a good method for designing practical systems.

1.4 STORED PROGRAM PROCESSORS

A *stored program processor* is an instruction-controlled digital system with a memory. The instruction-controlled system contains a processor and a control unit. The control unit is a synchronous sequential network, but the complete system is also a SSN. The memory unit is an electronic device that stores numbers. The numbers from the memory are sequentially supplied to the instruction-controlled system as instructions. A computer is one example of a stored program processor. However, many other commercial devices that are stored program processors are not computers.

Practical programmable devices are very similar to the instruction-controlled digital system illustrated in Figure 1-16. However, practical systems are much more complex and their design is beyond the scope of this book. Therefore, the discussion of programmable devices will use a block diagram approach.

Block Diagram

Let's examine the major functional pieces of a stored program processor using the block diagram in Figure 1-18. Later sections discuss the hardware that implements these blocks.

The clock

The stored program processor is a large synchronous sequential network that requires a clock signal to synchronize all its memory elements. The signal from the clock is a periodic pulse waveform; you can see it with an oscilloscope. To give meaning to the pulses, let's call them *ticks* to emphasize that they provide timing information. The typical clock will operate at one million ticks per second (1 MHz) or faster.

The clock circuit usually consists of some electronics and a quartz crystal. The crystal is a package containing a thin piece of quartz crystal with wires attached to opposite sides. Quartz is a piezoelectric material, which means that bending the material slightly makes a voltage across it. Similarly, applying a voltage to quartz will cause it to bend. A timing device is made from quartz by cutting a small wafer of it to the proper size so that it vibrates at a desired frequency. Because the piece of quartz is small and the material very stiff, the frequency is typically hundreds of kilohertz. This frequency is in the range of frequencies of computer clocks. When the proper electronic circuit uses a crystal, the vibrations of the quartz provide a voltage of known and very stable frequency. The electronic circuit provides energy to the crystal to enable it to continue vibrating. Very accurate watches also use quartz crystals to provide accurate frequencies.

The processor

The central block of the system is called the *processor*. The principal part of the processor is a combinational logic network, called the *ALU*, that manipulates binary numbers. Practical ALUs can perform dozens of different operations, but only one at a time. Usually ALUs can add, subtract, complement bits, position bits, and so on.

Registers in the processor hold the data numbers operated on or produced. Usually processors are parallel devices that work on multiple bit numbers. To better understand the discussion, think of 8-bit numbers because they are a common case. Signals from outside the processor control the operation the processor performs on the numbers. Any hardware involved in data transfer into or out of the processor is considered separate from the processor. Use the term *processor* for only the hardware that manipulates numbers.

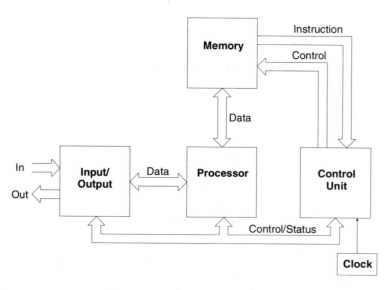

Figure 1-18   Stored program processor.

Some people call the processor the central processing unit, or *CPU*. Let's avoid the term *CPU* because it has different meanings to different people. For example, some people also use CPU to refer to the major processing hardware and control panel and display screen on mainframe computers. Unless you carefully define the term, avoid using CPU. Instead, use the term *processor*, because the job of this block in the stored program processor is to process information where that information is in the form of binary numbers.

The control unit

The control unit is a synchronous sequential network that sends control signals to the processor, memory, and other parts of the system. Interpret the bold arrow between the blocks as a path along which information flows from one block to another. This path is rather abstract at this point and may not be unique wires in actual hardware. An arrow head is shown pointing toward the control unit because it gets information on the status of operations performed by the processor.

The clock is shown connected to the control unit to emphasize that the control unit is a SSN, but timing signals are passed to other blocks also. For example, the processor has registers that are clocked.

The signals from the control unit tell the processor to manipulate numbers according to the algorithms built into the SSN. The control unit is instruction controlled; therefore, it can do more than one algorithm based on its design. Typical control units recognize several hundred different instruction codes. This level of complexity is practical because the control unit needs only a few states to control the processor for different operations. However, the control unit can perform only those instruction algorithms that are hard-wired into the control unit.

The memory

The addition of a memory device to the instruction-controlled system greatly increases the capability of the system. The memory, which is a collection of registers, holds binary numbers used for two distinct purposes. The first type of number is instruction code numbers used by the control unit. The second type is data numbers that are associated with the processor.

Instruction code numbers.   The memory holds instruction code numbers that are sequentially accessed by the control unit under control of the control unit. The control unit will get an instruction code from the memory, carry out that operation, and then return to the memory for further instructions. Very large, complex algorithms can be carried out as a series of small operations that are built into the control unit. However, the added capability due to the memory increases the complexity of the digital system very little.

Figure 1-18 shows a control path from the control unit to the memory. The control signals select the correct register and control the memory. The path from the memory to the control unit carries instruction code numbers. The block diagram does not specify how many bits this path can carry.

Program.   The sequence of instruction code numbers stored in the memory is called a *program*. Although the word *program* is routinely associated with computer programming, the word is more general. For example, when you go to the theater to see a play, you are given a program. The program describes the first act, the second act, and so on. The instruction codes in memory describe the first action, the second action, and so on.

Software.   If the memory technology allows people to change the numbers in the memory easily, then this electronic machine can be easily redesigned to do different jobs or different sequences. If the code numbers in the memory require only electronic information, the redesign of the unit can be done without any physical wiring changes at all. Because the numbers in the memory are rather abstract as opposed to hardware, the numbers are collectively called *software*.

The ability to change the operation of a digital system without changing the wiring makes stored program processors very important. Identical copies of the hardware can be manufactured under carefully controlled conditions so that the reliability and cost are very good. Then the operation of the hardware can be changed for different applications by putting a program into the memory.

Data numbers.   The memory also holds data numbers for the processor. The processor has enough registers to store only the numbers operated on by the ALU. Therefore, results generated by the processor are stored in the memory. The processor will later retrieve needed numbers from the memory.

Figure 1-18 shows a two-directional path between the processor and the memory for transferring data numbers. Transferring a number from the processor to the memory is called a *store* operation, and transferring a number from the memory to the processor is called a *load* operation. The control path from the control unit to the memory selects the memory register and controls the memory hardware.

Input/output

The input/output (I/O) block in Figure 1-18 includes any hardware that allows data transfer between the processor and hardware devices external to the stored program processor. People often call the devices outside the computer *the real world*.

Consider the real world that is the concern of this book. The stored program processor is primarily a control system. It is used inside consumer products, manufacturing machines, or laboratory instruments. The usual input devices are sensors, and the output devices are actuators. Common sensors are switches, temperature sensors, and oxygen sensors. For example, oxygen sensors monitor automobile engine exhaust to determine whether the proper amount of fuel is being used. Actuators include solenoids, indicator lights, and motor starters. The solenoid may move a valve such as a hydraulic valve or the fuel valve in a gasoline engine fuel injector. A motor starter may turn on the fresh air fans for the building you work in. Some other I/O devices include the switches operated by the keys of a personal computer, or the solenoids that drive the pins in the head of a dot matrix printer.

Most input/output devices need power electronics or signal conditioning electronics to interface to the signals in the computer. The power electronics is not properly part of the computer, but rather is called *interface electronics*. The computer only uses low-power-level logic signals.

The data path in Figure 1-18 between the I/O block and the processor is two-directional. Data numbers travel from the processor to output devices and from input devices to the processor over this path. The control unit controls the input/output section using the control path. The control unit selects which input/output device will communicate with the processor.

Instruction Operation

The complete operation of an instruction may require several ticks of the clock. In general, several machine cycles are necessary to complete the operation of an instruction as the control unit proceeds through its states. In the stored program processor, a complete instruction operation happens in two distinct phases or groups of ticks.

Fetch phase

During the *fetch phase*, a copy of the instruction code number is brought out of the memory and transferred to the control unit. In spoken language, say *the control unit goes to the memory to fetch the instruction*. Of course, the word *goes* is figurative—only electrical binary signals travel on wires. The control unit sends binary control signals to the memory to accomplish the fetch.

Usually the control unit will fetch the instructions from the memory in the physical order of the memory registers—one after another. Part of the fetch operation then is preparation for the next fetch.

The fetch phase of an instruction requires one or more ticks of the clock. The actual number of clock ticks depends on the design of the electronics.

Execute phase

During the *execute phase*, the instruction operation is carried out or executed. Execution means that the control unit sends signals to other parts of the machine to direct the operations specified by the instruction code. Instruction execution may affect only the processor, but it may also affect data transfers between the memory or I/O section and the processor. The execute phase of an instruction also requires one or more ticks of the clock.

In spoken language, people refer to the *execution* of an instruction, meaning both the fetch and execute phases together. The distinction of fetch and execute phases is most important to an understanding of how a stored program processor works.

The control unit will automatically continue to fetch and execute instructions until something stops it. People thus speak of *executing* programs or *running* programs.

In control system applications, usually the program will run forever, as long as the system is powered up and operational. Control programs run in loops so they continuously control devices connected to them.

Definition of the Instruction Set

The control unit of a given stored program processor can carry out only a finite number or set of operations. That is, the control unit cannot do an infinite number of different operations. Therefore, the control unit recognizes only a limited number of instruction code numbers. The group or set of valid code numbers that the control unit and processor, working together, can recognize and carry out is called the *instruction set* of the machine.

In many practical designs, the number of possible numbers that can be sent to the control unit exceeds its capabilities. That is, some of the numbers are invalid because the control unit was not designed to recognize them. The hardware will do something if it fetches an invalid number, but usually you will not know what it was! The instruction set includes only the valid codes recognized by the control unit.

General-purpose Computer

Suppose that you set out to design a stored program processor that could be used for many different applications. You might ask this question: What do I put into the processor and control unit to make a system that can do everything? Or more simply: What does everything include? Of course, one design may be better than another because it is cheaper, faster, or smaller. But the question here is, What is necessary to do everything?

The answer is a list of characteristics of a machine called the *general-purpose computer*. Here are the necessary operations that such a machine must have:

- *Arithmetic operations.*   The machine must be able to do some arithmetic. It could be very simple and only add, or very complex with hardware floating point operations.

- *Logical operations.*   The logical operations AND, OR, and COMPLEMENT are required. Many practical machines that are not computers are built with these logical operations, but they do not have any arithmetic capability.

- *Load and store operations.*   The machine must have the capability to store and retrieve many data numbers. The processor usually stores only a few numbers and does not qualify; a memory device is required. For those interested in computer control systems, this ability is essential to integration and differentiation, which are necessary for control system operation.

- *Testing and branching.*   More simply, this operation means the ability to make decisions. The hardware must first test a number and then respond to the test result. The response will be to change the order in which the instructions are used. This operation is called *branching*.

- *Input and output.*   The machine must be able to communicate with other devices, or it could only churn inside, accomplishing nothing useful.

Most computers that are called computers are general-purpose computers. However, knowing these characteristics will help you understand the instruction set of a computer. When

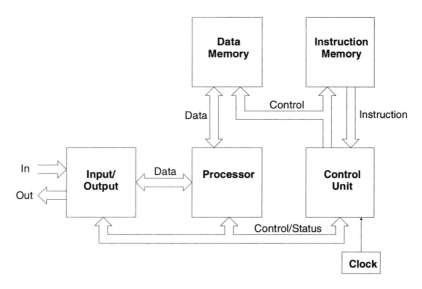

Figure 1-19 A Harvard architecture computer.

you are faced with learning several hundred instructions, learning is easier if you group the instructions into categories.

Computer architecture

The term *computer architecture* describes the overall design of a computer. The design shown in Figure 1-18 is often named the *Princeton architecture* or *von Neumann architecture* in honor of the professor who developed this design. One of its principal characteristics is the use of the memory. The memory holds both the instruction codes and the data numbers. This design allows the limited number of memory registers to be used most efficiently. One characteristic of the Princeton machine is seldom used in small computers. It can treat instructions as data numbers. Therefore, the program can change itself as it runs. Because microcomputers often use permanent memory to hold the program, using this feature in most microcomputer applications is not possible.

The principal alternative design is named the *Harvard architecture*. The Harvard design is diagramed in Figure 1-19. The difference between the Princeton and Harvard architectures is that the Harvard machine has two memories instead of one. One memory holds the instruction codes and the other holds data numbers. The two kinds of numbers are logically separated, but each memory must be large enough for its job. An advantage of this design is that the hardware can be made to operate faster because each memory operates independently.

Most small computers now have the Princeton architecture. However, the computers of the future will include more Harvard architecture machines because of the speed advantages.

1.5   MICROPROCESSORS AND MICROCOMPUTERS

Now let's look at how modern hardware is used to build the parts discussed on the block diagram. The many kinds of hardware in use have led to many terms to describe it. Because so many people and different companies are involved with microcomputer hardware, the terminology is naturally used somewhat differently by different people. To avoid confusion, the terminology introduced here is used consistently in this book. Usually, this terminology is consistent with Motorola terminology.

Integrated Circuit Technology

The development of integrated circuits has made the microcomputer possible. The integrated circuit may be the most significant technological development of the twentieth century because it makes so many other technologies possible.

Transistors

Let's begin with a brief discussion of transistors, because they are the building blocks of electronic circuits. Transistors are made from very pure silicon, which is a semiconductor material. Regions of the silicon are doped with a few atoms of another element to give it the proper characteristics to form transistors. The transistor is generally used as an electronic switch in digital circuits such as a computer. Tens of thousands of these switches are needed to build a practical computer. The transistors and their interconnections must be very small to be practical. One advantage of the integrated circuit is that many thousands of transistors can be interconnected in a very tiny device.

Integrated circuit construction

The construction of an integrated circuit begins with the manufacture of extremely pure silicon crystals. Wafers are cut from the crystalline silicon that are about one millimeter thick. These wafers are polished so the surface looks much like a black glass mirror. A smooth surface is necessary for the processes that follow to succeed.

The wafer is very pure silicon into which doping atoms must be introduced in microscopic-sized regions that are rigidly controlled. The doping atoms become part of the crystalline structure of the silicon. Doping atoms enter the silicon structure if they are introduced in a gaseous form at a high temperature—the silicon must be very near its melting point. However, doping atoms must only go into the silicon at the correct places.

In the most basic manufacturing process, the doping atoms are controlled by a coating that is placed on the surface of the wafer. The coating resists the entry of the doping atoms into the silicon. The coating is made with windows or openings that allow the atoms to enter at the correct place. The coating is applied as a photosensitive material that is exposed to ultraviolet light through a mask. The mask allows the light to expose only selected areas of the coating. The coating is then developed by a photographic process that removes or retains the coating as required, thus forming the window openings.

After the coating is in place, the wafer is passed through a furnace tube that contains the doping atoms. When the temperature and time of exposure are properly controlled, the doping atoms will enter the silicon and form part of a transistor or other device.

The complete wafer will require many passes through various furnaces with coatings in different patterns and different doping atoms. When all the processing is completed, a tiny area of the wafer will contain a complete electronic circuit. Usually hundreds of copies of this circuit will be made on the wafer simultaneously. When these circuits are separated into individual circuits, they are called *chips*. One process to separate the pieces of the wafer is to scribe the surface of the wafer with a diamond-pointed tool and then break it along the score marks—the wafer chips into pieces.

The chips containing complete functional circuits must be mounted in useful packages. The individual chips are usually a square of one to twenty millimeters on a side, although chip sizes change as technology changes. The package provides eight to several hundred electrical connections to the chip. The package also provides mechanical protection for the chip. The completed package is called an *integrated circuit*, or *IC*. The integrated circuit industry has standardized packages and their pin arrangements.

A simple integrated circuit contains a few hundred transistors, and a sophisticated microcomputer chip contains many hundreds of thousands of transistors. The cost of ICs ranges from a few cents each to many hundreds of dollars each.

The Microprocessor and Microcomputer

In the early 1970s, implementing large portions of a computer in integrated circuits became practical. Certainly ICs were used in computers before then, but only to do small functions. When the technology had progressed so that about 7000 transistors could be put into a single low-cost chip, the microcomputer era began. With such a chip, the processor and control unit of a relatively powerful computer can be built into a single integrated circuit called a *microprocessor*.

Be careful with the term *microprocessor*. Unfortunately, the computer business is subject to a wide range of interests and attracts uninformed people who use terminology incorrectly. Newspaper articles often contain errors of terminology in technical areas. The term *microprocessor* has been used to describe almost any computer-related item from a single integrated circuit to a complete computer system.

Definition of microprocessor

In this book, the term *microprocessor* means the electronics in a single integrated circuit that implements the processor and control unit of a computer. Often, this integrated circuit contains some minor support electronics for other parts such as the clock. If the integrated circuit contains other major devices, such as memory, the part of this larger chip that implements the processor and control unit is called the *microprocessor*. Figure 1-20 shows the microprocessor part of a computer.

The memory used with a microprocessor is also integrated circuits. A single IC is used if the required number of memory registers is small enough. If more memory is required,

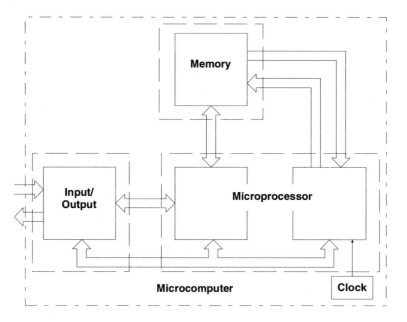

Figure 1-20   The microprocessor and microcomputer.

multiple ICs are used. Memory ICs are designed so that many copies of the same part can be added to make a larger memory.

The input/output section of the computer can be a single IC if relatively few I/O signals are required. As with memory ICs, the I/O ICs are designed so that many copies of the same IC can be added to make more signals.

The last part of the computer is the clock circuit. Most clock circuits use a quartz crystal as the time base.

Definition of microcomputer

The four major components of the computer—the microprocessor IC, one or more memory ICs, one or more I/O ICs, and the clock—are shown in Figure 1-20. Although a few other minor electrical components such as capacitors may be required, a working computer can be constructed from these four basic components. A single printed circuit board usually connects the ICs, making a computer called a *microcomputer*.

The single-chip microcomputer

As integrated circuit manufacturing technology improves, more and more transistors can be made in a single chip. Therefore, the range of choices of how to build computer hardware is great. One use of the improved technology has been to make bigger and better microprocessors with bigger and better memory chips. However, many applications don't require the greater power of the larger microprocessors. Another choice is to integrate the microprocessor, memory, I/O, and the clock electronics into the same chip to make a *single-chip microcomputer*.

The single-chip microcomputer is an ideal component for controlling mechanical and electrical devices, and it is used inside many consumer products. Because this chip controls the product, it is sometimes called a *microcontroller*. The names *single-chip computer* and *microcontroller* are interchangeable, although some companies prefer one over the other in their literature.

The processor and control unit part of the single-chip computer is called a *microprocessor*. Microprocessor is a reasonable name because the electronics from the microprocessor integrated circuit is incorporated into the single-chip computer. The quickly changing technology makes creating perfectly clear terminology difficult.

Identifying the individual parts of the single-chip computer from outside the IC package is impossible. Usually, you can identify an IC as a single-chip computer only from the part number printed on the package. People often buy equipment containing microcomputers without realizing that a computer is present.

Microcomputer Applications

The microprocessor and the single-chip computer began affecting the design of most electronic devices in the 1980s. The applications of these devices beyond the late 1990s will be so widespread that examples will be unnecessary. All engineers must learn the fundamentals of this technology. Most consumers already make daily use of products with computers embedded in them so they don't need to understand the computer themselves. Other products are more obviously computer based, but the product designers usually try to isolate the user from the details of the computer hardware and software.

Embedded control systems

A microcomputer is said to be *embedded* if it is inside a device that is not called a *computer*. The microcomputer provides a cost-effective and practical implementation of the electronics required by the device. Microcomputers provide sophisticated features to consumer products at low cost. The computer makes the products easy to use by people with a wide range of skills.

The following is a list of some common products that contain embedded microcomputers. Some of these microcomputers may be proprietary chips with limited features to reduce cost. Others use standard off-the-shelf parts that are available to the public.

Satellite TV receivers.   Satellite receivers not only have digital displays and digital image processing, they also call your service company to bill you for the movies you watch!

Microwave ovens.   Most microwave ovens can be programmed for various cooking times and power levels. Some ovens sense food temperature and change the program accordingly. An internal microcomputer controls these and other features.

Home heating thermostats.   Programmable thermostats control home temperatures and reduce energy consumption.

Automobiles. Most cars use a microcomputer to control the delivery of fuel and the spark timing for the engine. Microcomputer-controlled transmissions communicate with the engine controller to provide smooth transmission operation. Microcomputer-controlled anti-lock braking systems greatly improve the safety of cars and trucks. Electronic dashboard instruments contain a microcomputer.

Robotics. The revolution in manufacturing caused by robots and related manufacturing devices is a direct result of the use of microcomputers in the robots. Some robots use a separate microcomputer to control each axis of movement of the robot.

Personal computers

The name *personal computer*, or *PC*, originally described a small desktop or portable computer system intended for business and computation purposes. The name implies that a single person will use the computer system. However, the technology has grown in many directions so that the PC is sometimes connected to several terminals for a group of people to use; it may be too large to fit on the desktop or small enough to fit into a briefcase; and it may be used for playing games or for acquiring data from scientific experiments. Some very powerful personal computers are suitable for large computational jobs and are called *workstations* to distinguish them from other lower power devices.

Often people call a personal computer system a *microcomputer*, which confuses the meaning of the terminology. A personal computer is usually a collection of components that contain many microcomputers. The main computer box contains the central microcomputer, but other microcomputers are involved, too. The keyboard usually contains a single-chip computer. The disk drive interface probably contains a microcomputer. The hard disk drive probably contains a microcomputer. The display monitor interface may contain a microcomputer. The printer probably contains a computer. If the printer uses the PostScript language, the computer in the printer may be more powerful than the computer in the main box. Therefore, a personal computer system is a collection of many microcomputers as they are defined here.

Distinguishing the types of computers

The personal computer, or the even more powerful minicomputer, is a computer system purchased as a package. Even if some components are bought separately, they will be designed to be compatible with the system components. By contrast, a microcomputer is a collection of one or more integrated circuits on a circuit board that requires other devices such as a power supply to operate.

The personal computer user usually purchases software designed to meet a particular application need. Using the computer is a matter of learning to use the software.

The microcomputer user needs some knowledge of digital systems to apply microcomputer hardware effectively. Programming will probably be done at a level that directly controls the microcomputer hardware. The microprocessor manufacturer or a third party will provide software development tools, but the user will likely develop the applications software.

1.6 REVIEW

A broad range of topics have been overviewed in this introductory chapter. All of these will help you understand both microcomputer hardware and software. However, the section on number systems is the most important. Later chapters will use all the types of numbers introduced here without review. You should practice working with binary and hexadecimal numbers. The ability to visualize the bit patterns represented by hexadecimal numbers, and the ability to write a hexadecimal number to represent a binary number without hesitation are important when learning about microcomputers.

Always note the typeface used to display a number in this book. Hexadecimal numbers use a distinctly different typeface, named "Helvetica," from decimal numbers. This typeface makes them stand out without the need for other identifiers.

1.7 EXERCISES

1-1. Do the following calculations by changing the given decimal numbers to 8-bit two's complement numbers and then performing the indicated operation on the two's complement numbers. Do any two's complement overflows result?

(a) $(+23) + (-120) =$

(b) $(-20) + (-32) =$

(c) $(-44) + (+44) =$

(d) $(+15) + (+106) =$

(e) $(+47) - (+92) =$

(f) $(+127) + (-127) =$

(g) $(-1) + (-1) =$

(h) $(-47) - (-107) =$

(i) $(-56) + (-72) =$

(j) $(-2) + (-126) =$

1-2. Find the equivalent numbers.

(a) 11111101 two's complement $=$ _____ Hex

(b) 37 Hex $=$ _____ 2

(c) 01010010 BCD $=$ _____ 2

(d) 011000100 BCD $=$ _____ Hex

(e) 99 Hex $=$ _____ 10

(f) $10000011_2 =$ _____ 8

(g) $57_8 =$ _____ 10

1-3. Change the sign of each of the following two's complement numbers.

(a) 00000001

(b) 10111100

(c) 00001111

(d) 00000000

(e) 10000000

(f) 0111110000000001

(g) 11111111

(h) 11110000

(i) 01010101

(j) 10101100

1-4. Find the decimal equivalent value of each of the following two's complement numbers.

(a) 00000001

(b) 11111000

(c) 0000000000111111

(d) 00000000

(e) 1000000000000000

(f) 01111111

(g) 10000000

(h) 10000010

(i) 01000000

(j) 1111111111111111

1-5. Write each of the following numbers in hexadecimal format.

    (a) 11110000   (f)   11100111

    (b) 10110101   (g) 10011110

    (c) 00111101   (h) 01100111

    (d) 0110111100000000   (i)   10010000

    (e) 1010101101101100   (j)   0110111011101111

1-6. Add the 8-bit two's complement numbers 01011101 and 01101101 and make observations about the result.

1-7. Subtract the 8-bit two's complement number 00101111 from 11111100.

1-8. Subtract the 8-bit unsigned binary number 00101111 from 11111100.

1-9. Two 8-bit two's complement numbers are called A and B. When B is subtracted from A, a borrow from outside the numbers is necessary. From this information, what is the relationship, if any, between the numbers A and B?

1-10. Two 8-bit unsigned numbers are called C and D. When C is subtracted from D, a borrow from outside the numbers is necessary. From this information, what is the relationship, if any, between the numbers C and D?

1-11. What is the largest positive 16-bit two's complement number both in hexadecimal and decimal?

1-12. What is the smallest, or most negative, 16-bit two's complement number both in hexadecimal and decimal?

1-13. If a positive two's complement number is added to a negative two's complement number, under what circumstance will a two's complement overflow occur?

1-14. If you are using 8-bit two's complement numbers represented by hexadecimal numbers, which of the possible hex numbers represent positive numbers and which represent negative numbers?

1-15. Convert the 8-bit two's complement numbers 01011001 and 11000110 to the equivalent 16-bit two's complement numbers.

1-16. If a positive two's complement number is subtracted from a negative two's complement number, a two's complement overflow occurs if the answer is (positive, negative, other).

1-17. Subtract the two's complement number 01101010 from the two's complement number 10010001 and discuss the result.

1-18. Which numbers in each of the following pairs of two's complement numbers, represented in hexadecimal, are the algebraically greatest: 33 & 77, F0 & 0F, CC & DD, FF & 01?

1-19. The two's complement number for −1 is used frequently. Find the 8-bit, 16-bit, and 32-bit two's complement numbers for −1, and then write them in hexadecimal notation.

1-20. The hexadecimal representation of the 8-bit two's complement number equivalent to −1 is FF. Learn to form the hex numbers for −2, −3, and −4 quickly by counting downward from FF.

1-21. Test your skill at using hexadecimal numbers by determining mentally, not with a number table, the leftmost and rightmost bits of these hexadecimal numbers: 33, 91, C3, FE, 7FFD.

1-22. Add the two's complement numbers 11011101 and 1101100110001111.

1-23. What is the difference in meaning between the abbreviations KB and Kb?

1-24. What is the octal number equivalent to 4096_{10}?

1-25. Would representing 12-bit binary numbers with octal or hexadecimal numbers be more convenient for people?

1-26. Convert the unsigned binary numbers 01110011 and 11110000 to BCD numbers.

1-27. Is time more significant to a combinational logic network or a synchronous sequential network?

1-28. How many binary signals are necessary to represent a code for 25_{10} different items?

1-29. A thumbwheel switch generates four binary signals to represent a decimal number. When the switch is set to the number eight, three outputs are low and one is high. Sketch the four signals and label them with mixed-logic notation.

1-30. If a circuit uses binary signals of 0 volts and -12 volts, which of these voltages is the low level and which is the high level?

1-31. A signal wire is labeled START(L). Describe the apparent function of each voltage level.

1-32. A 2-input logic gate has low-asserted inputs and a high-asserted output when it performs the AND function. The wires connected to the inputs are labeled A(L) and $\overline{B}$(L). The output is labeled START(H). Should A and B be zero or one to make START true?

1-33. Design an EXCLUSIVE OR circuit using only gates as shown in Figure 1-4.

1-34. Explain the difference between the notation $\overline{PB1}$ used in a logic equation and the same notation on a wire in a Motorola manual.

1-35. Write a word description of each of the logic functions AND, OR, COMPLEMENT, and EXCLUSIVE OR.

1-36. Draw the logic symbol for a 2-input OR gate with high-asserted inputs and a low-asserted output. Also draw the AND symbol for this gate.

1-37. Make a voltage level table that describes a 2-input OR gate with low-asserted inputs and a high-asserted output.

1-38. A 3-input logic gate has low-asserted inputs and a high-asserted output when performing the AND function. Draw the logic symbol for this gate when it performs the OR function.

1-39. Design a new circuit to replace the one in Figure 1-5 with the following hardware changes. The switches both work like the lower switch in the figure, and the new light is high-asserted.

1-40. Why do the outputs of a synchronous sequential network depend on what the inputs were in the past?

1-41. Generally the faster a computer works, the better it is. Why not increase the speed of the computer's clock until the computer almost stops working, and then leave the clock at that setting?

1-42. Trace the operation of the synchronous sequential network in Figure 1-14.

(a) Assume that the network is in state Y1Y2=11 and the input X=0, and then sketch the circuit and label each wire with high or low to indicate its logic level. Compare your results to the state transition diagram in Figure 1-13.

(b) Next assume that one active clock transition takes place. Label the wires a second time with their new logic levels. Compare your results to the state transition diagram.

1-43. Change the state transition diagram in Figure 1-17 so that an instruction of 1 multiplies by three instead of by two.

1-44. Assume that a second input port and a transfer control signal are added to the system in Figure 1-16. The second input will transfer to register B the same way the first input transfers to register A. Modify the state transition diagram in Figure 1-17 so that an instruction of 1 inputs two two's complement signed numbers, and then puts the A number minus two times the B number in the output port.

1-45. Serial transfers are generally considered to be slower than parallel transfers because the same medium is assumed for each transmission. Could serial transmission of 8-bit numbers over a 5-meter fiber optic cable be faster than parallel transmission over a 5-meter copper cable?

1-46. When the microprocessor "fetches," does it get instruction numbers or data numbers from the memory?

1-47. Can the instruction set of a computer be changed after the computer is constructed?

1-48. How many embedded microcomputers would you estimate are in the typical personal computer system? Have you considered the hard drive, the video board, the network card, the printer, the modem, and other peripheral devices?

1-49. Look at the switch connections in Figure 1-5. The switches are likely to be some distance from the electronics making the signal wires from the switches quite lengthy and vulnerable to damage. Because a damaged wire is much more likely to get connected to the circuit common than to the voltage source, which of the two connections shown would provide the best electrical safety?

Chapter 2

Instruction Subset and Machine Language

Programming a microcomputer requires knowledge of its internal operation. High-level programming languages try to isolate the programmer from the internal operation of the computer. Most applications of microcomputers require unique hardware configurations that make use of a high-level language difficult. In such cases, microcomputer programming is usually done with a low-level language that requires intimate knowledge of the internal operation of the microprocessor. In particular, you must know and understand the operation of the hardware instructions.

This chapter introduces the fundamental operation of the microprocessor and the instructions that control the hardware. The examples use the Motorola 68HC11 microcomputer, although the fundamental concepts are the same for other devices, including those manufactured by other companies.

This chapter clearly demonstrates that hardware and software are not separate topics when learning microcomputers. Programming cannot be done well without fully understanding the registers in the microprocessor and the memory.

Not all the many instructions available in the 68HC11 are covered here. The advanced instructions are in Chapter 5 after the intervening chapters introduce some helpful programming tools.

Motorola Products

All the material in this chapter except Figure 2-1 and the double-precision instructions applies to the Motorola 6800, 6801, 6802, 6803, and 68701 chips.

2.1 THE 68HC11 COMPUTER OPERATION

The functions performed by machine instructions can be more easily understood if you understand the operation of the computer hardware. This section introduces the principal registers in the 68HC11 and explains their uses during instruction fetch and execution.

The Programming Model

You would find comprehending all the details of the microprocessor hardware and instructions at once very difficult. You must learn the many details a few at a time. To aid your learning, an abstract model of the most important microprocessor registers is used. Because this model provides enough detail to understand the fundamentals of programming, it is called the *programming model* of the computer. Usually a model of the memory registers accompanies the programming model. The programming model is useful only for learning and for writing programs because it does not provide any details of the electronics. Figure 2-1 shows the complete programming model of the 68HC11 microprocessor, although some of these registers will not be discussed until much later. You must think about only the registers discussed and ignore the others to avoid confusion. Remember that the programming model is an abstraction that cannot accurately relate how the hardware is constructed.

The memory model

A computer is a collection of registers and some other hardware that manipulate numbers and move numbers around among those registers. Some of the registers are in memory and others are in the microprocessor.

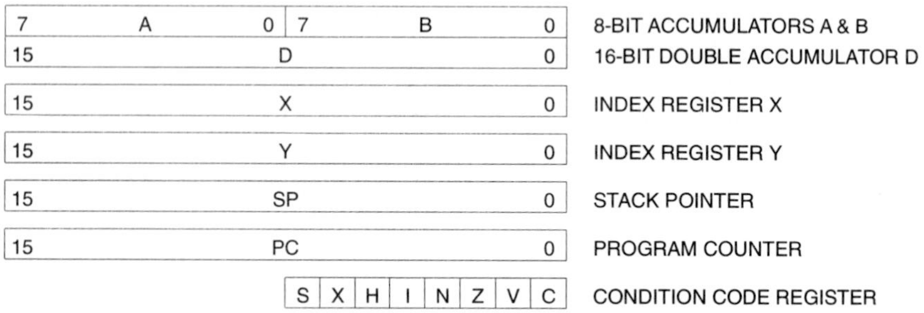

Figure 2-1 Motorola 68HC11 programming model.

The memory registers will be represented as shown in Figure 2-2. The numbers in the boxes represent the contents of the registers; that is, the bits stored in the flip flops. The numbers to the left of the boxes represent the addresses of the memory registers. Wherever a figure does not have numbers, you should assume that the number is unknown or is of no current interest. All the registers always contain numbers, but if the number is unknown, the box will be blank.

Usually a figure illustrates only a few memory registers, but they may have widely separated addresses. A break in the memory with dots in the register boxes indicates that some registers are not shown.

The size of the hexadecimal numbers represents the Motorola 68HC11 hardware correctly. The numbers in the memory registers are 8-bit numbers requiring two hexadecimal digits. The address numbers are 16-bit numbers shown as four hexadecimal digits. Because the addresses have 16_{10} bits, the 68HC11 can directly address 64-K memory registers.

The memory register with address 0000 is explicitly shown in many figures in this book so that you can determine the order of increasing addresses. In Figure 2-2, you can see that increasing addresses go from the top of the figure to the bottom. But this order would not be apparent if the addresses C239, C23A, and so on were not shown. The addresses are often not shown, meaning that any address could be used. For certain topics, figures illustrate memory, with increasing addresses going from the bottom to the top of the figure. Computer manuals sometimes have both types of figures on the same page. You can always determine the direction of a figure when address 0000 is shown.

In spoken language, the two ends of the memory are called the *top* and *bottom* of memory. Most people use the word *bottom* to describe the end of memory with address 0000 and the word *top* to describe the end with address FFFF. However, some people use the words *top* and *bottom* in the opposite sense.

An instruction format

Figure 2-3 illustrates one kind of instruction and its associated data number in memory. The example is the *Load Accumulator A* instruction. The mnemonic name of this instruction is LDAA. A *mnemonic* is an abbreviation or symbol designed to make the abbreviation easy to remember. The LDA means load accumulator and the final A specifies the A accumulator. In spoken language, we often refer to an instruction by spelling its mnemonic name.

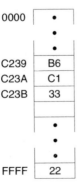

Figure 2-2 Memory model.

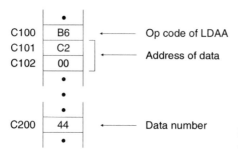

Figure 2-3 Example of an instruction code—Load Accumulator A.

The LDAA instruction has two parts. The first is the operation code byte B6 that specifies the operation to be performed by the microprocessor. The term *operation code* is usually shortened to *op code*. All instructions have an op code as the first byte.

The second part of the instruction is the 2-byte address C200 that specifies the memory register containing the data for this instruction. The instruction shown will copy the data number 44 into the accumulator A register in the microprocessor when the microprocessor executes it.

Not all instructions will contain an address like this instruction. Many more details of this instruction and others are covered in later sections.

The operation code of the instruction in the figure is the number B6 located at address C100. The address part of the instruction C200 is located at memory locations C101 and C102. The data value is the number 44 in the memory register with address C200. After the microprocessor executes this instruction, accumulator A will contain 44.

The microprocessor model

Figure 2-4 illustrates four of the microprocessor registers—only those registers needed for a particular discussion are shown. Later sections introduce the remaining registers in the microprocessor.

The size of the boxes in the figure and the size of the hexadecimal numbers correctly represent the number of bits in the registers. Therefore, the figure shows that the program counter register is a 16-bit register—also called a *double-byte register*. Drawing the program counter as two joined boxes emphasizes that it holds two bytes. The A accumulator register is an 8-bit register. Figure 2-4 has the same size box for the 8-bit A register as was used for an 8-bit memory register in Figure 2-3.

Accumulators. A general-purpose register in a microprocessor that holds a data number is called an *accumulator*. The data number in an accumulator may be the result created by an instruction—for example, the sum of two numbers. Also, the number in the accumulator may be operated on by an instruction.

The name *accumulator* may be confusing—the name also applies to the display mechanism on outdated mechanical adding machines that accumulated a total or sum. For now, just think of the number in the accumulator as a data number. Some people call the accumulator a *general-purpose register*, which is also not very informative.

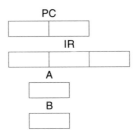

PC

IR

A

B

Figure 2-4   The microprocessor model.

The Motorola 68HC11 has two 8-bit accumulators called A and B as Figure 2-4 illustrates.

Caution! People are often inexact in using spoken language. They often say *A register* when describing the A accumulator register. Do not confuse the A register with the memory register with address 000A. Similarly, the spoken expression *the registers* refers only to the registers inside the microprocessor, not the memory registers.

Program counter.   The microprocessor register that holds the address of the next instruction to be fetched from memory is called the *program counter*. The Motorola 68HC11 has a 16-bit program counter labeled PC in Figure 2-4. The size of the program counter is the same as the size of the memory address. In spite of its name, the program counter is not a counter, and it does not count programs! The program counter is a register that holds addresses for use by the control unit in the microprocessor.

Instruction register.   The register in the microprocessor that holds an instruction code number as it is fetched from memory is called the *instruction register*. The instruction register is said to be *transparent* because its operation cannot be seen. The microprocessor uses the instruction register for its internal operations. Therefore, instructions cannot control the instruction register. Even laboratory test equipment cannot easily access the hardware in this register. The programming model usually does not show the instruction register for this reason.

Figure 2-4 shows the instruction register as a 3-byte register because the instruction in the figure requires three bytes. Remember that the programming model does not give an accurate representation of the electronics in the microprocessor.

Internal Computer Operations

Chapter 1 discussed the operation of a programmable device such as a computer using a block diagram. This section repeats that discussion in greater detail using the programming model. One typical instruction is traced through each step of the fetch and execute phases of its execution.

Figure 2-5 shows the programming model with a Load Accumulator A instruction in memory. Remember that the Load Accumulator A instruction copies the contents of the data

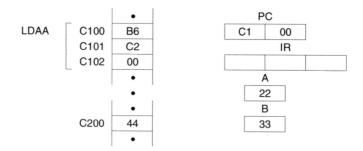

Figure 2-5   Programming model with an instruction in memory.

location addressed by the instruction into the A accumulator. The number that was originally in the accumulator is lost. This **LDAA** instruction occupies three bytes of memory.

Following the step-by-step operation of the computer requires a known starting point. Assume that some electronic device put the numbers shown into both the memory and the microprocessor registers—assume that it was done by magic if you prefer.

In Figure 2-5, the address of the op code of the instruction is in the program counter. The numbers in the A and B accumulators are meaningless—they were specified so you can see changes in the numbers. Previous instructions would normally leave their results in the accumulators.

The operation of the computer can now be traced one clock tick at a time. Because the ticks or cycles of the microprocessor clock synchronize the operation of the entire microcomputer, the register transfers are traced at each tick.

The term *tick* is a simplification of the 68HC11 internal clock that runs at one-fourth the crystal clock frequency.

Instruction fetch

The *fetch phase* of the instruction execution brings a copy of the next instruction code from memory into the microprocessor instruction register. The microprocessor enters the fetch phase each time it completes the previous instruction. When the microprocessor is turned on and starts running, the very first operation is a fetch.

The fetch of a Load Accumulator A instruction is traced in the following sections. LDAA was chosen because all programs use it, and it is easy to understand. Bold type in the figures directs your attention to the numbers discussed in the text and has no other meaning.

First tick of the instruction operation.   During the first tick or machine cycle of the fetch phase of an instruction:

- *The microprocessor sends the number in the program counter register to the memory as an address and tells the memory to read.* The program counter determines where an instruction will be fetched from, because it supplies the address to the memory. You should say that the program counter is *pointing* to the memory register as you might do with your finger. Figure 2-6 shows the program counter pointing at address C100. A hardware signal controls the memory read operation.

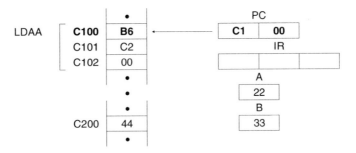

Figure 2-6   Beginning of the fetch phase of an instruction.

- *The memory sends a copy of the number at the specified address to the micropro-cessor, and the microprocessor puts the number into the left byte of the instruction register.* Part of the instruction code has now been fetched from the memory and put into the instruction register as Figure 2-7 shows.

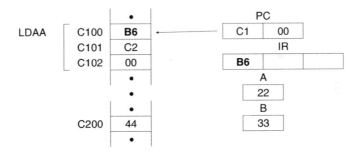

Figure 2-7   First byte of instruction has been fetched.

- *The microprocessor adds one to the number in the program counter and puts the result back into the program counter.* The program counter now contains the address of the next memory register in order. Figure 2-8 shows the result of this operation. Because the microprocessor adjusts the program counter this way, it will automat-ically fetch instruction bytes from the memory in order. When the current instruction is finished, the program counter will automatically point to the next instruction in order. Using a program counter eliminates the need for the instruction to specify the address of the next instruction, greatly simplifying the hardware.

- *The microprocessor determines if fetching should continue by examining the in-struction op code in the left byte of the instruction register.* If the instruction is more than one byte long, the fetch phase must continue. If not, the execute phase is entered. For this example, two more bytes are to be fetched, so the fetch phase must continue. The registers shown in the programming model don't change due to this step.

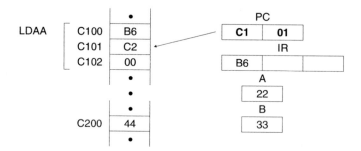

Figure 2-8   The program counter has been incremented after the first clock tick.

So far, the fetch phase has brought the first byte of the instruction into the microprocessor and adjusted the program counter to point to the next memory location. The microprocessor has determined that three bytes must be fetched for this instruction.

Second tick of the instruction operation.   Because the example instruction is three
bytes long, the fetch phase of the operation continues.

At the second tick of the clock:

- *The microprocessor sends the number in the program counter to the memory as an
address and tells the memory to read.*

- *The memory sends a copy of the number at the specified address to the microprocessor, and the microprocessor puts the number into the next byte of the instruction
register. Two bytes of the instruction code are now in the instruction register as
Figure 2-9 shows.*

- *The microprocessor adds one to the number in the program counter and puts the
result back into the program counter.* Figure 2-10 shows the registers after the
second tick of the clock.

After two ticks of the clock, the first two bytes of the instruction are in the instruction
register, and the program counter is pointing to the next location. The microprocessor already
knows that a third byte must be fetched; therefore, the fetch operation will continue.

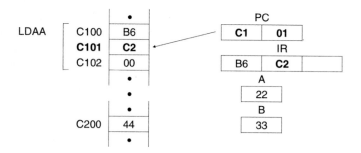

Figure 2-9   The second byte of the instruction has been fetched.

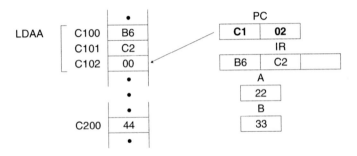

Figure 2-10   The program counter has been incremented after the second clock tick.

Third tick of the instruction operation.   The third tick continues the fetch operation getting the last byte of the instruction code in this example:

- *The microprocessor sends the number in the program counter to the memory as an address and tells the memory to read.*

- *The memory sends a copy of the number at the specified address to the microprocessor, and the microprocessor puts the number into the next byte of the instruction register.* The three bytes of the instruction code are now in the instruction register as Figure 2-11 shows.

- *The microprocessor adds one to the number in the program counter and puts the result back into the program counter.* Figure 2-12 shows the registers now that the fetch operation is complete.

In summary, during the fetch phase of this instruction, the three bytes of the instruction were put into the instruction register and the program counter was adjusted. The entire instruction code has been fetched from memory into the microprocessor; therefore, the fetch phase of the instruction operation is complete.

The program counter now contains the address of the next location in memory after this instruction code. The next location must contain the next instruction in the program. The design of the computer requires that instructions be in order in the memory.

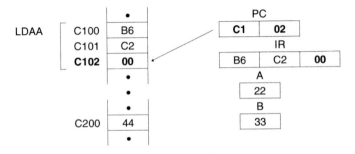

Figure 2-11   The third byte of the instruction has been fetched.

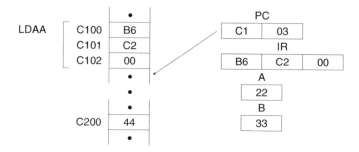

Figure 2-12   The fetch operation is complete after the third clock tick.

The program counter register is defined as a register that holds the address of the next instruction. You should understand that the contents of the program counter change during the fetch operation—the definition applies at the completion of the fetch phase of an instruction.

The instruction register holds the instruction code as it awaits the execute phase of the instruction. The fetch phase of an instruction never changes the accumulators or any memory registers.

Instruction execution

The microprocessor automatically enters the execute phase of instruction operation at the completion of the fetch phase. The *execute phase* carries out the function specified by the instruction code. The function of the Load Accumulator instruction used in the example is to transfer the data number to the accumulator.

Fourth tick of the instruction operation.   The execute phase of the Load Accumulator instruction requires only a single clock tick or machine cycle to do the following:

- *The right two bytes of the instruction register are sent to the memory as an address, and the memory is told to read.* The right two bytes of the instruction register contain the address part of the instruction code and thus point to the data number in memory.

- *The memory sends a copy of the number at the specified address to the microprocessor, and the microprocessor puts the number into the A accumulator.* Figure 2-13 shows that the single-byte data number for this instruction has been put into the accumulator. The number originally in the accumulator is lost.

During the execute phase, the microprocessor copied the addressed data number into the accumulator, destroying the original number there. The execute phase of this instruction does not affect the program counter. Only a few instructions affect the program counter during the execute phase.

Both the fetch and execute phases of the instruction operation are now finished. The microprocessor automatically begins a new fetch at the next tick of the clock.

After the instruction operation is finished, the number in the program counter specifies the address of the next instruction to be fetched. The number in the instruction register is no

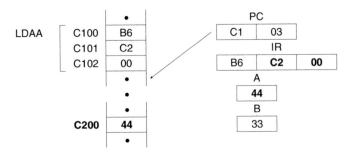

Figure 2-13   The data number has been put into the accumulator.

longer useful. The accumulators hold data numbers including the one that resulted from this instruction. Figure 2-14 shows the final state of the registers.

Instruction operation notation

A concise notation can describe the function performed by an instruction. Unfortunately, many varieties of notation are used in textbooks and computer manuals. Because this book uses Motorola hardware for the examples, it will use Motorola notation. But even this notation is a problem—the Motorola notation is not only somewhat incomplete and therefore sometimes incorrect, but Motorola has also changed notation from one manual to another. You must be careful when reading Motorola literature with this book. Look for the notation problems to avoid confusion.

A complete notation.   Let us begin the discussion of notation by using the LDAA instruction as an example. A word description of this instruction is: *The Load Accumulator A instruction transfers or copies the contents of the addressed memory register into the A accumulator.* This statement only describes the function performed; it does not explain the details of how the function is accomplished. For example, the actions of the program counter are not stated.

The phrase *the contents of* a register, meaning the number in the register, can be written symbolically. A common notation encloses the symbolic name of the register with parentheses;

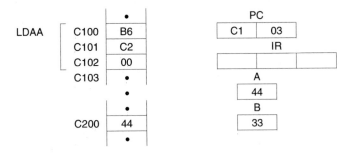

Figure 2-14   The result at the completion of the LDAA instruction after four clock ticks.

for example, (A) refers to the 8-bit number in the A accumulator. The spoken language for (A) is *the contents of A*.

Describing the contents of a memory register is more complicated. The memory register of interest depends on the address specified by the instruction. We could use the notation M to identify *the memory register addressed by the instruction*. Then the notation M identifies a memory register as the A identified the A accumulator. Adding parentheses gives the notation (M), meaning *the contents of the memory register addressed by the instruction*.

An arrow symbol denotes the transfer or copy operation, so the LDAA instruction is described by (M)→(A). This notation literally means that the contents of the A register becomes the contents of the memory register addressed by the instruction and that the previous number in the A register is lost. This brief notation describes the function of Load Accumulator completely but leaves out the details of how the microprocessor does it.

Motorola shortened notation.   The notation introduced so far is an excellent notation, but Motorola seldom uses it in its manuals! Instead, Motorola usually uses a shortened version of this notation. But the shortened version cannot correctly describe all the register transfers that occur.

Motorola solves this problem by returning to the longer notation whenever necessary. Therefore, two notation schemes are mixed, sometimes within the same statement, which may cause confusion. However, because this mixed notation is common, any additional confusion is avoided by following the most used version of Motorola notation. You must understand the notation problems or you will be confused when you read information about the instructions.

Here is the shortened version of the notation—the parentheses are left out! You are to infer from the context of the statement that the parentheses are there when they are needed. Fortunately, most of the time you will need to assume that the parentheses are there.

When using this new Motorola notation, the LDAA instruction is described by M→A. Because you already understand this instruction, understanding the notation is easy—you may even prefer it until a more difficult situation occurs. In the new notation, the M refers to the contents of the memory register addressed by the instruction, and the A refers to the contents of the A accumulator.

The LDAA instruction can now be described in three ways: (M)→(A), (M)→A, and M→A. All these mean the same thing, and all are used. Whenever reasonable, the shortest form of the Motorola notation is used in this book.

The instruction set table.   Let's look at a table that uses the Motorola notation to describe some more instructions that have the same format as the LDAA instruction discussed earlier. Figure 2-15 is a table that describes only a few instructions. A table for all the instructions of a microprocessor is called the *instruction set table*. A complete table for the 68HC11 is in Appendix A. You should avoid any information provided there beyond that in Figure 2-15 until much later.

The columns of the instruction set table are defined as follows:

- *Source Form.*   The mnemonic name of the instruction. A person writing a program on paper would write this name, so it is the source of the program.

Source Form	Operation	Boolean Expression	Machine Code		Bytes
			Op Code	Operand	
ADDA	Add Memory to A	A + M → A	BB	hh ll	3
LDAA	Load Accumulator A	M → A	B6	hh ll	3
STAA	Store Accumulator A	A → M	B7	hh ll	3
STOP	Stop program in trainer		3F		1

Figure 2-15   Part of the instruction set table.

- *Operation.*   This column contains a very short word description of the operation performed by the instruction.

- *Boolean Expression.*   This column has the logical description of the register transfers performed by the instruction written in Motorola notation.

- *Machine Code-Op Code.*   This column documents the instruction binary op code in hexadecimal format.

- *Machine Code-Operand.*   This column specifies the additional bytes beyond the op code that the instruction requires. The instruction format used so far requires two bytes for the address of the data. The address of the data is the operand. The hh designates the high byte of the address that is the second byte of the instruction code. The ll designates the low byte of the address that is the third byte of the instruction code.

- *Bytes.*   This column reiterates the total number of bytes of memory required by the instruction code. You can determine this number from the Machine Code column.

The instruction set table is in alphabetical order by mnemonic instruction name.

Now let's look at some details of the instructions shown in Figure 2-15, which are listed in alphabetical order. Then, using only these few instructions, an example program can be written:

- *ADDA.*   The ADDA instruction adds the number in the A accumulator to the number in the addressed memory location and then puts the sum in the A accumulator. The number in memory is unchanged, and the original number in A is lost. The numbers are 8-bit numbers. The addition is a binary addition, so the microprocessor does not distinguish two's complement signed numbers from unsigned numbers.

- *LDAA.*   The LDAA instruction was described earlier. However, the word *load* in an instruction name always means to copy a number from a memory register to a microprocessor register. Later you will find load instructions for other registers. The size of the number copied depends on the size of the destination register. Therefore, LDAA transfers an 8-bit number.

- *STAA.*   The STAA instruction copies the number in the A accumulator into the memory register addressed by the instruction. The previous number in the memory register is lost, and the accumulator is unchanged. The word *store* in an instruction name always describes the transfer of a number from a microprocessor register to a memory register. You will see many more kinds of store instructions in later discussions. The size of the number copied depends on the size of the source register.

- *STOP.* The *STOP* instruction is shown in italics, because this term is NOT the correct name of the instruction with an op code of 3F. The *STOP* instruction is used in Motorola trainers to stop a program. The 3F instruction will stop a program correctly only when used in a Motorola trainer. The real name and operation of this instruction will be discussed much later, because it is very complicated. The examples that follow use the *STOP* instruction to stop the program.

If you are familiar with the Fortran or Basic programming languages, you recognize the name STOP used here for the *STOP* instruction because those languages have a STOP that has the same purpose.

Instruction similarities

All instructions have fetch and execute phases that are similar. Therefore, you need not learn the detailed hardware operation for every instruction to write programs. The instruction set table does not describe all hardware operations; it only describes the functions performed by the instructions. Accordingly, this chapter continues at a lesser level of detail describing only the functions performed by the instructions.

Machine language programming

The binary numbers that form a program in the memory of the computer are called *machine language*. Machine language programming is the process of creating a sequence of binary instruction codes and then putting them into the memory to make a program. A program is just a series of instruction codes. A machine language program directly controls the hardware in the computer.

Let's look at a small machine language program that uses the instructions covered so far. This exercise demonstrates the computer operation during a complete program. The program adds two 8-bit numbers from memory registers to form an 8-bit sum that is put into a memory register. The program will do no more than add—nothing else such as error checking will be done. Therefore, the program adds either unsigned numbers or two's complement signed numbers.

Program loading.   The phrase *loading a program* means to put the instruction numbers and data numbers into the memory. Loading can be done by a person keying at a keyboard or by a computer program controlling a peripheral device that holds the numbers.

Figure 2-16 shows the add program loaded into memory. The numbers to be added, called *N1* and *N2*, are placed in memory locations C20B and C20C. The program must put the sum, called *SUM*, into location C20D. The number in location SUM was initially set to FF; when the program changes the answer, it will be obvious. You should always put in an

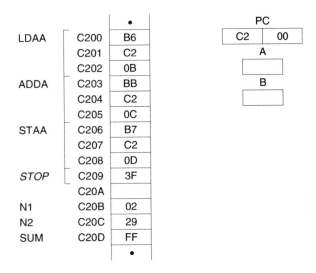

Figure 2-16   The first machine language program.

incorrect answer when testing a program—bad luck will cause the correct answer to occur accidentally when the program is incorrect.

Figure 2-16 shows the computer ready to run because the program counter contains C200—the address of the first instruction. The accumulators are blank because the numbers in them are unknown when the computer is started.

The locations—memory addresses—of both the instructions and data numbers in the figure have no significance. Both the instructions and data numbers could be located anywhere in the memory. The example does not use memory location C20A to show that the data numbers do not need to be next to the instructions. The instructions and data numbers may be widely separated in memory. The numbers in all other memory registers and microprocessor registers do not affect this program.

Program operation.   Let us follow the operation of the program in detail. The logic of the program is N1 + N2 → SUM.

Assume that the computer started running somehow—a person has pressed some buttons. A trainer computer requires a single command to start the computer. The sequence of the instructions in the program is as follows:

- *First instruction.*   Figure 2-17 illustrates the state of the computer after it has fetched and executed one instruction. The program counter is pointing to the next instruction at location C203. The first data number N1 has been loaded into the A accumulator. Notice that memory location C20B was not affected.

- *Second instruction.*   Figure 2-18 illustrates the computer after two instructions. The program counter is pointing to the next instruction at location C206. The two numbers, N1 in the A accumulator and N2 in a memory register, have been added

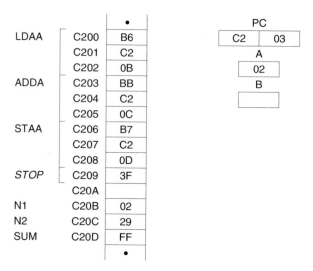

Figure 2-17   Machine language program—one instruction executed.

and the sum is in the A accumulator. Whatever number was in the A accumulator has been lost.

- *Third instruction.*   Figure 2-19 shows the result after three instructions. The program counter contains C209, so it is ready for the fetch of the fourth instruction. The sum has been stored into memory, and the number in memory location C20D was lost. The STAA instruction did not affect the contents of the A accumulator.

- *Fourth instruction.*   The last instruction, *STOP*, only stops the computer and does not alter any registers in the microprocessor or memory.

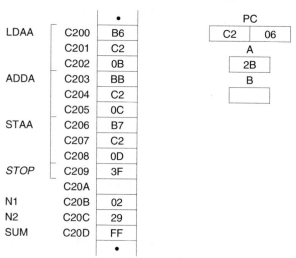

Figure 2-18   Machine language program—two instructions executed.

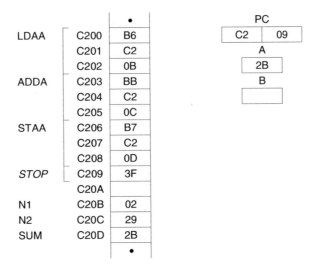

Figure 2-19   Machine language program—three instructions executed.

Problems and common errors

Many kinds of mistakes could be made in loading and running the example program in Figure 2-16. Here is a list of some of the more likely errors and the kinds of consequences that you should expect. The variety of possible actions that could occur is almost limitless, but learning that so many things could go wrong is valuable.

Here is a list of common errors:

- *The program counter is loaded with an incorrect address.* The computer will use whatever numbers it encounters as instructions, which will probably lead to chaos. As an example, suppose that the program counter was loaded with C201, which is only one binary bit different from the correct value. The first op code fetched will be C2 instead of B6. From here on, the program operates incorrectly.

- *The STOP instruction is left out.* Assume further that the numbers in locations C209 and C20A are not 3F. The computer will use the numbers starting at address C209 as instructions—probably executing the data numbers as instructions. What happens will depend upon the numbers encountered and hence is unpredictable.

Sometimes the computer will continue using garbage from memory until it accidentally runs into some good program and it starts operating accordingly, which can be very confusing! You intend to run your program, and some other program runs instead.

- *An instruction code is entered incorrectly.* Certainly an incorrect code will make the program work incorrectly. But suppose that the number at location C208 was entered as 08. Then the address part of the STAA instruction would be C208. When this program is run, the sum of two numbers will be stored at location C208. The incorrect sum FF is left at location C20D. Regardless, the program will run to the *STOP* instruction, giving the impression of correct operation until you find the wrong answer.

But the problem is more serious than that. Storing the result at C208 modifies the STAA instruction. Therefore, the program will do something different when it is run a second time. After that, the program will do the same thing every time as long as the data numbers are not changed. Usually programs must be run several times before you can have any confidence that they are working correctly. And even then, you should be cautious.

2.2 INSTRUCTIONS AND ADDRESSING MODES

Determining how the instructions will access data is a major consideration in the design of a microprocessor. How the instructions access data significantly affects the use and programming of the microcomputer. In particular, if the data number for an instruction is in memory, the microprocessor will form a memory address to access it. If the data number is in a register inside the microprocessor, a memory address is not needed.

The address formed by the microprocessor as part of the instruction execution is called the *effective address*. The microprocessor uses the effective address to control the memory. In some designs, the effective address may be formed from several different numbers.

Some instructions can tell the microprocessor to form the address of the data several ways. The various ways are called the *addressing modes* of the computer. An addressing mode is a particular way that the microprocessor forms the effective address of a data number in memory.

The Motorola 68HC11 has six addressing modes. You will find that certain categories of instructions have a subgroup of the possible modes. Learning the instruction set so that you can program the computer is simplified by first learning the addressing modes.

This section introduces a few new instructions to illustrate the addressing modes. This section also introduces all the fundamental addressing modes of the Motorola 68HC11 and gives a brief description of their uses. Later you will become aware of the tradeoffs in using various addressing modes. Only the basics are covered here; minor variations are covered later after you have some experience.

Extended Addressing

The Motorola 68HC11 uses 16-bit addresses. *Extended addressing* means that the complete 16-bit address of the data is in the instruction code. The address can be any number from 0000 through FFFF. All instructions with extended addressing have the format that Figure 2-20 shows.

The name *extended* that Motorola chose for this mode has little meaning at this point. Some other companies call an identical addressing mode *absolute addressing*. Regardless of the name, the idea is that the complete address of the data is in the instruction code.

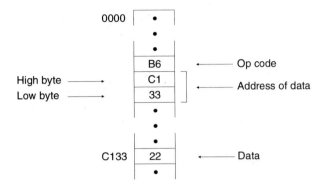

Figure 2-20   Memory format of extended addressed instructions. The example is a Load Accumulator A instruction.

All instructions with extended addressing have two distinct parts—an op code and an address. The instruction works the same at any location in memory, so Figure 2-20 does not specify addresses for the instruction. The data number also can be anywhere in the memory; the example shows it at location C133.

Some possible locations for the instruction make no sense, but the hardware won't care and will operate correctly. For example, the op code could be at FFFF and the address part at addresses 0000 and 0001. Because FFFF is the highest possible memory address, the program counter will wrap around to zero as the instruction is fetched. Certainly this operation should be avoided.

Another example, although somewhat more devious, is an instruction that addresses its op code as the data number. Both examples are *very bad programming practice*, but the hardware in the computer doesn't care!

Example

The example in Figure 2-16 illustrates extended addressed instructions. Extended addressing is the simplest addressing mode, so it was used for the instructions in the first programming example.

Memory map and address range

A visual technique makes understanding the implications of the addressing modes easier. A *memory map* is a diagram of the whole possible memory that indicates uses of groups of registers. The memory map displays all possible memory even though hardware may not be installed for all the registers. Individual registers cannot be shown, so only an outline of the memory areas is used. A memory map conveniently shows the range of registers accessible to an instruction with a certain addressing mode. Many instructions cannot access all the memory registers.

Figure 2-21 is a memory map for the extended addressing mode. It shows all possible locations of the data accessible to a given instruction by bracketing that memory region. The

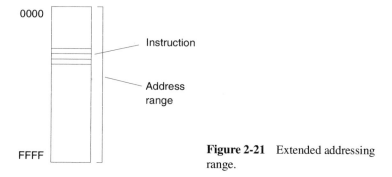

Figure 2-21   Extended addressing range.

number of bytes the instruction occupies and the location of the instruction are also shown. The address of the instruction is not specified to indicate that the instruction can be anywhere in the memory.

The memory map is very simple for the extended addressing mode because it can access the entire memory. Other addressing modes are more complicated because only portions of the memory can be accessed.

Direct Addressing

Suppose that the data numbers for the example in Figure 2-16 were moved to locations 0010 through 0012. Figure 2-22 shows the new program—look carefully at the instruction codes. The address parts of all instructions in the figure that reference memory have 00 as the most-significant byte. These zero bytes lead one to think that a compromise could be made to shorten the length of the instruction. A shorter instruction would be valuable. It would use less memory, and it could be fetched faster. Now we are looking at the performance aspects of the computer. Shorter and faster instructions would improve the performance or apparent speed of the computer without the greater expense of faster hardware.

Motorola made a second addressing mode called *direct addressing* to gain these advantages. In this mode, the most-significant byte of the address is 00 by default and can be left out of the instruction code. A new op code is necessary to specify this way of forming the address of the data.

Direct addressed instructions require only two bytes of memory. The first byte is an op code, as it is for all instructions. The second byte is the least-significant or low byte of the address of the data. The high byte of the address of the data always defaults to 00.

Figure 2-24 shows the memory format of the direct addressed instructions. Addresses are not shown for the instruction, because it can be anywhere in memory.

Using direct addressed instructions imposes the limitation that the data numbers must be placed in the address range 0000 through and including 00FF. The benefits of this addressing mode outweigh the limitations. Because direct addressing is very practical, most 68HC11 programs will have some data numbers in the direct addressing range. The instructions will be put elsewhere in the memory to conserve the valuable direct addressing range.

		•
N1	0010	02
N2	0011	29
SUM	0012	FF
		•
		•
		•
LDAA	C200	B6
EXT	C201	**00**
	C202	10
ADDA	C203	BB
EXT	C204	**00**
	C205	11
STAA	C206	B7
EXT	C207	**00**
	C208	12
STOP	C209	3F
		•

		•
N1	0010	02
N2	0011	29
SUM	0012	FF
		•
		•
		•
LDAA	C200	96
DIR	C201	10
ADDA	C202	9B
DIR	C203	11
STAA	C204	97
DIR	C205	12
STOP	C206	3F
		•
		•
		•
		•

Figure 2-22 Modified program with data numbers at new locations.

Figure 2-23 Program using direct addressed instructions.

Example

Figure 2-23, shown next to Figure 2-22, shows the program to add two numbers using direct addressed instructions. Compare this program to the equivalent program using extended addressed instructions in Figure 2-22. These figures are shown side-by-side to make this comparison easy. You will see that the program with direct addressed instructions is three bytes shorter. The program uses less memory and requires three fewer clock ticks for fetching the instructions.

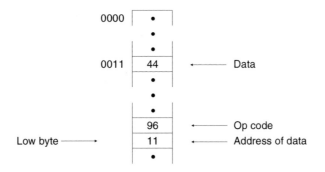

Figure 2-24 Memory format of direct addressed instructions. The example is a Load Accumulator A instruction.

Memory pages and addressing range

The addressing range of direct addressing is limited by its 8-bit address. A 16-bit address is needed to specify the whole address range. The same limitation applies to other addressing modes where not enough bits are available for the address. The limited range of memory that can be accessed is called a *page* of memory.

Some people call direct addressing *zero page addressing* because the high byte of all the addresses is 00. You will even find this name in some Motorola manuals.

The memory map in Figure 2-25 shows the direct addressing range and the 2-byte instruction. You can easily visualize the page of memory. The imaginary line between locations 00FF and 0100 marks the end of the page—nothing is physical about this line.

Direct addressing is a compromise between the size of the addressing range and speed of program execution. The range of addresses for direct addressing is only 0000 through 00FF, not the entire memory as with extended addressing. Now you can see why extended addressing has its name—its range is extended or greater than the direct addressing range.

Inherent Addressing

Some instructions operate on data in microprocessor registers only. A memory address is not needed, so only the op code part of the instruction is necessary. Instructions requiring only an op code have *inherent addressing*. The op code specifies the operation and the internal microprocessor register or registers. Inherent addressing really designates the lack of a memory addressing mode.

The ABA instruction has inherent addressing. ABA adds the B accumulator to the A accumulator and then puts the sum into the A accumulator. The B accumulator is not changed. The mnemonic name ABA is easier to remember if you read it as *add B to A*.

Look at the instruction set table in Appendix A for more details. A memory map is not appropriate to this addressing mode.

Example

Figure 2-26 shows a program to add two numbers using the ABA instruction. The logic of this program is slightly different from the version in Figure 2-23. You should see that the

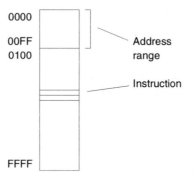

Figure 2-25 Direct addressing range.

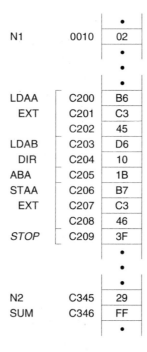

		•
N1	0010	02
		•
		•
		•
LDAA	C200	B6
EXT	C201	C3
	C202	45
LDAB	C203	D6
DIR	C204	10
ABA	C205	1B
STAA	C206	B7
EXT	C207	C3
	C208	46
STOP	C209	3F
		•
		•
		•
N2	C345	29
SUM	C346	FF
		•

Figure 2-26   Program with inherent addressed instruction.

two accumulators hold the numbers to be added. The program puts the sum into memory and leaves the second number in the B accumulator. Leaving a number in a register at the completion of a program segment is useful in certain programming strategies.

The example also uses some instructions with extended addressing, direct addressing, and inherent addressing. This example illustrates that instructions with different addressing modes may be used in the same program.

Implied addressing

The Motorola 68HC11 and many of Motorola's other products have inherent addressing. Strangely, some of Motorola's manuals call inherent addressing *implied addressing*. If you have a manual that uses the name *implied addressing*, just substitute the name *inherent*, because it is referring to the same addressing mode.

Double-byte Data

Some registers in the microprocessor hold 16-bit or double-byte numbers. Therefore, some instructions, such as a load instruction for one of these registers, must address double-byte numbers in memory.

Double-byte numbers in the Motorola 68HC11 are always stored in two consecutive memory registers with the high byte at the lower address and the low byte at the higher address.

This ordering convention for the bytes is called the *big-endian* convention. Some computers use the *little-endian* convention where the low byte is at the lower address and the high byte is at the higher address. Still other computers can use either convention.

An instruction contains only the address of the first or high byte of the data. The microprocessor automatically accesses the second byte of the data at the next address. The instruction op code specifies that the data number is two bytes long. The instruction format for direct and extended addressed instructions is the same for both single- and double-byte data numbers.

A double-byte register

The programming model in Figure 2-27 includes the double-byte register called *index register X*. The instruction register will no longer be shown in the figures because its operation should now be apparent. The index register will be used to demonstrate addressing of double-byte numbers.

The purpose of the X register is discussed later. At this point, only the functions of the load and store instructions for X will be considered. They are named LDX and STX. In spoken language, say *load X* and *store X*. The X implies an index register even though the type of register is not in the name as it is in the load accumulator instruction. An example of the LDX instruction is in the next section.

More notation

Describing the function of the LDX instruction in spoken language is simple—it loads a double-byte number from memory into the X register. In contrast, describing the function performed by the LDX instruction in precise written notation is difficult. It is difficult because the data number is two bytes long and only one of the bytes has an address in the instruction.

Here is the notation used in the instruction set table in Appendix A. It says that the operation of LDX is M:(M + 1) → X. As discussed earlier, multiple meanings of the notation are used together. The symbols to the left of the arrow designate the double-byte number that will be transferred to register X. The double-byte number is formed from two 8-bit or single-byte numbers that are *concatenated*—they are placed end-to-end. The colon indicates concatenation.

The M to the left of the colon designates the contents of the memory register that is addressed by the instruction. This use of M is the same as before. But the M on the right side

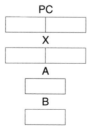

Figure 2-27 Programming model including the X index register.

of the colon designates the address, not the contents, of the memory register specified by the instruction. On the right side of the colon, one is added to the address to form the address of the next byte higher in memory. Finally, the parentheses around the M + 1 designate the contents of that memory location. The two bytes from the two consecutive memory locations are concatenated and transferred to X. The byte at address M is put in the left half of X, and the byte at address M + 1 is put in the right half.

The notation is inconsistent and confusing! Leaving out the necessary parentheses to make the notation shorter causes the problem. The address of the memory register and the contents of that memory register cannot be distinguished properly.

Figure 2-28 illustrates the LDX instruction with extended addressing. The notation M refers to the contents of location C234, which is the number 56. The notation (M + 1) refers to the contents of location C234 plus one, which is the number 78. The index register is loaded with 5678.

Immediate Addressing

To understand immediate addressing, you need the concept of a constant. A *constant* is a data number that never changes unless a new program is written. Running the program will never change the constant number. Constants may relate to physical quantities like the number π, or to data organization within the program such as table spacing.

A constant can be put inside an instruction code. The addressing mode that places a constant data number inside the instruction is called *immediate addressing*. Immediate addressed instructions do not contain an address. Figure 2-29 shows that they have an op code and a data number.

When immediate addressed instructions are fetched, the data number is fetched along with the op code. At the completion of the fetch phase, the data is immediately available. No more clock ticks are necessary to return to memory to get the data number.

Immediate addressed instructions require fewer clock ticks than similar extended or direct addressed instructions. Also, the instruction and data numbers together require fewer bytes of memory. So even if a constant is needed many places in a program, immediate addressed instructions will be shorter and faster than equivalent instructions with other addressing modes.

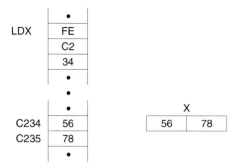

Figure 2-28   Extended addressed instruction with double-byte data.

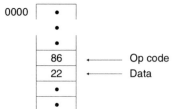

Figure 2-29 Memory format of immediate addressed instructions that have single-byte data numbers.

Example

Figure 2-29 shows a Load Accumulator A instruction with immediate addressing that loads the single-byte number 22 into the accumulator. Figure 2-30 shows the LDX instruction that loads the number 1234 into the X register. The formats of both instructions should look familiar to you. Both instructions have an op code followed by an operand.

The operand for immediate addressed instructions is a data number—the data number is inside the instruction. The LDAA instruction has a single-byte data number because the A accumulator holds single-byte numbers. The LDX instruction has a double-byte data number because the X register is two bytes long. Because the notation is difficult, you should verify that the LDX does the operation M:(M + 1) → X.

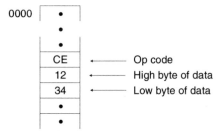

Figure 2-30 Memory format of immediate addressed instructions that have double-byte data numbers.

Self-modifying programs

Consider another viewpoint of the constants contained in immediate addressed instructions. Sometimes an instruction must store a number that will be used by another instruction. In this case, immediate addressing will not be useful. Here is the problem. The immediate data value could be changed only by another instruction storing the new number inside the immediate addressed instruction—a very poor programming practice in the microcomputer environment. *A program must never change instruction codes as the program runs.* Self-modifying programs are impractical and very difficult to debug. However, the computer hardware can access the data inside the immediate addressed instruction using other instructions.

To avoid such problems within a single instruction, store instructions don't have immediate addressing. If such an instruction were possible, that single instruction would make a self-modifying program—the store immediate instruction would store the data into itself!

You can verify that store instructions do not have immediate addressing by scanning the instruction set table in Appendix A.

The discussion above discourages writing programs that change the data within instructions. Similarly, never write programs that modify instruction op codes as they run.

Data ownership

No other instruction can use the data number within an immediate addressed instruction. *The constant within the immediate addressed instruction is owned by that instruction and no other instruction may use it.* Good programming practice requires this viewpoint. By contrast, many different instructions usually reference the data numbers used by direct and extended addressed instructions.

Addressing range

Figure 2-31 illustrates the addressing ranges for immediate addressed instructions. The range is quite limited because it includes only the space in the instruction for data. It is interesting that the data number moves to a different location if the instruction is moved to a new location. In contrast, the data stays in fixed locations when direct or extended addressed instructions are moved to new locations.

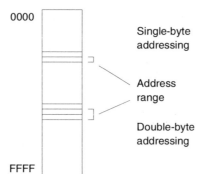

Figure 2-31   Immediate addressing range for single-byte and double-byte data instructions.

Indexed Addressing

Indexed addressing is much more complicated than the addressing modes discussed before. Let us begin by looking at the mechanics of how indexed addressing works. A later section explains the need for this addressing mode.

Indexed addressed instructions contain an op code and an offset byte. The instruction uses an index register in forming the effective address; that is, the address of the data. Figure 2-32 illustrates the format of these instructions using the LDAA instruction and the X index register.

An indexed addressed instruction forms the effective address by adding the offset byte from the instruction to the number in the index register—the index register is unchanged. Therefore, the effective address of the data number is the sum of two numbers. The number

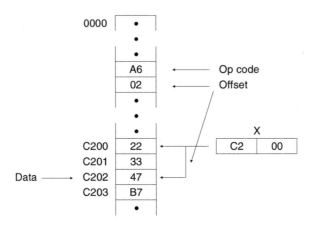

Figure 2-32   Memory format of indexed addressed instructions. The example is a Load Accumulator A instruction.

in the index register is a double-byte number, and the offset in the instruction code is a single-byte number. Both are unsigned numbers. Consequently, the lowest address for the data is the same as the number in the index register. The highest address is the number in the index register plus FF. Once again, FF is the largest unsigned number—not minus one!

The offset number inside the instruction is a constant. Changing the offset requires writing a new program that changes the instruction code. On the contrary, other instructions in the program probably will change the number in the index register.

Example

Figure 2-32 shows a way to visualize the indexed addressing mode using a specific LDAA instruction. Think of the index register X as pointing to the memory register with address C200. The index register is a *pointer* because it contains an address.

The data number is in the memory register that is offset in the higher direction by the offset distance, which is 02 for the instruction in the figure. So the instruction in the figure would load the A accumulator with the number 47, which is at address C202. The C202 is formed by adding C200 from the index register to 02 from the instruction.

The LDAA instruction does not change the number in the index register. Other indexed addressed instructions, such as LDX, will change the index register after it has been used to form the address.

Memory pages and addressing range

Indexed addressed instructions access a page of memory, because the offset in the instruction is only eight bits long. The index register determines the location of the page. The page includes those memory registers offset by 00 through FF from the address in the index register. Because the index register is 16 bits long, it can point to any address in the whole possible memory. So indexed addressed instructions can access the entire memory.

Figure 2-33 illustrates the indexed addressing range as the offset varies while the number in the index register remains unchanged. The size of the indexed addressing page is the same as the size of the direct addressing page. One viewpoint of indexed addressing is that it is direct addressing with a movable page. Both instruction formats have an op code and one byte for forming the effective address of the data.

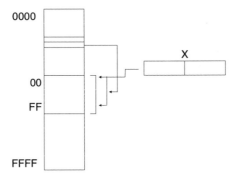

Figure 2-33 Indexed addressing range as offset varies from 00 to FF.

Purpose

Here is a preview of the use of index addressing. Many programs need to change the address part of instructions as the program runs, but the instruction codes must not change. Indexed addressing requires part of the effective address to be in the index register in the microprocessor. The contents of that index register can be changed by the program. The effect of altering the index register is to alter the effective address of the instruction. The instruction code including the offset byte remains unchanged.

Relative Addressing

The 68HC11 branch instructions have only program relative addressing, and only branch instructions have program relative addressing. In the 68HC11, program relative addressing and branch instructions go together. You must learn about branch instructions to make sense of program relative addressing—the two topics are inseparable.

Branch instructions make decisions by either altering the program flow or not altering the program flow. The decisions are two-way binary decisions. Information in the condition code register in the microprocessor determines whether a branch will occur. You must understand the condition code register to use the branch instructions correctly.

Condition code register

The 8-bit microprocessor register that holds test results is called the *condition code register*. The individual bits in this register report individual test results. Therefore, you need to examine the bits of this register one at a time. The other registers in the microprocessor are thought of as containing 8-bit or 16-bit numbers. The condition code register is used bit by bit.

Bit numbers.   The bits within a register are numbered. The numbering scheme is applicable to all registers. However, it is especially useful in discussing the condition code bits and the tests related to them. Figure 2-34 shows the bit numbers for an 8-bit register. The bits are always numbered starting with 0 at the right and continuing to the left while counting in decimal. Therefore, bit 15 is the left bit of a 16-bit register.

The origin of this numbering scheme may help you remember it. If the number in the register is a weighted number, the weights of the positions starting at the right are 2^0, 2^1, 2^2, etc. The bit numbers are the exponents for these weights.

Bit numbers are particularly useful in verbal communications. If you speak of *bit 1* rather than the *second bit*, you will avoid confusion. The second bit could be the second bit from the right or the second bit from the left.

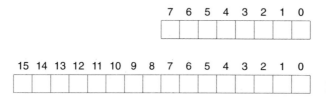

Figure 2-34   Register bit numbers.

Bit names.   The bits of the condition code register are referenced very frequently, so they have been given word names. Also, the word names are shortened to single-letter names. The instructions in a program will use the bits in the condition code register in many different ways. The word names help to identify the bits in terms of their uses.

Unfortunately, assigning names is difficult because not all instructions use the bits the same way. A name that is appropriate to one instruction is incorrect for another instruction. However, only one name can be assigned to each bit. So don't assume too much when you learn the names of the bits. You must interpret the names according to the operation performed by the instruction you are considering.

Condition code bits.   Figure 2-35 shows the condition code register with both the bit word names and the bit letter names. Some bits in this register indicate test results while others control microprocessor hardware. The S, X, and I bits are hardware control bits that are discussed in later sections. Ignore these three bits when considering branch instructions. The remaining five bits—H, N, Z, V, and C—indicate test results although they may not be those you expect from the bit names.

The microprocessor performs certain tests automatically each time an instruction is executed. The microprocessor does these tests without the need for instructions directing the tests. Programs ignore the test results when they are not needed. Because the microprocessor does not need extra time to perform the tests, they cost nothing in terms of performance.

Usually, the test is performed on the result generated by an instruction. When a test is performed, the condition tested for is either true or false. A true test result is indicated by 1 in a condition code bit, and a false test result is indicated by 0.

The names of the condition code bits fit the ADDA instruction best, so it will be the example. Figure 2-36 shows the addition of two binary numbers where number N1 is in the

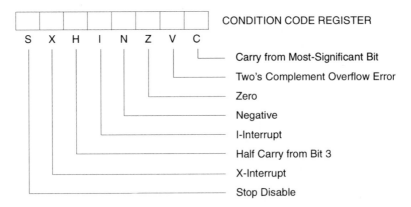

CONDITION CODE REGISTER

S   X   H   I   N   Z   V   C

— Carry from Most-Significant Bit
— Two's Complement Overflow Error
— Zero
— Negative
— I-Interrupt
— Half Carry from Bit 3
— X-Interrupt
— Stop Disable

Figure 2-35   Condition Code Register bit identifiers.

A accumulator and number N2 is in memory. The instruction puts the sum in the A accumulator. The ADDA instruction affects the five condition code bits that report test results as follows:

- *C—The C bit indicates the result of testing for a carry from the most-significant or left bit.* One indicates a carry occurred; zero that no carry occurred. C is 1 in the example because a carry from bit 7 occurred. The C bit is not significant if the addition was done on two's complement signed numbers. However, for unsigned numbers, the carry means the sum was too big for the register.

- *V—The V bit indicates the result of testing for a two's complement overflow error.* The example had such an error because adding two negative numbers gave a positive sum. So, V is 1 in the example. For the V bit to give significant results, the numbers must be two's complement signed numbers. If your numbers are unsigned, ignore the V bit.

- *Z—The Z bit indicates the result of testing the result of the instruction for zero.* The result must contain all zero bits for Z to be 1 or true. Z is 0 in the example because

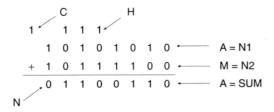

```
        C           H
1        1  1  1
    1  0  1  0  1  0  1  0  ◄──── A = N1
+   1  0  1  1  1  1  0  0  ◄──── M = N2
    ─────────────────────
    0  1  1  0  0  1  1  0  ◄──── A = SUM
N
```

x	x	1	x	0	0	1	1	Resultant CCR
S	X	H	I	N	Z	V	C	Unknown = x

Figure 2-36   Condition code bits following ADDA instruction execution.

the result that went to A was not zero. The Z bit confuses many people who expect it to be 0 when the result is zero.

- *N—The N bit indicates a negative result from the instruction.* In effect, it is a copy of the left bit of the result. N is 0 in the example because the sum of the numbers is positive. The N bit is only significant for two's complement numbers. If your numbers are unsigned, ignore the N bit.

- *H—The H bit indicates a carry halfway through the 8-bit number; that is, it indicates a carry from the bit 3 position.* H is 1 in the example. The H bit is generally used only with BCD numbers.

Condition code bit notation.   Not all instructions affect all the condition code bits. Some bits are not meaningful or are not affected during the execution of some instructions. Motorola uses a notation to indicate the effects an instruction has on the condition code bits. Table 2-1 lists this notation. The instruction set table in Appendix A uses the same notation.

The notation - means that the bit is unaffected by the instruction. The instruction does not change the condition code bit under any circumstance. Sometimes an instruction that controls a condition code bit is followed by an instruction that does not affect that bit. Then the next instruction after those two uses the information left in the condition code bit. Therefore, leaving the bit unaffected is useful.

An instruction may always put a 0 or a 1 in a condition code bit. The most obvious example is an instruction designed to set a condition code bit to 1. Certainly that instruction will put a 1 into the bit.

The notation ↕ means, in informal language, *the bit works like you think it ought to work!* More formally, the bit is affected by the instruction, although you may have to interpret the result. An example is the C or carry bit test for a subtraction instruction. The subtraction cannot cause a carry, but you can extrapolate the meaning carry to the meaning borrow. Indeed, the C bit indicates whether a borrow from outside the number occurred during a subtraction. But the C bit is not given a new name when it is used with a subtract instruction—it is still called the C bit. Look at the SUBA instruction in the instruction set table in Appendix A.

Caution! People often assume they know how the condition code bits work from the name of an instruction. This assumption is dangerous because the condition code bits are complicated and don't work as expected for all instructions. *Always use the instruction set table to determine how the condition code bits work.* Many programming errors occur because people assume too much about condition code bits.

TABLE 2-1   CONDITION CODE NOTATION

Symbol	Operation
-	Bit is unaffected by this instruction.
0	Bit is always cleared to 0 by this instruction.
1	Bit is always set to 1 by this instruction.
↕	Bit is set or cleared depending on instruction.

Branch instruction format

Branch instructions allow a program to make decisions. A branch instruction may alter the program flow based on the information the condition code register provides. A branch instruction is always used with another instruction that leaves test results in the condition code bits.

A branch instruction, when it branches, alters the contents of the program counter register. If it doesn't branch, the program counter is not altered, and the next instruction follows immediately after the branch instruction. If the program counter is altered, the next instruction is fetched from the effective address created by the branch instruction.

The 68HC11 has many different branch instructions that respond to almost any decision to be made. A later section covers the details of the branch instructions.

The effective address.   Remember that the branch instructions have only program relative addressing which is illustrated in Figure 2-37. A program relative addressed instruction forms the effective address by adding the relative offset byte of the instruction to the program counter. The sum is put into the program counter changing the address that it contains. This addressing mode, unique to the branch instructions, is similar to indexed addressing. Relative addressing uses the program counter as a kind of index register.

The effective address of a relative addressed instruction depends on the location of the instruction in memory. The effective address formed is relative to the location of the instruction—the same instruction code will form different addresses if it is located at different places.

Here are some important details about the effective address. The effective address is formed during the execute phase of the instruction. The program counter will have been incremented during the fetch phase. The program counter already points to the next instruction before the execute phase begins. During the execute phase, the program counter may be altered. If not, the next instruction follows in order because the program counter was already pointing at it. If it is altered, the relative offset byte is added to the incremented program counter.

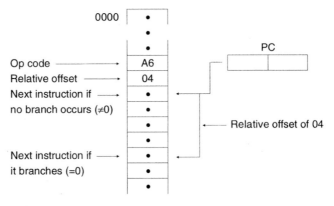

Figure 2-37   Memory format of program relative addressed branch instructions. The example is a Branch on Equal to Zero (BEQ) instruction.

To repeat, the program counter is not pointing to the op code of the branch instruction at the time the effective address is formed. Some manuals imply that it is by referring to the program counter plus a number, such as PC + 2. Don't be confused by this notation.

The BEQ instruction.   Figure 2-37 illustrates the BEQ—branch on equal to zero—instruction and program relative addressing. The BEQ instruction examines the Z bit in the condition code register to determine whether to branch. The condition of the Z bit depends on an instruction executed sometime before the BEQ. If Z = 0, the previous instruction had a result that was not zero, so the next instruction will follow the BEQ. If Z = 1, the last result was zero and the BEQ branches to the location determined by the relative offset.

Look at the BEQ in the instruction set table in Appendix A. The Boolean expression column describes only the use of the condition code bit. The ? Z = 1 means the branch occurs if this question has a true answer.

Branch instruction example

The program example in Figure 2-38 demonstrates the operation of a branch instruction. The program determines if the data number in memory location C444 is zero. If the data number is zero, the program stops at address C208. If the data number is not zero, the program stops at address C205. The program examines the data number by loading it into the A accumulator with a LDAA instruction. The LDAA makes the Z bit respond. The BEQ instruction then either does or does not branch based on the information in the Z bit.

Addressing range and pages

The relative offset byte in a branch instruction is an 8-bit *two's complement signed number*. The offset can be either positive or negative. The examples so far have positive numbers, so you could not detect that the offset is signed.

The 8-bit size of the offset causes a page of memory for the addressing range of the branch instructions. The instruction reaches locations with both higher and lower addresses

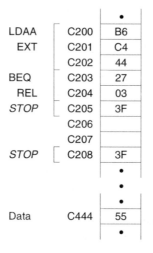

Figure 2-38   Example using a branch instruction.

than that of the instruction. Adding a positive offset generates a higher address. Adding a negative offset generates a lower address.

Figure 2-39 illustrates the memory map for the program relative addressed branch instructions. The page is located with the instruction approximately in the middle—the exact numbers depend on how you count. Some people start counting at the next instruction, and others start at the op code of the branch instruction.

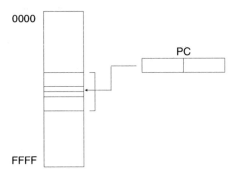

Figure 2-39   Program relative addressing range.

Unconditional and long branches.   The branch instructions discussed so far are called *conditional branches* because they branch only on the correct conditions. The 68HC11 also has some unconditional branches that don't depend on the condition code bits. One of these is the *branch always instruction*, BRA, that has program relative addressing. Another is the *jump instruction*, JMP, that also branches unconditionally but has extended and indexed addressing. The name *jump* clearly sounds different from *branch*. Using different names helps you remember that JMP does not have relative addressing.

The JMP can branch to any location in the whole memory. Long conditional branches to any location in memory are accomplished with two instructions, a branch and a JMP.

Relative addressing idiosyncrasies

The way the effective address is formed may bother you. Think about adding the 16-bit number in the program counter to the 8-bit relative offset. The address in the program counter is an unsigned number. Addresses increase from 0000 at the low end of memory to FFFF at the high end of memory. Address FFFF is not called minus one!

The relative offset in the instruction is a two's complement signed number. So the microprocessor apparently adds a 16-bit unsigned number to an 8-bit signed number. Normally, you would not expect useful results from such an addition.

However, the microprocessor does calculate the correct effective address. For this purpose only, think of the address in the program counter as a two's complement number. Treating the address as a two's complement number is possible because the program counter is only 16 bits long—incrementing the program counter when it contains FFFF gives 0000. Next, extend the 8-bit signed offset to a 16-bit signed offset by copying the sign bit eight times. The microprocessor gets the correct address by adding these two 16-bit numbers. Of course, the correct effective address is now interpreted as an unsigned number.

Relative addressing problems

Some effective addresses that can be formed by branch instructions are not useful. For example, a relative offset of FF or minus one is incorrect—it causes a branch to the offset part of the branch instruction. Similarly, an offset of minus two is incorrect because it makes an infinite loop when the instruction branches to itself.

Caution! Be careful to use the correct offset. Errors cause branches to incorrect locations, so errors may crash the program. The error is worse if the program doesn't crash because you may not realize that an error occurred. If you are hand coding a program, you can get the correct offset by counting from the branch instruction to the destination point. Start counting at the next register after the branch instruction with the number zero. If the destination is at a higher address, count in the positive direction. If it is at a lower address, count in the negative direction.

2.3   ADDRESSING MODE SUMMARY

The 68HC11 has six basic addressing modes. The fundamentals of these have been introduced in this chapter. A later chapter introduces some advanced instructions with more complicated addressing that is based on the fundamental modes.

The following summarizes the characteristics and uses of the six fundamental addressing modes. Figure 2-40 is a graphic presentation of the memory formats of each mode.

- *Inherent addressing*—the instruction code has only an op code without an operand such as an address. The instruction operates only on microprocessor registers.

- *Immediate addressing*—the instruction code is an op code and a constant data number. The data number is a single- or double-byte number as needed. The data value should be used only by the immediate instruction.

- *Direct addressing*—the instruction code is an op code and the low byte of the address of the data. The high byte of the address defaults to 00. The instruction can address only memory locations 0000 through 00FF. This range of memory is called the *direct addressing page.*

- *Extended addressing*—the instruction code is an op code and a double-byte address that can access the whole memory from address 0000 through FFFF.

- *Indexed addressing*—the instruction code is an op code and a single-byte unsigned offset. The offset is added to an index register to form the address of the data. The data is in a page extending from 00 to FF, higher than the location pointed to by the index register. The double-byte index register can point to any location in the whole memory. The index register is not changed when the offset byte is added to it.

- *Relative addressing*—the instruction code is an op code and a single-byte signed relative offset. The offset is added to the program counter to alter it if the instruction branches. Program relative addressing is available only to branch instructions. The

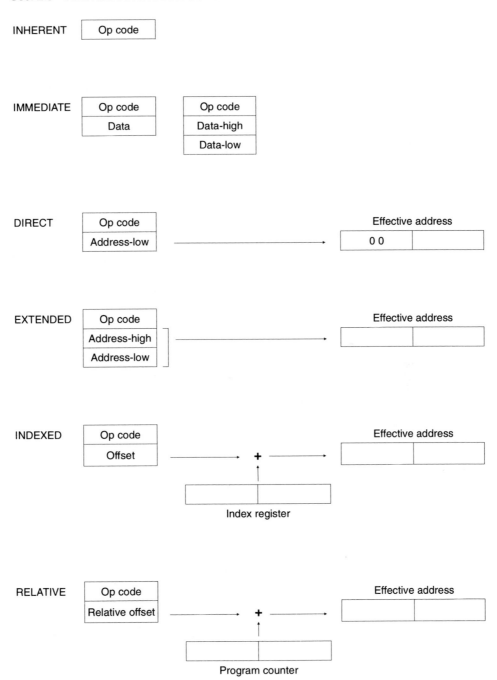

Figure 2-40   Memory formats of instructions with the six fundamental addressing modes.

branch instruction is located approximately at the middle of the addressing page that extends to relatively higher and lower addresses.

You will find knowing the memory formats of the various addressing modes useful because it will help you visualize how a program looks in memory.

2.4   AN EXPANDED REPERTOIRE

You will now see many more instructions that have the basic addressing modes. This section covers many details in a brief discussion. This discussion is based on the assumption that you understand the instruction formats and the addressing modes. Once you know these basic categories, you can learn more instructions very quickly. You probably will need to review these instructions several times so that you can remember what instructions are available. Then you will be able to write programs using these tools.

The Instruction Set Table

You will constantly refer to the instruction set table in Appendix A. That table is the principal tool of the programmer. It documents the programming details required to write programs. Figure 2-15 showed only some columns of the table. Figure 2-41 is a reproduction of a few rows of the complete table.

Each column of the instruction set table in Figure 2-41 is described next.

- *Source Form*—lists the instruction mnemonic and specifies any operands required besides the instruction op code. The operand abbreviation designates the type of operand required as follows:

(opr)	data or data address for an instruction that references memory
(rel)	relative offset of program relative addressed branch instruction

- *Operation*—a very short word description of the function performed by the instruction. The description is incomplete for some instructions, so don't depend on it alone. The description for CLC, *clear the carry bit*, is straightforward.

- *Boolean Expression*—the detailed description of the register transfers that specify the function of the instruction. The ABX instruction adds the 16-bit number in X to 00 concatenated with the 8-bit number in the B accumulator, and then puts the sum in the X register.

- *Addressing Mode*—the first item is the microprocessor register used by the instruction when one is used, or it is blank otherwise. The second item is the memory addressing mode. Some inherent addressed instructions use two microprocessor registers. The registers can be ascertained from the instruction mnemonic and are not listed as is true for the ABX instruction.

Motorola 68HC11 Instruction Set

Source Form	Operation	Boolean Expression	Addr. Mode	Machine Code Op Code	Machine Code Operand	Bytes	Cycles	S	X	H	I	N	Z	V	C
•	•	•	•	•	•	•	•	•	•	•	•	•	•	•	•
ABX	Add B to X	X + 00:B → X	INH	3A		1	3	-	-	-	-	-	-	-	-
•	•	•	•	•	•	•	•	•	•	•	•	•	•	•	•
ADDA (opr)	Add Memory to A	A + M → A	A IMM	8B	ii	2	2	-	-	↕	-	↕	↕	↕	↕
			A DIR	9B	dd	2	3								
			A EXT	BB	hh ll	3	4								
			A IND,X	AB	ff	2	4								
			A IND,Y	18 AB	ff	3	5								
•	•	•	•	•	•	•	•	•	•	•	•	•	•	•	•
CLC	Clear Carry Bit	0 → C	INH	0C		1	2	-	-	-	-	-	-	-	0
•	•	•	•	•	•	•	•	•	•	•	•	•	•	•	•
LDX (opr)	Load Index Register X	M:(M + 1) → X	X IMM	CE	jj kk	3	3	-	-	-	-	↕	↕	0	-
			X DIR	DE	dd	2	4								

Figure 2-41 A few rows of the instruction set table.

- *Op Code*—the instruction op code for the corresponding addressing mode. Some advanced instructions in the 68HC11 have 2-byte op codes, so two columns are under this heading. Look at the ADDA instruction with both IND,X and IND,Y addressing in the instruction set table.

- *Operand*—the operand bytes that follow the op code. The abbreviations have the following meaning:

ii	8-bit immediate data
dd	low byte of a direct address
hh ll	high and low bytes of an extended address
ff	unsigned 8-bit offset in indexed addressed instruction
jj kk	high and low bytes of 16-bit immediate data
rr	signed 8-bit relative offset in branch instruction

- *Bytes*—the number of bytes of memory occupied by the complete instruction code. The op code and operand columns specify the bytes required, so the bytes column contains redundant information. For example, the LDX IMM has an op code and two bytes of data for a total of three bytes.

- *Cycles*—the number of E-clock cycles required by the Motorola 68HC11 to fetch and execute the instruction. Other Motorola products have some of the same instructions as the 68HC11, but the number of cycles may differ slightly. The figure shows that the ADDA EXT instruction discussed earlier requires four clock cycles.

- *Condition Codes*—specifies the operation performed on each condition code bit during the execution of the instruction. The instruction has the same effect on the condition code bits regardless of the addressing mode, so the table lists only the condition codes once for each instruction. The condition code notation from Table 2-1 is reproduced here:

-	Bit is unaffected by this instruction.
0	Bit is always cleared to 0 by this instruction.
1	Bit is always set to 1 by this instruction.
↕	Bit is set or cleared depending on instruction.

As the condition codes column shows, the CLC clears the C bit. The LDX affects the N and Z bits, clears the V bit, and does not affect the C bit. For LDX, N is a copy of bit 15 of X.

A Tour through the Basic Instructions

The following discussion provides a tour through the instruction set table in alphabetical order. A few comments and examples are included along the way. Only basic instructions that relate to the A and B accumulators, the X index register, and the condition code register are discussed in this chapter. Chapter 5 discusses the other instructions.

As you look through the instruction set table reading the following comments, you should look for inconsistency. The designers of microprocessors make compromises to simplify the hardware and reduce cost. These compromises often lead to inconsistent features in the instructions. The 68HC11 is very consistent, but not completely so. For example, some instructions will not have all the addressing modes. Similarly, an instruction will be available to operate on the A accumulator, but the parallel equivalent for the B accumulator doesn't exist.

You should find each of the following instructions in Appendix A as you proceed through this tour of the instruction set:

ABA   *Add B to A.*   Does binary addition of the accumulators; the addition is correct for either unsigned numbers or two's complement signed numbers. The C bit is meaningless for two's complement numbers, while the V bit is meaningless for unsigned numbers. No parallel instruction puts the result into the B accumulator.

ABX   *Add B to X.*   Add B accumulator to X index register and put the answer in the X register. ABX is only useful for unsigned numbers such as addresses. Notice that the condition code bits do not respond to this arithmetic operation. No parallel instruction subtracts B from X, nor adds A to X.

ADCA, ADCB   *Add memory with carry to accumulator.*   Same as the ADD instruction except that it also adds the carry bit. These instructions are used for multiple precision addition of either unsigned or signed numbers.

ADDA, ADDB   *Add memory to accumulator.*   These instructions are the fundamental instructions for adding. The condition code bits work as
their names suggest. Notice that only the 8-bit add instructions
ABA, ADCA, ADCB, ADDA, and ADDB affect the H bit. Any
other instruction that affects H only copies it.

BCC   *Branch on carry clear.*   Branch if the C bit is clear (0), and
don't branch otherwise. None of the branch instructions affect
any of the condition code bits. The branch conditions of BCC
are opposite those of the BCS instruction. Each 68HC11 branch
instruction has a parallel instruction with the opposite branch
condition.

BCS   *Branch on carry set.*   Branch if the C bit is set (1), and don't
branch otherwise. BCS is most often used to branch to an error
routine when an unsigned number operation gives an incorrect
result.

BEQ   *Branch on equal to zero.*   Branch if the result of the previous
instruction was zero; otherwise, don't branch. Notice that this
instruction is not branch if the Z bit is clear (0)! Don't interpret
BEQ to mean branch on equal unless you are comparing two
numbers. See the section on comparison branches.

BMI   *Branch on minus.*   Branch if the result of the previous instruction was minus or negative; otherwise, don't branch. The numbers operated on by the previous instruction must be two's
complement signed numbers. Unsigned numbers can't be
negative! Using the word *minus* instead of *negative* in the name
of this instruction is strange. Apparently this name *BMI* avoids
conflict with the name *BNE* used for another instruction.

BNE   *Branch on not equal to zero.*   Branch if the result of the previous instruction was not equal to zero; otherwise, don't branch.
Notice that the branch occurs if Z bit is clear (0). Don't interpret the mnemonic BNE as branch on not equal unless you are
comparing two numbers. See the section on comparison
branches for more discussion.

BPL   *Branch on plus.*   Branch if the result of the previous instruction was plus or positive; otherwise, don't branch. The numbers
operated on by the previous instruction must be two's complement signed numbers. Unsigned numbers can't be positive, because they don't have a sign bit.

BRA   *Branch always.*   Unconditional branch or branch always. A
BRA should never follow immediately after a conditional

branch, because that just reverses the branch condition. Other branches are available that will do the job in one instruction.

BRN	*Branch never.* This instruction does nothing except waste time and space! It is effectively a 2-byte NOP instruction.
BVC	*Branch on overflow clear.* Branch if the V bit is clear (0), and don't branch otherwise.
BVS	*Branch on overflow set.* Branch if the V bit is set (1), and don't branch otherwise. BVS is most often used to branch to an error routine if a two's complement overflow error has occurred.
CBA	*Compare B to A.* Subtract the B accumulator from the A accumulator and throw the answer away! The name *compare* means that the difference from a subtraction is discarded so that the accumulators are not changed. The condition code bits are affected, so this instruction will always be followed by a branch instruction that responds to the comparison. Carefully note the order of subtraction, because the opposite order gives different results.
CLC	*Clear the carry.* Clear, zero, the C bit in the condition code register.
CLR, CLRA, CLRB	*Clear memory or accumulator.* Put the number 00 into the specified register. Zero is used so frequently that an instruction was provided to create it. A LDAA IMM instruction could be used instead of CLRA, but it requires two bytes instead of one. The CLR does not have direct addressing. No clear instructions exist for any other registers.
CLV	*Clear overflow.* Clear, zero, the V bit in the condition code register.
CMPA, CMPB	*Compare accumulator to memory.* Subtract the contents of a memory register from the accumulator and discard the answer so that the accumulator is unchanged. Makes the condition code bits respond. See CBA.
COM, COMA, COMB	*Complement memory or accumulator.* Complement all the bits of the specified register by changing 0s to 1s and 1s to 0s. Notice that the operation is done by a subtraction. The carry bit is always set to 1. The COM does not have direct addressing.
CPX	*Compare X to memory.* Subtract a double-byte number in memory from the X index register and discard the answer; the X register is unchanged. Makes the condition code bits respond.

DEC, DECA, DECB   *Decrement memory or accumulator.*   Subtract one from the number in the specified register. Very useful because adding and subtracting one is the essence of counting. Notice that the carry bit is unaffected, which leads to many programming errors, so be careful! The DEC does not have direct addressing—an apparent design compromise because direct addressing would be used frequently if it were available.

DEX   *Decrement X.*   Subtract one from the number in the index register X. Only the Z bit is affected; however, the X register usually holds addresses making the other condition code bits of little use. X is not a data register in the sense of an accumulator, although it is used for some limited data manipulation. Notice that this instruction is the first one introduced in this book that gives a 16-bit arithmetic result instead of an 8-bit result.

INC, INCA, INCB   *Increment memory or accumulator.*   Add one to the number in the specified register. See DEC.

INX   *Increment X.*   Add one to the number in the index register X. See DEX.

JMP   *Jump.*   An unconditional branch to any location in memory. Does not affect any condition code bits. Does not have direct addressing.

LDAA, LDAB, LDX   *Load accumulator or index register.*   Load a number from memory into the specified register. Although a carry cannot occur, the C bit is unaffected rather than set to 0. See ADCA for an example of why the C bit works this way.

NEG, NEGA, NEGB   *Negate memory or accumulator.*   Change the sign of the two's complement number in the specified register. Changing the sign is sometimes called *taking the two's complement of a number.* Notice that the operation is done by a subtraction. Changing the sign of 00 gives the correct result 00 with V set to 0. Changing the sign of 80 gives the correct result 80 with V set to 1. The NEG does not have direct addressing.

NOP   *No op.*   Do no operation, just waste time and space. NOP is not used in final programs, but it is helpful while debugging new programs. You can remove an instruction from your program without moving the remaining instructions to new locations. Just replace the undesired instruction by the correct number of NOPs.

SBA   *Subtract B from A.*   Subtract the B accumulator from the A accumulator and put the difference in the A accumulator. No parallel instruction puts the result into B, or subtracts the A accumulator from the B accumulator.

SBCA, SBCB

Subtract with carry.   Subtract memory register and carry bit from an accumulator and put the difference into the accumulator. These instructions are used for multiple precision subtraction of either signed or unsigned numbers. See ADCA.

SEC, SEV

Set the carry or overflow.   Set the carry bit C or the overflow bit V to 1. Affects no other condition code bits. May be used to force a test condition to a known value.

STAA, STAB, STX

Store accumulator or index register into memory.   Store the contents of the specified microprocessor register into memory. The carry bit is unaffected rather than made 0 even though a carry cannot occur. These instructions do not have immediate addressing, because it would not be useful—an IMM store would make a self-modifying program.

SUBA, SUBB

Subtract memory from accumulator.   Subtract the addressed memory register from the accumulator and put the difference into the accumulator. May be used to subtract either unsigned or two's complement signed numbers. A common misconception is that the two's complement signed numbers cannot be subtracted! When working with 8-bit numbers, a 1 in C indicates an error for unsigned numbers while a 1 in V indicates an error for signed numbers. The H bit is not affected by any subtraction instruction.

TAB, TBA

Transfer A to B or B to A.   TAB transfers or copies the A accumulator to the B accumulator, and TBA transfers B to A. After the TAB instruction is executed, both the A and B accumulators contain the number originally in A. The 68HC11 lacks an instruction to exchange accumulators A and B.

TAP

Transfer A accumulator to condition code register.   Transfer the A accumulator into the processor condition code register. *Caution!* Don't use this instruction unless you are an advanced programmer. It affects all the condition code register bits including those that control the hardware. The TAP may alter the hardware operation in undesirable ways.

TPA

Transfer condition codes to A.   The mnemonic came from "transfer the processor condition code register into the A accumulator." This instruction is rarely needed in programs because the branch instructions respond to the condition code bits adequately. No parallel instruction transfers to B.

TST, TSTA, TSTB   *Test memory or accumulator.*   Test the specified register and
                         make the condition code register respond by subtracting 00.
                         The register tested is unchanged. The TST is a convenience that
                         is used with branch instructions. The instruction set table im-
                         plies that the V and C bits don't work, but they are affected
                         correctly.

You now know enough instructions to start writing useful programs in machine language. Chapter 5 introduces a few more instructions. However, the majority of the instructions in most practical programs are included in this list.

You will need to expend some effort to remember all the instructions introduced here.
But as you use them to write a few programs, you will learn them easily. Because each
instruction has been designed to do a needed function, it will easily come to mind as you
need that function. However, practice at writing programs is necessary.

The Comparison Branch Instructions

The branch instructions covered so far are relatively easy to understand because they involve
only a single condition code bit. The comparison branches are more complex because they
use two or more condition code bits to decide whether to branch. The microprocessor applies
a logic relationship to the bits to determine the branch condition. Generally, you do not need
to learn the relationship, because the name of the instruction tells you what the instruction
does. The name of the instruction describes the relationship between two numbers that are
being compared.

Branch with subtraction

*All the branches discussed here require a subtraction operation immediately preced-
ing the branch.* Don't try to circumvent this requirement! If you do, the instruction may
appear to work correctly for your immediate data numbers, but other numbers may lead to
errors.

The reason that the subtraction is necessary is that these instructions all compare two
numbers. The essence of the comparison is the subtraction. The subtraction instructions that
are appropriate for comparisons include the following: CBA, CMPA, CMPB, CPX, SBA,
SUBA, and SUBB. For example, SBA subtracts the B accumulator from the A accumulator,
so the A accumulator holds the minuend and the B accumulator holds the subtrahend. A few
more subtraction instructions are introduced later.

As you read the descriptions of the comparison branches, think about the words *greater*
and *less* versus *higher* and *lower*. Motorola was consistent in the naming conventions for
the branch instructions. The consistent names will help you remember whether the instruction relates to two's complement signed numbers or unsigned numbers. The words *greater*
and *less* are used with signed numbers; greater means more positive and less means less
positive. The words *higher* and *lower* are used with unsigned numbers; higher means bigger
and lower means smaller.

Here are the comparison branches:

BGE

Branch if greater than or equal to.   Causes a branch if (N is set AND V is set) OR (N is clear AND V is clear). The branch will occur if and only if the *two's complement signed* number represented by the minuend (e.g., accumulator A) was greater than or equal to the *two's complement* number represented by the subtrahend (e.g., M).

Caution! Some Motorola manuals describe this instruction as *branch if greater than or equal to zero*. This name comes from applying the instruction name to the difference from the subtraction rather than from the numbers being compared.

BGT

Branch if greater than.   Causes a branch if (Z is clear) AND ((N is set AND V is set) OR (N is clear AND V is clear)). The branch will occur if and only if the *two's complement signed* number represented by the minuend (e.g., accumulator A) was greater than the *two's complement* number represented by the subtrahend (e.g., M).

Caution! Some Motorola manuals describe this instruction as *branch if greater than zero*. This name comes from applying the instruction name to the difference from the subtraction rather than from the numbers being compared.

BHI

Branch if higher.   Causes a branch if (C is clear) AND (Z is clear). The branch will occur if and only if the *unsigned* number represented by the minuend (e.g., accumulator A) was higher (bigger) than the *unsigned* number represented by the subtrahend (e.g., M).

BHS

Branch if higher or same.   BHS is a second name for the BCC instruction. Causes a branch if (C is clear). The branch will occur if and only if the *unsigned* number represented by the minuend (e.g., accumulator A) was higher (bigger) than or equal to the *unsigned* number represented by the subtrahend (e.g., M).

BLE

Branch if less than or equal to.   Causes a branch if (Z is set) OR ((N is set AND V is clear) OR (N is clear AND V is set)). The branch will occur if and only if the *two's complement signed* number represented by the minuend (e.g., accumulator A) was less than or equal to the *two's complement signed* number represented by the subtrahend (e.g., M).

Caution! Some Motorola manuals describe this instruction as *branch if less than or equal to zero*. This name comes from applying the instruction name to the difference from the subtraction rather than from the numbers being compared.

BLO

Branch if lower. BLO is a second name for the BCS instruction. Causes a branch if (C is set). The branch will occur if and only if the *unsigned* number represented by the minuend (e.g., accumulator A) was lower (smaller) than the *unsigned* number represented by the subtrahend (e.g., M).

BLS

Branch if lower or same. Causes a branch if (C is set) OR (Z is set). The branch will occur if and only if the *unsigned* number represented by the minuend (e.g., accumulator A) was lower (smaller) than or equal to the *unsigned* number represented by the subtrahend (e.g., M).

BLT

Branch if less than. Causes a branch if (N is set AND V is clear) OR (N is clear AND V is set). The branch will occur if and only if the *two's complement signed* number represented by the minuend (e.g., accumulator A) was less than the *two's complement signed* number represented by the subtrahend (e.g., M).

Caution! Some Motorola manuals describe this instruction as *branch if less than zero.* This name comes from applying the instruction name to the difference from the subtraction rather than from the numbers being compared. The BLT instruction is not equivalent to the BMI instruction!

Do not be concerned by two's complement overflow errors that occur during the subtraction that precedes the branch instruction. The branch instructions will correctly account for the information in the V bit. To emphasize this point, you should demonstrate to yourself that a subtraction of two's complement numbers followed by the BGT instruction does not perform the same function as the subtraction followed by BPL. The numbers that cause two's complement overflows will not work correctly with BPL.

Using BEQ and BNE for comparisons

The instructions BEQ and BNE can compare two numbers to see if they are equal or unequal. These instructions were used earlier to test if one number was zero or not zero. If they are to do a comparison, they must be preceded by a subtraction operation just like the other comparison branches. But now the meaning of the instruction will be interpreted as follows:

BEQ

Branch on equal. Causes a branch if (Z is set). The branch will occur if and only if the number represented by the minuend (e.g., accumulator A) was equal to the number represented by the subtrahend (e.g., M). The two numbers may both be two's complement signed numbers or they may both be unsigned numbers.

BNE

Branch on not equal.   Causes a branch if (Z is clear). The branch will occur if and only if the number represented by the minuend (e.g., accumulator A) was not equal to the number represented by the subtrahend (e.g., M). The two numbers may both be two's complement signed numbers or they may both be unsigned numbers.

The BEQ and BNE instructions cause some confusion because they are used two different ways. To complicate matters, some Motorola manuals describe BEQ as *branch on equal to zero* and others describe it as *branch on equal* without distinguishing the two different uses.

Applications of Complex Instructions

Many instructions were designed for specific applications. After you have some programming experience, the need for these specialized instructions is obvious. Therefore, the following programming examples demonstrate the uses of some of these instructions.

Multiple precision arithmetic

A program that does multiple precision addition requires the ADCA instruction. The 68HC11 instructions you have seen operate on 8-bit numbers. When larger numbers are needed, they will be made up of 8-bit single-precision pieces. The example program that follows adds two double-precision or double-byte numbers from memory and stores the double-byte sum in memory. The same technique as used in the example applies to larger numbers.

Figure 2-42 will help you visualize the algorithm for the example. The numbers all reside in 8-bit registers, so the figure shows sample 8-bit numbers. The registers are arranged to appear as if 16-bit numbers are being added. In the figure, the numbers 44 and C1 are added to obtain 05 and a carry from bit 7. Then the carry, the 82 and the A3 are added to obtain the sum 26 and a carry.

The program example.   Figure 2-43 shows the program for the double-precision addition. The LDAB, ADDB, and STAB instructions add the least-significant eight bits. The STAB instruction does not affect the C bit, so the carry information generated by the ADDB is not lost.

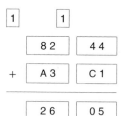

Figure 2-42   Adding two double-precision numbers.

		•
N1	0010	82
	0011	44
N2	0012	A3
	0013	C1
SUM	0014	FF
	0015	FF
		•
		•
		•
LDAB	C200	D6
DIR	C201	11
ADDB	C202	DB
DIR	C203	13
STAB	C204	D7
DIR	C205	15
LDAA	C206	96
DIR	C207	10
ADCA	C208	99
DIR	C209	12
STAA	C20A	97
DIR	C20B	14
		•

Figure 2-43 Double-precision add program module.

Next, the most-significant eight bits of the numbers and the carry from the least-significant bits are added. Accounting for the carry could be done by branching on carry set to a program module that adds one, but using the ADCA instruction is much easier. The ADCA adds two 8-bit numbers and the C bit. This one instruction does the job if the carry information is correct.

When the ADCA adds the most-significant eight bits, the N, Z, V, and C bits are affected. The information in these bits describes the 16-bit result. The C bit indicates whether an addition error occurred if unsigned numbers are used. The V bit indicates whether an addition error occurred for two's complement signed numbers. Similarly, the N and Z bits refer to the 16-bit sum. So a branch instruction that follows this addition correctly responds to the 16-bit operation.

Multiple branch conditions

Many programs require multiple decisions. Figure 2-44 illustrates the testing of a two's complement signed number to see whether it is negative, zero, or nonzero positive. Remember that zero is positive in the two's complement number system. Be careful to distinguish zero from nonzero positive numbers. This example additionally shows that the branch instructions do not affect the condition code bits for good reasons.

The program uses a TST instruction to set the N and Z bits in the condition code register. These bits provide the information needed for the decisions. The BEQ instruction

		•
TST	C200	7D
EXT	C201	C5
	C202	55
BEQ	C203	27
REL	C204	03
BPL	C205	2A
REL	C206	02
STOP	C207	3F
STOP	C208	3F
STOP	C209	3F
		•

Figure 2-44 Program module with multiple branch paths for negative, zero, and nonzero positive numbers.

branches to the routine that handles the zero case; the routine is just a *STOP* instruction in the figure. Because the BEQ does not affect the condition code bits, the BPL uses the same condition code results for a second branch. If the BPL does not branch, the number tested must have been negative, and a separate test for negative is unnecessary. The example program stops at address C207 for a negative number, at address C208 for a zero number, and at address C209 for a nonzero positive number.

Double-precision Instructions

The 68HC11 has a few instructions that operate on double-byte numbers in a general way. All these instructions require a double-byte accumulator. To accommodate this need, the A and B accumulators are used together as a single 16-bit accumulator called the *D accumulator*. The programming model in Figure 2-1 illustrates the D accumulator with the A accumulator on the left and the B accumulator on the right. The A accumulator holds the most-significant byte, and the B accumulator holds the least-significant byte of the double-byte number in D. The D accumulator is not an additional register, but only a different way of using the A and B registers. For example, bit 15 of the D accumulator is bit 7 of the A accumulator. The basic instructions that use the D accumulator are as follows:

ADDD   *Add memory to D.*   Add a double-byte number from memory to the D accumulator.

LDD   *Load accumulator D.*   Load a double-byte number from memory into the D accumulator. Notice that the name is not LDAD.

STD   *Store accumulator D.*   Store a double-byte number from the D accumulator into memory. Notice that the name is not STAD.

SUBD   *Subtract memory from D.*   Subtract a double-byte number in memory from the D accumulator.

All instructions that use the D accumulator make the condition code bits respond to a 16-bit number. The number is always the result obtained from the execution of the instruction. The branch instructions may be used following the double-precision instructions. For example, two two's complement signed 16-bit numbers can be compared using the instruction sequence LDD, SUBD, and BGT.

2.5   MACHINE LANGUAGE PROGRAMMING EXAMPLE

You now have enough background to write a program that does a significant job. This section discusses a program that brings together many instructions, addressing modes, programming techniques, and other concepts. The program will be very small, but it should make you think about how good programs should be written.

Minimum Requirements of a Good Program

You should think about the characteristics that are desirable in a good program. This task is easier if you have had some programming experience regardless of the kind of computer or language used. The ideas considered here are most basic.

- *The program must do the same thing every time it is run.* You should not need to reenter any data or instructions to run the program again. It must perform the same function on the same data producing the same results every time.

- *The program must not modify itself.* That is, the instructions cannot be changed by the program. A program that changes itself and then returns itself to the original state before finishing is also unacceptable. Programs that don't change themselves are called *pure procedure* programs.

- *The program must work correctly for all reasonable data values.* For example, a valid data value of zero is often overlooked in program design.

- *The instructions and data numbers should be separated in memory.* Any data that must be entered to run the program or data values that change while the program runs must be separated from the instructions. They should not be mixed in one area of memory so that the instructions have to branch over data values. The two areas in memory are called the *program section* and the *data section*.

- *The program should adapt easily to different sets of data.* For example, a program that works on a table of numbers must easily adapt to tables at any reasonable location and of any reasonable length. Writing a new program for each table is not practical.

- *The program should start at the first instruction and stop at the last instruction.* Programs that are scattered around in memory are difficult to understand.

Many more considerations are involved in writing sophisticated programs. The ideas presented here provide a good starting point for the first significant program. Chapter 4 presents more information on good programming practices.

A Classic Example: Copy a Table

A table or list is a series of data numbers grouped together in memory. For example, a manufacturing facility may have many identical automated manufacturing machines. The computer could monitor the number of parts each machine makes by counting the parts as they exit the machine. Each number in a table may represent the number of parts manufactured by a different machine. All the data numbers together represent the production of the manufacturing facility. Microcomputer programs for control and data acquisition applications frequently use tables.

Let's look at a practical and useful program that copies a table of numbers from one place in memory to another. The goal is to have two copies of the same data numbers. For example, these two tables may be used in a machine-monitoring application. The first table may represent production counts for the last hour. At the end of an hour, a copy is made to be used by a report program that prints the data, and the entries in the first table are set to zero. While the printer is printing the data from the second table, the first table is updated with new production data.

Assume that the number of numbers in the tables will be small so that the length of the table can be represented by a single-byte number. Similarly, assume that the entries in the tables are single-byte numbers. Figure 2-45 illustrates what the program should do to tables of four entries.

A brainstorming session

The copy-a-table program could be written many different ways. Let's begin with a brainstorming session to evaluate various approaches.

Copy-a-table instruction.   You could browse through the instruction set table until you find an instruction that copies a table! Then the program would consist of that one instruction. Of course, you are not surprised that the 68HC11 does not have such an instruction, although some microprocessors do. One reason you would not expect to find such an instruction is that it would require at least two addresses. One address would specify the

	•		•
C110	11	C110	11
C111	22	C111	22
C112	33	C112	33
C113	44	C113	44
	•		•
	•		•
	•		•
C130	37	C130	11
C131	02	C131	22
C132	7A	C132	33
C133	C1	C133	44
	•		•

Figure 2-45   Memory contents before (left) and after (right) running the copy-a-table program. The source table is at address C110, and the destination table is at address C130.

location of the first table; the second address, the location of the second table. All the instructions you have seen only have one address; so it is unlikely that such an instruction exists, and it doesn't.

Copy a memory byte.   The next approach you may consider is an instruction that copies one number from one memory register to another. This instruction could be used repeatedly to copy the entire table. The 68HC11 does not have this instruction either, for it too would require two addresses. However, more sophisticated microprocessors do have such instructions.

Load and store.   A simpler and more practical approach is to use the LDAA and STAA instructions. The LDAA can get a data number from the first table, and the STAA can store it into the second table. The execution of two instructions will be required to move each data number, but the job can be done. Figure 2-46 shows part of a program to copy the table shown in Figure 2-45 using this approach.

Let's evaluate the program. Certainly, the program is very simple and easy to understand. The program also will run very quickly because it does nothing more than the required operations. It also meets most of the minimum requirements discussed earlier. However, adapting it to different tables is not very easy. The program must be rewritten if any change, such as length or location, is made to the tables. Also, the length of the program will be unacceptable for long tables. This approach is not very good as a general solution, but it is useful for very short tables.

A loop.   The next approach uses the LDAA and STAA instructions, but it requires that the same two instructions be executed repeatedly, once for each entry in the table. Repeating or iterative programs are called *loops*. A loop does its function and then branches back to use the same instructions over and over. In the copy-a-table program, the loop will copy one table entry each time around the loop. When all the entries have been copied, the program will break out of the loop and stop.

•	•	•
LDAA	C010	B6
EXT	C011	C1
	C012	10
STAA	C013	B7
EXT	C014	C1
	C015	30
LDAA	C016	B6
EXT	C017	C1
	C018	11
STAA	C019	B7
EXT	C01A	C1
	C01B	31
LDAA	C01C	B6
•	•	•

Figure 2-46   Part of a copy-a-table program using LDAA and STAA.

One problem here is that the LDAA and STAA instructions that do the copying contain addresses. The addresses must change each time around the loop or the same entry would be copied each time. That is, the same instructions must copy a different entry each time around the loop. To meet the requirement that the instructions don't change due to program execution, the LDAA and STAA instructions must have indexed addressing. Indexed addressing allows a program to modify the effective address of an instruction without modifying the instruction code. Indexed addressing makes writing practical looping programs for the 68HC11 possible.

A basic looping program

A looping copy-a-table program has several distinct parts. The parts include a means of keeping track of the next entry to be copied and a means to stop the program from looping when all entries have been copied. The looping program will be much more complex than the in-line program in Figure 2-46. It also will run much slower, but it will be better in almost every other way.

The pointer.   The index register X will be used to point to the table entry to be copied. It does this by holding the address of the entry in the first table that will be picked up by the program. It also will be used to determine where to store the entry into the second table. As the program copies individual entries, the pointer will be adjusted forward one location each time around the loop.

But the pointer must point to the first location of the first table as the loop begins running the first time. The pointer, the X register, must be initialized to contain this address before the loop starts.

Pointers must always be initialized before they can be used by the program loop. The initialization must be done by the program—not by a person entering a number every time the program is to be run.

The counter.   The loop in the program will repeatedly copy table entries, but it must stop looping when all entries are copied. The program can stop looping simply by counting the entries as they are copied. When the number of entries that have been copied equals the length of the table, the loop can exit.

A number called a *counter* will be put into a register before the loop starts. The counter will be tested and adjusted each time around the loop. When the number in the counter reaches a limit, the loop will exit.

The counter can be in any appropriate register. The two basic choices are an accumulator and a memory register. Generally, you have two accumulators and thousands of memory registers, so the memory register is the most common choice. Using a memory register as a counter is easy, because most computers, including the 68HC11, have instructions to increment or decrement a memory register.

A counter must always be initialized before it can be used by a loop. The program must do the initialization; you cannot expect a person to enter a value each time.

The counting method.   Now consider some ways the program could use the counter register. For example, the table in Figure 2-47 shows four entries. The counter could be

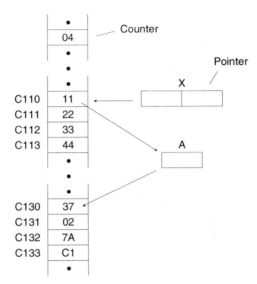

Figure 2-47   Pointer and counter used by copy-a-table loop.

initialized to negative four, the program could add plus two each time around the loop, and then stop at a limit of plus four. The counter could be initialized to an unsigned seven and decremented by one to the limit of three. Similarly, the counter could be initialized to an unsigned four and decremented to zero. Many other ways can be used to count from a starting point to an ending point. However, some ways to use the counter are better than others.

In choosing a counting method, you probably will choose to use an unsigned number in the counter. A signed number doesn't add anything useful over an unsigned number, but it does make the numbers more difficult to read and interpret. Also, the range of possible numbers is twice as great if an unsigned number with the same number of bits is used. In addition, a negative counter value doesn't have any obvious physical meaning.

Next, in choosing the counting method, you probably will choose to start the counter at the number of table entries, and then count to zero by ones. Detecting whether a number is zero is very easy because the computer hardware automatically detects zero and indicates it with the Z bit. Testing for any other value requires more instructions and more complication. Likewise, subtracting one from a number is easy because the hardware has a decrement instruction. Therefore, counting will not be done by twos or any other number besides one. All microprocessors have decrementing and zero testing capabilities because they are essential to counting.

Furthermore, decrementing an unsigned number by ones makes the number in the counter equal to the number of loops yet to go. When you test the program, decrementing the counter makes interpreting the number in the counter easy.

The code.   Figure 2-48 lists the instructions for a program that copies a table using a loop. The operation and design of the program should be studied carefully. The program has some design problems that can be overcome, but then the program would be somewhat longer. The next section discusses these design deficiencies.

Address	Contents	Instruction	Description
•	•	•	•
0030	C1		Address of first table
0031	10		
0032	04		Table length
0033	xx		Working counter
•	•	•	•
•	•	•	•
C010	DE	LDX DIR	Initialize pointer to first table
C011	30		
C012	96	LDAA DIR	Get initial counter value
C013	32		
C014	97	STAA DIR	Store initial counter value into working counter
C015	33		
C016	27	BEQ REL	Branch to *STOP* if working counter equals zero
C017	0A		
C018	A6	LDAA IND,X	Get next entry from first table
C019	00		
C01A	A7	STAA IND,X	Put entry into second table
C01B	20		
C01C	08	INX	Advance pointer to the next table entry
C01D	7A	DEC EXT	Count down the working counter
C01E	00		
C01F	33		
C020	20	BRA REL	Go around again
C021	F4		
C022	3F	*STOP*	Stop the program
•	•	•	•

Figure 2-48   Copy-a-table program.

The data section of the program was put at addresses 0030 through 0033 so that it would be in the direct addressing range. This decision leads to good program design that uses the capabilities of the addressing modes well. The program section includes locations C010 through C022. The only consideration in choosing these addresses was to put the program outside the direct addressing range to allow easy program expansion. If more data locations are needed in the direct addressing range when the program is changed, the instructions will not have to be moved.

Here is a step-by-step discussion of the program. Look at the instructions and comments in Figure 2-48 as you read the following:

- *Initialize the pointer.* The LDX instruction at address C010 loads the pointer, the X index register, with the address of the first entry of the first table. The double-byte address is supplied by the data section at locations 0030 and 0031. You can see that the first table is at address C110 because that is the number loaded into X.

- *Initialize the counter.* Because the counter will change as the program runs and will be zero when the program stops, it must be initialized at the beginning of the program. If it were not initialized, the program would copy zero entries the next time it runs. The program must do the initialization. The data section holds the initial value of the counter, the table length, that must be copied to a working counter. The working counter will change as the program runs, but the initial value will not. The LDAA instruction at address C012 gets the initial counter value, 04, and the STAA instruction at address C014 puts it into the working counter. So the example will copy four numbers. Notice that an instruction to copy one number from one memory location to another would be useful here.

- *Break out of the loop if all entries have been copied.* Breaking out of the loop at this point is not intuitive. Though hindsight is required, consider using the BEQ instruction that branches on equal to zero. The BEQ at address C016 uses the Z bit to determine its actions. The STAA instruction that precedes the BEQ stored a number into the working counter and made the Z bit respond. Check the instruction set table to confirm that STAA affects the Z bit. Because the working counter controls the exit from the loop, this loop appears correct so far. For example, if the data section said the length of the table was zero entries, the loop would correctly exit before doing anything. If the number of entries is not zero, the program continues in sequence. The relative offset in the branch cannot be determined until the program is completed, because it is doing a forward reference. That is, the instruction must branch to a location farther ahead in memory.

- *Copy one entry.* The LDAA and STAA instructions with indexed by X addressing at addresses C018 and C01A copy the entry. The LDAA gets the entry from the first table into the A accumulator using the X register with an offset of zero as the pointer. The STAA puts the entry into the second table using the X register with an offset of 20 as the pointer. The offset of 20 determines the location of the second table; namely, address C130. The effect of the offset 20 is to allow the X register to point to both tables. A later section discusses the many limitations of this approach. The approach was chosen for this example both to make it simple and to demonstrate some problems.

- *Advance the pointer to the next entry.* The loop must be prepared for the next time around. When it runs again, it must copy the next entry, not the first entry again. So, the pointer must be adjusted by incrementing X with the INX instruction at address C01C. The effect is to advance the pointers to both tables. Notice that the INX must follow the copying of an entry so that the initial value in the pointer is the first address of the table. Setting the initial address of the table to one lower address and putting the INX before the copy would be confusing.

- *Count off one entry copied.* The working loop counter is decremented by the DEC at address C01D. The Z bit is affected by the DEC instruction, so the Z bit indicates whether the working counter has reached zero. Notice that the DEC instruction that is used has extended addressing because the DEC instruction cannot have direct addressing.

- *Go around again.* The end of the loop sends control back to the beginning of the loop to test the working counter. Notice that the BRA instruction at address C020 does not affect the condition code bits. After branching back to the BEQ, the Z bit still contains the information from the DEC instruction. The DEC decremented the working counter, so the BEQ will respond to the correct information. Also notice the negative relative offset in the BRA that causes a branch to a lower address. A JMP instruction could be used instead of the BRA, but the BRA is shorter and faster. Because most loops are short, the BRA is almost always used.

- *Stop the program.* The *STOP* instruction at address C022 is reached by the BEQ after the loop is completed.

This program is much more complicated than the one in Figure 2-46. It will run several times slower. Besides copying an entry, it also must adjust a pointer and a counter, test the counter, and loop back.

A slower and complicated program may be considered a disadvantage. In contrast, this program is much better if the table is large, if the table must be moved to different locations, or if the table length is changed. Each of these modifications can be done by relatively minor changes. For practical applications, this program would be far better than the one in Figure 2-46.

Limitations and problems

The copy-a-table program in Figure 2-48 has many limitations and a major design error. Here are some comments about the program:

- *Location of tables.* The tables may be located anywhere in memory provided the second table starts at an address 20 locations higher than the start of the first table. Changing this spacing between the tables requires changing the offset of the STAA IND,X instruction—the program must be changed! Also, the range of spacings is 00 through FF, which is not a very large range; this range also limits the length of the tables. A long table with a small spacing will cause the second table to overlap the first table. Similarly, the second table must always be at a higher address than the first table because the offset of the STAA IND,X is an unsigned number.

- *Size of the tables.* The maximum length of the tables is FF entries because the working counter is a single-byte number. The minimum table size is zero. Setting the size to zero in effect disables the program from working. This size may be useful during the debugging stage, especially if the copy-a-table program is a small module of a larger program.

- *Insignificant changes in the program will make it malfunction.* The order of the program parts is very dependent on the programming approach. For example, if the pointer was advanced after decrementing the working counter, the program would no longer work. The problem is the dependence of the BEQ instruction on the Z bit condition left by the DEC instruction. Similarly, if the pointer was initialized after the initialization of the working counter, the BEQ would not work correctly for the same reason. Both defects could be remedied by putting a TST instruction immediately before the BEQ to test the working counter. Then the other parts could be arranged in any order without problems. The program would be a better program.

- *Some data is in the program section instead of the data section.* This is a serious conceptual as well as practical problem. The offset 20 in the STAA IND,X instruction determines the spacing between the tables. The number 20 is a data number that should be put into the data section and then used as a data number. Changing the table spacing in the current program requires that the program be rewritten. Several ways to change the program to correct this problem are available, but all require a more complex program design.

You should always examine your programs for the kinds of details discussed here. Correcting such defects as a program is written is easier than changing it at a later time. In particular, you should ask yourself what changes are likely to be made to the program some time in the future. Plan ahead for these changes.

An improved program

The program in Figure 2-49 is an improvement over the program in Figure 2-48. It removes the most serious limitation—the data is now all in the data section. Also, the spacing between the tables can be changed without writing a new program. The spacing is now a data value stored in the data section of the program.

Here are some observations about the new program:

- Both the data section and the program section were moved to new locations to encourage you to investigate the changes in the instructions that are necessary.

- The order of the program parts was changed.

- The working counter is tested before the branch instruction so further changes will have no consequence on the BEQ instruction.

- The pointer to the first table must be saved and restored each time around the loop.

The limitations that remain all involve restrictions of instruction addressing modes or the size of the numbers. For example, the maximum spacing between the tables is still FF. You should investigate this improved program to determine the limitations that remain.

2.6   REVIEW

You have now seen all fundamental addressing modes and a subset of the 68HC11 instructions. You have also seen machine language programming where a person looks up the binary

Address	Contents	Instruction		Description
•	•	•		•
0020	C2			Address of first table
0021	20			
0022	20			Spacing between tables
0023	04			Table length
0024	xx			Working counter
0025	xx			Save pointer
0026	xx			
•	•	•		•
C100	DE	LDX	DIR	Get address of first table
C101	20			
C102	DF	STX	DIR	Store in save pointer location
C103	25			
C104	96	LDAA	DIR	Get initial counter value
C105	23			
C106	97	STAA	DIR	Store initial counter value into working counter
C107	24			
C108	D6	LDAB	DIR	Get table spacing
C109	22			
C10A	7D	TST	EXT	Test working counter
C10B	00			
C10C	24			
C10D	27	BEQ	REL	Branch to *STOP* if working counter equals zero
C10E	11			
C10F	DE	LDX	DIR	Restore pointer to first table
C110	25			
C111	A6	LDAA	IND,X	Get next entry from first table
C112	00			
C113	3A	ABX	INH	Adjust pointer to second table
C114	A7	STAA	IND,X	Put entry into second table
C115	00			
C116	7A	DEC	EXT	Count down the working counter
C117	00			
C118	24			
C119	DE	LDX	DIR	Get pointer value
C11A	25			
C11B	08	INX		Advance pointer to the next table entry
C11C	DF	STX	DIR	Store in save pointer location
C11D	25			
C11E	20	BRA	REL	Go around again
C11F	EA			
C120	3F	*STOP*		Stop the program
•	•	•		•

Figure 2-49   An improved copy-a-table program.

instruction codes in the instruction set table and then puts the codes into memory through a manual process.

Programming in machine language is, at least, very tedious. A computer system and an assembler program can help you program by providing the binary numbers. The program will do the work of looking in the instruction set table. The next chapter discusses assemblers and assembly language programming.

2.7 EXERCISES

2-1. Rewrite the program in Figure 2-16 to add three numbers together. The first instruction must be kept at the same address; this means that the data numbers must be moved. The problems that result from this move are common to all computer programming.

2-2. Rewrite the program in Figure 2-23 so that the instructions are moved to address C400 but the data numbers remain at the same locations.

2-3. Trace the execution of the program in Figure 2-26 by writing down the contents of the program counter, instruction register, A and B accumulators, and memory register C346 after each instruction is executed.

2-4. Find the values in the H, N, Z, V, and C bits after each instruction in Figure 2-26 is executed.

2-5. Modify the program in Figure 2-26 to add three numbers together. Experiment with different addressing modes and programming approaches.

2-6. What is the machine code for a single instruction that undoubtedly puts zero in the B accumulator (two answers!), and another instruction that puts zero into memory location C111?

2-7. The BHS instruction is normally used with (signed, unsigned, BCD, other) numbers.

2-8. If a negative number is subtracted from a negative number, the V bit will be 1 if the answer is (positive, negative, V will never be 1).

2-9. Is the two's complement overflow bit affected by the TAB instruction?

2-10. The complete hexadecimal instruction code for a single instruction that adds the C bit to the A accumulator is _____.

2-11. List all branch instructions that must follow a subtraction operation under normal circumstances.

2-12. Which branch instruction must immediately follow a TSTA instruction to cause a branch only if the number in the A accumulator is a nonzero unsigned number?

2-13. What single instruction will set the C bit? The N bit?

2-14. Write a program module to exchange the numbers in the A and B accumulators.

2-15. What is the hexadecimal code for a BRA instruction that branches to itself? What is the effect of executing this instruction in a larger program?

2-16. Find all the known condition code bits if a DEC instruction operates on the hexadecimal number 80.

2-17. Find all the known condition code bits if a NEG instruction operates on the hexadecimal number 80 or the number 00.

2-18. How do branch instructions affect the condition code bits? Why?

2-19. Compare the meaning of the C and V bits immediately after an addition instruction has been executed. Does any difference occur if unsigned or two's complement numbers are used?

2-20. How does subtracting two two's complement numbers compare to changing the sign of the subtrahend and adding? Do you get the same results in the condition code bits?

2-21. Are the condition codes the same if one is added to the A accumulator with the ADDA instruction versus the INCA instruction?

2-22. If you wanted to branch on the sign of a number in memory, is it better to use LDAA or TST followed by the branch instruction?

2-23. If the X register contains FFFF when an INX instruction is executed, the result in the C bit is (1, 0, unknown)_____.

2-24. Two unsigned numbers are loaded into the A and B accumulators and then subtracted with an SBA instruction. If the program must branch only when the number in the B accumulator is lower than the number in the A accumulator, use the _____ instruction following the SBA.

2-25. Compare the BLO instruction to the BCS instruction.

2-26. List all instructions that do addition of any kind on 16-bit numbers.

2-27. List all instructions that do subtraction of any kind on 8-bit numbers.

2-28. If the sequence of instructions in Figure 2-50 is run starting at address D100, the *STOP* instruction at address D10A will be reached if the number at address 0020 is _____ or it is in the range _____ through and including _____.

2-29. If the sequence of instructions in Figure 2-51 is run starting at address D100, the *STOP* instruction at address D10C will be reached if the number at address 0012 is in the range _____ through and including _____ or it is in the range _____ through and including _____.

2-30. Do you get the same results from the NEG instruction as from COM followed by INC?

2-31. List all instructions that have extended addressing but do not have direct addressing.

2-32. In forming an instruction with indexed by X addressing, what must the offset byte be if the index register contains D333 and the instruction must access data at location D347?

2-33. Find the hexadecimal code for a BRA instruction at address C144 that must branch to address C156; and another at C199 that must branch to C188.

2-34. The complete hexadecimal instruction code for a single instruction that increments the D accumulator is _____.

2-35. Bit 12 of the D accumulator is also bit _____ of the _____ accumulator.

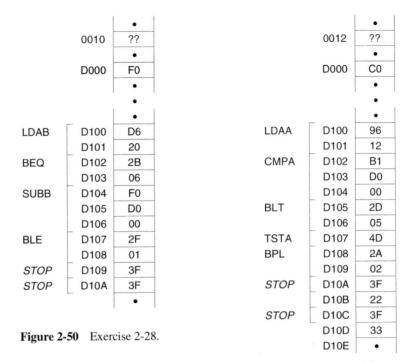

Figure 2-50   Exercise 2-28.

Figure 2-51   Exercise 2-29.

2-36. Modify the program in Figure 2-43 to subtract two two's complement double-precision numbers. Use the SBCA instruction. Your program must stop at one location if the subtraction was correct, and at another if incorrect.

2-37. Modify the program in Figure 2-44 to accomplish the same goal with the branch instructions in a different order—put the BPL first. Repeat this exercise with a BMI as the first instruction.

2-38. Modify the program in Figure 2-48 to use the B accumulator as the working counter.

2-39. Refer to Figures 2-52 and 2-53. Design and write new sequences of instructions that do the same job as those in the figures, but use the D accumulator in a good way to do the job.

2-40. Consider the program in Figure 2-48. What advantage, if any, is achieved by copying the table starting at the highest address and working to the lowest address?

2-41. In Figure 2-48, would it make any difference if the LDX DIR instruction were moved to immediately after the STAA DIR instruction?

2-42. Discuss the use of indexed addressing for the LDX instruction that is the first instruction in the program in Figure 2-48.

2-43. What is the length (bytes) of the longest table the program in Figure 2-48 can correctly copy?

2-44. Modify the program in Figure 2-48 by adding a TST instruction immediately before the BEQ instruction.

	•	
0010	??	
0011	??	
	•	
	•	
	•	
CLRB	D100	5F
ADDB	D101	DB
	D102	11
ADDB	D103	DB
	D104	11
STAB	D105	F7
	D106	D1
	D107	17
LDAB	D108	C6
	D109	00
BCC	D10A	24
	D10B	01
INCB	D10C	5C
ADDB	D10D	DB
	D10E	10
ADDB	D10F	DB
	D110	10
STAA	D111	F7
	D112	D1
	D113	16
STOP	D114	3F
	•	

Figure 2-52   Exercise 2-39.

	•	
0030	??	
0031	??	
	•	
	•	
	•	
CLRA	D100	4F
LDAB	D101	D6
	D102	31
ADDB	D103	CB
	D104	01
ADCA	D105	99
	D106	30
BNE	D107	26
	D108	03
TSTB	D109	5D
BEQ	D10A	27
	D10B	04
STAA	D10C	97
	D10D	30
STAB	D10E	D7
	D10F	31
STOP	D110	3F
	•	

Figure 2-53   Exercise 2-39.

2-45.   Modify the program in Figure 2-48 so that the table locations are specified by the addresses of their first entries. Then the tables can be anywhere in memory. The data section will contain the addresses of the two tables and the single-byte length of the tables. Use only instructions introduced in this chapter.

2-46.   Assume that the program in Figure 2-49 is modified and run using the four entry tables in Figure 2-47. After the program stops, find the numbers in the following registers: A, B, X, memory locations 0024 and 0025. Assuming that the *STOP* instruction does not change the condition codes, what will the values in H, N, Z, V, and C be?

2-47.   An unsigned minuend is in the B accumulator, and an unsigned subtrahend is in the A accumulator. Will the instruction sequence SBA, NEGA put the correct difference in the A accumulator?

Chapter 3

Assemblers and Assembly Language

When writing machine language programs, you must continually refer to the instruction set table to find the instruction codes. The process is quite tedious and mechanical. You first decide what instruction you want and what addressing mode is appropriate. Then you follow the Source Form column in alphabetical order until you find the mnemonic name of the desired instruction. Next you look down the Addressing Mode column to find the row for your desired addressing mode. Now look across the row and finally you have the op code. If you don't remember the operand format, you continue horizontally across the row to find the format. At this point, you probably write the op code on paper and add the required operand. After completing this process for all the instructions in your program, you enter the codes into the memory of your computer with some appropriate hardware.

After you perform these steps a few times, you will decide that this work is boring and quite routine. It is a job best done by a computer! This conclusion leads us to the idea of assembly language and assemblers.

Assembly language is a symbolic representation of the instructions and data numbers in a program. A program called an *assembler* translates the symbols to binary numbers that can be loaded into the computer memory. The name *assembly language* apparently comes from the operation of the assembler program. The assembler puts together or assembles the

Motorola Products

All material in this chapter applies to the Motorola 6800, 6801, 6802, 6803, and 68701 chips as well as the 68HC11.

complete instruction code from the op code and operand. This chapter discusses the Motorola assembly language for the 68HC11 and the basic operation of an assembler.

The material in this chapter applies to any vendor's assembler that uses Motorola 68HC11 assembly language. When you purchase an assembler, be careful to determine if the assembler is compatible with standard Motorola language. Some assemblers will generate correct binary code, but the syntax and symbolic statements could be quite different from the Motorola language.

3.1   THE ASSEMBLY PROCESS

When you use assembly language to program a microcomputer, the assembler program will usually run on a computer other than the microcomputer. The computer system that runs the assembler program is called the *assembly system*. The assembly system requires sufficient resources, such as printers and disk drives, to carry out all the functions of the assembler program. The computer may be a personal computer system, a minicomputer system, or a mainframe computer system. Although the principles involved are the same in all cases, the way you access each of these systems differs somewhat. To simplify the discussion and yet represent common systems, assume that the assembly system is a personal computer using floppy disks as the storage medium. Assume that a direct connection runs from the assembly computer to the microcomputer.

When the internal machine language of the assembly system computer is different from the language of the microcomputer, the assembler program is called a *cross assembler*. This name recognizes that the two computers are different, and nothing more. Be careful when purchasing an assembler program; you must specify both the computer it runs on and the computer for which it generates code.

The following discussion is an overview of the assembly process. It presents the process from the development of the symbolic program to loading the binary numbers into the microcomputer. The details of the symbolic language are discussed later in this chapter.

The Editor, Assembler, and Loader Programs

The ultimate goal of the assembly process is to put the binary instruction codes and binary data numbers that are your program into the memory of your microcomputer. The microcomputer is called the *target computer*. All work that precedes putting the program into memory is aimed at the target computer.

The assembler program reads a symbolic source module that it translates into a binary object module. The *source module* is a physical entity, such as a disk file, that contains all the characters that make a symbolic program. The symbolic program is called the *source code*. The *object module* is a physical entity, such as a disk file, that contains the binary numbers that will be loaded into the memory of the target computer. The binary numbers in the object module are called the *object code*.

The word *source* correctly implies that the source code is the origin or beginning of the program. The object or purpose of the assembly system is to generate the binary codes, so the name *object code* for these binary codes is appropriate.

A *load module* is a physical entity, such as a disk file, that can be read by a loader program. A *loader program* reads the load module and places the binary numbers into the memory of the target computer. Some assembler programs generate object modules that are also load modules, so the name *load module* is sometimes an alternative to the name *object module*. However, other systems may require an intermediate program, sometimes called a *linker*, to convert the object module into a load module.

Creating the source module

Figure 3-1 illustrates the development of the source module. The characters of the source program are coded into some form of memory storage device that is compatible with the assembly system. Many common devices such as floppy disks, hard disks, or magnetic tape can be used. The examples in this book refer to floppy disks because they are the most common.

The source module is usually created by an editor program. An *editor program* can create, edit, store, and retrieve text characters from a disk file. The programmer will type the symbolic characters for the source program, change them as needed, and then direct the editor to make a disk file and write the characters into the file. The file can be retrieved later by the editor if further changes are needed.

Usually, the editor program will run on the assembly system computer. A programming editor is similar to the more common word processor program. However, the editor has functions devoted specifically to editing computer programs, so it is generally better than a word processor for this job.

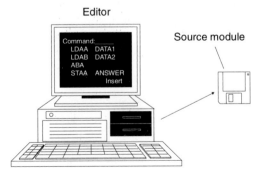

Figure 3-1 Creating the source module with an editor.

Learning to use an editor is easy because you can see the text change on the computer screen as you type. The editor also will print a paper copy of the text.

You should learn most of the commands available in your editor program. Usually the commands were put there because they are immediately useful to editing programs. Sometimes people are content to learn only how to insert and delete single characters. If you intend to do any serious work with an editor, the time spent to learn most of the editor features will be saved many times over as you use the editor.

Creating the object module, load module, and listing

Figure 3-2 illustrates the development of the load module and listing. The assembler program creates the object module from information contained in the source module. Another program may be required to convert the object module into a load module. The information in both the object module and load module is put into some form of memory storage device such as floppy disk, hard disk, or magnetic tape. The medium must be compatible with the assembly computer system and the loader hardware and software. Assume the use of floppy disks as an example.

The printed *listing* shows all the source and object code. Usually, the listing will be printed on paper. However, some assemblers will allow you to print to disk, which saves the listing in a disk file.

Running the assembler program is usually very easy. You tell it the name of the source file and which devices will receive the object code and listing, and it does the work. Generally, any information provided by the assembler will be in the printed listing.

Loading the program

The object code stored in the load module must be loaded into the target computer memory. The physical form of the load module determines how this process is done. For

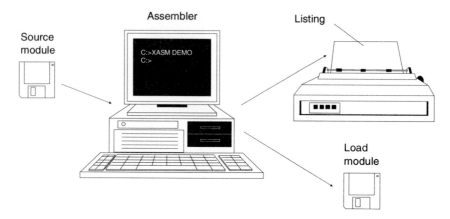

Figure 3-2   Creating the load module and listing with an assembler.

example, if the load module is a floppy disk, the target computer must be interfaced to an appropriate floppy disk drive.

Using a direct-wired connection between the assembly computer and the target computer to transfer the load information is common. A program in the assembly system transmits the binary numbers from the object file over the wires. A program in the target computer, called a *loader program*, receives the numbers and places them into the memory at the proper locations. Of course, this approach requires the proper hardware interfaces in both computers and compatible software. Figure 3-3 illustrates this way of loading a program.

Another common way to transfer the object code to the target computer is the use of a memory integrated circuit. The assembly system will have a device and software to put the binary numbers into a permanent memory integrated circuit. The integrated circuit is then plugged into the target computer. One advantage here is that the target computer does not need a loader program and peripheral device. However, the error checking that the loader would do is lost. This approach is very practical for distributing mass market software such as video games. The memory integrated circuit is made into a cartridge that plugs into the video game computer.

The Source Module, Load Module, and Listing

Information associated with the assembler consists of three distinct parts: the source module, the object or load module, and the listing. The source module is the input to the assembler, which then creates the object module and the listing. The assembler may directly create a load module, or another program may convert the object module to a load module.

The information in a source module

The *source module* contains the entire source program in symbolic form. The format of the information must be compatible with the assembler program that must read it. Several types of symbolic information are discussed here:

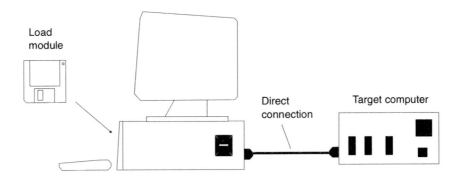

Figure 3-3 Loading the program into the memory of the target microcomputer.

- *Symbolic instructions, data, and addresses.*   Your assembly language program will use the instruction mnemonic names provided in the instruction set table. You will create your own symbols to represent data numbers and addresses.

- *Directives.*   *Directives* are symbols that control and direct the operations of the assembler program. Another name for a directive is *pseudo op*; this name comes from the similarity between the symbolic op codes of the target computer instructions and the symbolic control codes for the assembler program.

- *Comments.*   *Comments* are statements added by the programmer to explain the program. These statements are printed on the listing, but they do not affect the object code in any way.

The information in a load module

The *load module* will be read by the loader program to put the binary numbers into the memory of the target computer. The format of the information in the load module must be compatible with the loader program that must read it. Here is what it contains:

- *Binary instructions and data.*   The binary codes for your instructions and data numbers are in the object module. The symbols that were used in the source program are not in the object module, because they have been translated into binary numbers. Regenerating the source program by reading the object module is not possible.

- *Load information.*   Any information needed by the loader program is provided by the assembler program in the load module. For example, the loader is given the beginning load address for a series of binary numbers. The loader program then puts the numbers at consecutive locations beginning at the load address.

- *Error-checking information.*   Many chances for error occur from the time the assembler generates binary codes until the loader puts the numbers into memory. For example, exposure of a disk to magnetic fields could change the numbers in the load module. Therefore, a way of assuring correctness for the numbers loaded into the target computer is necessary.

The usual method of assuring correctness requires a checksum. The assembler program uses an algorithm to create a number from all the other numbers put into the load module. The algorithm may be as simple as adding. Then the checksum number is put into the load module. The loader program will follow the same algorithm as it reads the numbers from the load module. After finishing reading, it compares its result to the checksum in the load module. If the checksums match, the loader assumes no errors. Notice that errors could still be in the data even if the checksums match, because the algorithm will give the same checksum for many different numbers. However, missing an error is very unlikely to happen. If the checksums don't match, the loader indicates an error.

The information on the listing

Sample *listings* are illustrated later in this chapter, but here is an overview of the information found on the listings made by assemblers:

- *The source code.*   All characters in the source module are printed on the listing. Some assemblers will automatically adjust spacing so that the program is in neat columns. Others print the characters exactly as they were created by the editor.

- *The object code.*   The binary codes created by the assembler program are printed in hexadecimal format. Also, the memory addresses used by those numbers are printed beside them.

- *The symbol table.*   The symbol table lists all the user symbols that the programmer created in the source program. The table usually includes the numerical values of the symbols that the assembler determined. Some assemblers also add a cross-reference table to the symbol table. The cross-reference includes the listing line numbers of all lines where each user symbol was used within the source program.

- *Error messages.*   Errors detected by the assembler program are usually indicated within the listing near the place where the error was detected. The error message may be an English statement or a code number that must be found in a manual. The assembler will indicate the number of errors found near the end of the listing. The number of errors is the first item to look at when reading a listing. If errors are found, continuing with the load process has no value. The assembler finds syntax errors such as misspelled instruction names. It may also detect incorrect addressing modes, multiple use of the same memory location, and other programming errors. Unfortunately, it cannot determine if your program works as you want it to work!

Some assemblers allow the user to suppress various information such as the symbol table or even the entire listing. Usually, you should print all this information and use it to help you write programs.

3.2   THE MOTOROLA ASSEMBLY LANGUAGE

This section lists and describes the characteristics of an assembly language source statement and defines the standard Motorola language. Examples illustrate each statement. Some more extensive and comprehensive examples are in the next section. The information included in this chapter is a subset of all possible source statements because not all the 68HC11 instructions have been presented. However, you will easily learn the omitted information in later chapters.

Many companies sell assembler programs for Motorola products. Different assemblers have different characteristics and features such as the directives that control the assembler. Therefore, one assembler may assemble a source module without error while another may indicate a large number of errors and produce no useful code when assembling the same source.

All useful 68HC11 assemblers accept standard Motorola notation for the instructions. This section covers a notation that works with all assemblers for the standard Motorola 68HC11 language. Most assemblers recognize extensions to the language beyond those statements in this section. Therefore, the material covered here is incomplete for any particular assembler.

Source Statement Content

Assembly language programming requires that you create symbolic source statements. An assembly language *source statement* is a single line of a source program. A source statement may contain various numbers and symbols. The numbers and symbols can be connected by operators to make expressions.

Spaces or blanks in the source program are important. Spaces serve as delimiters or separators between groups of other characters. Therefore, spaces cannot be embedded at arbitrary places in the source statement as in some other computer languages. If you try to embed a space, the assembler will treat the characters preceding the space and the characters following the space as two separate items.

Numbers

A source statement may contain numbers specified by several different number bases. The assembler translates all numbers to 16-bit binary numbers, and then other commands determine the number of bits to be used.

The examples in Table 3-1 show the different number types. A hexadecimal number is written with a preceding dollar sign. A binary number has a preceding percent sign. However, the decimal number requires no special character.

A common error is the omission of the dollar sign from a hexadecimal number. If the result is a valid decimal number, the assembler treats it as such. However, if the hexadecimal number begins with a letter, the assembler treats the result as a symbol.

The examples in the table show that the assembler makes negative two's complement signed numbers when a number has a minus sign. No other designation of type of number is needed. Also, leading zeros are not necessary in the source statement.

TABLE 3-1   NUMBER EXAMPLES

Number Type	Source Example	Binary Number Formed
Decimal	10	0000000000001010
Decimal	−100	1111111110011100
Hexadecimal	$10	0000000000010000
Hexadecimal	$ACBD	1010110010111101
Binary	%1001	0000000000001001

User symbols

User symbols are groups of characters that represent memory addresses or data values. The assembler assigns numerical values to the user symbols as part of the assembly process. Because you will create the user symbols, you should choose symbols that are useful to the reader of the assembly language source program.

In contrast to user symbols, the assembler has a predefined or known symbol for each instruction. Because these symbols have been chosen to make them easy to remember, they are called *mnemonics*. For example, LDAA is the instruction mnemonic for the load accumulator A instruction.

User symbols must meet the following minimum requirements to be acceptable to the assembler program:

- *A symbol consists of one to six alphanumeric characters.* The first character must be alphabetic. Some assemblers allow longer symbols with other characters, but all assemblers accept symbols as specified here. If a symbol is too long, most assemblers truncate the extra characters, which may cause confusing program errors that are usually not detected by the assembler. For example, an assembler that allows only six character symbols may treat both ABCDEFGHI and ABCDEFXYZ as the symbol ABCDEF.

- *A user-defined symbol may not consist of the single characters A, B, X, or Y.* The assembler has already defined these symbols to represent the microprocessor registers. Although most assemblers allow it, you should not define symbols that are identical to instruction mnemonics. For example, LDAA as a user symbol may be confusing.

- *Each user symbol must be unique.* The assembler will generate an error message for doubly defined symbols. For example, the same symbol cannot refer to the addresses of two different memory locations.

Here are some valid user symbols: TIME, INITCT, TEMP3, and AA. The following are incorrect user symbols: 34ABC, 125, and A.

Expressions

Symbols, numbers, and combinations of symbols and numbers separated by the arithmetic operators +, −, *, and / are called *expressions*. A single number or a single symbol is the simplest expression. Spaces cannot be embedded within an expression.

Inexperienced programmers should avoid using complicated expressions. Probably, all programmers should avoid complicated expressions!

An example of an expression is ABC+1. It means to add one to the value of the symbol ABC. So, if the value of ABC is 5, the value of the expression is 6. Expressions that add one to a symbol are used frequently.

The meaning of an expression depends upon its application. Good names for the symbols makes understanding the expression easier.

Source Statement Format

A *field* is a part of a source statement with a particular meaning. Each source statement contains up to four fields of information. These fields—from left to right—are the label, operation, operand, and comment fields. Some fields may be left blank. The fields have variable lengths and are delimited by spaces.

The items within the fields must be separated by at least one space. You may use more spaces without changing the meaning of the statement. Some assemblers print the listing with your spaces included while others arrange columns independent of your spaces.

The maximum length of a source statement in the listing depends on the assembler. If your line length exceeds the assembler limit, the assembler will do something to accommodate the extra characters. Usually the result is not very good. For example, the excess characters may be printed at the beginning of the next line.

In addition, the length of the source statement may cause the listing line to exceed the line length accepted by the printer or to be greater than the paper width. All these problems should be resolved before you make a source program.

Label field

A *label* is a user symbol that represents the memory address of the labeled item. A symbol is called a *label* when it represents an address. A symbol is made a label by placing its first character in column one of the source statement. The label field extends from column one to the first space character. Placing the symbol at column one tells the assembler to assign the value of the current address to the symbol. The assembler keeps the current address in its location counter.

The EQU directive discussed later overrides the assignment of an address value to the symbol starting in column one by assigning its own value. A symbol defined by EQU is usually not an address and then is not called a *label*.

Column one defines the label; therefore, the label field is always to the left of the operation column. This arrangement mimics the memory diagrams in Chapter 2 that show the address to the left of the memory registers containing instruction codes.

Column one of the source statement is special in that it defines labels. However, most statements do not need a label. Put a space in column one if no label is required.

Here is an example of a source statement with a label:

```
        START        CLRA
```

where the symbol START is the label with the S in column one.

Operation field

The next field to the right of the label field, called the *operation field*, contains an instruction mnemonic or an assembler directive. The operation field can be any length and starts at any column beyond column one or beyond the label. You can use spaces to align statements.

The allowed instruction mnemonics are defined on the instruction set reference for the microprocessor in Appendix A. The instruction set table therefore is both an assembly language reference and a machine language reference. The source form column of the instruction set table also shows the assembly language format for each instruction.

In this example of a source statement:

           START   CLRA

the CLRA instruction mnemonic is in the operation field beyond the label START. One space between the T and C is required, and the others are ignored.

Operand field

The content of the *operand field* varies depending on the operator used in the operation field. If the operator is a directive, the operand may be nothing or it may be an expression that is appropriate to that directive. If the operator is an instruction mnemonic, the operand will be nothing, a symbol, a number, or an expression with an addressing mode character. Table 3-2 illustrates the instruction operands and their formats for instructions introduced so far. A few more formats will be covered in later chapters. Table 3-3 is a series of examples of instructions illustrating each addressing mode.

No spaces are allowed in the operand field. Be careful because extraneous spaces may cause your program to be incorrect, while the assembler interprets the statement as correct.

Carefully notice that the instruction formats for direct, extended, and relative addressing are the same. The assembler can easily determine that an instruction uses relative addressing, because this mode is only for branch instructions. However, many instructions have both direct and extended addressing. Nothing in the assembly language statement explicitly determines which addressing mode should be used.

The assembler must choose between the direct and extended addressing modes based on the value of the operand. If the operand is an address in the range 0000 through 00FF, the assembler selects direct addressing. If not, the assembler selects extended addressing.

Immediate addressing is indicated by the # character. Because immediate addressing means that a constant number is in the instruction, using the symbol # that means number makes the format easy to remember. Similarly, indexed addressing is indicated by ,X at the end of the operand, which is easy to remember. Be careful not to omit the # or ,X because the result will be a correct statement to the assembler that will not cause an error message, but it will change the meaning to a different addressing mode.

TABLE 3-2   INSTRUCTION OPERAND FORMATS

Addressing Mode	Operand Format	Operand Type
Inherent	none	none
Immediate	#operand	data number
Extended	operand	address
Direct	operand	address
Indexed	operand,X	offset
Program relative	operand	address

TABLE 3-3   INSTRUCTION FORMAT EXAMPLES

Addressing Mode	Example 1	Example 2
Inherent	INCA	TAB
Immediate	LDAB #40	LDX #$0100
Extended	STAA LENGTH	LDAA UPLIMIT
Direct	SUBA COUNT	ADDD CORRECT
Indexed	NEG 5,X	LDD FACTOR,X
Program relative	BVS OERROR	BHI NEXT

Comment field

The *comment field* is the remainder of the line to the right of the operand field. The comment field may be left blank. Otherwise, you will put messages or comments to document the program in this field.

The assembler ignores comments when generating object code—it only prints the comments on the listing. The length of the line that can be printed by the assembler is limited by the assembler, the printer, and the paper width. Long comments will cause problems. You must determine the limits inherent to your assembly system.

Spaces delimit the fields, so the comment field starts after the first space after the operand field. Be careful to avoid putting spaces in the operand field because a space makes the rest of the operand a comment.

As an example of a comment, the statement

```
        READY        TST FLAG      CAREFUL! THIS IS TRICKY
```

has a comment to the right of the instruction.

The Basic Assembler Directives

Directives are symbolic commands in the source program that control the operation of the assembler program. Directives are placed in the operation field although they do not direct the assembler to generate instruction codes.

The following sections describe the basic, most frequently used Motorola directives. Most commercial assemblers have many more directives to give the programmer greater control over the assembler. Unfortunately, no easy way to standardize the directives among companies is available, so a given source program cannot be used with all assemblers. The basic directives presented here work with all assemblers that follow the Motorola assembly language format.

In some of the descriptions of directives that follow, square brackets enclose optional quantities. The brackets are not part of the statement—they indicate only that some quantities are optional. Optional items can be omitted entirely. Also, lowercase names distinguish various quantities from the directive name in uppercase letters. Use your values for anything in lowercase.

Memory allocation

Specifying how certain areas of memory will be used is called *memory allocation*. Some memory allocation directives determine where numbers, whether they are instructions or data, will be placed in the memory of the target computer. Other directives determine the amount of space or the number of registers allocated for data numbers.

ORG directive.   The *origin directive*, ORG, puts an address into the assembler location counter. The assembler program uses the location counter to track the current address as statements are assembled. The source statements that follow create numbers that are assembled into memory locations at the address specified in the location counter. Each time the assembler generates a new byte, it also adds one to the location counter. Therefore, the original number in the location counter determines the origin or beginning point for loading numbers into the memory of the target computer. The format of the statement is:

ORG   expression

where the expression usually is a hexadecimal address.

Normally, at least two ORG statements will be in a program listing—one to specify the beginning of the data section, and one for the beginning of the program section. For example, the statement

ORG $C800

tells the assembler the following statements will specify numbers to load into consecutive memory registers beginning at address C800.

Most assemblers initialize the location counter to zero if they don't find an ORG directive in a source program. Be careful to avoid having two ORG statements that each cause code to be assembled at the same memory locations. Some assemblers will warn you if this happens. If this error is made, usually the last numbers to be loaded into the conflicting locations will be the final numbers in the memory of the target computer.

RMB directive.   *RMB* is an acronym for *reserve memory bytes*. The format of the statement is

[label]   RMB   expression

where the expression determines a number the assembler adds to the location counter. Adjusting the location counter causes the assembler to skip the number of bytes determined by the expression. In effect, RMB reserves a block of memory registers for use elsewhere in the program. The loader does not affect the memory bytes skipped by an RMB. Some assemblers leave the contents column of the listing blank to indicate that nothing happens to these registers during loading.

The label on an RMB refers to the first of the bytes reserved. For example, the statement

WORKCT   RMB   2

reserves a double-byte in memory that probably is a working counter. The instructions that initialize and adjust the counter will use the label WORKCT to access it.

The RMB directive is used here because running the program must initialize the working counter. Loading a program into memory should not initialize the working counters.

Number formation

The assembler must be directed to generate a number and put it into memory whenever a data number is required. The directives that generate numbers also determine such details as the number of bits allotted to the number.

Use the number formation directives to generate data numbers only. Never use them to generate instruction code numbers. Always use instruction mnemonics for instructions to avoid confusion.

FCB directive.   *FCB* is an acronym for *form constant byte*. The FCB directive creates single-byte numbers and puts them into consecutive memory registers. The format of the statement is

        [label]   FCB   expression[, expression, ...]

where the expressions determine the values of the numbers. The most common expression is a single number. The label refers to the first of the numbers specified by the expressions. For example, the statement

        TABLE   FCB $27,12,LENGTH+1,%10

creates a table of four single-byte data numbers beginning at the address TABLE. An expression with a symbol specifies the third number. The symbol LENGTH was defined elsewhere in the source. Notice that you cannot label any bytes in the table beyond the first. If you need such labels, use separate FCBs for each data number.

The FCB truncates each number it creates to its least-significant eight bits. Some assemblers warn you if the most-significant eight bits of the numbers are not all zeros.

FDB directive.   *FDB* is an acronym for *form double-byte*. The FDB directive creates double-byte numbers and puts them into consecutive memory registers. The high or most-significant byte is at the lower address and the low byte is at the higher address. The format of the statement is

        [label]   FDB   expression[, expression, ...]

where the expressions determine the values of the numbers. The label refers to the high byte of the first double-byte number specified by the expressions. For example, the statement

        LARGE   FDB $C3D

creates a single double-byte number where LARGE represents the address of the 0C byte. Notice that you cannot label the low byte of a double-byte number. If you need a label on each byte, use two FCB directives to form the double-byte number. An alternative is the expression LARGE+1 that refers to the low byte of the double-byte number.

Symbol definition

Symbols represent either addresses or data numbers. The assembler cannot distinguish the two types of numbers. The only distinction is the means of defining the symbol. Always define address labels by placing the first character in column one. Always define data symbols with a directive in the Motorola assembly language.

EQU directive. The *EQU directive* equates a symbol to a numerical value. The symbol must have its first character in column one. The format of the statement is

symbol   EQU   expression

where the symbol is assigned the value of the expression. The EQU overrides the normal assignment of the symbol by directly specifying its value. In other words, even though the symbol is in column one, it is not assigned a value from the location counter. For example, the statement

PI   EQU   31416

creates a symbol PI and assigns it the value 31416_{10} in the assembler symbol table. Symbols defined by EQU are usually data values rather than addresses or labels.

The value of the expression is always a 16-bit number. The use of the symbol elsewhere in the program determines whether eight or 16 bits are needed.

The EQU directive affects only the assembler symbol table. It does not directly affect the memory of the target computer. If the symbol PI was never used anywhere else in a program, the memory of the target computer would contain no evidence that the value 31416_{10} exists.

Assembler control

Some directives control how the assembler program works and therefore do nothing to the object code generated by the assembler. Most such directives are for the convenience of the programmer and can be ignored. However, the following control directives are essential to use of the assembler.

END directive. The *END directive* marks the end of the source program and needs no label or expression. The END statement tells the assembler it has reached the last source statement. The assembler will not read any more source lines after the END directive. If any additional source statements are beyond the END directive, they will be ignored. When the assembler encounters the END, it proceeds to generate the object module.

If the END is omitted, the error recovery action taken depends upon the particular assembler program in use.

The END directive is not the same as the *STOP* instruction in the program. *STOP* is an instruction that controls the target microcomputer. END is a directive that controls the assembler program.

Asterisk directive in column 1. An asterisk in column one is a directive that tells the assembler the entire line is a comment. The assembler prints the characters following the asterisk on the listing, but ignores them for everything else. The asterisk character has this special meaning only in column one. For example,

* THIS LINE IS A COMMENT

is a line that is entirely comment. The normal meanings of the source statement fields have been overridden.

Asterisk or $ directive.   The assembler treats an asterisk anywhere other than column one or in a comment as another directive that is equivalent to a $ *directive*. This directive tells the assembler to substitute the current address for the symbol ∗ or $. A common usage is to set a label equal to the current address.

An example of the use of the ∗ directive follows:

```
              ORG      $C100
FIND          EQU      *
              LDAA     SEARCH
```

The label FIND is assigned the value C100 because that is the number specified by the ORG statement on the previous line. Also, FIND is the label on the LDAA instruction, which is assembled at address C100. This way of labeling the LDAA instruction is useful if the program is changed frequently, because the label does not need to move from line to line.

3.3   EXAMPLES FROM THE ASSEMBLER

The first example here includes a source listing and the corresponding assembler listing. The source listing was made with a programming text editor. The assembler listings for the examples were made with a commercial cross assembler, but anything unique to that assembler has been removed from the figures. Therefore, the listings are good examples for almost any assembler.

Some Good Assembly Language

The first example in this section is nothing more than a collection of assembly language statements. It is not a useful program! The example includes a variety of statements to show what the assembler does. A realistic program example follows later.

The source listing

A listing of source statements is shown in Figure 3-4. An editor program printed the source listing from the contents of the source module. Refer to the figure as you read the following comments:

- Box 1 points to a comment line. The asterisk directive in column one defines this line as a comment. All program listings should have a title section to identify the program. Comment lines create the title section and separate various parts of the listing. Remember, a comment does not affect the target computer.

- Box 2 points to a line that defines the symbol CONST and assigns it the value 17_{10}. This line does not directly affect the memory of the target computer. EQU statements are usually put at the beginning of the source module so that the symbol definitions are easy to find.

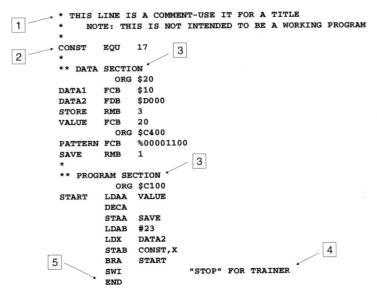

```
1 →    * THIS LINE IS A COMMENT-USE IT FOR A TITLE
       *      NOTE: THIS IS NOT INTENDED TO BE A WORKING PROGRAM
       *
2 →    CONST    EQU   17        3
       *
       ** DATA SECTION
                  ORG $20
       DATA1    FCB   $10
       DATA2    FDB   $D000
       STORE    RMB   3
       VALUE    FCB   20
                  ORG $C400
       PATTERN FCB    %00001100
       SAVE     RMB   1              3
       *
       ** PROGRAM SECTION
                  ORG $C100
       START    LDAA  VALUE
                DECA
                STAA  SAVE
                LDAB  #23
                LDX   DATA2
                STAB  CONST,X                     4
        5       BRA   START
                SWI              "STOP" FOR TRAINER
                END
```

Figure 3-4   Assembly language source listing.

- Box 3 lines are comments that are titles for major sections of the listing. The data section is a collection of statements that allocate memory and define data numbers. The program section is the list of instructions in the program. The titles mark the listing to make reading it easier. Sometimes, additional directives tell the assembler to print these parts of the program on separate pages.

- Box 4 points to a comment added at the end of a source statement to clarify that statement. In this book the SWI instruction is called the *STOP* instruction because that is its function in a Motorola trainer and some other trainers, but the assembler must be given the correct instruction name. The comment to the right of the instruction should help you recognize the purpose of this instruction. Use the space to the right of instructions for comments that explain unusual uses for instructions. This space is not very useful for documenting the function of the program.

- Box 5 points to the END directive that tells the assembler that the source module contains no more source statements. All source listings have exactly one END directive, and it always is the last statement.

Some statements in this source program have been indented. The instructions were indented to avoid the label defining column one. The ORG statements were indented an extra amount to make the listing easier to read.

The assembler listing

Figure 3-5 shows the listing that the assembler printed. A few lines were removed from the figure. They contained additional titles, disk file identifiers, blank lines, etc.

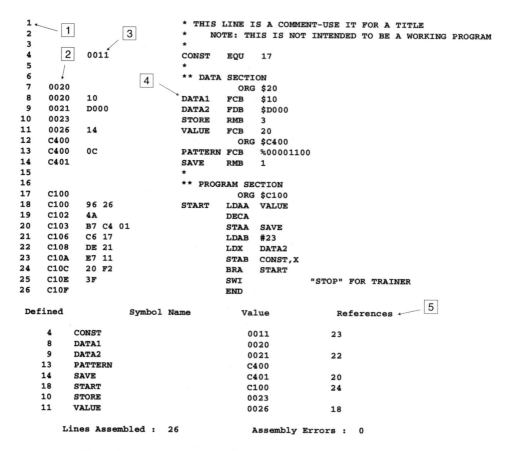

```
1                                    * THIS LINE IS A COMMENT-USE IT FOR A TITLE
2        1              3           *      NOTE: THIS IS NOT INTENDED TO BE A WORKING PROGRAM
3                                   *
4        2    0011                  CONST    EQU    17
5                                   *
6                          4        ** DATA SECTION
7    0020                                    ORG    $20
8    0020    10                     DATA1    FCB    $10
9    0021    D000                   DATA2    FDB    $D000
10   0023                           STORE    RMB    3
11   0026    14                     VALUE    FCB    20
12   C400                                    ORG    $C400
13   C400    0C                     PATTERN  FCB    %00001100
14   C401                           SAVE     RMB    1
15                                  *
16                                  ** PROGRAM SECTION
17   C100                                    ORG    $C100
18   C100    96 26                  START    LDAA   VALUE
19   C102    4A                              DECA
20   C103    B7 C4 01                        STAA   SAVE
21   C106    C6 17                           LDAB   #23
22   C108    DE 21                           LDX    DATA2
23   C10A    E7 11                           STAB   CONST,X
24   C10C    20 F2                           BRA    START
25   C10E    3F                              SWI            "STOP" FOR TRAINER
26   C10F                                    END
```

```
 Defined           Symbol Name        Value         References        5

     4      CONST                    0011              23
     8      DATA1                    0020
     9      DATA2                    0021              22
    13      PATTERN                  C400
    14      SAVE                     C401              20
    18      START                    C100              24
    10      STORE                    0023
    11      VALUE                    0026              18

        Lines Assembled :  26            Assembly Errors :  0
```

Figure 3-5   Assembler listing for a collection of assembler statements.

Layout of the listing.   The assembler listing has several distinct parts. Each part is particularly useful for a certain task. Refer to the boxes on Figure 3-5. The following comments describe the listing by referencing the boxes:

- Box 1 points to the column of decimal numbers that identify the lines on the listing. It is most convenient to talk about the listing by referring to these line numbers.

- Box 2 identifies the address column of the listing. These hexadecimal addresses refer to memory locations in the target computer.

- Box 3 identifies the columns that relate to the memory contents of the target computer. However, some numbers in this column are not memory contents as you will determine from the context of the line. All numbers in the contents columns are hexadecimal numbers. Because some instructions have five bytes, a large space is needed.

- Box 4 identifies the source statements that were read from the source module by the assembler program. The assembler reproduces the source characters on the listing. Therefore, most people don't print the source listing using the editor. Printing the assembler listing and using it as a source listing is almost as easy and certainly more useful.

- Box 5 points to the beginning of the user symbol table. All user symbols that the assembler found are listed here with the values that the assembler determined for them. This symbol table also includes a cross-reference.

Always remember that the assembler created anything printed on the listing that was not in the source. You can better understand the listing if you know what created the information on it.

The assembler listing in detail.   Two software programs (namely, an editor and an assembler) are necessary to create the assembler listing. The program that the listing represents will be put into the target computer. Aspects of all three of these programs are contained in the listing. Consider these programs as you read the following comments about the assembler listing in Figure 3-5. The listing is referenced by line number. Also look at Figure 3-6 to see the contents of memory after the load module is loaded.

- *Line 4.*   The EQU directive creates the user symbol CONST and assigns it the value 17_{10}. The only thing affected by this statement is the assembler symbol table. Nothing in the target computer is affected, so the address column of the listing is left blank by the assembler. The contents column contains the hexadecimal number 0011, which is the 16-bit equivalent of 17_{10}. Be careful! Just because the 0011 is printed in the contents column does not mean that it will be loaded into the target computer. Remember that the contents column may contain numbers that do not represent numbers in the memory of the target computer.

- *Line 7.*   The ORG directive tells the assembler the first memory location to be used by the items generated by the following statements. The numerical value usually is a hexadecimal number as illustrated here. The number 0020 was chosen to illustrate some data numbers in the direct addressing range of memory. Instructions referencing these numbers should have direct addressing. Indenting the ORG statement two character positions makes the listing easier to read. Notice that the contents column of the listing is blank to indicate that the memory of the target computer is not affected.

- *Line 8.*   The FCB creates a single-byte number 10 as shown by the contents column. This byte is at address 0020 due to the ORG statement on line 7. Line 8 also defines the address label DATA1. The symbol DATA1 is assigned the value 0020 as the symbol table shows.

  After loading, the number 10 that was created by the FCB is at location 0020 as shown in Figure 3-6.

	•	C106	C6
0020	10	C107	17
0021	D0	C108	DE
0022	00	C109	21
0023		C10A	E7
0024		C10B	11
0025		C10C	20
0026	14	C10D	F2
	•	C10E	3F
C100	96		•
C101	26	C400	0C
C102	4A	C401	
C103	B7	C402	•
C104	C4		
C105	01		

Figure 3-6   Memory contents after loading the program in Figure 3-5.

- *Line 9.*   The FDB directive forms the double-byte number D000 that goes into the next two bytes at addresses 0021 and 0022. The listing shows both bytes on one line. Consequently, the address on the next line is two higher than the address on this line.

  Locations 0021 and 0022 contain D000 in Figure 3-6.

- *Line 10.*   This RMB directive tells the assembler to allocate or reserve three memory locations. RMB does not specify any value to be loaded into these locations, and the memory of the target computer will not be altered when the program is loaded. To indicate that nothing will be loaded, the space in the contents column is blank. Because three bytes were allocated, the address on the next line is three higher than on this line. Notice the label STORE appears to be meaningful to the application because the reserved bytes can only be used to store numbers.

  Locations 0023 through 0025 are blank in Figure 3-6 to indicate that the contents are unknown. These locations were not affected by the loader.

- *Line 11.*   The FCB directive forms the hexadecimal number 14 that is equivalent to the decimal number 20. The label VALUE is assigned the numerical value 0026.

  Location 0026 contains the number 14 in Figure 3-6.

- *Line 12.*   A new ORG statement can be used anywhere in the source program. Here some data numbers are put into locations beyond the direct addressing range because the address C400 is higher than 00FF.

- *Line 13.*   This FCB directive generates the number 0C that was specified in binary format. Binary format is especially useful when the individual bits of the number are important to the application. Leading zeros are not necessary, but here they help you visualize the bit pattern in the number.

Location C400 contains 0C in Figure 3-6.

- *Line 14.*   The RMB directive reserves one memory location. The assembler assigned the label SAVE the value C401 as you can see in the symbol table.

Location C401 is shown blank in Figure 3-6 to indicate that the contents are unknown.

- *Line 17.*   Another ORG determines the location of the program section. Usually, an ORG will be at the beginning of both the data and program sections. The addresses specified by the ORGs do not have to be in numerical order.

- *Line 18.*   The first instruction in the program is labeled START to make finding the starting point of the program easy. The LDAA instruction here has direct addressing. Because the assembly language format of this LDAA refers to either direct or extended addressing, you can only determine the addressing mode by looking at the instruction code the assembler generated. Because the instruction is two bytes long, it must have direct addressing. If the instruction had been three bytes long, it would have had extended addressing.

Locations C100 and C101 contain the instruction code 96 26 in Figure 3-6.

- *Line 19.*   The DECA instruction has only inherent addressing and requires only the one byte 4A in location C102.

- *Line 20.*   This instruction format could specify either direct or extended addressing. By examining the contents columns, the 3-byte form indicates extended addressing. The assembler determined that the value of SAVE is C401 by looking in the symbol table. Therefore, the address is outside the direct addressing range, and the assembler chose extended addressing.

Locations C103, C104, and C105 contain the 3-byte instruction code B7 C4 01 in Figure 3-6.

- *Line 21.*   The data values used in immediate addressed instructions are constants; therefore, putting the actual numerical value in the source statement is often reasonable. If the same constant must be used in more than one place, a symbol for it should be defined by an EQU directive. Then changing the EQU statement changes all occurrences of the constant. Here the constant 17 was specified as decimal 23.

Locations C106 and C107 contain the instruction code C6 17 in Figure 3-6.

- *Line 22.*   This LDX instruction references a double-byte number at location DATA2 using direct addressing.

- *Line 23.*   The symbol CONST specifies the constant offset part of this instruction. Symbols are commonly used for offsets. Be careful! The offset is an 8-bit unsigned number, so the value of the symbol must fit into eight bits. Many times this offset will be zero. Then, put the actual number into the source statement.

- *Line 24.* The BRA instruction has only program relative addressing. The assembler determined the relative offset to be F2, a negative number, that causes the BRA to branch to a lower address.

- *Line 25.* The SWI instruction stops the program when used in a Motorola trainer. The function of this instruction is described later.

- *Line 26.* The END directive tells the assembler that no more source code needs to be read. The address C10F printed in the address column has no significance.

Reading an assembler listing and looking at all the details are quite tedious, but necessary to learn what the assembler does. After you learn the details, you will read the listing by focusing only on the parts you need.

The symbol table. The symbol table lists all information the assembler has about the user symbols that were defined in the program. Here are descriptions of the columns and messages included in the symbol table:

- *Defined*—the line number of the listing line where the symbol was defined. For example, the symbol PATTERN was defined on line 13. Only one line number is possible for each symbol on a correct listing.

- *Symbol name*—the names of the user symbols defined in the source module. You can often find programming errors by looking for misspelled symbol names. Sometimes the erroneous symbol is a hexadecimal number that has the dollar sign missing. The large blank area to the right of the symbols provides space for long symbol names.

- *Value*—the values of the user symbols in hexadecimal number format. The assembler determines these values from the source statements. For example, the value of SAVE is C401 because the ORG said to start assembling at address C400, and then one byte was assembled before the label SAVE was defined.

- *References*—the line number of the lines where each symbol was used or referenced. A long program may reference a symbol hundreds of times, making this list very long. In this example, the symbol START was referenced on line 24 by the BRA instruction.

- *Lines assembled*—the number of source lines that the assembler read from the source module. This number must match the number of lines that the editor created. If the numbers don't match, probably an erroneous END statement is in the source.

- *Assembly errors*—the number of errors detected by the assembler. Unless this number is zero, proceeding with the existing program has no value. It will not load and run correctly. Do not be discouraged if the number of errors is very large. For example, having approximately one error per line as an average may not seem possible. Large error numbers usually occur when many references are made to the same erroneous symbol. Similarly, one error may cause several other errors, so fixing the one real error may resolve many others.

The symbol table in this example is only marginally useful because this program listing is so short. Practical programs often have listings that are several hundred pages long. Then, the symbol table is very valuable for finding symbols, their values, and the instructions that reference them.

Some Bad Assembly Language

Assemblers have many features that enable programmers to create good, useful programs. The same features also can create programs that seem to defy logic. Sometimes programmers use assembler features in clever ways that cause problems for others using the program. This section expresses opinions on some bad ways to use assembly language.

The bad example

Figure 3-7 is a listing containing a collection of statements designed to illustrate some bad ideas. The statements are all valid assembly language statements. The assembler finds no errors. But the organization, ease of use, readability, and statement formats are problems.

As you read the following comments, you should envision a real program, not the small sample shown in the figure. Actual programs usually require a few hundred pages for the listing instead of a few dozen lines. Longer programs magnify the importance of the problems discussed here.

Bad documentation.   Look at the documentation the comments on the listing provide. Most of the documentation is useless as the following points out:

- *Line 1.*   The title is inadequate for an actual program. A good title section will describe the program function, memory requirements, revision history, author, date, and other pertinent items.

- *Line 3.*   The comment on line 3 was too long, so the assembler wrapped part of it to the next line in an unsatisfactory manner. This comment is very difficult to read.

- *Line 5.*   The relevance to the program of the symbol CONST should be stated in the comment field.

- *Line 16.*   The data section ends at line 15, and the program section begins at line 16. No headings or blank lines separate the two parts. Therefore, finding the start of the program on the listing is difficult. The program and data blur together.

- *Line 18.*   Putting good comments that describe the function of the program in this listing is not possible because it has no reasonable function—it's only an example of statements. However, a problem occurs in the comment on line 18. The comment does not attempt to describe why the STAA instruction is in the program. It only restates the obvious by telling what the machine instruction does. To make matters worse, the comment is incorrect! The STAA instruction stores the contents of the

```
 1                              * THIS LISTING DEMONSTRATES SOME BAD IDEAS
 2                              *
 3                              *    NOTE: THIS IS NOT INTENDED TO BE A USEFU
                                L PROGRAM
 4                              *
 5              0015            CONST   EQU    $15
 6                              *
 7                              ** DATA SECTION
 8                              *
 9    0020                              ORG 32
10    0020    10                MARY    FCB    $10
11    0021    3946              BOB     FDB    $ABCD/3+2
12    0023    36                ABCDEF  FCB    CONST+BOB
13    0024                      LDAA    RMB    1
14    0025    0A                A       FCB    $A
15    0026    EC                TWOONES FCB    %11101100
16    0027    96 24             START   LDAA   LDAA
17    0029    4A                        DEC    A
18    002A    97 45                     STAA   $45      LOAD THE A REGISTER
19    002C    C6 20                     LDAB   #MARY
20    002E    D7 25                     STAB   ABCDEF+2
21    0030    DE 15                     LDX    CONST
22    0032    E7 20                     STAB   MARY,X
23    0034    20 F5                     BRA    START+4
24    0036    3F                        SWI             "STOP" FOR TRAINER
25    0037                              END
```

Defined	Symbol Name	Value	References	
14	A	0025		
12	ABCDEF	0023	20	
11	BOB	0021	12	
5	CONST	0015	12	21
13	LDAA	0024	16	

Figure 3-7 Assembler listing with bad ideas.

accumulator—it does not load the accumulator. *Incorrect comments are worse than no comments.* Always strive to write the most clear, most informative, and most useful comments.

Whenever you write a program for any serious use, you should print a listing and then put it aside for a few days. Later, read your comments to see whether they make sense. Often you will find that your comments and documentation can be improved.

Bad labels.   Most of the address labels are poorly named. The programmer must usually be very creative to select good names for the program labels. Here are some specific problems:

- *Lines 10 through 12.*   People's names as labels may be cute, but they are unlikely to be helpful. The label ABCDEF is equally poor, because it conveys no useful information.

- *Lines 13 and 14.*   The label LDAA is the same as an instruction mnemonic and is very confusing. The label A is easy to confuse with the accumulator A and may cause errors in some assemblers.

- *Line 15.*   The label TWOONES seems to mean the number contains two ones, but it doesn't. Labels that try to identify the characteristics of a data number will be troublesome because the data value is likely to change.

Sometimes people conclude that a label on every assembly language line is a good idea—it isn't. Only use labels where they are needed. The reader must examine any unnecessary labels; this extra effort wastes time. Probably the only label that is useful when the program doesn't use it is the label on the first instruction of the program—the starting address.

Bad numbers.   The format of a number should match its application. Normally, numbers with different bases should not be mixed in expressions. Here are some examples from Figure 3-7:

- *Line 9.*   The ORG statement specifies an address in decimal format. Addresses should always be hexadecimal numbers. The assembler printed the hexadecimal address in the address column, but the ORG statement is still difficult to read.

- *Line 11.*   The assembler calculates a number in a very complicated way. Generally such calculations should be avoided. If the calculation is necessary, include parentheses in the expression so that the precedence of operations is unmistakable. Also, be careful to understand exactly how the assembler does integer division.

Some assemblers will use a default value of zero if a number is omitted. Generally avoid doing this; always state the number exactly as you want it.

Bad expressions.   Most expressions containing more than one symbol are difficult to read and may lead to programming errors. Avoid all complicated expressions. An expression that adds one to a symbol is the most useful and hence most common expression. Here are some problems in Figure 3-7:

- *Line 12.*   The expression CONST+BOB generates a constant number for the FCB directive. The expression uses the address label BOB, which implies a 16-bit number, but the FCB creates only a single-byte number. Adding the symbol CONST to the truncated address is even more confusing. Expressions should use symbols that represent consistent quantities.

- *Line 20.*   The STAB instruction stores a number at an address that is 2 higher than the label ABCDEF. An offset from a label, like the 2, is dangerous. For example, executing the STAA instruction will store to location 0025, which destroys a constant. This instruction must be an error. Probably some changes were made in the data section. A change moved the location where this instruction was supposed to store its data. When you change the data section of a program, finding every instruction that references every data location and correcting how they address the data are very difficult. A much better solution here is to label the location where the number is to be stored and use that label in the STAB instruction. This problem would not occur if the STAB instruction contained a single label instead of an expression.

- *Line 23.* The instruction on line 23 branches in a strange way. The expression START+4 evaluates to 002B, which is not the address of an instruction. Probably the program was changed by removing a 1-byte instruction, which now makes the BRA instruction incorrect. A change in some unrelated part of the program makes the BRA instruction incorrect. The problem would not occur if the expression START+4 were replaced by a label for the correct branch location.

Almost always a complicated expression can be replaced by a single number and a good comment to explain the number.

Bad instructions.   Instruction statements should be written to be simple and easy to read. Here are some problems in Figure 3-7:

- *Line 16.* The instruction LDAA LDAA is correct but quite confusing. People have difficulty separating the two meanings of the same symbol. The problem, of course, is caused by the choice of address label.

- *Line 17.* The instruction on line 17 probably is an error. Because line 14 has a label A, this instruction probably intended to decrement location A using extended addressing. However, the assembler could not distinguish the label from the A accumulator, and it generated an inherent addressed instruction of code 4A that decrements the A accumulator. The assembler does not consider this instruction to be an error. Instead, it is a feature that allows people to add a space before the accumulator designator in the instruction mnemonics. The space may improve the readability of the line.

- *Line 18.* The address operand for this STAA instruction is address 0045. Even if the program was moved to a different place, or the data number was moved, this instruction will always store to address 0045. Using a number for the address defeats much of the power of the assembler because the assembler is not determining the address. Also, when changes are made to the program, finding this statement in the listing to determine what is wrong will be difficult. *Never use a number as the address part of an instruction.* Always use a symbol so that the assembler can change the address for you as you modify your program.

- *Line 19.* The instruction on line 19 is probably an error. It is loading the low byte of an address into an accumulator, which is unlikely to be useful.

- *Line 21.* The instruction on line 21 is probably an error, or at least bad practice. The address operand CONST was defined on line 5 using an EQU directive. Using the EQU is a poor way to define this address. The LDX instruction stores a data number at address 0015, but this is not obvious. Labeling the item at location 0015 and using that label in the LDX instruction is better. Then the assembler listing will have an entry in the address column for address 0015 and the use of that location will be apparent. Also, mistakes causing two statements to use the same address will be unlikely. Using the EQU to define addresses defeats much of the usefulness of the assembler. In contrast, sometimes you must link a program to another program that was assembled separately. Then you must use the EQU to define the address label from the other program.

- *Line 22.*   The instruction on line 22 has a poorly chosen offset. The symbol MARY was defined as an address label on line 10. Using an address that could also be an offset is unlikely. This instruction can illustrate a good use of the EQU directive. The required offset is a constant that can be defined by an EQU using a meaningful label to name the offset. If the offset must be changed later because the program is changed, a change in the EQU statement reliably changes all occurrences of the offset. Because the same offset usually occurs in many instructions throughout a program, using a symbol for an offset is practical.

Extraneous statements.   The example in Figure 3-7 is not a real program, but it illustrates the following problems anyway:

- *Lines 11 and 15.*   The program never uses the data stored at locations BOB or TWOONES. Never put items in a program that are not used—they waste resources and are very confusing.

- *Line 24.*   The program never uses the SWI instruction on line 24. When editing a program, be careful not to leave unused instructions in the program.

Bad symbol table.   Apparently part of the symbol table was discarded. Some people ignore the symbol table; therefore, some assemblers have an option to suppress printing of the symbol table. Always print and use the symbol table—it is very useful. It will help you to find problems and to understand the program.

Assembly Language Copy a Table

A program was written in Chapter 2 to copy a table from one place in memory to another place in memory. That example, in Figure 2-48, is machine language that was coded by hand—you looked at the instruction set table to find the codes. Now let's look at the same program written in assembly language.

The assembler listing is Figure 3-8. The program is identical to the one in Chapter 2 including some programming errors. You can compare the contents columns of the figures to see that the assembler generates the same codes. Of course, using the assembler is much easier than hand coding the program.

The documentation in the assembler listing is similar to that in Chapter 2. It could be improved somewhat. Here are some comments on the program listing:

- *Lines 1 through 6.*   These comments title and otherwise document the program.

- *Lines 7 through 10.*   These lines create a user symbol and assign it a value. The comment to the right of the EQU statement gives a brief explanation of the purpose of the symbol. The best place for symbol definitions is at the beginning of the listing so that the reader knows the symbols before encountering them in the program.

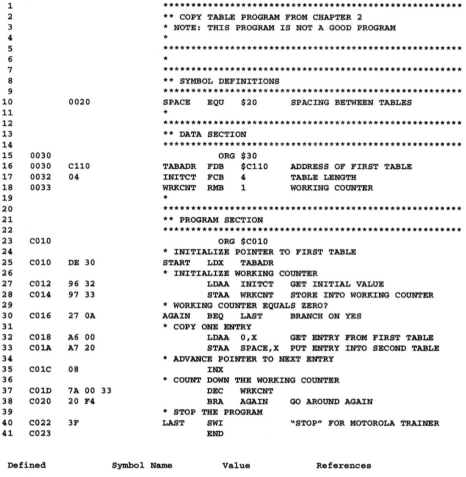

```
1                            ******************************************************
2                            ** COPY TABLE PROGRAM FROM CHAPTER 2
3                            * NOTE: THIS PROGRAM IS NOT A GOOD PROGRAM
4                            *
5                            ******************************************************
6                            *
7                            ******************************************************
8                            ** SYMBOL DEFINITIONS
9                            ******************************************************
10          0020             SPACE   EQU   $20      SPACING BETWEEN TABLES
11                           *
12                           ******************************************************
13                           ** DATA SECTION
14                           ******************************************************
15   0030                            ORG   $30
16   0030   C110            TABADR  FDB   $C110    ADDRESS OF FIRST TABLE
17   0032   04              INITCT  FCB   4        TABLE LENGTH
18   0033                   WRKCNT  RMB   1        WORKING COUNTER
19                           *
20                           ******************************************************
21                           ** PROGRAM SECTION
22                           ******************************************************
23   C010                            ORG   $C010
24                           * INITIALIZE POINTER TO FIRST TABLE
25   C010   DE 30           START   LDX   TABADR
26                           * INITIALIZE WORKING COUNTER
27   C012   96 32                   LDAA  INITCT   GET INITIAL VALUE
28   C014   97 33                   STAA  WRKCNT   STORE INTO WORKING COUNTER
29                           * WORKING COUNTER EQUALS ZERO?
30   C016   27 0A           AGAIN   BEQ   LAST     BRANCH ON YES
31                           * COPY ONE ENTRY
32   C018   A6 00                   LDAA  0,X      GET ENTRY FROM FIRST TABLE
33   C01A   A7 20                   STAA  SPACE,X  PUT ENTRY INTO SECOND TABLE
34                           * ADVANCE POINTER TO NEXT ENTRY
35   C01C   08                      INX
36                           * COUNT DOWN THE WORKING COUNTER
37   C01D   7A 00 33                DEC   WRKCNT
38   C020   20 F4                   BRA   AGAIN    GO AROUND AGAIN
39                           * STOP THE PROGRAM
40   C022   3F              LAST    SWI            "STOP" FOR MOTOROLA TRAINER
41   C023                           END
```

```
Defined          Symbol Name        Value          References

   30      AGAIN                     C016           38
   17      INITCT                    0032           27
   40      LAST                      C022           30
   10      SPACE                     0020           33
   25      START                     C010
   16      TABADR                    0030           25
   18      WRKCNT                    0033           28        37

        Lines Assembled :  41           Assembly Errors :  0
```

Figure 3-8   Assembler listing of the copy-a-table program.

● *Lines 11 through 15.*   The ORG statement locates the data section in the direct addressing range. The title identifying the data section makes the listing easier to read. Sometimes the data definitions are put on separate pages of the assembler listing for the same reason.

- *Lines 16 through 18.*   These statements allocate memory registers for the data numbers. The symbols used as address labels suggest the use of the associated registers. Comments to the right further document the purpose of each data value. The numbers 4 and 1 on lines 17 and 18 are in decimal format. Use decimal format for small numbers that are the same in both decimal and hexadecimal because they are easier to read.

- *Lines 19 through 22.*   These comments title the program section. The blank comment lines improve the readability of the listing by providing white space.

- *Line 23.*   The ORG locates the program section. The program section usually is outside the direct addressing range to conserve that space for data values.

- *Lines 24 and 25.*   The LDX initializes the X index register as the table pointer. The addressing mode of this instruction is direct because the label TABADR was given a value in the direct addressing range. The comment titling this module of the program was put on a separate line. The comment could be put to the right of the instruction, but the technique used here makes the listing easier to read. Using a separate line for the comment requires more paper to print the listing, but the small extra cost is justified. The label START is not used by the program, but it helps the reader understand where the program starts running.

- *Lines 26 through 28.*   The working counter is initialized. The comments in Figure 2-48 are to the right of the instructions. But, in the assembly language listing, the function of the instructions is described by line 26 for clarity. Extra comments to the right of the instructions further explain the instructions.

- *Lines 29 and 30.*   The loop begins here. The label AGAIN was chosen to suggest that something would branch to this place again and again. *The only reason to have a label on an instruction, other than the first instruction, is to support a branch instruction elsewhere in the program.* Names like AGAIN, NEXT, and BACK help the reader recognize loops. Chapter 4 discusses better documentation for loops.

- *Lines 31 through 33.*   The program copies one entry from the first table to the second table using two indexed addressed instructions. The comments at the right explain the details. The offset in the STAA instruction was specified symbolically to suggest its effect; namely, that it determines the spacing between the tables.

- *Lines 34 and 35.*   Now that an entry has been copied, this instruction advances the pointer to the next entry. The comment is much longer than the instruction, but don't be bothered by this. The instruction does a very important function in the program. The importance justifies the size of the comment.

- *Lines 36 through 38.*   The program decrements the working counter and ends the loop. The fact that the DEC instruction affects the Z bit of the condition code register is important, but this fact is not documented on the listing. The reader must know how the instructions work. Also, the loop end is primarily documented by the comment on line 34. The label AGAIN in the BRA instruction also suggests the end of the loop. Chapter 4 shows how the loop can be better documented. The technique here makes identifying the loop difficult.

- *Lines 39 and 40.*   The instruction on line 40 stops the program execution. The label LAST was chosen to suggest the location of the last instruction to be executed in the program. *The STOP instruction should be the last instruction in the listing for good readability.* The comment on line 40 tells you the SWI instruction does the *STOP* function in the Motorola 68HC11 trainer.

- *Line 41.*   The END directive tells the assembler the end of the source has been reached. The address printed in the address column is not useful.

Now consider the symbol table in Figure 3-8. The user symbols that were defined in the program are listed in alphabetical order. The following comments point out some details that are in the symbol table:

- The symbol table shows that the symbol AGAIN was defined on line 30. On line 30 of the listing, you see that AGAIN starts in column one of the source line that defines it as an address label.

- The symbol table shows that the value assigned to AGAIN is C016. The C016 in the address column of listing line 30 confirms this.

- The symbol AGAIN is referenced on line 38 according to the cross-reference. Indeed, line 38 contains the BRA AGAIN instruction where the symbol determines the address part of the instruction. Notice the difference between a reference to a symbol and the definition of that symbol. A symbol can only be defined (that is, created and assigned a value) by an EQU statement or by usage as a label. Both techniques require the symbol to start in column one.

- The symbol SPACE was assigned the value 0020 by the EQU directive. An equal sign to the left of the value column indicates that an EQU directive was used for this symbol.

- The symbol table indicates that two references to the symbol WRKCNT are on lines 28 and 37. Some symbols are referenced hundreds of times in large programs.

Incorrectly spelled symbol names are a very common source of errors that can be recognized from the symbol table. If a symbol is doubly defined, many assemblers will list each value and the cross-reference will tell you where the multiple definitions are located. The symbol table can save you the effort of paging through a long listing when errors occur.

When you look at an assembly language program listing, generally you should first look to see if assembly errors are reported.

Next, look at the symbol table. You may be able to solve a problem in seconds using the information in the symbol table. Remember that many people make things difficult by ignoring the symbol table.

3.4   REVIEW

Why use an assembler? You have seen the copy-a-table program both hand coded and assembled by an assembler. You can develop the program either way. So why use an assembler? Certainly, the assembly system is costly. A computer system with a printer is necessary. Both an editor and an assembler program are required. Hand coding requires only an inexpensive instruction set table.

First, using an assembler makes developing new programs faster and easier for you. This alone justifies the cost of the assembly system. The assembler generates instruction codes very quickly. During program development, the program will be modified many times. The hand-coding method requires a large amount of time to develop any practical program.

Second, the assembler does error checking. Although most of the errors detected are simply typing errors and misspelled words, the assembler also detects errors in the use of instructions. Errors involving addressing modes and incorrect addressing ranges can be detected by the assembler. Similarly, use of the same locations in memory for more than one purpose can be detected. In large programs developed by a group of people, this error detection is very useful.

Third, the assembler documents the program in the best possible way. Both the source and object code are printed on paper and stored in computer-readable form. The assembly computer system becomes a very sophisticated filing system for storing the program.

The assembler frees you from routine work so that you can be more creative and productive in writing programs.

3.5   EXERCISES

3-1. Write an assembly language statement to generate a LDD instruction that undoubtedly puts the number 3041_{10} into the accumulator.

3-2. The last statement in an assembly language program is always _____.

3-3. Why not always use the FCB directive instead of the RMB directive?

3-4. In Figure 3-8, if line 23 was changed to ORG $C022, what would be different about the program? Will it copy the table?

3-5. Fill in the missing parts (a) through (r) of the strange program in Figure 3-9. Some items have been overprinted with Xs to hide their values.

3-6. After completing question 3-5, assume that the object code was loaded into memory and run from location START to the *STOP* instruction. Then consider the following:
   (a)   The number that was stored in memory location 0030 is _____.
   (b)   The last time the BEQ instruction was executed, it (did, did not) _____ branch.
   (c)   The number in memory location 0104 is _____.
   (d)   The number in the X index register is _____.
   (e)   The number in the A accumulator is _____.
   (f)   The number in the B accumulator is _____.

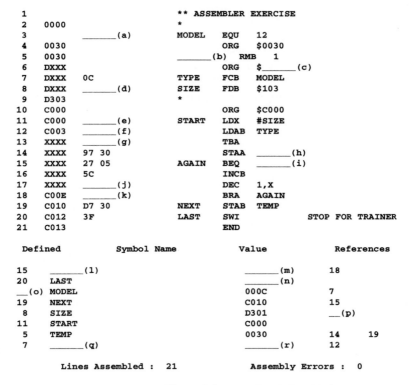

```
 1                              ** ASSEMBLER EXERCISE
 2   0000                       *
 3            _____(a)         MODEL   EQU    12
 4   0030                               ORG    $0030
 5   0030            _____(b)  RMB    1
 6   DXXX                               ORG    $_____(c)
 7   DXXX   0C                  TYPE    FCB    MODEL
 8   DXXX   _____(d)           SIZE    FDB    $103
 9   D303                       *
10   C000                               ORG    $C000
11   C000   _____(e)           START   LDX    #SIZE
12   C003   _____(f)                   LDAB   TYPE
13   XXXX   _____(g)                   TBA
14   XXXX   97 30                       STAA   _____(h)
15   XXXX   27 05               AGAIN   BEQ    _____(i)
16   XXXX   5C                          INCB
17   XXXX   _____(j)                   DEC    1,X
18   C00E   _____(k)                   BRA    AGAIN
19   C010   D7 30               NEXT    STAB   TEMP
20   C012   3F                  LAST    SWI            STOP FOR TRAINER
21   C013                               END
```

```
 Defined            Symbol Name          Value              References

 15       _____(1)                          _____(m)      18
 20       LAST                               _____(n)
 __(o)    MODEL                          000C               7
 19       NEXT                           C010               15
  8       SIZE                           D301               __(p)
 11       START                          C000
  5       TEMP                           0030               14      19
  7       _____(q)                          _____(r)      12
```

Lines Assembled :   21   Assembly Errors :   0

Figure 3-9   Exercises 3-5 and 3-6.

(g)   Would the binary code for the BEQ instruction change if the statement on line 10 were changed to ORG $DD00?

(h)   List the line numbers of all the instructions that would have changed binary codes if the statement on line 4 were changed to ORG $D000.

3-7.   Fill in the missing parts (a) through (o) of the strange program in Figure 3-10. Some items have been overprinted with Xs to hide them.

3-8.   After completing question 3-7, assume that the object code was loaded into memory and run from location START to the *STOP* instruction. Determine what numbers are in the A and B accumulators, and in the X index register when the program stops. What numbers in memory were changed by running the program?

3-9.   Write an assembly language program module to do the following: If the number in the A accumulator is −4 or greater and is +4 or less, replace it with zero, otherwise don't change the number. The numbers are two's complement numbers.

3-10.   Write an assembly language program according to the specifications in Exercise 2-45.

3-11.   Write an assembly language program that works on a table specified by the address of its first entry and its length in bytes. The program must search through the table from its beginning until it finds the first entry containing a specified single-byte value. The program then stops

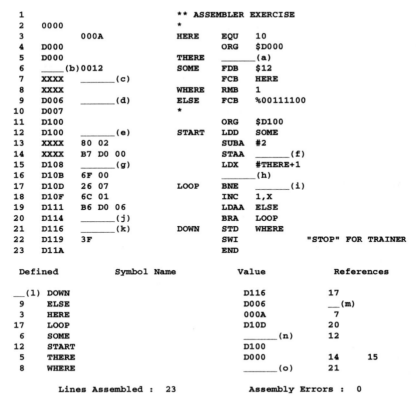

```
 1                                    ** ASSEMBLER EXERCISE
 2    0000                            *
 3              000A                   HERE      EQU     10
 4    D000                                       ORG     $D000
 5    D000                            THERE      _____(a)
 6    ____(b)0012                     SOME       FDB     $12
 7    XXXX      _____(c)                       FCB     HERE
 8    XXXX                            WHERE      RMB     1
 9    D006      _____(d)            ELSE       FCB     %00111100
10    D007                            *
11    D100                                       ORG     $D100
12    D100      _____(e)            START      LDD     SOME
13    XXXX      80 02                            SUBA    #2
14    XXXX      B7 D0 00                          STAA    _____(f)
15    D108      _____(g)                       LDX     #THERE+1
16    D10B      6F 00                            _____(h)
17    D10D      26 07                 LOOP       BNE     _____(i)
18    D10F      6C 01                            INC     1,X
19    D111      B6 D0 06                          LDAA    ELSE
20    D114      _____(j)                       BRA     LOOP
21    D116      _____(k)            DOWN       STD     WHERE
22    D119      3F                               SWI             "STOP" FOR TRAINER
23    D11A                                       END
```

Defined	Symbol Name	Value	References
__(1)	DOWN	D116	17
9	ELSE	D006	__(m)
3	HERE	000A	7
17	LOOP	D10D	20
6	SOME	_____(n)	12
12	START	D100	
5	THERE	D000	14 15
8	WHERE	_____(o)	21

Lines Assembled :   23   Assembly Errors :   0

Figure 3-10   Exercises 3-7 and 3-8.

with the relative position of the entry in the B accumulator and FF in the A accumulator. If the entry is not found, it will put 00 in the A accumulator.

3-12. Figure 3-11 is an assembly language listing of the improved copy-a-table program from Chapter 2. Consider each of the following:

(a) Study the instruction codes generated by the assembler to determine their addressing modes without looking them up in the instruction set table.

(b) If only the statement on line 9 was changed to ORG $C000, which instructions would have different addressing modes, and what are those modes?

(c) If only the statement on line 19 was changed to ORG $C000, which instructions would have different addressing modes, and what are those modes?

3-13. Modify the program in Figure 3-11 so that it adds all the entries of the first table to the corresponding entries of the second table, and stores the sums in the second table.

3-14. Write an assembly language program that reports whether two tables, specified by the addresses of the first entries and the length in bytes, have identical entries.

```
 1                                    *********************************************************
 2                                    ** IMPROVED COPY TABLE PROGRAM FROM CHAPTER 2
 3                                    *
 4                                    *********************************************************
 5                                    *
 6                                    *********************************************************
 7                                    ** DATA SECTION
 8                                    *********************************************************
 9      0020                                    ORG   $20
10      0020    C220                  TABADR   FDB   $C220     ADDRESS OF FIRST TABLE
11      0022    20                    SPACING  FCB   $20       SPACING BETWEEN TABLES
12      0023    04                    TABLEN   FCB   4         TABLE LENGTH
13      0024                          WRKCNT   RMB   1         WORKING COUNTER
14      0025                          WRKPNT   RMB   2         WORKING POINTER SAVE
15                                    *
16                                    *********************************************************
17                                    ** PROGRAM SECTION
18                                    *********************************************************
19      C100                                   ORG   $C100
20                                    * INITIALIZE POINTER TO FIRST TABLE
21      C100    DE 20                 START    LDX   TABADR
22      C102    DF 25                          STX   WRKPNT
23                                    * INITIALIZE WORKING COUNTER
24      C104    96 23                          LDAA  TABLEN    GET TABLE LENGTH
25      C106    97 24                          STAA  WRKCNT    STORE INTO WORKING COUNTER
26                                    * GET TABLE SPACING
27      C108    D6 22                          LDAB  SPACING
28                                    * WORKING COUNTER EQUALS ZERO?
29      C10A    7D 00 24              AGAIN    TST   WRKCNT
30      C10D    27 11                          BEQ   LAST      BRANCH ON YES
31                                    * COPY ONE ENTRY
32                                    *     GET POINTER TO FIRST TABLE
33      C10F    DE 25                          LDX   WRKPNT
34                                    *     GET ENTRY FROM FIRST TABLE
35      C111    A6 00                          LDAA  0,X
36                                    *     POINT TO SECOND TABLE
37      C113    3A                             ABX
38                                    *     PUT ENTRY INTO SECOND TABLE
39      C114    A7 00                          STAA  0,X
40                                    * COUNT DOWN THE WORKING COUNTER
41      C116    7A 00 24                       DEC   WRKCNT
42                                    * ADVANCE POINTER TO NEXT ENTRY
43      C119    DE 25                          LDX   WRKPNT
44      C11B    08                             INX
45      C11C    DF 25                          STX   WRKPNT
46      C11E    20 EA                          BRA   AGAIN     GO AROUND AGAIN
47                                    * STOP THE PROGRAM
48      C120    3F                    LAST     SWI             "STOP" FOR MOTOROLA TRAINER
49      C121                                   END
```

Defined	Symbol Name	Value	References			
29	AGAIN	C10A	46			
48	LAST	C120	30			
11	SPACING	0022	27			
21	START	C100				
10	TABADR	0020	21			
12	TABLEN	0023	24			
13	WRKCNT	0024	25	29	41	
14	WRKPNT	0025	22	33	43	45

```
      Lines Assembled :  49          Assembly Errors :  0
```

Figure 3-11   Exercises 3-12 and 3-13.

Chapter 4

Program Structure and Design

Programming is a topic seldom discussed in engineering and microcomputer textbooks. Many authors assume that people automatically know how to program once they understand what computer instructions are. Such an assumption is unrealistic. Programming is a complicated process—a skill that is most easily developed with guidance from an experienced and thoughtful mentor.

Because this chapter discusses programs and programming, let's begin by defining the following words:

- *Program*—the sequence of instructions, and the associated data values, that the computer hardware uses to carry out an algorithm—the step-by-step procedure required to do something.

- *Programmer*—a person who creates new programs and modifies or maintains existing programs.

- *Programming*—the act of creating programs, which includes initial design and planning, documenting, coding into a language, testing, and debugging.

147

- *Software*—the programs and related information used by a computer. The word *software* is often used as a synonym for *program*.

These definitions provide us with some common ground for discussing programming. Don't confuse these definitions with job titles or department names. For example, the person employed as a programmer may have quite different responsibilities from those stated in the definition.

4.1 THE HARD COLD FACTS

Programming is usually taught by discussing the tools used to make programs. The writer assumes that the reader will see the light and understand why these tools are important. But often the light doesn't shine! So let's begin by looking at some issues related to programming before getting to the tools and techniques.

What Does Software Cost?

Software is expensive. Industry studies show that labor cost for program development is about one to one and a half hours per machine instruction. This work is performed by expensive skilled labor. Most useful programs require a minimum of several thousand instructions!

If you are a college student, you certainly believe that software is expensive. Think of the time and effort you put into your last program. You probably felt that the minor task that the program performed was not worth the effort!

Many on-the-job people have personal computers that they have never programmed to do a custom job. Even when advanced application programming languages are available, the effort required to write a custom program is not worthwhile. Usually, a program is purchased that is good enough to meet the needs. Also, a large software industry exists to supply people with software tailored to specific applications.

Don't be deluded by the highly sophisticated personal computer program that costs a few hundred dollars or less. Mass market programs selling at that price depend solely upon a large volume of sales to justify their existence. Such programs often cost many hundreds of thousands of dollars to develop. Marketing costs drive the total cost even higher.

What about Software Quality?

The quality of much software and most documentation is poor. Even after extensive testing, software frequently has bugs. Some people say that all programs have at least one bug.

Programmers use the term *bug* to describe a mistake because the word *mistake* devastates their egos. Bugs are not only costly to the supplier, but also to the customer, who relies on the software to do a job.

You might think that good testing and correction will overcome the problem of bugs. Experienced people say that *testing proves the presence of bugs, not the absence of bugs.*

Seldom does testing prove the reliability of software because not every possibility can be tested. Because testing has not been very successful, many companies now distribute pre-production copies of software to selected users. They hope to find the bugs before wide-spread distribution complicates the issue. Despite this effort, bugs frequently show up later.

Testing and debugging of software usually take more time than the original planning and writing. Consequently, most new software is completed later than the planned schedule. Software projects commonly require two to three times the estimated time that was allotted.

After software has been used for a while, changes are often desired or necessary. Changes can easily introduce new bugs, and the changes are often very difficult to make. Experience at modifying software has led some companies to the following rule: If more than five percent of a program must be changed, throw the program away and start over!

How could software quality be so troublesome? Besides the complexity of most software, poor quality is partly due to the way people learn about computers. They must start by learning what a computer is, what instructions are, and what programming is. These are all very detailed efforts that focus the person's attention away from the final goal. As the person learns more, the number of details increases. Unless people make a concerted effort to see the bigger picture, they can become trapped in the habit of seeing only details. The result is poor-quality software. Some people go through an entire career without making an effort to see the final goal.

Programming Is Hard Work

Some people like programming. Others don't! To those who like programming, it is the best expression of creativity. To those who dislike programming, it is tedious, boring work.

Regardless of your feelings about programming, it is a difficult and exacting job. Programs with mistakes usually are nearly worthless. The pressure to be correct is over-whelming. However, no magic will remove the work from programming.

Starting on the Right Path

Let's make your life easier. Programming can be made easier, and you can be more productive if you approach the task properly. Studies of programming and related activities suggest ways to make programming more cost effective and less work, and the programs produced more reliable. This chapter discusses many ideas that have been put into practice and have proved to be useful.

Trial and error is the least effective way to write software. Deliberate and careful design approaches using good tools are necessary for true long-term success that includes maintenance and enhancement of the software.

Two good software development techniques—called *structured programming* and *top/down design*—are widely accepted. They are the starting points for good software design and implementation. They are not the answer to all problems. They are starting points. This book will encourage the use of both of these techniques. All examples in later sections adhere to these principles.

Structured programming strongly influenced the design of most high-level programming languages developed since about 1970. Before then, people did not understand the programming problem. Since then, even the hardware architecture of computers has been influenced. Top/down design has also changed language design for the better. Many modern languages enforce or encourage these ideas in your programs.

Assembly language, as discussed earlier in this book, has total flexibility. Assembly language does nothing to enforce good programming. Only the discipline of the programmer enforces good programming. So the programmer must understand and embrace the techniques to gain the benefits that structured programming and top/down design promise.

4.2   PROGRAM DESIGN—WHAT'S IMPORTANT

Programming in any language is a human activity. Errors frustrate people, and successful working programs cause great joy. Building programs and running a computer system are usually enjoyable activities. Many people work long hours at their home computers as a hobby.

Contrary to this view is the pain of detailed design work that goes into complex software. Program coding requires very exacting work.

Some people who use hobby computers call themselves *hackers*. This name implies that they use trial and error to get their programs to work as they wish. Hackers usually do little planning or detailed design of their software.

Programmers' Goals

People usually have goals when they start to write a program. Assume that they are not programming as a hobby, but they have serious reasons to do the job. These people may be college students, working engineers, software developers, or consultants. *Programmer* is a useful name for such individuals, although the job title of programmer may imply quite different duties. Once again, the term *programmer* as used here means the person—usually persons—who designs and writes a program.

Programmers have similar goals as they set out to write programs. Different applications will affect the goals, so assume that microcomputer hardware as discussed in this book is adequate for the application. The list of possible goals is probably very long, so consider only the following five:

- *Write the shortest program.*   The shortest program means the program that occupies the fewest bytes of memory and yet does the required task. The reasons for this goal are many. In product engineering or personal computer applications, the program length may be important because using less memory hardware reduces costs. College students may gain status with other students if their programs are shorter when doing the same assignment. Making the program fit existing resources could be useful if adding more memory is not practical. Many considerations including cost may limit memory size, especially in time-sharing applications.

- *Write the fastest program.* The fastest program is the one that runs in the shortest time while doing the required task. In product applications, such as automobile engine control, quick processing of complex algorithms is necessary to make the engine run correctly. Personal computers are being used for ever-larger applications, but they must quickly interact with the persons operating the computer. The fees for using some time-sharing computers are based on the run time of the program.

- *Write an easily understood program.* People, including the person writing the program, will need to understand the function of the instructions in the program. Understanding is necessary for changing the program, fixing bugs, or coping with changes in the hardware. Complex programs are more difficult to understand than simple programs. The program algorithm may affect the ease of understanding of a program, so the programmer may choose the algorithm accordingly.

- *Write an easily modified program.* Modifying a program means changing it to cope with changes in the environment. For example, a monitoring computer for several automatic manufacturing machines must monitor more machines if production must increase. The engine control computer for an six-cylinder engine will require changes if it is to control a four-cylinder engine. A personal computer program will require changes to use new features of an upgraded operating system. Each of these cases involves changes in an existing and functioning program. These examples clearly imply that an entirely new program is not necessary.

- *Meet the schedule.* Meeting a schedule is usually the goal of the programmer's supervisor or client, or of a college professor. The schedule specifies the date when the software user can do productive work with the program. A contract may dictate the schedule. Release of a new model of a product will enforce a deadline. Remember that microcomputers often are part of a product, and late release will give the competitor's products an advantage. To meet the schedule, people often forget all the other goals discussed and press on as fast as possible.

You probably can add several other goals to this list. What were your goals the last time you wrote a program?

What We Are Working With

The goals of programmers must be put into correct perspective in view of practical market conditions. The cost and performance of microcomputer hardware have changed greatly since microcomputers were first developed. Here are some accomplishments of the integrated circuit and computer manufacturers since the first hardware appeared:

- *Speed.* Microprocessor speed has increased more than a factor of 100.

- *Cost.* The cost of chips for equivalent performance has decreased more than a factor of 1000.

- *Memory size.*   Usual memory sizes have increased from a few kilobytes to a few dozen megabytes.

- *Secondary storage.*   Secondary storage has evolved beyond slow and small floppy disks with a capacity of a few hundred kilobytes. Hard and optical disks and CD ROMs (compact disk read only memory) have hundreds of megabytes to gigabytes of capacity and are much, much faster than the original floppy disks.

The enormous improvement in computer hardware has also increased its complexity. The complexity of software development has similarly increased. Sometimes additional computing power simplifies the programmer's job through more advanced software, but the complexity is still there.

In product engineering applications with embedded computers, the computer is part of a product and even the electronics will influence the programmer's task. Development software cannot easily hide the hardware complexity, although that is the intention.

The effect of increased complexity has been an ever-increasing cost of software development. Modification and long-term maintenance of software are large problems. The success of microcomputers has made enhancements a necessary marketing device. Competitors will be enhancing their products too!

Assessing the Goals Based on Reality

Let's evaluate each of the programmer's goals in view of the reality of the market and the technology available. Here is some discussion and some opinions about each goal:

- *Write the shortest program.*   This goal emphasizes expensive human resources to reduce the cost of inexpensive hardware. Human resources are an expensive part of product development, and people write programs. Automated factories build computer hardware, so it is low in cost. Thus, this goal is of little practical value. Even if many copies of the program will be used, making deliberate efforts to write short programs is seldom practical.

  Effort toward writing short programs also encourages programming tricks to cut memory usage. Tricks usually lead to problems sometime in the lifetime of a product. In contrast, selecting a good algorithm that is particularly efficient in using memory would be practical.

- *Write the fastest program.*   With the high and ever-increasing speed of microcomputer hardware, writing the fastest program is not an important goal for most applications. A big related problem, if speed is important, is knowing how to write the program for fast execution before writing begins. Similarly, determining what to optimize for speed after completing a program that is too slow is difficult. Usually, rewriting an entire program to optimize its speed is impractical anyway.

  The speed of a program must be determined under realistic control situations. One testing approach, called *profiling*, is to take data on the frequency of usage of various parts of the program. You find out how often each part of the program runs under realistic circumstances. The usage data reveals which parts have a significant effect on the performance. You then optimize those significant parts. For example, doubling the speed of a part that accounts for one percent of the total processing time does little good. Doubling the speed of a part that accounts for 40 percent of the time makes a large overall improvement.

The realistic goal is to get a good program working, and then optimize the parts that will make a difference. This goal accomplishes overall cost-effectiveness. Deliberate effort to make a fast program from the beginning is usually not practical.

The execution speed of a program is almost independent of the length of the program. Long programs are not necessarily slower than short ones. For example, suppose that a short program is written as loops inside loops. An equivalent long program is a sequence of instructions without loops. The short program will certainly run slower than the long one.

- *Write an easily understood program.* The word *understood* implies understanding by people—including both the programmer and others. Better understanding results from simple straightforward problem solutions as opposed to complex and clever solutions.

Never use programming tricks. For example, never use the op code of an instruction as the data value for another instruction. Clever tricks may slightly shorten or speed up a program. On the other hand, using expensive human resources to save inexpensive computer resources is not cost-effective. Therefore, an easily understood program without tricks is a principal goal of realistic programmers.

- *Write an easily modified program.* Modification of programs happens frequently both during development and after completion. Making a program easy to modify requires planning and effort during the design. However, this small effort is cost effective because it saves very large efforts when making the inevitable changes. Successful programmers focus on long-term goals and not just immediate success.

- *Meet the schedule.* Many people in the software industry accept late delivery of software as normal, because it happens so frequently. Users of personal computers have coined the term *vaporware* to describe advertised software that is not available for purchase. Good software delivered late is often not profitable.

The best way to meet this goal is to organize carefully and plan for changes. Software written to ease understanding and modification is necessary. Forget about writing short or fast executing programs. Forget about clever programming tricks. Emphasizing the correct goals will get the job done as fast as possible.

The purpose of this section is to emphasize that you should write programs that are easy to understand and easy to modify. No significant effort should go into writing a program so that it is fast or short. This statement does not mean that silly things should be done. It means that little value is derived from spending serious effort to make short or fast programs.

This direction is somewhat contrary to human nature. Many people feel that much of the fun of programming is in finding clever solutions to programming problems. Instead, that creativity should be applied to bigger problems than to the minute details of a program.

4.3 PRACTICAL PROGRAMMING

A set of guidelines will emphasize the important goals in the creative process of programming. Guidelines cannot cover all circumstances and guarantee success—success requires good judgment. The ideas presented here are completely compatible with the goals of easily understood and easily modified programs:

- *Don't use a single resource for multiple purposes.* When an item is used for multiple purposes, compromises will usually be necessary in the use of that item. Instead, create a new resource for each independent purpose. Separate resources will sometimes require a slightly longer program, but the advantages are worth it. For example, don't use a loop counter to form addresses in a pointer. The two uses should be independent of each other; then a change in one use won't affect the other.

- *Use no intimate knowledge of the hardware configuration.* For example, do not use unimplemented bits of a register as default values; they may change later with hardware revisions. Similarly, moving program modules to new memory locations should have no effect on the correctness of the program.

- *Keep the instructions and data separated.* The data numbers should not be scattered between the instructions. Not only is understanding such a program difficult, but the program will have unnecessary instructions to branch over the data numbers.

- *Use tables.* Collect program parameters and related data together in tables or structures. If the program is designed to interact with a person, provide software that will easily modify the tables.

- *Only put constant data within the instructions.* To change data within instructions, you must have intimate understanding of the instructions in the program. Only when a change in the data also requires the program to be rewritten should data be put inside instructions. For example, immediate and indexed addressed instructions contain constants.

- *Avoid tricks.* Clever use of instructions usually leads to solutions that are not easy to enhance and alter. Tricks usually lead to greater cost.

- *Don't write self-modifying programs.* Programs that change their instructions as they run are very difficult to understand, debug, and modify. If the computer has permanent memory that can't change, as many microcomputers do, the program won't work at all if it must modify itself.

- *Use the instructions in the instruction set well.* The instruction set might be somewhat inconsistent, so you might prefer to avoid some instructions because they are rarely used. However, the design probably includes odd instructions because they make certain functions easy to do. Use all the available resources.

You should always actively strive to write good programs. Don't concentrate only on solving the problem at hand. You will find problem solving will be easier when you search for the simple, straightforward solution instead of the complex and elegant solution.

4.4 FLOWCHARTING

A tool to help organize the programmer's thoughts is valuable when designing and writing programs. Organization will aid in promoting the proper goals. Organizing your thinking with a design tool helps you to include details that you might miss otherwise.

Computer programs are algorithmic processes—they carry out large jobs one small step at a time. Most people understand algorithmic processes best when they are illustrated graphically. Several graphical tools exist, but the one most widely used is the flowchart.

A *flowchart* is a diagram made from several standard symbols connected by flow arrows. The flowchart symbols represent actions to be taken. The arrows direct the reader to follow the progress of the actions of the program.

The flowchart is both a documentation device and a design tool. At the design stage, the programmer usually sketches the flowchart on paper and changes it repeatedly as the design progresses. When the design and program coding are complete, the flowchart becomes a documentation device because it clearly illustrates what the program does and how it does it. Both the programmer and others use the flowchart to understand a program.

Flowchart Symbols

Flowcharts are made from several standard symbols. Let's consider a subset of the standard symbols and their meanings. Additional specialized symbols are available, but they often complicate the flowchart and add little in the way of clarity.

Primary flowchart symbols

The primary symbols are necessary for making most flowcharts. With them, you can make any flowchart. All other symbols are optional.

Flow lines and flow arrows. Flow lines with arrow heads guide the reader through the other symbols in the correct order. Several flow lines may come together at an intersection point, and then one line continues from that point as shown in Figure 4-1. Several flow arrows cannot leave an intersection point though, because the reader would not know which line to follow.

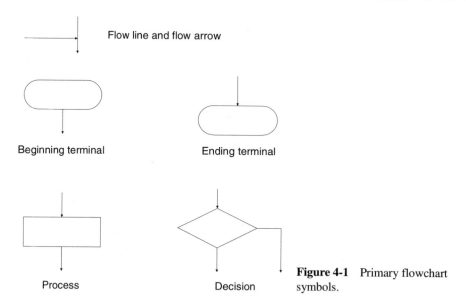

Flow line and flow arrow

Beginning terminal

Ending terminal

Process

Decision

Figure 4-1 Primary flowchart symbols.

Only a single flow line should terminate on a symbol. If two or more flow lines must come together at a symbol, the flow lines should come together before the symbol with only a single flow line terminating at the symbol. This design leads to better clarity.

Terminal.   The flowchart begins and ends with a unique symbol called a *terminal* as shown in Figure 4-1. The beginning terminal tells the reader where to start reading the flowchart. Put the name of the program module associated with this flowchart on the terminal symbol. The ending terminal indicates where the computer stops processing instructions in this program module. The label on the ending terminal is usually END or RETURN.

Process.   The process symbol is a rectangular box that means do something or process some information. Figure 4-1 shows the process symbol. Write the name or description of the process inside the symbol. Most of the symbols in flowcharts are process symbols.

Each process box identifies a group of instructions that together carry out a particular function. This function has a distinct beginning point and a distinct ending point. Hence, the symbol has a single flow arrow entering it and a single flow arrow leaving it. This simple yet important observation, one flow arrow entering and one flow arrow leaving, is discussed further later.

Decision.   The choice between two alternative flow paths is the essence of a decision. Figure 4-1 illustrates the diamond-shaped symbol for a decision. A short word description of the decision followed by a question mark identifies the symbol.

The computer makes binary choices, so the fundamental decision symbol has two flow arrows leaving it and one arrow coming to it. The paths going out usually have two opposite labels such as YES/NO, UP/DOWN, or TRUE/FALSE. These words answer the question inside the symbol. Decisions made up of more than two alternatives require several fundamental decisions.

Secondary flowchart symbols

The symbols included here are commonly encountered. However, their use is optional and usually unnecessary.

Connector.   Two connectors containing the same identifier indicate the same point in a broken flow line—they connect two parts of a broken flow line. The connector symbol shown in Figure 4-2 is a small circle. Connectors usually contain a letter as an identifier. Each connector will have either one flow arrow coming into it or one flow arrow going out of it. Connectors are seldom used. You should avoid connectors whenever possible because reading broken flow lines is difficult.

Off-page connector.   Figure 4-2 shows the off-page connector symbol—it connects two parts of a broken flow line that are on different sheets of paper. *The off-page connector is never needed and should never be used!* You will understand this very strong statement after reading the next two major sections.

Flowchart Example

A simple flowchart that includes all the primary symbols is shown in Figure 4-3. The symbols are not labeled, because the figure illustrates only the symbols and not a program. However, the logic of the algorithm can be understood even without labels.

Look at the figure carefully to see the individual flowchart symbols. Here are some characteristics of this flowchart that you should observe:

- The flowchart begins with a single terminator symbol.

- Each process box has a single flow arrow coming into it and a single flow arrow leaving it.

- The decision symbol has exactly two flow arrows leaving it.

- When the two flow arrows come together, they meet before the ending terminator so that only one arrow meets the ending symbol.

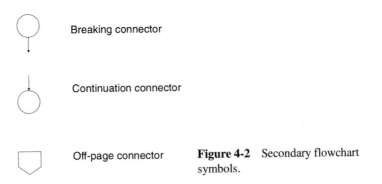

Breaking connector

Continuation connector

Off-page connector

Figure 4-2   Secondary flowchart symbols.

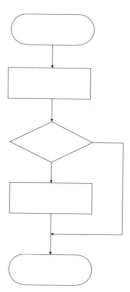

Figure 4-3   An example flowchart.

- The arrow heads clearly show the flow of the algorithm represented by the flowchart.

- The flowchart has a single ending terminator.

Using the flowchart symbols as described here for all flowcharts is good practice. Most of these characteristics should seem simple and obvious. However, when an algorithm becomes very complex, and hence its flowchart becomes complex, forgetting these simple rules is easy.

4.5   STRUCTURED PROGRAMMING

One requirement of good programming is simplicity. If programs are very complex, they are difficult to write, to modify, and to understand.

Programs consist of building blocks that flowchart symbols represent. Fortunately, we need only three symbols to make flowcharts: the terminal, the process block, and the decision. This concept seems simple enough; however, connecting these building blocks with flow arrows can be done in many ways.

Undisciplined programmers write *spaghetti* programs. The flowchart for a spaghetti program has the flow lines and boxes entangled in ways to make changes and understanding almost impossible. Figure 4-4 illustrates such a flowchart. Try to follow all the possible paths through the flowchart. Some processes have three or four different paths leading to them.

The effort to cope with spaghetti programs led to research to determine whether fundamental program building blocks exist. The research revealed only three fundamental ways of connecting the flowchart symbols. These three ways of connecting the symbols, called

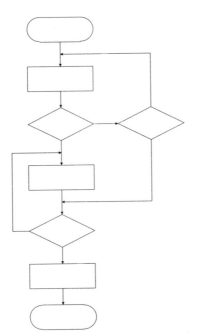

Figure 4-4   The flowchart for a spaghetti program.

program structures, are enough to build all possible programs. The discipline now called *structured programming* was the result of the research. The purest form of structured programming requires the building of programs from the three fundamental program structures only.

Experience has shown that structured programming leads to better quality software and to more cost-effective software. Of course, programs are only correctly structured if programmers make them structured, or the computer language enforces structuring. A programmer cannot make an unstructured program look structured on a flowchart. The programmer must design a structured program and then write the program by following the design. A programmer cannot write a spaghetti program, get it to work by trial and error, and then make a structured flowchart for it.

Structured programming is just one step in designing good programs and does not guarantee high-quality programs. It is just one necessary ingredient in successful software engineering.

Fundamental Program Structures

The names of the three fundamental program structures are SEQUENCE, IF-THEN-ELSE and DO-WHILE. Look at the three fundamental structures as represented by the flowchart in Figures 4-5 through 4-7. Observe that each structure has a single beginning or entry point and a single exit point. Similarly, each process box on a flowchart has a single entry point and a single exit point. Never draw any additional flow arrows entering or leaving a process box.

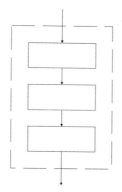

Figure 4-5 The SEQUENCE structure.

Each structure is itself a process—drawing a rectangular process box around each structure is correct. The fundamental program structure is the basic building block of programs. Structured programming advocates building programs from only these fundamental structures. The phrase *correctly structured program* describes a program built only from fundamental structures.

SEQUENCE structure

The *SEQUENCE structure* is simply several process boxes strung together one after another. Figure 4-5 shows that this series of boxes begins at one point and ends at one point. Remember this structure by saying that it does one thing after another in a sequence.

IF-THEN-ELSE structure

Figure 4-6 illustrates the *binary decision structure* called *IF-THEN-ELSE*. The decision chooses between two alternative processes. The flow arrows must merge at one point after the processes on the two sides. If these two sides do not come back together, you get spaghetti because control is transferred outside the structure.

A common form of this structure has a process that does nothing on one side. The effect is a structure, sometimes called *IF-THEN*, that does something or bypasses it.

Remember the IF-THEN-ELSE structure by saying that it tests a condition, and IF the condition is satisfied, THEN do something, otherwise do something ELSE.

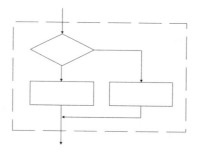

Figure 4-6 The IF-THEN-ELSE structure.

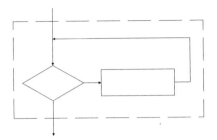

Figure 4-7   The DO-WHILE structure.

DO-WHILE structure

Figure 4-7 illustrates the iterative or *loop structure* called *DO-WHILE*. The internal process or body is repeatedly executed until the decision determines that the process should not execute again. That is, it loops until the decision exits the loop.

You can use any kind of condition for the decision. The decision may be based on a number. Alternatively, it could depend upon a complex series of input/output operations. Usually, the body of the loop will do something to alter the condition so that the loop exits eventually. If the exit never occurs, the loop is an *infinite loop*, meaning that it never ends.

If you are familiar with other kinds of loops, you can easily distinguish the DO-WHILE loop because the DO-WHILE has the decision *before* the body of the loop. The decision may cause the loop to exit without the body of the loop executing even once.

Remember the DO-WHILE structure by saying that it will test for a condition, and then DO something over and over WHILE the condition exists.

Extended Program Structures

A few other structures, called *extended structures*, meet the general requirements of structured programming. A variety of names have been created for these extra structures. You should look for the fundamental characteristics of structures of other names because they may be the same structures you already know.

Many high-level languages have some extended structures. People also write assembly language versions, although these extra structures are not necessary—the three fundamental structures will carry out all tasks. However, extra structures are sometimes more convenient for the programmer or for the language compiler.

DO-UNTIL structure

The *DO-UNTIL loop* structure shown in Figure 4-8 is similar to the DO-WHILE. The main difference is that the decision to exit the loop is placed after the body of the loop rather than before.

A serious limitation of the DO-UNTIL is that the body of the loop always executes at least once. Because the body executes before the decision, the decision cannot prevent the execution of the body at least once.

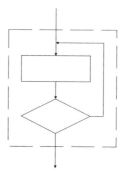

Figure 4-8   The DO-UNTIL structure.

Practical programs often need to execute the body of a loop zero times. So the DO-UNTIL loop is impractical as a general solution. The DO-WHILE loop has no such limitation; it is more flexible and adequate for all programs.

The DO-UNTIL loop can be troublesome if the programmer is not careful. If the loop decrements a numeric counter down to zero to end the loop, an initial counter value of zero causes problems. Probably the loop will execute the maximum number of times that the counter can specify—decrementing a zero value, if unsigned numbers are used, results in the largest possible value. The problem happens because the loop decrements the counter before the decision.

The usual solution to this problem is another decision before the loop that checks for this special case. The decision bypasses the loop to avoid the problem of an initial zero value. This decision adds extra code to the program that is unnecessary and confusing.

The DO-UNTIL loop does reflect normal thought to some extent. That is, people usually think about doing something first and then about making a decision based on the results. By contrast, experienced programmers think of a loop structure and how to get out of the loop. So the DO-WHILE loop seems normal to experienced programmers and odd to others.

IN-CASE-OF structure

Figure 4-9 shows the multiple decision structure with more than two alternatives called *IN-CASE-OF*. The first distinguishing feature of this structure is that one alternative for each decision is to do nothing. The second feature is that the program exits the structure after a single test is satisfied and the corresponding process is executed.

The IN-CASE-OF structure can be made from fundamental structures. Figure 4-10 shows how multiple IF-THEN-ELSE structures are made. A table lookup technique for making the IN-CASE-OF structure is also possible.

The IN-CASE-OF structure is just a program module that has a name because it occurs so often in practical programs. It does meet the basic requirement of a structure in that it has one beginning and one ending point.

Some high-level programming languages have a CASE structure to make programming a little easier. Also, the language compiler may be able to optimize the machine language for fast execution.

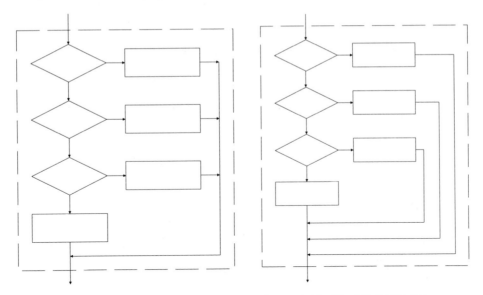

Figure 4-9 The IN-CASE-OF structure.

Figure 4-10 The IN-CASE-OF structure made from IF-THEN-ELSE.

Identifying Structured and Unstructured Programs

The flowchart shown in Figure 4-11 represents a structured program. Dashed lines enclose the fundamental structures. The design uses structures inside structures to build the overall flowchart and, eventually, the whole program. Notice that you can study the structure of the program without even knowing the names in the symbols.

To check a flowchart for correct structure, begin by drawing boxes over the flowchart to enclose program structures at all reasonable places. If all the boxes have only a single entry point and a single exit point, the flowchart meets a major requirement of structured programming.

Enclosing a collection of spaghetti with a box and then claiming to have a structured program is not legitimate! On the other hand, if each box contains only fundamental structures, you have met all the requirements.

Look at the unstructured spaghetti flowchart in Figure 4-4. Try to draw boxes on the flowchart that enclose program structures that each have a single entry point and a single exit point. Your boxes will reveal that the program modules have multiple entry and exit points.

Programs represented by unstructured flow charts usually contain unnecessary branch or jump instructions. Therefore, you can usually identify an unstructured program from its listing by looking at the use of the branch and jump instructions.

Some high-level programming languages have a branch statement called *GOTO*. Some companies encourage their employees to use *GOTO-less* programming to avoid jumping around in the program. Their goal is to encourage structured programming. However, those

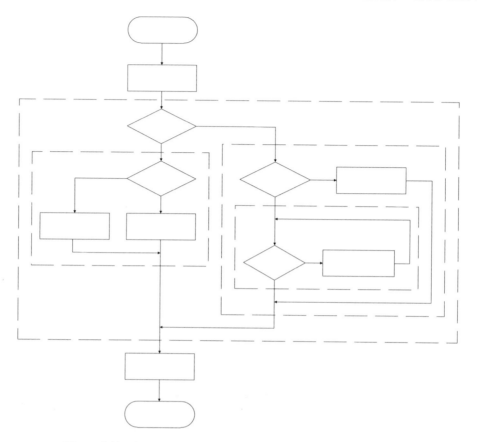

Figure 4-11   A structured flowchart made from fundamental structures.

languages with a GOTO usually require the GOTO statement to make all the fundamental structures. In contrast, other high-level programming languages lack the equivalent of a GOTO statement eliminating any chance for unstructured programs.

Making Structured Loops

If you are unaccustomed to structured programming, making structured DO-WHILE loops may seem difficult. As an example, the loop in Figure 4-12 is clearly unstructured. The loop has two exit points. The problem is caused by the need to terminate the loop prematurely when the program detects an error. The decision in the example sends control out of the loop, effectively making a second exit point.

The alternative flowchart in Figure 4-13 easily handles the premature termination using a structured design. The technique requires that the termination condition for the loop be forced. Then the loop follows its normal course to the end and remains correctly structured.

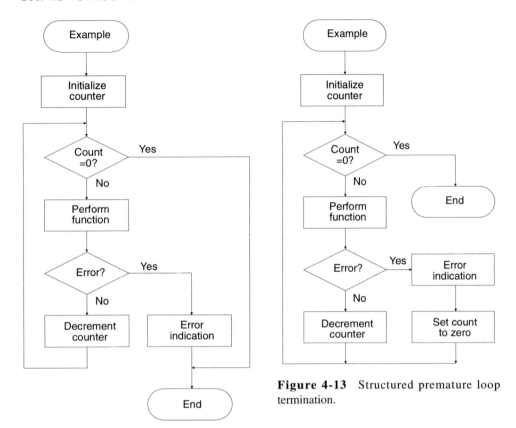

Figure 4-12   Unstructured premature loop termination.

Figure 4-13   Structured premature loop termination.

Complicated loops may need to exit due to several different conditions. These loops are easily implemented in fundamental structures by using a flag that determines whether the loop will exit. Outside the loop, the flag is set to an initial value that enables the loop to execute. Then, inside the loop, various program conditions alter the flag. The loop exits based on the value of the flag.

Sometimes, the value of the flag that causes the loop to terminate is used outside the loop. Figure 4-14 illustrates using such a flag to terminate a loop. After the loop in the figure exits, the flag has the value one or two depending on the condition that caused the exit.

A Troublesome Case

One problem often disturbs people when they begin using structured programming. The problem is an error that aborts the program. Aborting a program means to prevent it from going to its end. Figure 4-15 illustrates the case of a program that detects an error that prevents it from continuing.

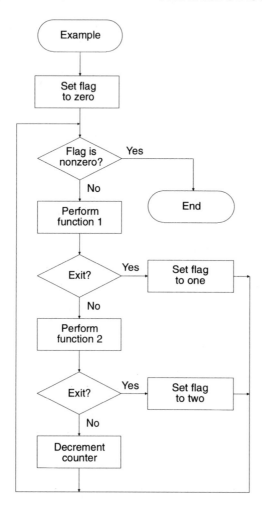

Figure 4-14 Using a flag to terminate a loop.

Aborting the program technically is a violation of structured programming because the program does not come to its normal end. However, aborting is quite practical to do, and it causes no problems. Specifically, aborting the program means to leave this execution of the program at some point and not return to continue from that point under any circumstance. Consequently, aborting the program prevents it from reaching the normal end of the program.

If a program is run again after being aborted, the new execution starts at the beginning of the program. It never resumes from the point where the program was aborted.

Starting again at the beginning is the reason that aborting a program is allowed within structured programming. Regardless of structuring, the data required by the program may be damaged by running only part of the program. However, running the entire program from the beginning will initialize the program and regenerate any data needed.

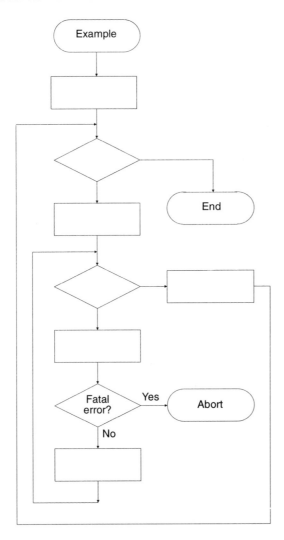

Figure 4-15 A troublesome case for structured programming.

Will You Use It?

If you have experience at writing spaghetti programs, you may decide that creating a structured program is unnecessary and nearly impossible. You will feel very limited by the requirements, which may at times seem artificial. If you persevere, you will learn that structured programming is neither impossible nor confining.

You will often discover that an unstructured program resulted from poor understanding of the problem. Experience shows that forcing a design into proper structure also forces the programmer to think carefully about the problem. When the design meets the requirements of structured programming, other problems are often solved. Thus, structuring is not just a programming tool, but it is also an aid to problem solving.

Sometimes people feel that making a structured flowchart limits what can be done in the program code. However, if you cannot solve the problem in the flowchart, you cannot solve the problem in the code.

Making Unstructured Programs Structured

Making an unstructured program correctly structured by simple changes is almost impossible. Generally, you must write an entirely new program. The logic of the unstructured program must be understood to design the structured equivalent. Therefore, illustrating an unstructured flowchart without labels and then making its equivalent is not possible. Instead, all the detail of the unstructured program must be understood.

Clearly, the only practical approach to structured programs is to design a structured program from the beginning. Then as changes are made, the structuring must be preserved. Experience will show you that this design is easier than it may seem. When a program is written in correctly structured form, changes are easier to make.

4.6   TOP/DOWN DESIGN

Top/down design is an old and simple idea that has a new name, because it is now applied to computer program design. Top/down design simply says to get the big picture first, and then consider finer and finer details as the design proceeds. The basic idea is to approach the design of a program by first identifying the major functional parts, or modules, of the program. This collection of major modules is called the *top level* of the design. Next, each of these parts is further broken into smaller parts, which together are called the *second level*. The second-level design is equivalent to the top-level module that it expands. The design continues until enough detail exists to write and document the program.

Detail is added to the design by adding more modules at lower levels. Altering the top-level modules to make more modules is not correct. Doing so would make too many modules, causing the designer to lose sight of the overall design. The result probably would be a spaghetti program.

For example, some second-level modules may be quite complex, so these modules are broken into parts on a lower third level. Again, the parts of the second level are not replaced, but each module is separately broken into parts.

Writing teachers recommend this approach for writing an English composition— another form of software. To write a composition, make an outline with the major sections identified first. Next, add subsections to the outline; the subsections are each equivalent to the major section that they are outlining. This expansion to additional detailed subsections continues until a complete outline results.

The table of contents of a book is an example of an outline of a written document. The design of the book you are reading was done this way. I decided upon the chapter contents by making the table of contents first. Then, I determined the sections of each chapter. As I wrote, smaller subsections were designed and added to a detailed table of contents, which served as an outline. The table of contents in the book contains only the upper levels of the detailed table of contents. Some changes in the design were made, but the major modules remained the same as in the original design.

Choosing Program Modules

You must decide how to break the design of a program into practical modules. A programmer must use judgment and experience to do this successfully. You must choose functional parts of the program to be modules. Each module should do something that is immediately recognizable by someone looking at the design. Only related functions should be together in a module.

For example, a program may gather some data and then print a report. The jobs of gathering the data and printing are two separate functions. Probably, these should be thought of separately in tackling the design. You will consider details such as scaling data values for unit conversions. These details can be thought about later, after all the top-level design is finished.

Top/Down Design Using Flowcharts

Figure 4-16 illustrates the top/down approach to computer program design using flowcharts. The first step in designing a program is making a top-level flowchart that will segment the total program into its major functional sections. Next, make a second-level flowchart for each process block that requires additional design work. You continue creating additional levels for complex modules. Your judgment determines when you have enough levels. Your creativity gets expressed and your talent is demonstrated when you make judgments about the design of the program.

The lower-level flowcharts are equivalent to the higher-level modules they represent. That is, the process blocks each have a single beginning point and a single ending point. Therefore, the expanded flowchart for a block also must have a single beginning point and a single ending point.

Relationship to structured programming

The requirement for a single beginning and single ending point is the same as one requirement of structured programming. Thus, structured programming is perfectly compatible with top/down design. Top/down design requires structured programming.

Number of levels

You must determine the number of levels required. Because this process is a creative design process, we can rely only on practical experience to provide guidelines or rules. However, some justification exists for the rules.

Maximum number of levels.   The first purpose of a flowchart is to display the function performed by each section of the program. The names placed on the symbols must

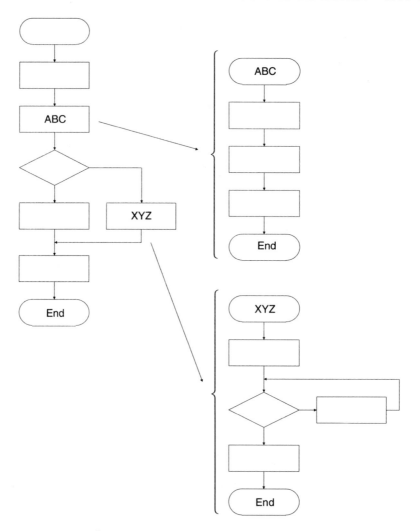

Figure 4-16   Top/down design using flowcharts.

describe these functions. The programming language statements will not appear on the flow-charts, because the language statements comprise the program implementation—not the program function. A listing of the language statements documents the program.

In other words, the flowchart should not have language statements on its symbols. Not placing language statements on the flowchart provides an effective maximum limit on the number of levels of flowcharts. Whenever the only choice is to put language statements on the flowchart, that flowchart is unnecessary.

Similarly, the flowcharts will be independent of the computer architecture and the hardware. Don't put hardware details on the flowchart.

Minimum number of levels.   At the other extreme, the minimum number of levels depends on the size of each flowchart. If your flowcharts have only a few levels, you may need to make large flowcharts to provide enough detail. However, large flowcharts destroy most of the advantages of top/down design. Each flowchart should be simple enough that you understand its entire function at a glance without extensive study. Most people can understand a flowchart after a quick glance if the flowchart has no more than ten symbols.

Therefore, a rule of thumb is to limit flowcharts to ten symbols. A ten-symbol flowchart is practical because it will easily fit on a single 8 1/2 by 11-inch sheet of paper, making the documentation convenient. Furthermore, *you will never need the off-page connector symbol*, because your flowcharts will each fit on one page. An earlier section encouraged you never to use the off-page connector because your flowcharts should fit on one page.

Top/Down Design Summary

Here is a summary of the guidelines for top/down design using flowcharts:

- Make a top-level flowchart first, then make second-level flowcharts, and then other levels if needed. Then start coding. Never start writing code without designing this way first.

- Run a working program as soon as possible to check the design. Don't be concerned if incomplete or dummy modules make the function of the program nonsense.

- Design each program module using structured programming, and then code them in a structured fashion according to the flowchart.

- Put comments in the program that match the flowcharts. Put the comments in the source as you create the source module, not after it has been assembled and tested.

- Use some commenting technique to show the program structure in the listing.

- Put functional names on the flowchart symbols. Don't make any flowcharts so detailed that only language statements can be put on the flowchart.

- Make flowcharts that are independent of the computer architecture and hardware whenever possible. For example, the flowcharts should contain no mention of addressing modes or microprocessor registers.

- Limit the maximum size of each flowchart to ten symbols. This guideline limits the complexity of the flowchart, and the flowchart will fit on one sheet of paper.

- Design each flowchart to meet the requirements of structured programming.

These guidelines will help you write good programs, but they do not guarantee success. Only careful and diligent work by a knowledgeable person will result in high-quality software.

4.7 STRUCTURED TOP/DOWN ASSEMBLY LANGUAGE

Assembly language programmers have total flexibility in writing and documenting their programs. The quality of their work varies from excellent to worthless. The following two examples illustrate these extremes. The comments will help you evaluate the ideas presented. Do not assume that the approaches used in the good program are the only good ways to write and document programs. Instead, form your own opinion and use your creativity to make improvements.

A Good Program

The sample program that follows provides an illustration of the techniques of structured programming and top/down design. First, structured flowcharts demonstrate top/down format even though the program is very short and simple. Next, the assembly language program listing illustrates the use of many instructions, addressing modes, and programming techniques. The documentation on the listing is adequate for most purposes, and it is based on the assumption that the reader has knowledge of the Motorola assembly language.

Example program specifications

The following program searches a table in memory for negative numbers. It both counts and sums these negative numbers and then indicates whether all numbers in the table are negative. The table may represent the inventory of certain parts, and the negative numbers are orders that could not be filled.

The table is specified by the address of its first entry (lowest address) and the length of the table. The table contains single-byte signed numbers. The length (unsigned), negative count, and sum values are double-byte numbers to allow for large tables and large sums. The program does not test for two's complement overflows.

Flowcharts for the example

Flowcharts for this example are shown in Figure 4-17. Though this program is very simple, the top/down design idea is useful. Each flowchart defines a specific function that is independent of the other parts. The title of each flowchart matches exactly the title of the higher-level process that is being expanded.

Documentation on the assembler listing

Documenting the program structure in the listing is very useful. The programmer should use some creativity in deciding upon a format. The format used in the example is only one possibility. It was chosen primarily to make the listing match the flowcharts closely. Here are the details of the format used in the listing:

- *Major sections of the listing.*   The title block, data section, and program section of the listing are labeled with headings surrounded by lines of asterisks. The asterisks

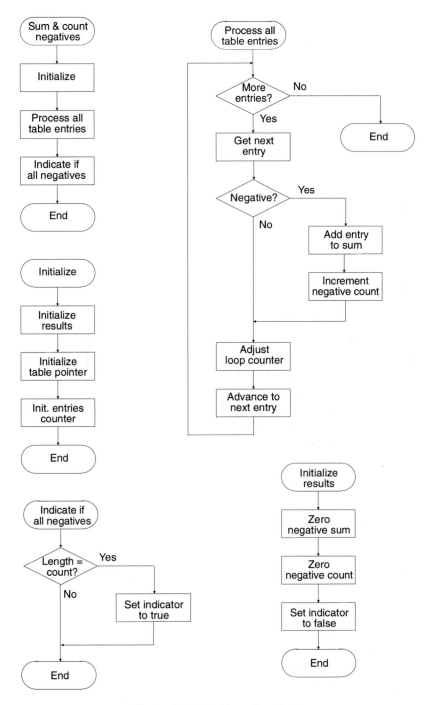

Figure 4-17   Top/down flowcharts.

make the headings easy to find. Printing each of these sections on separate pages of the listing is also common.

- *Top-level format.*   The top-level flowchart titles are surrounded by dashed lines as Figure 4-18 shows on lines 36 through 38. The titles on the flowchart must closely match those on the listing. All in-line programming for this level should be between its title and the title of the next section on the listing that is at the same level. In other words, the program should not jump around on the listing with logical parts of this section within other sections. All the code for the Initialization module should be on the listing before the title for the Process All Table Entries section.

- *Second-level format.*   The titles of the second-level flowcharts have headings without dashed lines as Figure 4-18 shows on lines 39, 47, and 49. These comments mark off the second-level modules. Although the headings for the second level are less prominent than those for the top level, you can easily see that the second-level modules are contained within the top-level module.

- *Third-level format.*   Indenting lower-level comments is common practice as Figure 4-18 shows on lines 40, 43, and 45. Indenting makes these comments less prominent on the listing and implies that these modules are within the second-level module.

- *Implementation specific comments.*   The last set of comments is at the right of the instructions as Figure 4-18 shows on lines 63 and 64. These comments are not flowchart titles; instead, they explain details of how the instructions were used in writing the program. Sometimes, such comments include specifics of a particular computer architecture and the particular programming approach used. This information is not on the flowchart, because the flowchart describes only the function of the program. Notice that the comment on line 63 was continued on line 64 with two periods that approximate an ellipsis to imply the continuation. The area to the right of instructions is also the perfect place to explain tricky uses of instructions. The explanation is on the line with the instruction, making it easy to understand.

Documentation is an area open to the creativity of the programmer. Some companies attempt to enforce consistency among programmers by mandating the format of the listing.

Editing comments

Because programs change frequently, the comments also will change frequently. You should plan for changes in your comments. The approach used in the example makes these changes easy. For example, changes often require moving sections of the program to new locations. An editor program can easily move a block of lines. If the comments are formatted as shown here, the instructions can be changed or moved without changing the comments. If a change requires that an instruction with a detailed comment at the right be moved, the instruction and its comment can be moved together easily. Not moving the specific comments with the instruction is difficult!

```
 1                              *****************************************************
 2                              ** SUM AND COUNT NEGATIVE TABLE ENTRIES
 3                              ** INDICATE IF ALL NEGATIVE
 4                              *
 5                              * THIS PROGRAM SUMS THE NEGATIVE ENTRIES IN A TABLE,
 6                              * COUNTS THE NUMBER OF NEGATIVES, AND INDICATES IF
 7                              * ALL ENTRIES ARE NEGATIVE.
 8                              *
 9                              * THE TABLE IS SPECIFIED BY ITS ADDRESS AND LENGTH.
10                              * TABLE ENTRIES AND INDICATOR ARE SINGLE-BYTES. THE
11                              * LENGTH, SUM, AND NEGATIVE COUNT ARE DOUBLE-BYTES.
12                              * TWO'S COMPLEMENT OVERFLOW CHECKING IS NOT DONE.
13                              *
14                              * AUTHOR: GENE H. MILLER
15                              *
16                              * REVISION HISTORY: VER2.1  5/28/1998
17                              *
18                              *****************************************************
19                              *
20                              *
21                              *****************************************************
22                              ** DATA SECTION
23                              *****************************************************
24     0010                              ORG  $10
25     0010     TABADR  RMB   2       PUT ADDRESS OF TABLE HERE
26     0012     TABLEN  RMB   2       NUMBER OF TABLE ENTRIES
27     0014     NEGSUM  RMB   2       PROGRAM PUTS SUM HERE
28     0016     NEGCNT  RMB   2       PROGRAM PUTS COUNT HERE
29     0018     ALLNEG  RMB   1       00=FALSE, FF=TRUE
30     0019     WRKCNT  RMB   2       WORKING LOOP COUNTER
31                              *
32                              *****************************************************
33                              ** PROGRAM SECTION
34                              *****************************************************
35     C100                             ORG  $C100
36                              *---------------------------------------------------
37                              * INITIALIZE
38                              *---------------------------------------------------
39                              * INITIALIZE RESULTS
40                              *      ZERO NEGATIVE SUM
41     C100  CC 00 00     START   LDD   #0000
42     C103  DD 14                STD   NEGSUM
43                              *      ZERO NEGATIVE COUNT
44     C105  DD 16                STD   NEGCNT
45                              *      SET INDICATOR TO FALSE
46     C107  7F 00 18             CLR   ALLNEG
47                              * INITIALIZE POINTER TO TABLE BEGINNING
48     C10A  DE 10                LDX   TABADR
49                              * INITIALIZE NUMBER OF ENTRIES COUNTER
50     C10C  DC 12                LDD   TABLEN
51     C10E  DD 19                STD   WRKCNT
52                              *---------------------------------------------------
53                              * PROCESS ALL TABLE ENTRIES
54                              *---------------------------------------------------
55                              * MORE ENTRIES?
56     C110  DC 19        LOOP    LDD   WRKCNT    TEST WORKING LOOP COUNTER
57     C112  27 1B                BEQ   NEXT      BRANCH ON NO
58                              * GET NEXT ENTRY
59     C114  E6 00                LDAB  0,X
60                              * NEGATIVE?
61     C116  2A 0D                BPL   POSITV    BRANCH ON NO
```

Figure 4-18   Assembly language listing for a good program.

```
62                               * ADD NEGATIVE ENTRY TO SUM
63    C118    4F                         CLRA           EXTEND SINGLE-BYTE NEGATIVE
64    C119    43                         COMA           ..TO DOUBLE-BYTE NEGATIVE
65    C11A    D3 14                      ADDD    NEGSUM
66    C11C    DD 14                      STD     NEGSUM
67                               * INCREMENT NEGATIVE COUNTER
68    C11E    DC 16                      LDD     NEGCNT
69    C120    C3 00 01                   ADDD    #1
70    C123    DD 16                      STD     NEGCNT
71                               * ADJUST LOOP COUNTER
72    C125    DC 19            POSITV    LDD     WRKCNT
73    C127    83 00 01                   SUBD    #1
74    C12A    DD 19                      STD     WRKCNT
75                               * ADVANCE TO NEXT ENTRY
76    C12C    08                         INX
77    C12D    20 E1                      BRA     LOOP
78                               *-----------------------------------------------------------
79                               * INDICATE IF ALL NEGATIVES
80                               *-----------------------------------------------------------
81                               * LENGTH = COUNT?
82    C12F    DC 12            NEXT      LDD     TABLEN
83    C131    93 16                      SUBD    NEGCNT
84    C133    26 04                      BNE     LAST      BRANCH ON NO
85                               * INDICATE TRUE
86    C135    86 FF                      LDAA    #$FF
87    C137    97 18                      STAA    ALLNEG
88    C139    3F               LAST      SWI               "STOP" FOR MOTOROLA TRAINER
89    C13A                               END
```

Defined	Symbol Name	Value	References			
29	ALLNEG	0018	46	87		
88	LAST	C139	84			
56	LOOP	C110	77			
28	NEGCNT	0016	44	68	70	83
27	NEGSUM	0014	42	65	66	
82	NEXT	C12F	57			
72	POSITV	C125	61			
41	START	C100				
25	TABADR	0010	48			
26	TABLEN	0012	50	82		
30	WRKCNT	0019	51	56	72	74

Lines Assembled :   89   Assembly Errors :   0

Figure 4-18   Continued.

A Bad Program

Sometimes people don't recognize the problems caused by poorly written and poorly documented programs. When they know every detail of a program, they have difficulty seeing other people's problems. Usually, the problems become clear to them when they study another person's program.

You should study the example bad program in Figure 4-19 to observe some bad practices you should avoid. This program does the same job as the program in Figure 4-18. Notice how simple things become very confusing!

Here are some general problems with this program and the assembler listing shown in Figure 4-19:

```
 1                                    * AN UNACCEPTABLE SUM NEGATIVE TABLE ENTRIES
 2                                    * PROGRAM WRITTEN IN VALID ASSEMBLY LANGUAGE
 3                                    * THAT WORKS CORRECTLY
 4                                    *
 5           0013                     R       EQU     $13
 6    0010                                    ORG $10
 7    0010                                    RMB     2           ADDRESS OF TABLE
 8    0012                            TL      RMB     1           TABLE LENGTH
 9    0013                                    RMB     2
10    0015                            TL1     RMB     1
11    0016                            WCNT1   RMB     1
12    0017                            WCNT    RMB     1
13    0018                            TEMP    RMB     2
14    001A                            NEG     RMB     2
15    001C                            *
16    C000                                    ORG $C000
17    C000    7F 00 13                        CLR     R           INIT RESULT TO ZERO
18    C003    96 13                           LDAA    R
19    C005    97 14                           STAA    R+1
20    C007    16                              TAB
21    C008    DD 1A                           STD     NEG         ZERO NEGCNT
22    C00A    7F C0 24                         CLR     ALL
23    C00D    DE 10                           LDX     $10         POINT TO TABLE
24    C00F    96 15                           LDAA    TL1
25    C011    D6 12                           LDAB    TL
26    C013    D7 16                           STAB    WCNT1
27    C015    97 17                           STAA    WCNT
28    C017    7D 00 16            L           TST     WCNT1       TEST LOOP COUNTER
29    C01A    26 16               TH          BNE     XYZ         AND BRANCH
30    C01C    96 17                           LDAA    WCNT1+1
31    C01E    27 06                           BEQ     E
32    C020    20 10                           BRA     XYZ
33    C022    FF                  F0          FCB     $FF
34    C023    01                  ONE         FCB     $001
35    C024                        ALL         RMB     2
36    C026    96 12               E           LDAA    TL
37    C028    D6 15                           LDAB    TL1
38    C02A    93 1A                           SUBD    NEG
39    C02C    26 03                           BNE     F
40    C02E    7A C0 24                        DEC     ALL
41    C031    3F                  F           SWI                 "STOP"
42    C032    E6 00               XYZ         LDAB    0,X         GET VALUE
43    C034    2B 02                           BMI     RST
44    C036    20 15                           BRA     BILL
45    C038    DB 14               RST         ADDB    R+1         ADD TOTAL
46    C03A    D7 14                           STAB    R+1
47    C03C    F6 C0 22                        LDAB    F0
48    C03F    D9 13                           ADCB    R
49    C041    D7 13                           STAB    R
50    C043    CC 00 00                        LDD     #0
51    C046    5C                              INCB
52    C047    D3 1A                           ADDD    NEG
53    C049    D7 1B                           STAB    NEG+1
54    C04B    97 1A                           STAA    NEG
55    C04D    08                  BILL        INX                 INCREMENT IND
56    C04E    D6 17                           LDAB    WCNT
57    C050    F0 C0 23                        SUBB    ONE         DECREMENT LOCATIONS
58    C053    D7 17                           STAB    WCNT1+1     17 & 18
59    C055    24 C0                           BCC     L
60    C057    7A 00 16                        DEC     WCNT1
61    C05A    20 BE                           BRA     L+3
62    C05C                                    END
```

Figure 4-19   Assembly language listing for a bad program.

- *Program structure.* You cannot determine the structure of the program by just looking at the listing. Without some study, you probably can't even tell that a loop is in the program. The few comments that are placed to the right of the instructions are little help.

- *Data organization.* The data section seems to be separated from the program section, but closer study will reveal that some data is placed within the program at lines 33 through 35. Careful scrutiny will reveal that the double-byte data numbers are not all stored with the high and then low bytes in adjacent memory locations.

- *Documentation.* The program does not have a title and other identification. The comments that are provided are almost useless because they don't relate to each other well enough. The abbreviations are confusing. Many of the user symbols are poorly named and cause confusion rather than helping document the program.

Here are some further problems listed by specific line numbers on the listing:

- *Line 5.* The address of a data location is defined by the EQU directive, which is confusing. For example, can you quickly see that the label R is on a double-byte number?

- *Lines 8 and 10.* A double-byte number is stored in two noncontiguous locations. This separation of the two bytes of the number not only is confusing, but it prevents proper use of the instructions available in the computer.

- *Lines 11 and 12.* The labeling of the two instructions is confusing because they represent a double-byte number that is incorrectly stored with the high byte at the higher address.

- *Line 13.* Two bytes are reserved and assigned a label when they are never used in the program.

- *Lines 17 through 22.* The instructions here put zero into some memory locations several different ways when one straightforward way would be much better.

- *Line 23.* Never put the numerical address in an instruction. Use a label.

- *Lines 24 through 27.* The instructions here copy the table length to the working counter in a disorganized way. First, the program gets the low byte of the double-byte number into the A accumulator and the high byte into the B accumulator; this order is backward from the normal use of double-byte numbers with the D accumulator. Then, it stores the two bytes in the opposite order; the high byte is stored first, and the low byte is stored second. These little inconsistencies make following the program difficult. Probably, a better solution would be to use the D accumulator.

- *Lines 28 through 32.* The instructions test the double-byte working counter one byte at a time in an unstructured and disorganized manner. Avoid using multiple branches if possible. Here, the number could be loaded into the D accumulator to set the condition code bits.

- *Line 33.*   The number FF created here is a constant that should be in an instruction using immediate addressing.

- *Line 34.*   The number is specified with three digits, which is confusing. Either one or two digits would be much better.

- *Line 35.*   The RMB reserves two memory bytes, but only one of them is ever used by the program.

- *Line 40.*   Understanding would be much easier if an instruction sequence stored the number FF into location ALL. The use of the DEC here makes the function unnecessarily dependent upon another part of the program and requires the reader to learn about that other part.

- *Line 41.*   The program stops running due to this instruction that is not at the physical end of the program. This instruction is difficult to find while working with the program.

- *Lines 43 and 44.*   The branch technique on lines 43 and 44 is poor. The two instructions could be replaced by a single BPL BILL, which would be much better.

- *Lines 50 and 51.*   The number one is put into the D accumulator in a long and complicated way. The LDD instruction should load the number one.

- *Lines 53 and 54.*   The instructions here store the D accumulator in a difficult way. Use the available instructions well by using the STD instruction.

- *Lines 56 through 58.*   One byte of the working counter is decremented one way, and the other byte is decremented a second way. A better sequence would use the DEC instruction to decrement both bytes.

- *Line 61.*   Don't use expressions for branch addresses—label the branch location instead and use the label in the branch instruction.

Of course, this example was developed to illustrate a large number of problems. Probably no practical program would ever be this bad. However, all the problems illustrated have been seen by the author in other people's programs.

4.8   LARGE-SCALE TOP/DOWN DESIGN

Teams of people write most programs. Usually the size of a program, and sometimes its complexity, is more than one person can handle in a reasonable period. The team approach is only successful if careful coordination occurs between the software written by various people. Spaghetti programs are almost impossible to make work under team circumstances. The top/down design ideas also apply to programming team management.

The Top/Down Team

A programming team will, at least, consist of a team leader and several other programmers. The team leader is an experienced person who has designed software and written programs. The team leader will have the responsibility of designing the top level of the program. This person also may design the second level of the program if it is small enough. Often this design phase will require interactions between the team leader, the team members, and the program users. Both flowcharts and written materials document the design. The documentation of the interactions between program modules is particularly important.

Team members

The team leader gives the upper level flowcharts and documents to other team members. Each person will work on one section of the program. The flowcharts contain only structured modules. Therefore, the team leader is sure that the modules will fit together later, unlike those of spaghetti programs.

Each person now makes additional levels of flowcharts to expand the design into lower levels. Each person must meet the requirements of structured programming in designing new program modules.

A problem

Suppose that the procedure as described were to continue until everyone on the team completes the design of their sections. Then they do the actual coding of their program modules. When these modules are put together to make the final program, catastrophe will strike! The modules almost certainly won't go together correctly.

Certainly some design work has been incorrect. The program specifications likely have changed. Therefore, much of the design work and coding is useless because they must now be changed.

The problem occurs because the implementation phase is too late. Detailed coding was done before the design and specifications were proved.

Top/Down Implementation

Top/down implementation helps avoid problems when the separate program modules developed by different team members are brought together into a single program. The idea is to start writing and running the program soon after starting the design. Start writing the program even before designing the lower levels.

Stubs

At first, writing a program before the design is complete seems impossible, and to some extent it is. Consequently, some program modules developed early in the project will be dummy modules called *stubs*. Stubs don't do correct functions, because they are substitutes for the real code. The actual code will be designed and written later. A stub may contain some useful code, but some stubs contain no code at all. Certainly using stubs requires the judgment of a knowledgeable programmer.

An example using stubs

As an example of the use of stubs, consider writing a program that must read some temperature sensors and print a report. The report displays the data collected from the sensors. At the least, the program will have an input section for reading the sensors and a report section for generating the report.

The input program module.   The input program module does three major functions. First, it must read the temperature sensors. Second, it must use an algorithm or table lookup to correct the data for nonlinearity in the sensors. Third, it must scale the data values for useful units.

The temperature sensor program module will probably have three separate parts for these three functions. To write and test all the code for these will be time-consuming. It also may be impossible because the characteristics of the sensors may not yet be known. Also, the sensors may not yet be interfaced to the computer hardware, so they don't operate.

These problems should not be a deterrent to writing the program. Instead, the input program can contain a stub that generates phony data from the temperature sensors, a stub that linearizes the data in a phony way, and a stub to scale the data in a phony way.

The report program module.   The report program module will print a report on a printer. The report will need headings, neat columns, titles, time and date, and so on. However, all the details needed on the report probably are not known. The required data to be presented may not have been decided. However, these problems should not deter the programmer from writing the report program.

The report program can be written with stubs. The report may contain headings with little information, the time and data may be phony, and the arrangement on the page may be less than desirable. However, the program can be written and made to work.

After the report program module and the input program module are both completed, they can be put together to make a working program. Any bugs can be found easily because the code is still very simple—much of it is stubs.

The demonstration.   After the program modules have been completed using stubs, a complete working program can be demonstrated. The report generated will contain phony data created in the stubs in the input program. The organization of the report will also need improvements. Regardless of these details, a program has been written that can be demonstrated and evaluated. The team leader, the team members, and maybe the ultimate users of the program can see some results.

Probably this working system will do something that is not right—that is, incorrect, not just missing. Often people will not specify the desired results from a program completely. However, when they see it running, they say, "See, that is wrong." If what they see is based on stubs, changing the design is still relatively easy and little detailed work is lost. The program design can be changed, and the program can be demonstrated again.

Program testing

Running and demonstrating the program containing stubs checks the design, overall function, and initial coding of the program. The next step is to remove some stubs and replace them with correct modules. The best approach is to replace these one at a time so that finding the location of bugs will be easy. Each time a stub is replaced with actual code, a new aspect of the program will perform correctly. Eventually, the entire program will work correctly.

The top/down implementation technique tests most of the program modules to some extent as the program is built and the stubs are replaced. Testing done within the actual program, and not by a separate test program, is valuable. You are assured that each program module works in the context of the actual program. This built-in testing significantly reduces the need to write separate test programs.

As the program evolves toward the final design, it should be demonstrated to interested parties several times. If errors occur, they will be caught before further work is completed. Catching errors early reduces the effort needed to correct them. Likewise, if additional features are needed, adding them to the design is easier if the program is not yet completed.

Team member interactions

Usually, the team leader will merge new modules into the overall program and test their operation. The other team members do not need to interact with each other. Their individual parts of the program are largely independent of the other programmers' work. The reduced interaction improves the productivity of the team.

Replacing a team member will have little effect on the other team members. New people added to the team need learn only about their part of the program—they do not need to learn all the program.

A major problem of adding new people is the nonproductive time spent informing them about the work completed. Top/down design reduces this time, because the modules are designed to be independent of each other—the design specifications are developed before the modules are written.

The results

Top/down design applied to team programming has been very successful. The goal is to get a system working before coding all the program. The working program demonstrates the design and some working code even if the data is phony. When the entire project is completed, everyone will have confidence that the program works correctly; it has been working and demonstrated from the early stages of design.

The top/down approach to team management is possible only if you have correctly structured programs. Otherwise, the jumping around in spaghetti programs causes different programmers' modules to interact at an intimate level. Changing one person's module will affect another person's module; getting the program to function correctly will be difficult.

Studies of practical programming teams have shown that a combination of structured programming methods and top/down design methods leads to successful projects. Of course, these are techniques that do not exclude other approaches. Usually, some bottom/up testing of low-level modules is also done.

4.9   SMALL-SCALE TOP/DOWN DESIGN

An individual programmer also gains from the use of top/down design and top/down imple-mentation. Don't try to design or write all of a program at once. Organize your efforts by making top/down structured flowcharts, and then code the program from the flowcharts. At first, code only the higher-level modules freely using stubs for the lower-level modules. Get your program running as soon as possible so that you can do some testing. You may even leave major portions of the program out of the design by making empty stubs. They can be added later. Likewise, you may write the program for hardware that is readily available, and then convert it later for the specific hardware required. Avoid taking on too much at once.

To use top/down design and top/down implementation effectively, you need good editor and assembler programs. The editor will make changes such as replacing a stub easy. After each change, immediately assemble the new program and run it to see the results. Your program will evolve to completion rapidly.

4.10   REVIEW

Good programming approaches have been demonstrated throughout this chapter. Examples have been used to illustrate both good and bad programs. This book may be unique in that it illustrates and labels bad programs. Of course, deciding whether a program is good or bad is a matter of judgment. Not everyone will agree. Therefore, this chapter has given reasons for the judgments that were made. When you write programs, you too will be making such judgments continually. You must decide what is important and set your goals accordingly.

The design of software is an activity that affords people the opportunity to be creative to an extent possible in few other endeavors. Their results can be elegant and beautiful problem solutions or ugly and costly catastrophes.

4.11   EXERCISES

4-1. Name the three fundamental program structures. Sketch a flowchart for each structure.

4-2. If a loop processes information before checking whether it should exit the loop, is it a DO-WHILE or a DO-UNTIL structure?

4-3. Figure 4-20 is a structured flowchart that contains only the three fundamental structures. Iden-tify all the structures by drawing boxes around each structure.

4-4. If an assembly language program listing for a structured program has documentation as de-scribed in this chapter, what is the implication of two consecutive lines containing comments?

4-5. Consider an assembly language program listing for a program that has documentation as de-scribed in this chapter, and has a comment line followed by several lines of program code. What is implied if the first line beyond the comment does not have an address label, but the second line does have an address label?

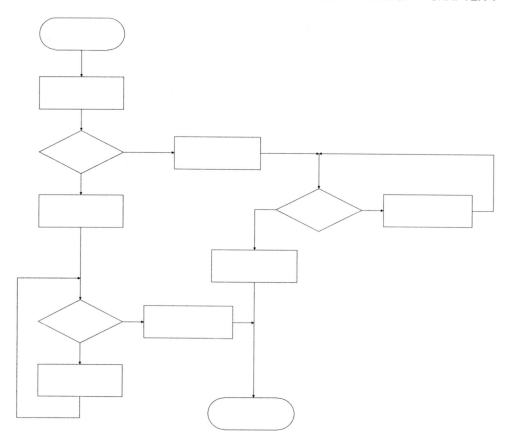

Figure 4-20   Exercise 4-3.

4-6. Make a single-level structured flowchart that represents the following algorithm:
Input a number. If the number is greater than 200_{10}, stop the program. Sum the integer numbers from 1 to (and including) the number, then print the number and the sum and stop the program.

4-7. Make top/down structured flowcharts for the following program:
Input a positive number. Print a report that lists the squares and cubes of the numbers from zero to the number read. Limit the input number to 100_{10} by printing an error message for incorrect numbers and ending the program. After the report is finished, request a YES or NO for repeating the program and respond accordingly. Include messages and headings for the report.

4-8. Make a single-level structured flowchart for the program in Exercise 3-11.

4-9. Make a single-level structured flowchart for the program in Exercise 3-14.

4-10. Make a new structured flowchart using only the three fundamental structures that is equivalent to the incorrectly structured flowchart in Figure 4-21.

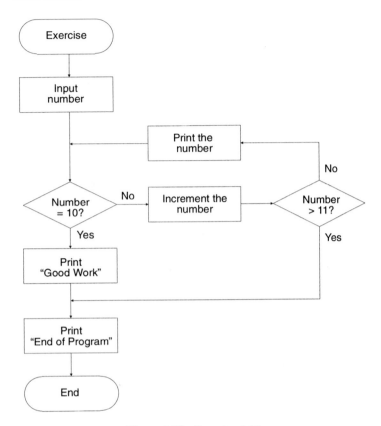

Figure 4-21   Exercise 4-10.

4-11.   Find a program written in a high-level language such as Basic, Fortran, or C. Study the program to determine whether it is built from the three fundamental program structures.

4-12.   Make a new structured flowchart using only the three fundamental structures that is equivalent to the incorrectly structured flowchart in Figure 4-22.

4-13.   Make a new structured flowchart using the three fundamental structures and the DO-UNTIL structure that is equivalent to the flowchart in Figure 4-22.

4-14.   Modify the program given in Figure 4-18 by making new flowcharts so that it sums and counts both the positive and negative entries found. Your program must be properly structured using only the three fundamental structures.

4-15.   Write an assembly language program from your flowcharts developed for Exercise 4-10. Your program must follow the flowcharts, and the listing must be well documented.

4-16.   Name the program structure that starts on line 13 of Figure 5-16.

4-17.   Name the program structure that starts on line 38 of Figure 5-17.

4-18.   Name the program structure that starts on line 44 of Figure 5-17.

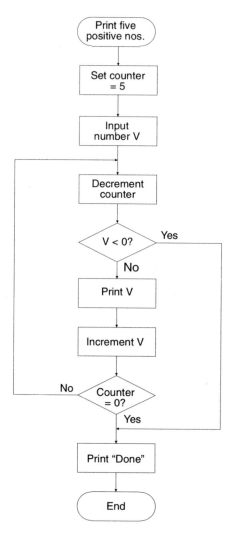

Figure 4-22   Exercises 4-12 and 4-13.

4-19.   Name the program structure that starts on line 51 of Figure 5-18.

4-20.   Name the program structure that starts on line 56 of Figure 5-18.

Chapter 5

Advanced Assembly Language Programming

The more complicated Motorola 68HC11 instructions that were avoided in previous chapters are introduced in this chapter. Many of the more complex instructions were designed to solve particular programming problems. This chapter discusses both the problems and the use of the more complicated instructions to solve the problems. All examples presented use assembly language, so memory diagrams are usually omitted.

5.1 MORE INDEXING

The 68HC11 has a Y index register that was avoided in previous chapters. The function of the Y index register is the same in all ways to the function of the X index register. The instruction set has a parallel instruction using the Y index register for every instruction that in any way uses the X register.

The 68HC11 has many instructions that relate to the index registers. Therefore, many instruction op code numbers are needed for these instructions.

Motorola Products

Many 68HC11 instructions discussed in this chapter are not available in the Motorola 6800, 6801, 6802, 6803, and 68701 chips. For example, the 68HC11 has a Y index register that the other chips do not have. Therefore, the 68HC11 has many more instruction codes than the other processors. However, compatible programs written for these other processors will run on the 68HC11.

Consequences of a Large Instruction Set

Many Motorola 8-bit microprocessors do not have a Y index register. When the Y index register is included, the required number of instruction op codes exceeds the number of combinations that fit into an 8-bit number. Therefore, some 68HC11 instructions have op codes that are double-byte numbers.

Prebytes

All 68HC11 instructions that in any way relate to the Y index register have double-byte op codes. A few other 68HC11 instructions not associated with Y also have double-byte op codes. To make talking about double-byte op codes easy, Motorola describes an added op code byte that precedes the normal single op code byte. The added byte is called a *prebyte*. Motorola literature refers to a *prebyte* and an *op code byte* rather than a *double-byte op code*.

Performance considerations

Using a prebyte makes the instructions that relate to the Y register one byte longer in length. Consequently, they take one clock cycle longer to fetch from memory. The result is a performance penalty to using the Y index register. Normally, you should use the X index register for all applications that require only a single index register. If you need two index registers, use both X and Y because the program will be better than if you use only the X index register.

Op Code Maps

A table or map of the op code numbers is helpful in understanding the assignment of instruction code values. Appendix A shows the op code maps for the 68HC11.

The instruction set of the 68HC11 has three different prebyte values; therefore, four maps are needed. Each of the 256_{10} numerical values on a map corresponds to a box. The box either names the instruction that corresponds to its numerical value or indicates that the code is invalid. Most of the codes were chosen orderly, so the columns of the maps correspond to particular addressing modes. Because the four maps together have room for 1024_{10} possible codes and only 307_{10} of them are valid instructions, most of the codes are invalid.

Index Register Exchange Instructions

The instructions for changing the index registers covered so far allow you to load, store, increment, decrement, or add the B accumulator to the index registers. Here are two instructions that make manipulating the index registers in many other ways easy:

XGDX, XGDY

Exchange the D accumulator with the X or Y index register. These instructions exchange or swap the 16-bit numbers in the two registers. None of the condition code bits are affected.

Usually, index registers hold addresses and act as pointers. For simple manipulation of tables, the load, store, increment, and decrement the index register operations are sufficient. However, more complex calculation of addresses is impractical. The exchange instructions XGDX and XGDY allow the program to calculate addresses easily. Move the number in the index register to the D accumulator; then operate on the number using instructions that affect the A, B, and D accumulators. Finally, move the result back to the index register.

The exchange instructions also make temporarily saving the D accumulator easy. Use the exchange instruction to save the D accumulator in an index register while using D for other purposes. Using an exchange is easier and faster than storing the D accumulator in memory.

Another Instruction for D

The compare accumulator D to memory instruction, CPD, is unusual. It has only the basic addressing modes, yet it has a prebyte even when Y is not involved. You can see this easily on op code map page 3 in Appendix A.

CPD

Compare accumulator D to memory.   Subtract a double-byte number in memory from the D accumulator and discard the answer so that the accumulator is unchanged, but make the condition code bits respond.

Notice the inconsistent naming of the CPD, CMPA, CMPB, CPX, and CPY instructions.

5.2   BIT AND BYTE MANIPULATION

All instructions discussed so far have operated on single-byte or double-byte data numbers. These instructions treat the data as collections of bits that represent a number. This section considers manipulating or using the individual bits within a number and operating on a number as a collection of bits.

Shift Operations

Several instructions can logically shift the bits in an accumulator or a memory register. Shifting means to move a collection of bits to an adjacent position. The shifting of a number is useful as a general programming tool. The meaning of these instructions depends on the context in which they are used.

Rotate instructions

The rotate instructions shift all the bits of an 8-bit register and the carry bit circularly one bit position. The effect is to alter a 9-bit number. Look at the instruction set table in Appendix A as you read about these instructions. Here is a list of the rotate instructions:

ROL, ROLA, ROLB   *Rotate left memory byte or accumulator A or B.* Figure 5-1 illustrates this operation. The name *rotate left* means move bit 7 to the C bit and move the C bit to bit 0 as the other bits are shifted.

ROR, RORA, RORB   *Rotate right memory byte or accumulator A or B.* Figure 5-1 illustrates this operation. The name *rotate right* means move bit 0 to the C bit and move the C bit to bit 7 as the other bits are shifted.

The rotate instructions only move the bits one position. The 68HC11 has no multiple position rotate instructions. However, the inherent addressed versions only require one byte of memory. So, four rotates of an accumulator—the maximum number needed for an 8-bit accumulator—only require four bytes of memory. More than four rotates are unnecessary because fewer rotates in the opposite direction do the same job.

The instruction sequence CLC ROLA, or the single instruction ASLA, both multiply by two for both two's complement signed and unsigned numbers. Following this multiply, the C and V bits correctly indicate overflow information.

Shift instructions

The shift instructions move all the bits of a register one position and insert a 0 at the input end of the register. The bit at the output end of the register moves to the carry bit.

Look at the instruction set table in Appendix A as you read about these instructions. Carefully examine the graphical representations of the instruction operations.

Here is a list of the shift instructions:

ASL, LSL,   *Arithmetic or logical shift left memory byte or accumulator A,*
ASLA, LSLA,   *B, or D.* Each instruction has two different names—use the
ASLB, LSLB,   name, either arithmetic or logical, that fits the application of the
ASLD, LSLD   instruction in the program. Figure 5-1 illustrates the operations with an 8-bit register. Those instructions operating on the D accumulator operate on 16-bit numbers.

LSR, LSRA, LSRB,   *Logical shift right memory byte or accumulator A, B, or D.*
LSRD   Each instruction has a single name because the arithmetic meaning, divide by two, applies to unsigned numbers but not to two's complement numbers. Figure 5-1 illustrates these operations with an 8-bit register.

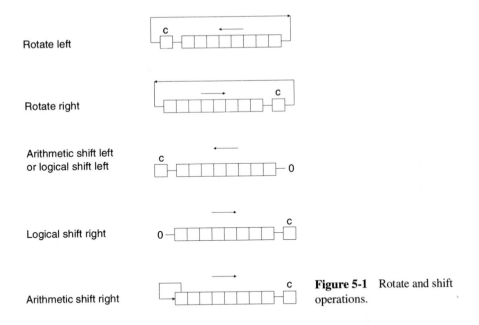

Rotate left

Rotate right

Arithmetic shift left
or logical shift left

Logical shift right

Arithmetic shift right

Figure 5-1   Rotate and shift operations.

ASR, ASRA, ASRB   *Arithmetic shift right memory byte or accumulator A or B.* These instructions divide an 8-bit two's complement signed number by two. They are useful for little else. Figure 5-1 illustrates the operation of these instructions.

The shift instructions only move the bits one position. The 68HC11 has no multiple position shift instructions.

Division by two using ASR.   Table 5-1 illustrates successive divisions by two of some two's complement numbers. The ASR instruction can do division by two. Carefully look at the resulting numbers that are all correct! The ASR instruction uses an integer division

TABLE 5-1   SUCCESSIVE DIVISION BY TWO USING ASR

Binary	Decimal	Binary	Decimal
00001010	+10	11110110	−10
00000101	+5	11111011	−5
00000010	+2	11111101	−3
00000001	+1	11111110	−2
00000000	0	11111111	−1
00000000	0	11111111	−1

technique called *floored division*. Floored division rounds the integer quotient to the floor or most negative number. A characteristic of floored division is that a quotient cannot be zero if the original number is negative. Be careful using ASR because floored division is different from the integer division used by some high-level programming languages. For example, Fortran does not do floored division.

Double-precision instructions.   Double-precision shift instructions are only available in the 68HC11 for the D accumulator. It has no double-precision shift instructions that operate on a number in memory. When a program must operate on 16-bit numbers, the double-precision instructions provide significant speed improvement over the 8-bit instructions.

To shift a double-byte number in memory, consider using a logical shift followed by a rotate instruction. These instructions use the C bit so that multiple precision shifting is easy.

For review, the 68HC11 has instructions that perform the double-precision operations load, store, add, subtract, and logical shift left and right on the D accumulator.

Read-modify-write instructions

The rotate and shift instructions that operate on a memory register are called *read-modify-write* instructions. To rotate the register, the microprocessor must read the number from memory into the microprocessor, rotate the bits including the C bit, and then write the result back to the memory register. Therefore, read-modify-write instructions are complex. They use more clock cycles than you might expect.

Logical Operations

Instructions are available for the logical operations AND, OR, COMPLEMENT, and EXCLUSIVE OR. Table 5-2 defines these functions using truth tables. All logical instructions operate on 8-bit numbers. However, you probably would not want to AND eight pairs of bits at once as part of an application program! Instead, the logical instructions manipulate bits. Refer to the instruction set table in Appendix A as you read the following.

TABLE 5-2   LOGIC FUNCTIONS

AND			OR			COMPLEMENT		EXCLUSIVE OR		
X	Y	F1	X	Y	F2	X	F3	X	Y	F4
0	0	0	0	0	0	0	1	0	0	0
0	1	0	0	1	1	1	0	0	1	1
1	0	0	1	0	1			1	0	1
1	1	1	1	1	1			1	1	0

$$F1 = X \wedge Y \qquad F2 = X \vee Y \qquad F3 = \overline{X} \qquad F4 = X \veebar Y$$

Bit picking

The normal use of the AND instruction is to pick selected bits from an 8-bit number. Here is the operation of the related AND and BIT instructions:

ANDA, ANDB
And memory byte to accumulator. These instructions AND the corresponding bits in the addressed memory register and the accumulator register and put the result in the accumulator. The N, Z, and V condition code bits are affected.

BITA, BITB
And memory byte to accumulator and discard the result. These instructions AND the corresponding bits in the addressed memory register and the accumulator register and discard the result without altering the accumulator. The N, Z, and V condition code bits are affected.

You can see the effect of the AND instruction by looking at pairs of rows of the truth table in Table 5-2 for the function. Look at the first two rows of the AND table in Table 5-2 and observe that 0 ANDed with anything is 0. From the third and fourth rows, observe that 1 ANDed with something is the same thing.

This viewpoint helps in understanding how the AND instruction can pick bits from a number. The result from the AND operation contains a 0 at each bit position corresponding to a 0 in the mask. Therefore, a 0 effectively blocks or masks the original bit from the result and a 1 allows the original bit to appear in the result.

Using the AND instruction.   Here are two examples of the ANDA instruction where A designates the number in the A accumulator and M the number in the addressed memory register:

$$
\begin{array}{ll}
01010101 & \text{A} \\
\wedge\ 11110000 & \text{M} \\
\hline
01010000 & \text{A}
\end{array}
\qquad\qquad
\begin{array}{ll}
01011111 & \text{A} \\
\wedge\ 00000100 & \text{M} \\
\hline
00000100 & \text{A}
\end{array}
$$

In the first example, the four 1s in the memory register allow the bits 0101 from the A accumulator to show in the result. The four 0s in the memory register block the other bits from the result by making the result bits 0s. In the second example, only one bit of the original A accumulator shows in the result, and the other bits are 0s.

To continue this viewpoint, let's call the number in the memory register a *mask* because it lets some bits show through and blocks others. Unfortunately, the word *mask* is used to describe several other kinds of numbers.

Testing hardware signals.   Figure 5-2 illustrates a memory register with bits that represent the conditions of some push button switches. Somehow, hardware and software read the conditions of the push button switches and put the bits into the memory register labeled IN1. When a push button switch is pushed, a 1 is in the memory bit. When the push button is released, a 0 is in the bit.

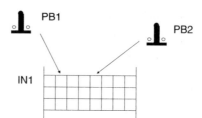

Figure 5-2   Memory bits representing push button switch conditions.

Let's look at a program that executes one program module if both push buttons PB1 and PB2 are pushed, and another program module if they are not both pushed. Figure 5-3 is the assembly language source for the program. Here is the use of each instruction:

- *LDAA.*   The first LDAA instruction copies the conditions of the switches from the memory register labeled IN1 into the A accumulator. Of course, the accumulator has eight bits, not just the two of interest. Assume that other devices such as switches control the other bits.

- *ANDA.*   The first ANDA instruction masks the bit pattern in the A accumulator so that only bit 6 shows in the result. Location MASK1 contains the mask bit pattern that selects bit 6. The AND instruction makes the other seven bits of the A accumulator all 0s. The result in A represents the condition of switch PB1. If the number in A is zero, the push button PB1 is not pushed. If the number in A is nonzero, then the push button PB1 is pushed. Observe that the A accumulator is zero or nonzero regardless of which bit PB1 uses. If you connected PB1 to a different bit, you would only need to change the mask.

- *BEQ.*   If the result in the A accumulator is zero because PB1 is not pushed, you know that the two push buttons are not both pushed. So the BEQ instruction sends program control directly to the module labeled NOT. Testing PB2 would have no value in this case.

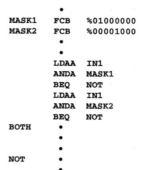

Figure 5-3   Source program for ANDing two push button switches.

- *LDAA.*   The second LDAA instruction restores the switch conditions to the A accumulator because the AND instruction changed the accumulator.

- *ANDA.*   The second AND instruction uses MASK2 to let the PB2 condition show in the result and to zero the other bits in the A accumulator. If the number in A is zero, the push button PB2 is not pushed. If the number in A is nonzero, then the push button PB2 is pushed.

- *BEQ.*   If the result in the A accumulator is zero because PB2 is not pushed, the BEQ sends program control to the module labeled NOT. If the A accumulator is not zero, then both push buttons are pushed and control goes to the program module labeled BOTH.

The example above ANDs two bits by testing each bit individually and branching to the correct place. Logical operations with bits are usually done this way, but some variations on the technique can be used. For example, a mask that selects both PB1 and PB2 simultaneously leaves one of four possible patterns in the accumulator. The program then would test for the desired pattern.

Also consider using the BIT instruction in the example in Figure 5-3 to avoid changing the accumulator during the first masking operation. Each approach has good points and bad points.

Bit packing

Selected bits of a register can be forced to known values. *Bit packing* means to force known bit values into certain bits of a register without changing the other bits. Both the AND and ORA instructions can force bits.

Normally, you will use the ORA instruction to put 1s into selected bits of an 8-bit number. Here is the operation of the ORA instruction:

ORAA, ORAB   *Or memory byte to accumulator A or B.* These instructions OR the corresponding bits in the addressed memory register and the accumulator register; the result is put into the same accumulator.

You can see the effect of the ORA instruction by looking at pairs of rows of the truth table for the OR function. Look at the first two rows of the OR table in Table 5-2 and observe that 0 ORed with something is the same thing. From the third and fourth rows, observe that 1 ORed with something is always 1.

Using the ORA instruction to pack 1s.   Here are two examples of the ORA instruction where A designates the number in the A accumulator and M the number in the addressed memory register:

01010101	A		01100000	A
∨ 11110000	M		∨ 00000100	M
11110101	A		01100100	A

In the first example, the 1s in the memory register force the corresponding bits in the A accumulator to 1s while the 0s leave the other bits unchanged. In the second example, the ORA changes only one bit of the original number in the accumulator.

The number that selects the bits to be operated on is called a *mask* as it was for the AND instruction. The meaning is very different for the two applications, but the name is the same. You will find some additional meanings for the word *mask* in later sections, so be careful to avoid being confused by the different but related meanings.

Using the AND instruction to pack 0s.   The following two examples use the AND instruction differently from the earlier example. Here the mask, the number in memory, is opposite the earlier mask because selected bits are forced to 0s.

Look at these examples:

$$
\begin{array}{ll}
01010101 & \text{A} \\
\wedge\ 11111011 & \text{M} \\
\hline
01010001 & \text{A}
\end{array}
\qquad\qquad
\begin{array}{ll}
01111111 & \text{A} \\
\wedge\ 11111101 & \text{M} \\
\hline
01111101 & \text{A}
\end{array}
$$

The masks used in these examples contain mostly 1s. Usually, only one or a few bits are forced to 0s in practical applications.

Logical instruction example.   The example in Figure 5-4 expands the example in Figure 5-2 to include the control of a light by a memory bit. Figure 5-5 is the source listing of a program that turns on the light if both push button switches are pushed, and turns off the light otherwise. The program uses two additional masks for controlling the light.

Notice that the program can change only the selected bit that controls the light. If other bits in the register labeled OUT control other devices, changing them will cause errors.

Bit reversing

The EXCLUSIVE OR instruction, named *EOR*, can reverse or complement selected bits of a register. Here is the operation of this instruction:

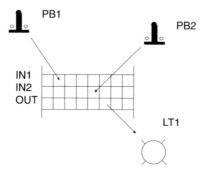

Figure 5-4   Memory bits representing push button switch and light conditions.

```
                              •
                   MASK1    FCB    %01000000
                   MASK2    FCB    %00001000
                   MASK3    FCB    %00000100
                   MASK4    FCB    %11111011
                              •
                              •
                   LDAA    IN1
                   ANDA    MASK1
                   BEQ     NOT
                   LDAA    IN2
                   ANDA    MASK2
                   BEQ     NOT
          BOTH     LDAA    OUT
                   ORAA    MASK3
                   STAA    OUT
                   BRA     NEXT
          NOT      LDAA    OUT
                   ANDA    MASK4
                   STAA    OUT
          NEXT       •
                     •
```

Figure 5-5   Source program for controlling light with two push button switches.

EORA, EORB   *Exclusive or memory byte to accumulator A or B.* These instructions EXCLUSIVE OR the corresponding bits in the addressed memory register and the accumulator register. The result is put into the same accumulator.

You can see the effect of the EOR instruction by looking at pairs of rows of the truth table for the EXCLUSIVE OR function. Look at the first two rows of the truth table in Table 5-2 and observe that 0 EORed with something is the same thing. From the third and fourth rows, observe that 1 EORed with something results in the opposite value.

Bit set and clear instructions

The 68HC11 has instructions that will set or clear selected bits in a memory register. These instructions contain a mask byte that selects the bits to be operated on. The meaning of the name *mask* in these instructions is different from its meaning when used with the AND and ORA instructions. Here is the operation of the bit set and clear instructions:

BSET   *Set all bits in a memory byte that correspond to 1s in the mask.* The BSET instruction ORs the mask from inside the instruction to the addressed memory byte and puts the result into the memory byte.

BCLR   *Clear all bits in a memory byte that correspond to 1s in the mask.* The BCLR instruction ANDs the complement of the mask from inside the instruction to the addressed memory byte and puts the result into the memory byte.

The mask byte selects the bits that the instruction operates on. For all instructions that contain a mask, 1s in the mask enable the operation and 0s disable the operation. Masks that are all 1s or all 0s are seldom used. If the mask contains all 0s, the instruction does nothing.

A BSET instruction with a mask of all 1s is useful, but a BCLR instruction with a mask of all 1s may be replaced by a CLR instruction.

Usually, you will think of the mask as a means to select the bits that the instruction will operate on. However, sometimes you must understand how the instruction uses the mask in computing the result. In particular, some hardware control problems will require this understanding.

Addressing modes for bit instructions. The BSET and BCLR instructions each have two addressing modes for a single instruction. No other instructions in the 68HC11 have two addressing modes. These bit instructions need a memory address of the register to be operated on and a mask byte to select the bits.

The instruction code contains the constant mask byte. Therefore, the mask byte is accessed by immediate addressing. The mask byte makes these instructions one byte longer than expected. Look at the instruction set table for the memory format of these instructions.

The bit instructions can use either direct addressing or indexed addressing to access the memory byte to be operated on. An instruction that can use only direct or indexed addressing has an unusual combination of addressing modes. In particular, the extended addressing mode is not available to these instructions.

To specify the addressing mode for the instructions easily, the instruction set table lists only the addressing mode for the memory byte to be operated on. Because the mask is always accessed by immediate addressing, the instruction set table does not show the addressing mode for the mask.

The assembly language format. The assembly language for the bit instructions has a new format to accommodate the mask byte, which makes the instruction statement longer. The format is the same as for other direct or indexed addressed instructions, except the mask byte is added at the end of the statement after a comma. For example, the statement

```
        EX1             BSET      LIGHTS,$10
```

sets bit 4 of the memory location labeled LIGHTS using direct addressing. The value of the mask byte is specified numerically or symbolically; however, the # symbol for immediate addressing is not used.

The order of the symbols in the assembly language—namely, op code, address, and mask—is the same as the order of the instruction bytes in memory. Therefore, the instruction set table gives you the order of the symbols.

Bit testing and branching

Practical programs frequently need to test one or more memory bits and then branch on the outcome of the test. The 68HC11 has instructions that combine the testing and branching operations into one instruction. Here are the bit test-and-branch instructions:

BRSET   *Branch if all the bits in a memory byte that correspond to 1s in the mask are set.* The BRSET instruction ANDs the mask from inside the instruction to the complement of the addressed memory byte and branches if the result is zero.

BRCLR   *Branch if all bits in a memory byte that correspond to 1s in the mask are clear.* The BRCLR instruction ANDs the mask from inside the instruction to the addressed memory byte and branches if the result is zero.

As in the BSET and BCLR instructions, the mask in the BRSET and BRCLR instructions selects the bits in the memory byte to be tested. All selected bits must meet the required condition for the branch to occur. The most common mask contains only a single 1, so one bit is tested.

Addressing modes for branch on bit instructions.   The BRSET and BRCLR instructions each have three addressing modes for a single instruction. No other instructions in the 68HC11 have three addressing modes. These two instructions need the address of the memory register to be operated on, a mask byte to select the bits in that memory register, and an offset for the relative addressed branch operation.

The instruction code contains the constant mask byte. Therefore, the mask byte is accessed by immediate addressing. The memory byte tested is accessed by either direct or indexed addressing. The branching operation uses program relative addressing the same as the other branch instructions. The multiple addressing modes make these instructions either four or five bytes long. The instruction set table shows the memory format for these instructions.

The assembly language format.   The assembly language format for these instructions is unique because of the required addressing modes. Be careful that your assembler uses this format because some assemblers require the index register at the end of the statement. The format here is the same as the BSET and BCLR instructions with a relative address symbol added at the end. For example, the statement

          EX2   BRCLR   1,Y,MASK1,NEXT

creates an instruction using indexed by Y addressing with an offset of 1 to access a memory register. The bits in the register that correspond to the 1s in the value of the symbol MASK1 are tested. If all those bits are 0s, then the instruction branches to location NEXT; otherwise execution continues at the next instruction.

The order of the symbols in the assembly language statement is the same as the order of the instruction bytes in memory. Therefore, the order of the hexadecimal codes in the instruction set table implies the order of the symbols.

Bit instruction example

Figure 5-6 is the assembly listing for a program that uses the bit instructions. The program tests two memory bits that represent push button switch conditions. If both push buttons are pushed, it turns on a light by putting 1 in a memory bit. Otherwise, it puts a 0 in the bit.

The program uses the same hardware as shown in Figure 5-4, and it performs the same function as the program in Figure 5-5. Look carefully at the hexadecimal codes to see the mask bytes and the multiple addressing modes.

```
 1                              *********************************************************
 2                              ** DEMONSTRATE BIT INSTRUCTIONS
 3                              **
 4                              * TURN ON LIGHT IF BOTH PUSH BUTTON SWITCHES
 5                              * ARE PUSHED, AND TURN IT OFF OTHERWISE
 6                              *
 7                              *********************************************************
 8                              *
 9                              *********************************************************
10                              ** SYMBOL DEFINITIONS
11                              *********************************************************
12         0040                 SW1       EQU    %01000000   SWITCH 1 MASK
13         0008                 SW2       EQU    %00001000   SWITCH 2 MASK
14         0004                 LT1       EQU    %00000100   LIGHT 1 MASK
15                              *
16                              *********************************************************
17                              ** DATA SECTION
18                              *********************************************************
19   D000                                 ORG    $D000
20   D000                       IOAREA    RMB    1           BEGINNING OF INPUT/OUTPUT AREA
21   D001                                 RMB    1           THE MEANING OF THESE LOCATIONS
22   D002                       IN1       RMB    1           ..IS HARDWARE DEPENDENT
23   D003                       IN2       RMB    1
24   D004                       OUT       RMB    1
25                              *
26                              *********************************************************
27                              ** MAIN PROGRAM
28                              *********************************************************
29   C100                                 ORG    $C100
30                              * CONTROL LIGHT CONTINUOUSLY
31   C100  CE D0 00             START     LDX    #IOAREA   POINT X AT IOAREA
32   C103  1E 02 40 05                    BRSET  IN1-IOAREA,X,SW1,TSTSCND
33   C107  1D 04 04                       BCLR   OUT-IOAREA,X,LT1
34   C10A  20 0C                          BRA    NEXT
35   C10C  1E 03 08 05          TSTSCND   BRSET  IN2-IOAREA,X,SW2,LTON
36   C110  1D 04 04                       BCLR   OUT-IOAREA,X,LT1
37   C113  20 03                          BRA    NEXT
38   C115  1C 04 04             LTON      BSET   OUT-IOAREA,X,LT1
39   C118  20 E6                NEXT      BRA    START
40   C11A                                 END
```

Figure 5-6   Listing for program using bit instructions.

5.3   ARITHMETIC OPERATIONS

The arithmetic operations covered so far are addition, subtraction, increment, decrement, and multiply or divide by two. The 68HC11 has several more arithmetic instructions that significantly improve the execution speed of programs that do arithmetic.

Multiplication

Programs that implement control systems frequently use multiplication. These programs often need fast execution. The multiply instruction MUL provides very fast multiplication when compared to using a software algorithm to calculate the product. The instruction set table contains further information on the MUL instruction.

Here is the operation of the multiplication instruction:

MUL   *Multiply the A accumulator by the B accumulator and put the product into the D accumulator.* The multiply instruction multiplies two 8-bit unsigned numbers to make a 16-bit unsigned product. Bit 7 of the product goes to the C bit. The original numbers in the accumulators are lost because the product of the numbers goes to D.

Carefully note that the MUL instruction can only multiply unsigned numbers. The 68HC11 does not have a two's complement number multiplication instruction. Be careful to use the correct type of numbers because the result will be incorrect with signed numbers, except in certain special cases.

Arithmetic

One use of the multiply instruction is to multiply two 8-bit numbers to obtain the product of the numbers as an arithmetic result. The biggest product of two 8-bit unsigned numbers is FE01, so there is no possibility of a carry that could affect the C bit. Instead, bit D7 of the result goes to the C bit. Then the C bit can be used for rounding. For example, suppose that the product is scaled by using only its most-significant eight bits. If C is set to a 1, it means that the least-significant eight bits are one-half or greater. So, C set to 1 indicates that the 8-bit result should be rounded up to the next highest number.

A program can multiply two's complement signed numbers by converting them to positive numbers before using the MUL instruction. The program must compute the sign of the answer and correct the product as required. The overhead of correcting the sign is easily overcome by the speed of the MUL instruction.

Shifting

If the multiplier is a number containing a single 1, the product from the MUL instruction contains the multiplicand shifted to the position of the 1 in the multiplier. If only the most-significant eight bits of the product are kept, the result is the multiplicand shifted to the right. If the least-significant eight bits are kept, the multiplicand is shifted to the left.

The following examples illustrate using MUL for shifting:

```
        11111111    A                    10110101    A
     ×  01000000    B                 ×  00001000    B
    0011111111000000    A:B          0000010110101000    A:B
```

The examples both have a number to be shifted in the A accumulator. The B accumulator holds the shift mask. In the first example, the result in the A accumulator is the number shifted right two places while the result in the B accumulator is the number shifted left six places. In the second example, the number is shifted right five places in A and left three places in B. One instruction can shift the number in the A accumulator from zero to seven places with this technique.

Division

The 68HC11 has two integer division instructions that give both a quotient and a remainder after dividing. They use the D accumulator and the X index register. The 68HC11 has no equivalent instructions for the Y register. Here are the divide instructions:

IDIV, FDIV   *Divide two 16-bit numbers giving a 16-bit quotient and a 16-bit remainder.* All numbers are 16-bit unsigned numbers. The dividend is the D accumulator, the divisor is the X register, the quotient goes to the X register, and the remainder goes to the D accumulator. The integer divide IDIV expects the quotient to be one or higher. The fractional divide instruction FDIV expects the quotient to be smaller than one.

Carefully select the correct divide instruction because the results obtained from the two instructions are very different.

Arithmetic

The *integer divide instruction IDIV* expects the quotient to be one or higher. The *fractional division instruction FDIV* expects the quotient to be smaller than one. The instruction assumes that the binary points of the dividend and the divisor are at the same relative position within the numbers.

The FDIV gives results with the binary points at the left end of the registers. If the quotient is higher than FFFF, or the divisor is zero, the quotient is set to FFFF and the remainder is indeterminate. The C bit indicates divide by zero and the V bit indicates a division overflow—the division was not a fraction, because the denominator was lower or the same as the numerator.

The IDIV instruction gives results with the binary points at the right end of the registers. If the quotient is too big, the case when dividing by zero, the quotient will be FFFF and the remainder is indeterminate. The C bit indicates divide by zero.

Table 5-3 shows the results of dividing several different numbers. Study the examples to see the effects of a division overflow.

Shifting

The division instructions shift a number to the right or left when the divisor contains a single 1. However, using MUL for shifting is usually better because the divide instructions put the quotient into the X register, which is less convenient. The MUL instruction also executes much faster than the divide instructions. The last examples in Table 5-3 illustrate using the divide instructions for shifting.

BCD Operations

The only instruction in the 68HC11 specifically designed for binary-coded decimal numbers is the DAA instruction. Here is the function of the DAA instruction:

TABLE 5-3   DIVISION INSTRUCTION EXAMPLES

Fractional Divide FDIV				Integer Divide IDIV			
D/X; X = Quotient, D = Remainder				D/X; X = Quotient, D = Remainder			
Before (point same)		After (point left)		Before (point same)		After (point right)	
D	X	.X	.D	D	X	X.	D.
0000	0005	0000	0000	000A	0005	0002	0000
000A	0005	FFFF	0000	000B	0005	0002	0001
0005	000A	8000	0000	0005	000A	0000	0005
0005	0009	8E38	0008	0005	0000	FFFF	0005
0001	FFFF	0001	0001	FFFF	0001	FFFF	0000
0123	1000	1230	0000	1234	0100	0012	0034

DAA   *Decimal adjust the A accumulator.* The DAA instruction correctly decimal adjusts the A accumulator immediately following
only an ADDA, ADDB, ADCA, ADCB, or ABA instruction. The
DAA corrects any digit that is out of decimal range following
the addition and affects the C bit for a BCD result.

The DAA instruction corrects to BCD by adding one of the numbers 00, 06, 60, or 66
to the A accumulator. The algorithm to choose the correct number is complex and depends
on the initial values in the C and H bits. The DAA instruction is the only instruction in the
68HC11 that uses the H or half-carry condition code bit.

5.4   THE STACK

Practical programs frequently need to save data generated by a program module. A program
module that executes later in the program will retrieve the data. Once the data is retrieved,
a copy of the saved data is often no longer needed. Using separate memory registers for each
number that is saved and retrieved often requires many registers for large programs. In
addition, the organization of these storage places is difficult, making programs more complicated than necessary.

A *stack* is a means of temporarily storing and retrieving numbers in a way so that it
reuses the same memory registers for the storage. The stack does not solve all problems of
data storage, but it is very practical for common programming operations.

Stack Operation

A mechanical analogy will help you learn how the stack operates. The stack is a very dynamic
device. One common mechanical stack is the plate rack used in some cafeterias. Figure 5-7(a)

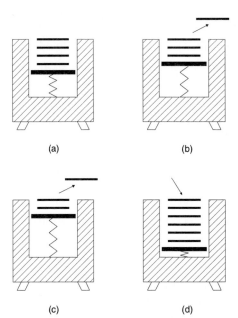

(a)   (b)

(c)   (d)

Figure 5-7   Plate rack analogy to computer stack operation.

illustrates a cross-section of a plate rack containing some plates. As hungry customers go through the cafeteria line, each person takes a plate off the top of the rack. When a plate is removed, a mechanism moves the plates upward so that the next plate is now at the top of the rack. Figures 5-7(b) and 5-7(c) illustrate this process. The next customer now can easily access the next plate that is at the top of the rack. When the dishwasher puts clean plates into the rack, the mechanism moves downward. The top plate is again at the top of the rack, as Figure 5-7(d) illustrates.

If you consider the usage of the plates, you see that the last plate put into the rack is the first plate removed from the rack. The rack is a *last-in-first-out* device, or a LIFO. Notice that a similar effect results from stacking the plates on a table. Then each customer would reach to the top of the pile of plates to get a plate.

A stack in the computer is usually a region of memory allocated by the program. Some computers have registers dedicated to the stack. The 68HC11 has both hardware and instructions for easy use of a stack in memory.

Stack Hardware

A register in the 68HC11 called the *stack pointer* controls the operation of the stack. The stack pointer automatically points to the memory location that is the next available stack location as numbers are stored and retrieved from the stack. The number in the stack pointer is a memory address; that is the reason this register is called a *pointer register*.

Stack Instructions

Several instructions can manipulate the stack pointer register and the contents of memory pointed to by the stack pointer. The stack instructions were designed for very specific purposes and you should use them only for those specific purposes. Using these instructions for other purposes usually leads to poorly designed programs.

Here are all the stack-related instructions:

INS, DES
Increment or decrement the stack pointer. These instructions respectively add one and subtract one from the number in the stack pointer register.

LDS, STS
Load or store the stack pointer register. The load the stack pointer instruction creates a stack by placing an address in the stack pointer register. The store the stack pointer instruction is seldom used or needed.

PSHA, PSHB
Push the A or B accumulator onto the stack. These PSH instructions store a byte from the accumulator to memory at the address specified by the stack pointer and then decrement the stack pointer by one. The addressing mode is called *inherent*, but it is a kind of indexed addressing that uses the stack pointer as the index register.

PSHX, PSHY
Push two bytes from the X or Y index register onto the stack. These PSH instructions store the double-byte number in the index register to memory at the address specified by the stack pointer and then decrement the stack pointer by two. The addressing mode is called *inherent*, but it is a kind of indexed addressing that uses the stack pointer as the index register.

PULA, PULB
Pull a byte from the stack and load it into the A or B accumulator. These PUL instructions increment the stack pointer by one and then load a byte into an accumulator from memory at the address in the stack pointer. The addressing mode is called *inherent*, but it is a kind of indexed addressing that uses the stack pointer as the index register.

PULX, PULY
Pull two bytes from the stack and load them into the X or Y index register. These PUL instructions increment the stack pointer by two and then load a double-byte number into an index register from memory at the address in the stack pointer. The addressing mode is called *inherent*, but it is a kind of indexed addressing that uses the stack pointer as the index register.

TSX, TXS, TSY, TYS
Transfer the stack pointer plus one to the index register or transfer the index register minus one to the stack pointer. These

instructions adjust the number transferred to overcome the automatic adjustment of the stack pointer when it is used for accessing the stack.

Be careful to understand the difference between the stack pointer and the stack. The stack pointer is a register in the microprocessor that holds the memory address of the next available memory byte in the stack. The stack is the collection of memory registers allocated to hold temporary data.

The stack can be any size provided enough memory is installed into the computer. Most programs require a small stack, generally fewer than 50_{10} bytes. Because the stack pointer holds a 16-bit address, your stack can be anywhere in the memory space. Generally, the stack will be outside the direct addressing range. Placing the stack in the direct addressing range provides no advantage, so placing it elsewhere frees that part of memory for use by direct addressed instructions.

The stack pointer is the last microprocessor register in the 68HC11 to be introduced. You have now seen all the registers that were shown on the programming model in Figure 2-1 in Chapter 2.

Stack Example

A program designed to illustrate the operation of the stack is Figure 5-8. The program creates a stack, creates two data numbers in the accumulators, and then exchanges the two numbers using the stack. A detailed description of the program and the stack operation follows.

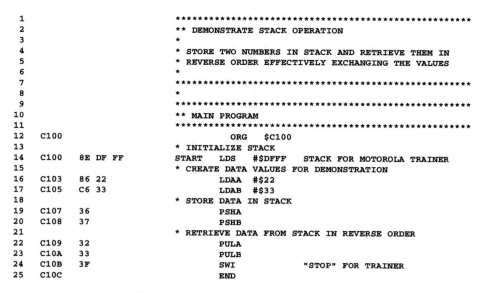

```
 1                          ********************************************************
 2                          ** DEMONSTRATE STACK OPERATION
 3                          *
 4                          * STORE TWO NUMBERS IN STACK AND RETRIEVE THEM IN
 5                          * REVERSE ORDER EFFECTIVELY EXCHANGING THE VALUES
 6                          *
 7                          ********************************************************
 8                          *
 9                          ********************************************************
10                          ** MAIN PROGRAM
11                          ********************************************************
12   C100                            ORG     $C100
13                          * INITIALIZE STACK
14   C100   8E DF FF        START   LDS     #$DFFF    STACK FOR MOTOROLA TRAINER
15                          * CREATE DATA VALUES FOR DEMONSTRATION
16   C103   86 22                   LDAA    #$22
17   C105   C6 33                   LDAB    #$33
18                          * STORE DATA IN STACK
19   C107   36                      PSHA
20   C108   37                      PSHB
21                          * RETRIEVE DATA FROM STACK IN REVERSE ORDER
22   C109   32                      PULA
23   C10A   33                      PULB
24   C10B   3F                      SWI               "STOP" FOR TRAINER
25   C10C                           END
```

Figure 5-8 Program to demonstrate the stack.

Figure 5-9 shows the stack and the microprocessor registers at each step of the program. The figure shows the memory with higher addresses at the top of the diagram rather than at the bottom, as in previous examples. This order helps in visualizing the stack because the stack builds downward to lower addresses. Drawing the memory this way implies downward movement.

Figure 5-9(a) shows the memory and microprocessor registers before the program is run. The memory chosen for the stack is at address DFFF and below. The figure shows all the registers without numbers initially because the numbers are unknown. Here is the program description:

- *Lines 14 through 17.* The LDS instruction on line 14 loads an initial address into the stack pointer register, effectively creating the stack. The initial address locates the highest address used by the stack. Generally, programs will have one and only one LDS instruction because stack operations are automatic after the stack is

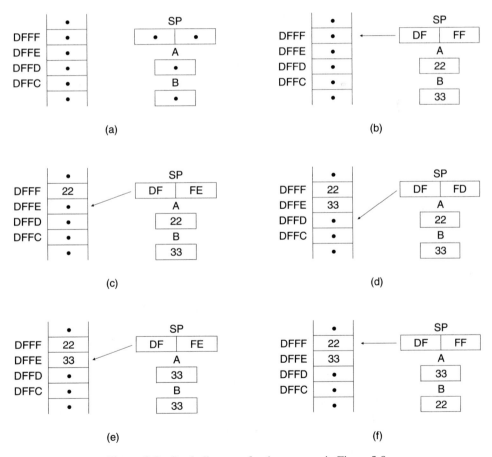

Figure 5-9   Stack diagrams for the program in Figure 5-8.

initialized. The next available location in the stack is address DFFF, and the stack is now empty. Next, the instructions on lines 16 and 17 load the accumulators with numbers for the demonstration. The stack diagram in Figure 5-9(b) illustrates the conditions after these instructions have executed.

- *Line 19.*   The PSHA stores the 22 from the A accumulator into memory at the address specified by the stack pointer, namely, address DFFF. Then the PSHA automatically decrements the stack pointer so that it now contains the address DFFE. The stack diagram now looks like Figure 5-9(c). Notice that the stack pointer always points to the next available stack location. We say that the data number has been *pushed onto the stack.* Also note that the PSHA instruction is only one byte long because the stack pointer holds the address for it.

- *Line 20.*   The PSHB pushes the 33 from the B accumulator onto the stack. The stack diagram now looks like Figure 5-9(d). The next available location in the stack is now at address DFFD. Two numbers are now in the stack.

- *Line 22.*   The PULA instruction first increments the number in the stack pointer to DFFE. Second, the PULA loads the number 33 from memory location DFFE into the A accumulator. The stack diagram now looks like Figure 5-9(e). The PULA instruction removed the number 33 from the stack so that only one entry is left in the stack. We say that a data number has been *pulled from the stack.* You should observe that the last number put into the stack is the first number taken out of the stack.

- *Line 23.*   The PULB instruction pulls the number 22 from the stack into the B accumulator. The stack diagram now looks like Figure 5-9(f). The stack is now empty. You can tell that it is empty because the number in the stack pointer is now the same as its initial value.

At the end of this program, the original numbers in the A and B accumulators have been exchanged. This very simple program stores data in memory. However, no memory locations were permanently devoted to this program module. After it finishes executing, no data locations remain allocated. Other program modules can now use the same memory in the stack.

Using the Stack

You may wonder whether data numbers remain in the memory locations used by the stack. For the example program in Figure 5-8, the numbers 22 and 33 do remain in locations DFFF and DFFE after this program runs. However, the only reasonable point of view is to say that they are gone. Two numbers were pushed onto the stack and two were pulled from the stack, so it must be empty! You must disregard the numbers that remain in the memory registers. To do otherwise defeats the benefits of the stack and leads to elusive programming bugs.

Similarly, a rule to follow when using the stack is *always pull the same number of bytes that were pushed*. To do otherwise is nonsense. If more bytes are pushed than pulled, some bytes are forever left in the stack and are of no value to the program. Likewise, if bytes are pulled that were not pushed, the program will be using garbage numbers obtained from locations outside the stack. The 68HC11 has no hardware that can detect when the stack is used improperly.

Finally, you must determine the maximum depth of your stack and allocate enough memory for it. Be careful that your stack does not overwrite the program at the maximum depth of the stack.

5.5   SUBROUTINES

A *subroutine* is a reusable program module. A main program will call upon it at several locations to do some task. Therefore, you only need to write the subroutine code once instead of each time it is needed. However, the memory saved due to subroutines is often less important than the improved program organization that results from the use of subroutines.

Subroutine Concepts

A subroutine is a program module that is logically separate and independent of a main program. To use the subroutine, the main program will transfer program control to the subroutine. The subroutine does its function and then returns control to the main program. The subroutine is independent of the main program, so changes in the main program do not require changes in the subroutine.

The subroutine is not only a powerful programming device, but it is also a useful conceptual tool. It helps in designing programs because you can think about the function performed by the subroutine without considering the mechanics of doing that function.

Well-written subroutines are perfectly compatible with the structured programming and top/down design ideas. Subroutines were used to achieve many of the advantages of structured programming long before structured programming was formulated.

Instructions for Writing Subroutines

The transfer of control from a main program to a subroutine and from a subroutine back to the main program is very easy. The instruction set includes instructions to do these functions in cooperation with the stack hardware. Here are the instructions:

JSR, BSR   *Jump or branch to subroutine and save the return address in the stack.* These instructions transfer control to the subroutine after pushing the program counter onto the stack. The JSR and BSR instructions differ only in their addressing modes. During the execute phase, these instructions push the double-byte number in the program counter on the stack. The program counter was already incremented during the fetch phase, so the number that is pushed is the address of the next instruction—it is the return address.

RTS   *Return from subroutine by retrieving the return address from the stack.* The RTS instruction pulls a double-byte number from the stack and puts it into the program counter.

To use subroutines, your program must define a stack so that the program counter containing the return address can be saved and restored. The body of the subroutine may use the stack to store temporary data. However, when the return point of the subroutine is reached, the stack must be returned to its condition at the entry to the subroutine. If the stack is not returned to its entry condition, the RTS instruction will obtain an erroneous return address. The body of the subroutine must remove or pull as many bytes from the stack as it adds or pushes to it.

Using the subroutine instructions

Figure 5-10 illustrates the instructions to send control to a subroutine and to return control to the main program. In the example, the first JSR instruction jumps to the subroutine at address C400 and puts the return address C103 on the stack. At the end of the subroutine, the RTS sends control to the main program by pulling the address C103 from the stack into

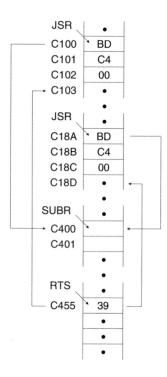

Figure 5-10   Using the subroutine instructions.

the program counter. The second JSR instruction at address C18A performs similarly putting the address C18D on the stack. Notice that this time the subroutine must return to address C18D, which is different from the first return address. Because the subroutine must return to a different address for each call, the address must be saved at the time of the call. The reason that the stack is used is because it temporarily saves the subroutine return address.

You should notice that the stack now has two different purposes. First, a program may save data numbers on the stack. Second, a subroutine will save the return address on the stack. Take care to avoid misusing the two kinds of numbers because both will be in the stack at once. Furthermore, in many practical programs, knowing the position of each number in the stack at any particular time is difficult if not impossible.

The First Subroutine Example

The subroutine in Figure 5-11 demonstrates the subroutine instructions with an example. The subroutine exchanges the low and high four bits of the A accumulator. The main program passes a data value to the subroutine in the A accumulator. The subroutine returns with the result in the A accumulator.

Here is a description of the program details:

- *Line 16.*   Loading an address into the stack pointer creates a stack. The use of subroutines always requires a stack.

- *Line 18.*   The main program gets the data value to be passed to the subroutine into the A accumulator. The main program is passing a parameter to the subroutine.

- *Line 19.*   The JSR instruction sends control to the subroutine and puts the return address C108 onto the stack. The JSR does not affect the accumulators, index registers, or condition code register. The information in those registers is effectively passed to the subroutine even if it's not needed. The stack pointer now contains DFFD.

- *Lines 23 through 25.*   This documentation is the minimum required for all subroutines.

- *Lines 30 through 37.*   The body of the subroutine is similar to any other program module. It performs a function on some data; here, it's the data in the A accumulator. Running the body of this subroutine does not alter the stack or stack pointer.

- *Line 41.*   The RTS instruction returns control to the main program without affecting the A accumulator. The result in the accumulator is passed to the main program.

- *Line 20.*   The main program continues running at address C108 because the RTS instruction removed that address from the stack and put it into the program counter. After the RTS, the stack pointer contains DFFF and the stack is empty.

Running the subroutine changes the contents of both the A accumulator and the condition code register from what they were when the JSR instruction was encountered. The

```
 1                             ******************************************************
 2                             ** SUBROUTINE DEMONSTRATION
 3                             * PASS VALUE IN A ACCUMULATOR
 4                             *
 5                             ******************************************************
 6                             ** DATA SECTION
 7                             ******************************************************
 8      0010                            ORG   $10
 9      0010    46             VALUE    FCB   $46      DATA FOR DEMONSTRATION
10                             *
11                             ******************************************************
12                             ** MAIN PROGRAM
13                             ******************************************************
14      C100                            ORG   $C100
15                             * INITIALIZE STACK
16      C100    8E DF FF       START    LDS   #$DFFF   STACK FOR MOTOROLA TRAINER
17                             * DEMONSTRATE SUBROUTINE OPERATION
18      C103    96 10                   LDAA  VALUE    GET DATA
19      C105    BD C1 09                JSR   SWAPA
20      C108    3F                      SWI            "STOP" FOR MOTOROLA TRAINER
21                             *
22                             ******************************************************
23                             ** SUBROUTINE SWAPA
24                             * SWAP HIGH AND LOW 4 BITS OF A ACCUMULATOR
25                             * MODIFIES A,CC
26                             ******************************************************
27                             *----------------------------------------------------
28                             * SWAP BITS
29                             *----------------------------------------------------
30      C109    48             SWAPA    LSLA
31      C10A    89 00                   ADCA  #0
32      C10C    48                      LSLA
33      C10D    89 00                   ADCA  #0
34      C10F    48                      LSLA
35      C110    89 00                   ADCA  #0
36      C112    48                      LSLA
37      C113    89 00                   ADCA  #0
38                             *----------------------------------------------------
39                             * RETURN FROM SUBROUTINE
40                             *----------------------------------------------------
41      C115    39                      RTS
42      C116                            END
```

Figure 5-11 Main program and subroutine example.

other microprocessor registers were not affected. Line 25 documents these changes because the writer of the main program must realize that the subroutine has made these changes.

Flowcharting Subroutines

The flowchart for a subroutine module differs only slightly from the flowchart for any other structured module. A structured subroutine will have a single beginning and a single ending point; therefore, it is a process. A rectangular process box on the main flowchart represents the subroutine. The subroutine is on a lower level of the top/down flowcharts. The flowchart for the subroutine is independent of the main program as is any lower-level flowchart. To provide some distinction for the subroutine, usually the word RETURN labels the terminator instead of the word END.

Figure 5-12 illustrates the symbol that represents the subroutine on the main program. Usually, a subroutine name represents the function of the subroutine. The name is used as a label in the assembly language program. Most people put the name of the subroutine at the top of the process box. The same name labels the beginning terminal symbol of the subroutine.

Parameter Passing

The difficult part of writing subroutines is choosing and implementing a means of passing parameters or data values between the main program and the subroutine. As a practical matter, the parameters are placed either in microprocessor registers or in memory. The parameter passed can be either the actual data value or a reference to the data value. A *reference* is the memory address of a data value.

When writing subroutines, many details must be considered, including deciding where to put the data in memory and deciding how to access it. Therefore, many different techniques can be used for parameter passing. The following sections discuss and illustrate with examples several different parameter-passing techniques. Understanding these techniques

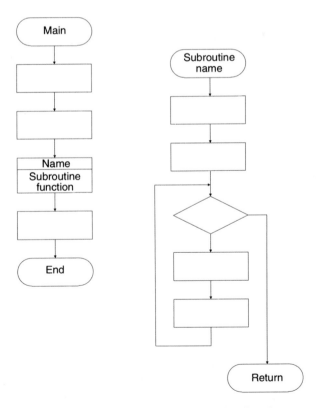

Figure 5-12   Main program and subroutine flowcharts.

at the assembly language level will help you understand the characteristics of high-level languages, especially the C-language that is widely used.

Usually, looking at the flowchart will not tell you the parameter-passing technique. The flowchart describes the function of the program and not the programming details. Therefore, the flowcharts are omitted for most of the subroutine examples that follow. The emphasis here is on the technique rather than the subroutine function.

A Bad Subroutine Example

Before looking at some good ways to pass parameters to and from subroutines, let's look at an example designed to illustrate the problems caused by bad techniques. The example is a trivial program of no real value except to illustrate some problems. Of course, the same problems occur in useful subroutines.

The listing for the example is Figure 5-13. The main program puts a number into location NUMBER and jumps to a subroutine that modifies the number in a trivial way. After control returns to the main program, it puts the result from the subroutine into an answer location. This process is done twice for different numbers. Here are some problems in the bad example program:

- *Lines 54, 57, and 59.*   No parameter passing occurs between the main program and the subroutine. Instead, the subroutine directly accesses the location labeled NUMBER to get data and to store the answer. The subroutine is not independent of the main program because it uses a data location owned by the main program. For example, if the main program was changed so that a location was no longer labeled NUMBER, the subroutine could no longer be used—the subroutine also would need changes. You could argue that the subroutine is not a subroutine but just a strange module of the main program. Similarly, if the main program is moved to a new location in memory, the subroutine must be assembled again for the new location. This requirement prevents separate assembly of the main program and the subroutine, which is a useful technique.

- *Lines 21, 25, 29, and 33.*   The main program must put the data for the subroutine into the special location labeled NUMBER each time it calls the subroutine. Then, the main program must get the result from the location NUMBER each time the subroutine returns. The main program cannot choose to use an arbitrary location each time it calls the subroutine. The main program is doing extraordinary things to accommodate the bad subroutine. One advantage of using subroutines is reusing them at many places in a program. However, the extraordinary setup in this example must be repeated each time the main program calls the subroutine. This repetition causes inefficient use of memory with nothing gained in return.

- *Lines 18 and 36.*   The main program was using the B accumulator before calling the subroutines. However, the BAD subroutine changed the B accumulator without restoring it. Consequently, the main program will malfunction at line 36. At the least, the subroutine documentation should list any registers that the subroutine modifies.

```
 1                                    *********************************************************
 2                                    ** DEMONSTRATE BAD SUBROUTINES
 3                                    *
 4                                    *********************************************************
 5                                    *
 6                                    *********************************************************
 7                                    ** MAIN PROGRAM
 8                                    *********************************************************
 9      C100                                    ORG    $C100
10                                    *-------------------------------------------------------
11                                    * INITIALIZE STACK
12                                    *-------------------------------------------------------
13      C100    8E DF FF              START   LDS    #$DFFF     STACK FOR MOTOROLA TRAINER
14                                    *-------------------------------------------------------
15                                    * USE BAD SUBROUTINES
16                                    *-------------------------------------------------------
17                                    * MAKE DATA IN MAIN PROGRAM
18      C103    C6 11                         LDAB   #$11
19                                    * SET UP DATA FOR SUBROUTINE
20      C105    86 22                         LDAA   #$22
21      C107    B7 C1 25                      STAA   NUMBER
22                                    * CALL SUBROUTINE TO ADD TWO
23      C10A    BD D0 05                      JSR    ADDTWO
24                                    * USE RESULT FROM SUBROUTINE
25      C10D    B6 C1 25                      LDAA   NUMBER
26      C110    B7 C1 27                      STAA   ANSWER2
27                                    * SET UP DATA FOR SUBROUTINE
28      C113    86 33                         LDAA   #$33
29      C115    B7 C1 25                      STAA   NUMBER
30                                    * CALL SUBROUTINE TO ADD TO POSITIVE
31      C118    BD D0 00                      JSR    BAD
32                                    * USE RESULT FROM SUBROUTINE
33      C11B    B6 C1 25                      LDAA   NUMBER
34      C11E    B7 C1 28                      STAA   ANSWER3
35                                    * STORE MAIN PROGRAM DATA
36      C121    F7 C1 26                      STAB   ANSWER1
37      C124    3F                            SWI               "STOP" FOR MOTOROLA TRAINER
38                                    *
39                                    *********************************************************
40                                    ** DATA SECTION
41                                    *********************************************************
42      C125                          NUMBER  RMB    1
43      C126                          ANSWER1 RMB    1
44      C127                          ANSWER2 RMB    1
45      C128                          ANSWER3 RMB    1
46                                    *
47                                    *********************************************************
48                                    ** SUBROUTINES BAD, ADDTWO, ADDONE
49                                    * INCREMENT POSITIVE TWO'S COMPLEMENT NUMBERS, OR
50                                    * ADD ONE OR TWO TO ANY NUMBER
51                                    *********************************************************
52      D000                                  ORG    $D000
53                                    * SUBROUTINE BAD--POSITIVE NUMBER?
54      D000    F6 C1 25              BAD     LDAB   NUMBER
55      D003    2A 03                         BPL    ADDONE     BRANCH ON YES
56                                    * SUBROUTINE ADDTWO
57      D005    7C C1 25              ADDTWO  INC    NUMBER
58                                    * SUBROUTINE ADDONE
59      D008    7C C1 25              ADDONE  INC    NUMBER
60      D00B    39                            RTS
61      D00C                                  END
```

Figure 5-13   An example of bad subroutines.

- *Lines 48, 54, 57, and 59.* The apparent single subroutine is really three different subroutines merged together with different entry points. The subroutine is incorrectly structured because it does not have a single entry point and a single exit point. Although this structure makes shorter subroutines, it is confusing and complicated. Changing one subroutine affects the other subroutines as well.

Almost everything about the bad subroutine causes problems in practical programs. The example given is very simple, and you may think that the points mentioned are of little importance. However, when you write large practical programs, you will discover that the problems discussed are significant.

Some Basic Subroutine Design Considerations

Answering the following questions while writing a subroutine will help you to write good programs:

- *Is your subroutine independent of the main program?* As a test of independence, determine whether the main program and its data can be moved to new memory locations without changing the subroutine. Also, can the assembly language labels in the main program be altered without any corrections to the subroutine? You should not need to assemble the subroutine again after you change the main program.

- *Is your subroutine written in correctly structured code?* A minimum requirement is that your subroutine may have only a single entry point and a single exit point. Always make the first instruction on the subroutine listing the first instruction executed in the subroutine. Likewise, always put the RTS instruction at the physical end of the subroutine listing.

- *Have you determined what data must be passed to and from the subroutine?* You must determine where the data is, how large the numbers are, how many numbers must be passed, and in what order they are to be passed.

- *Does your subroutine restore any microprocessor registers it changed?* If some registers are not restored, the subroutine must have documentation explaining what changes have occurred.

- *Must the main program do much setup to make the subroutine work?* Usually, the parameter-passing technique should fit the task done by the subroutine. Then, the main program will do little to set up the call to the subroutine. Use of the subroutine will fit into the main program in a natural way. Furthermore, if the main program calls the subroutine at many places, the number of memory locations used by the instructions that set up the subroutine may be unreasonable. A technique requiring little or no set up may be needed.

- *Have you determined where the subroutine will put its local variables?* The numbers that the subroutine works on when doing its function are called *local variables.* Usually, the microprocessor registers will hold these. However, if memory is used, the location of the local variables is a significant decision.

The sections that follow discuss many good parameter-passing techniques. The good techniques will satisfy all the questions asked here.

5.6   SUBROUTINE PARAMETER PASSING TECHNIQUES

Parameter passing refers to the method by which data numbers are sent from the main program to the subroutine, or from the subroutine to the main program. Many different techniques are used, although the names for the techniques are not universally agreed upon. The two general categories of techniques considered here are usually named *call-by-value* and *call-by-reference*.

Call-by-Value Technique

The *call-by-value* technique of parameter passing requires the main program to give the subroutine copies of the data values. The subroutine does its function and then may give some data values back to the main program. The main program may keep the original data value in memory; but it does not give the subroutine access to the original data number, which prevents the subroutine from changing it.

The simplest call-by-value technique uses the microprocessor registers to hold the data values. The main program first places the numbers in the registers, and then executes the jump-to-subroutine instruction. Therefore, the subroutine has the data when it starts running. The subroutine then places the results into the microprocessor registers and returns to the main program. The results are then available to the main program when it continues running. Using microprocessor registers is the most common technique of passing parameters at the machine and assembly language levels of programming.

The main limitations on passing numbers in the microprocessor registers are the number and size of data values that can be passed. The Motorola 68HC11 can hold data values in both the accumulators and the index registers. The A and B accumulators can each hold one byte, and the index registers can each hold two bytes. Therefore, the registers can hold a total of six bytes. Six bytes are enough for many practical applications.

The designs of the main program and the subroutine must be coordinated so that the correct numbers are passed in the correct registers. Generally, the main program is designed first. The main program determines the calling sequence for the subroutine. Then the subroutine can be written to be compatible with the calling sequence.

Call-by-value example

The first subroutine example in Figure 5-11 uses the call-by-value technique of parameter passing—the A accumulator holds the value. The next example is a very simple subroutine that illustrates several more characteristics of call-by-value subroutines. In this example, the main program and the subroutine were assembled by separate runs of the assembler program. When you assemble a subroutine separately, you will get assembly errors if the subroutine is not independent of the main program.

Assembling the main and subroutine modules separately has several practical advantages. First, it is convenient because subroutines can be developed independently of the main program—possibly by different people. Second, several subroutines can be grouped together into a library. The library routines can be thoroughly tested to ensure correctness. Then, different main programs can use the library subroutines. The programmer can depend on working subroutines in the library without the effort of writing new subroutines.

If the subroutine and the main program are assembled separately, you must provide a means of linking them together. Generally, this linking is not difficult.

Flowcharts.   Flowcharts for the example are in Figure 5-14. The example subroutine provides a time delay by looping for a specified amount of time. The resolution of the time delay is 20_{10} milliseconds because an inner loop provides this delay. An outer loop delays a number of 20_{10}-millisecond periods. The number is determined by the 8-bit value passed from the main program. The 8-bit timing value is an unsigned number, so the maximum time delay is 255_{10} times 20_{10} milliseconds or a little more than five seconds. More

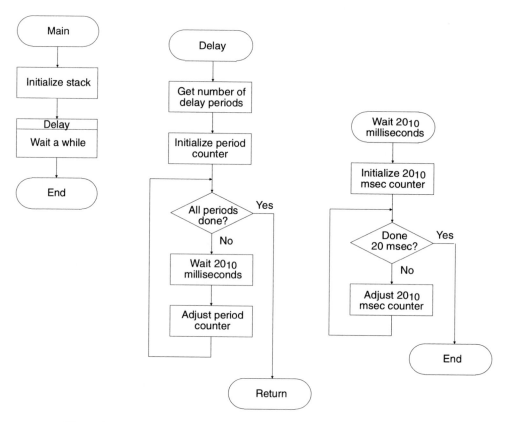

Figure 5-14   Flowcharts for DELAY subroutine and a main program to call it.

sophisticated ways can be used to provide timing, but this example illustrates the subroutine principles well.

The main program.   Figure 5-15 is the listing of the main program. The main program creates a stack, gets the time delay parameter into the A accumulator, and jumps to the subroutine. The statement on line 10 provides the link between the main program and the subroutine. Because the subroutine contains the label DELAY at the beginning of the subroutine, the assembler must know the value of DELAY to assemble line 32. It can't assemble the JSR instruction without knowing the address of the subroutine. Line 10 provides the address manually. The EQU directive tells the assembler the value of DELAY.

The subroutine.   The subroutine listing is in Figure 5-16. The top/down structured flowcharts show that the subroutine is a loop inside a loop. The following describes the listing in detail:

- *Lines 8 and 9.*   The purpose of the first program module is to get the parameter from the main program. In general, the parameter will be in a microprocessor register or in memory. This example uses the A accumulator, so this module has nothing to do and no code is necessary.

- *Lines 10 and 11.*   The next module initializes the working counter for the outer loop. The A accumulator is the working counter; it is already set, and no code is necessary.

- *Lines 12 through 25.*   This loop is the major loop that uses up time for the delay. The outer loop repeatedly uses the 20_{10}-millisecond loop.

- *Line 13.*   The period loop counter is tested to see if all the 20_{10}-millisecond delay periods are finished. The TSTA instruction is necessary because the BEQ instruction must have the correct Z bit information. The subroutine must not depend on the main program setting the condition codes correctly before jumping to the subroutine.

- *Lines 15 through 22.*   This 20_{10}-millisecond loop uses the index register as a double-byte working counter. You can calculate the number of loops required if you know the number of microprocessor clock cycles necessary to execute the loop. By looking at the instruction set table, you will find that each instruction in this loop requires three clock cycles. If N is the initial number in the X register, the number of clock cycles for this loop is 6+9N. If you know the clock rate, you can determine N. The Motorola trainer uses an 8.0-MHz crystal that gives an E-clock rate of 2.0_{10} MHz or 0.5_{10} microseconds per cycle. To get a loop time of 20_{10} milliseconds with this clock rate, the value of N is 115C.

- *Line 27.*   The RTS instruction gets the subroutine return address C108 from the stack and places it into the program counter, causing execution to continue in the main program. While the body of the subroutine is running, the stack pointer contains DFFD and the only number in the stack is the return address. After the RTS executes, the stack is empty.

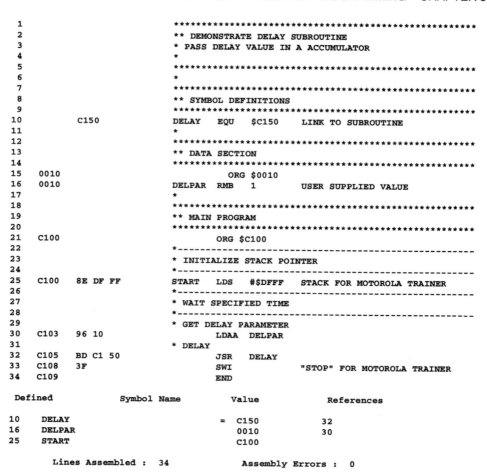

```
 1                              ***********************************************************
 2                              ** DEMONSTRATE DELAY SUBROUTINE
 3                              * PASS DELAY VALUE IN A ACCUMULATOR
 4                              *
 5                              ***********************************************************
 6                              *
 7                              ***********************************************************
 8                              ** SYMBOL DEFINITIONS
 9                              ***********************************************************
10          C150                DELAY   EQU   $C150     LINK TO SUBROUTINE
11                              *
12                              ***********************************************************
13                              ** DATA SECTION
14                              ***********************************************************
15   0010                              ORG $0010
16   0010                       DELPAR  RMB   1         USER SUPPLIED VALUE
17                              *
18                              ***********************************************************
19                              ** MAIN PROGRAM
20                              ***********************************************************
21   C100                              ORG $C100
22                              *-------------------------------------------------
23                              * INITIALIZE STACK POINTER
24                              *-------------------------------------------------
25   C100  8E DF FF             START   LDS   #$DFFF    STACK FOR MOTOROLA TRAINER
26                              *-------------------------------------------------
27                              * WAIT SPECIFIED TIME
28                              *-------------------------------------------------
29                              * GET DELAY PARAMETER
30   C103  96 10                        LDAA  DELPAR
31                              * DELAY
32   C105  BD C1 50                     JSR   DELAY
33   C108  3F                           SWI             "STOP" FOR MOTOROLA TRAINER
34   C109                               END
```

```
 Defined            Symbol Name        Value            References

10    DELAY                        =   C150               32
16    DELPAR                           0010               30
25    START                            C100

        Lines Assembled :   34          Assembly Errors :   0
```

Figure 5-15 Main program assembled separately from subroutine.

The DELAY subroutine is very simple because it only loops to use the time one data value specifies. However, it illustrates correct principles and is similar to most practical subroutines. It also demonstrates the separate assembly of the main program and subroutine, which is very common.

Position-independent code

The binary instruction codes that form the subroutine in Figure 5-16 have an interesting property. If you change the ORG statement on line 7 to specify any other address, such as ORG $0000, the binary codes will be the same. The binary codes that make up this subroutine can be placed anywhere in memory without alteration and they will work correctly. Therefore, this subroutine has *position-independent* or *binary relocatable* code.

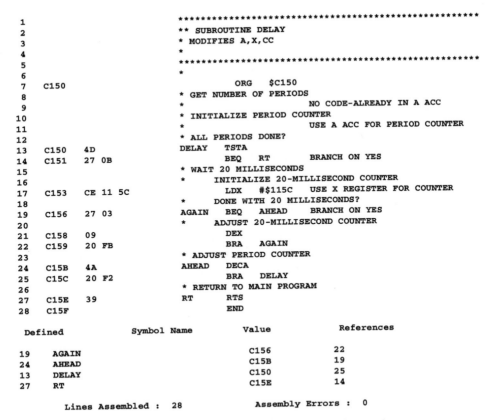

```
 1                            ********************************************************
 2                            **  SUBROUTINE DELAY
 3                            *  MODIFIES A,X,CC
 4                            *
 5                            ********************************************************
 6                            *
 7     C150                                 ORG    $C150
 8                            *  GET NUMBER OF PERIODS
 9                            *                          NO CODE-ALREADY IN A ACC
10                            *  INITIALIZE PERIOD COUNTER
11                            *                          USE A ACC FOR PERIOD COUNTER
12                            *  ALL PERIODS DONE?
13     C150   4D             DELAY   TSTA
14     C151   27 0B                  BEQ    RT         BRANCH ON YES
15                            *  WAIT 20 MILLISECONDS
16                            *      INITIALIZE 20-MILLISECOND COUNTER
17     C153   CE 11 5C               LDX    #$115C     USE X REGISTER FOR COUNTER
18                            *      DONE WITH 20 MILLISECONDS?
19     C156   27 03          AGAIN   BEQ    AHEAD      BRANCH ON YES
20                            *      ADJUST 20-MILLISECOND COUNTER
21     C158   09                     DEX
22     C159   20 FB                  BRA    AGAIN
23                            *  ADJUST PERIOD COUNTER
24     C15B   4A             AHEAD   DECA
25     C15C   20 F2                  BRA    DELAY
26                            *  RETURN TO MAIN PROGRAM
27     C15E   39             RT      RTS
28     C15F                          END
```

Defined	Symbol Name	Value	References
19	AGAIN	C156	22
24	AHEAD	C15B	19
13	DELAY	C150	25
27	RT	C15E	14

Lines Assembled : 28   Assembly Errors : 0

Figure 5-16   DELAY subroutine assembled separately from main program.

Position-independent code is useful in practical applications. For example, a company can manufacture a permanent memory-integrated circuit containing the binary numbers for a collection of position-independent subroutines. Purchasers of the memory can wire it into their computers without any consideration of the addresses that it will implement, and the subroutines will work correctly.

What makes the subroutine code position independent? In the 68HC11, the fact that the instructions in the subroutine do not use direct or extended addressing to reference other parts of the position independent program module is sufficient. Instructions with other addressing modes do not have addresses within the instructions, so they have no addresses to consider. For example, the branch instructions that have program relative addressing only have an offset inside the instruction. The branch instructions were designed with this addressing mode to make position-independent code possible. Remember though that only branch instructions have program relative addressing in the 68HC11.

In a later chapter, a very desirable type of subroutine called a *reentrant subroutine* is discussed. Because avoiding direct and extended addressed instructions that access data is necessary in reentrant subroutines, good practice avoids their use within all subroutines.

Avoiding direct and extended addressing encourages use of the stack to store temporary numbers.

All remaining subroutine examples in this chapter have position-independent code. In addition, they have *pure procedure* code because the instructions are not altered by using the subroutine. The name *pure procedure* implies that no numbers that are used as data values that change as the program runs are within the instructions.

Call-by-value in memory example

A constant data number can be passed to a subroutine as a value in memory. The value of the constant is determined when the program is loaded into memory. The constant is placed into memory immediately following the JSR instruction. A similar technique can be used for variable data when read/write memory holds the program and the data number.

The example in Figure 5-17 illustrates a subroutine that rotates only the A accumulator to the left a number of positions. The constant passed to the subroutine specifies the number of positions to rotate. An alternative is passing the number of rotates in the B accumulator, thus making the main program more complicated and using the B accumulator. In addition, passing the number of rotates in an accumulator treats the number as a variable rather than as a constant.

Here are some details of the program in Figure 5-17:

- *Line 22.*   The constant is placed immediately after the JSR instruction, thus passing it to the subroutine. The subroutine must return to address C109 in this example to avoid executing the data value as an instruction.

- *Lines 35 and 36.*   The subroutine will alter the X and B registers, so they are saved on the stack. Their values are restored at the end of the subroutine. The stack pointer now contains DFFA.

- *Line 38.*   The TSX instruction transfers the number in the stack pointer plus one to the X index register. The index register now contains DFFB, which is the address of the last byte stored in the stack. The index register points to the entry at the lowest address in the stack.

- *Line 39.*   The index register is loaded with the saved program counter. The JSR instruction saved the program counter in the stack. The saved address, C108, is the subroutine return address. The saved address is now in the X index register. The index register now points to the constant data value to be passed to the subroutine.

- *Line 40.*   The LDAB instruction gets the data value from the main program into the B accumulator for the subroutine to use. This step is the last in passing the parameter to the subroutine.

- *Lines 44 through 49.*   The body of the subroutine rotates the A accumulator the correct number of positions. This loop uses the value passed from the main program to determine the number of positions to rotate.

```
 1                              ***********************************************************
 2                              ** DEMONSTRATE SUBROUTINE--ROTATE A ACCUMULATOR LEFT
 3                              ** MULTIPLE POSITIONS
 4                              * PASS PARAMETER BY VALUE AS MEMORY CONSTANT
 5                              *
 6                              ***********************************************************
 7                              *
 8                              ***********************************************************
 9                              ** DATA SECTION
10                              ***********************************************************
11     0010                              ORG   $10
12     0010    06              VALUE     FCB   $06         DATA FOR DEMONSTRATION
13                              *
14                              ***********************************************************
15                              ** MAIN PROGRAM
16                              ***********************************************************
17     C100                              ORG   $C100
18                              * INITIALIZE STACK
19     C100    8E DF FF        START     LDS   #$DFFF      STACK FOR MOTOROLA TRAINER
20                              * ROTATE A ACCUMULATOR LEFT 4 POSITIONS
21     C103    96 10                     LDAA  VALUE       GET DATA
22     C105    BD C1 0A                  JSR   ROTAL
23     C108    04                        FCB   4           CONSTANT FOR POSITIONS
24     C109    3F                        SWI               "STOP" FOR MOTOROLA TRAINER
25                              *
26                              ***********************************************************
27                              ** SUBROUTINE ROTAL
28                              * ROTATE ONLY A ACCUMULATOR LEFT, CONSTANT FOLLOWING
29                              * JSR INSTRUCTION SPECIFIES POSITIONS--MODIFIES A,CC
30                              ***********************************************************
31                              *---------------------------------------------------------
32                              * INITIALIZE SUBROUTINE
33                              *---------------------------------------------------------
34                              * SAVE MAIN PROGRAM REGISTERS
35     C10A    3C              ROTAL     PSHX
36     C10B    37                        PSHB
37                              * GET NUMBER OF POSITIONS
38     C10C    30                        TSX
39     C10D    EE 03                     LDX   3,X
40     C10F    E6 00                     LDAB  0,X
41                              *---------------------------------------------------------
42                              * ROTATE A
43                              *---------------------------------------------------------
44     C111    5D              BACK      TSTB
45     C112    27 06                     BEQ   AHEAD
46     C114    5A                        DECB
47     C115    48                        ASLA
48     C116    89 00                     ADCA  #0
49     C118    20 F7                     BRA   BACK
50                              *---------------------------------------------------------
51                              * RETURN FROM SUBROUTINE
52                              *---------------------------------------------------------
53                              * ADJUST RETURN ADDRESS
54     C11A    30              AHEAD     TSX
55     C11B    6C 04                     INC   4,X
56     C11D    26 02                     BNE   RET
57     C11F    6C 03                     INC   3,X
58                              * RESTORE MAIN PROGRAM REGISTERS
59     C121    33              RET       PULB
60     C122    38                        PULX
61     C123    39                        RTS
62     C124                              END
```

Figure 5-17   Call-by-value in memory subroutine example.

- *Line 54.* The return address for the subroutine must be adjusted so that the subroutine returns to the location after the data value. The TSX instruction points the X index register to the last entry in the stack again. The saved return address can now be accessed.

- *Lines 55 through 57.* The return address saved in the stack is incremented by one. The first INC instruction adds one to the least-significant byte of the saved address. If a carry occurs, the second INC instruction adds one to the most-significant byte. Be careful when adjusting addresses. If the program doesn't check for a carry and increment the most-significant byte when necessary, the program will work for most addresses. However, when an address is adjusted that generates a carry, the program will fail. Such a bug could go undetected for a long time.

- *Lines 59 and 60.* The PUL instructions restore the main program numbers to the B and X registers from the stack.

- *Line 61.* The RTS instruction sends control to address C109 in this example by pulling this adjusted return address from the stack.

As in this example, subroutines using constant parameters often assume the character of a new instruction in the instruction set of the microprocessor. In this example, the subroutine acts as a rotate-the-A-accumulator-multiple-positions instruction. Some assemblers, called *macro-assemblers*, allow a new symbol to be defined for this pseudo-instruction. When the assembler encounters this new symbol, it automatically generates the jump to subroutine instruction and constant as required to call the subroutine.

Call-by-Reference Technique

Often the data values that the subroutine must work on are in memory. If so, the main program can pass the memory addresses of the data values to a subroutine. The addresses passed to a subroutine are called *references*, and the parameter-passing technique is named *call-by-reference*. The subroutine does the work to get the data from the locations referenced or addressed. Similarly, the subroutine can pass data back to the main program by directly storing into referenced memory locations. The main program gets the results from memory as it continues running. Usually, the main program will want the results in memory anyway, so it need do nothing to get the results.

Because the subroutine will have the addresses of the data numbers, you cannot prevent the subroutine from changing the original data. Any undesired changes in the data can make errors very troublesome and elusive. However, if you want the subroutine to modify the original data, call-by-reference is a good technique.

The call-by-value technique is commonly combined with the call-by-reference technique within the same subroutine. Each parameter should be passed by the technique best suited to its function. For example, a table is usually passed to a subroutine using the call-by-reference technique.

Call-by-reference in microprocessor register example

The reference or address of the data can be passed to the subroutine in a microprocessor register. In the 68HC11, references usually are passed in the index registers. With only two index registers, only two references can be passed this way. However, two references are enough for many practical subroutines.

Figure 5-18 is the listing of an example program that passes both a reference and a value to the subroutine. The example program uses only the microprocessor registers. A resultant value is passed back to the main program in a microprocessor register.

The example subroutine counts the number of zeros in a table of numbers in memory. The table is at the address specified by the reference. The value determines the length of the table. The subroutine returns the number of zeros in the table as a value. The main program and subroutine were assembled together to shorten the listing. Here are some details about the program:

- *Line 31.*   The value parameter is loaded into the A accumulator in preparation for transfer to the subroutine.

- *Line 32.*   The reference parameter is loaded into the X index register. Because a 16-bit address is sent to the subroutine, the data can be anywhere in memory.

- *Line 33.*   The JSR instruction does not alter the accumulators or index registers, so the parameters are effectively passed to the subroutine during this jump.

- *Lines 48 through 59.*   The data passed to the subroutine is used in the body of the subroutine. The result, the number of zeros in the table, is generated in the B accumulator.

- *Line 63.*   The RTS instruction does not alter the microprocessor registers. Therefore, the result is passed to the main program in the B accumulator during the return. Upon return to the main program, the values in the A, B, X, and condition code registers are different from the values at entry to the subroutine.

You should observe that the subroutine has pure procedure, position-independent code. In addition, the subroutine did not need to store any local data in memory to do its function.

Call-by-reference in memory example

The reference or address of a data value can be passed to the subroutine in the memory locations following the JSR instruction. This location is convenient and easily accessible for the reference, although the convenience of this location is not immediately apparent. Furthermore, in most applications, the location of the data will not change as the program runs. Therefore, the reference passed to the subroutine will be a constant determined when the program is loaded into memory.

The example in Figure 5-19 demonstrates passing parameters by reference in memory. All local variables are kept in the microprocessor registers. The subroutine does the same function as the example in Figure 5-18; only the parameter-passing technique is different. The subroutine searches a table of numbers and counts the number of zeros. The subroutine

```
 1                              *********************************************************
 2                              ** DEMONSTRATE SUBROUTINE--COUNT ZERO BYTES IN TABLE
 3                              * PASS VALUE IN A REGISTER AND REFERENCE IN X REGISTER
 4                              *
 5                              *********************************************************
 6                              *
 7                              *********************************************************
 8                              ** DATA SECTION
 9                              *********************************************************
10   0010                              ORG   $10
11   0010    04                LENGTH   FCB   4              DATA FOR DEMONSTRATION
12   0011                      ANSWER   RMB   1              ANSWER PUT HERE BY MAIN PROGRAM
13                             *
14   C300                              ORG   $C300
15   C300    45                TABLE    FCB   $45            TABLE TO BE SEARCHED FOR ZEROS
16   C301    00                         FCB   $00
17   C302    10                         FCB   $10
18   C303    00                         FCB   $00
19                             *
20                             *********************************************************
21                             ** MAIN PROGRAM
22                             *********************************************************
23   C100                              ORG   $C100
24                             *---------------------------------------------------------
25                             * INITIALIZE STACK
26                             *---------------------------------------------------------
27   C100    8E DF FF          START    LDS   #$DFFF   STACK IN MOTOROLA TRAINER
28                             *---------------------------------------------------------
29                             * COUNT NUMBER OF ZERO BYTES IN TABLE
30                             *---------------------------------------------------------
31   C103    96 10                     LDAA  LENGTH   GET TABLE LENGTH
32   C105    CE C3 00                  LDX   #TABLE   GET TABLE ADDRESS
33   C108    BD C1 0E                  JSR   CTZERO
34   C10B    D7 11                     STAB  ANSWER   SAVE RESULT FROM SUBROUTINE
35   C10D    3F                        SWI            "STOP" FOR MOTOROLA TRAINER
36                             *
37                             *********************************************************
38                             ** SUBROUTINE CTZERO
39                             * SEARCH TABLE OF (A) BYTES AT LOCATION (X), RETURN
40                             * NUMBER OF ZEROS AS (B)--MODIFIES A,B,X,CC
41                             *********************************************************
42                             *---------------------------------------------------------
43                             * SEARCH TABLE FOR ZEROS
44                             *---------------------------------------------------------
45                             * INITIALIZE ZERO COUNTER
46   C10E    5F                CTZERO   CLRB
47                             * AT END OF TABLE?
48   C10F    4D                AGAIN    TSTA
49   C110    27 09                     BEQ   RET      BRANCH ON YES
50                             * TABLE ENTRY ZERO?
51   C112    6D 00                     TST   0,X
52   C114    26 01                     BNE   AHEAD    BRANCH ON NO
53                             * INCREMENT ZERO COUNTER
54   C116    5C                        INCB
55                             * DECREMENT LOOP COUNTER
56   C117    4A                AHEAD    DECA
57                             * ADVANCE TO NEXT ENTRY IN TABLE
58   C118    08                        INX
59   C119    20 F4                     BRA   AGAIN
60                             *---------------------------------------------------------
61                             * RETURN FROM SUBROUTINE
62                             *---------------------------------------------------------
63   C11B    39                RET      RTS
64   C11C                               END
```

Figure 5-18 Passing parameters using call-by-value and reference in microprocessor registers.

needs the address of the table, the length of the table, and a place to return the number of zeros.

Here are some details about the program in Figure 5-19 and the memory diagram for it in Figure 5-20:

- *Lines 33, 34, and 35.*   The addresses of the memory locations that contain the table address, table length, and resulting zero count are placed following the JSR instruction by the FDB directives. The contents column shows these addresses to be C300, 0025, and C000. Figure 5-20(b) is a memory diagram that shows the JSR instruction and the references. From this figure, you cannot determine whether each reference is used to pass a value to the subroutine or for it to pass a result back. Sometimes a reference is used for both purposes. Observe that the main program

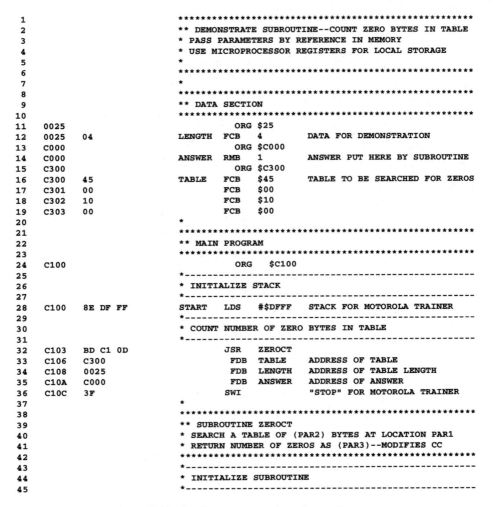

```
 1                          *******************************************************
 2                          ** DEMONSTRATE SUBROUTINE--COUNT ZERO BYTES IN TABLE
 3                          * PASS PARAMETERS BY REFERENCE IN MEMORY
 4                          * USE MICROPROCESSOR REGISTERS FOR LOCAL STORAGE
 5                          *
 6                          *******************************************************
 7                          *
 8                          *******************************************************
 9                          ** DATA SECTION
10                          *******************************************************
11    0025                              ORG   $25
12    0025    04            LENGTH   FCB   4          DATA FOR DEMONSTRATION
13    C000                              ORG   $C000
14    C000                  ANSWER   RMB   1          ANSWER PUT HERE BY SUBROUTINE
15    C300                              ORG   $C300
16    C300    45            TABLE    FCB   $45        TABLE TO BE SEARCHED FOR ZEROS
17    C301    00                       FCB   $00
18    C302    10                       FCB   $10
19    C303    00                       FCB   $00
20                          *
21                          *******************************************************
22                          ** MAIN PROGRAM
23                          *******************************************************
24    C100                              ORG    $C100
25                          *---------------------------------------------------
26                          * INITIALIZE STACK
27                          *---------------------------------------------------
28    C100    8E DF FF      START    LDS   #$DFFF    STACK FOR MOTOROLA TRAINER
29                          *---------------------------------------------------
30                          * COUNT NUMBER OF ZERO BYTES IN TABLE
31                          *---------------------------------------------------
32    C103    BD C1 0D               JSR    ZEROCT
33    C106    C300                    FDB    TABLE     ADDRESS OF TABLE
34    C108    0025                    FDB    LENGTH    ADDRESS OF TABLE LENGTH
35    C10A    C000                    FDB    ANSWER    ADDRESS OF ANSWER
36    C10C    3F                      SWI              "STOP" FOR MOTOROLA TRAINER
37                          *
38                          *******************************************************
39                          ** SUBROUTINE ZEROCT
40                          * SEARCH A TABLE OF (PAR2) BYTES AT LOCATION PAR1
41                          * RETURN NUMBER OF ZEROS AS (PAR3)--MODIFIES CC
42                          *******************************************************
43                          *---------------------------------------------------
44                          * INITIALIZE SUBROUTINE
45                          *---------------------------------------------------
```

Figure 5-19   Passing parameters by reference in memory.

```
46                              * SAVE MAIN PROGRAM REGISTERS
47   C10D   36          ZEROCT   PSHA
48   C10E   37                   PSHB
49   C10F   3C                   PSHX
50                              * GET TABLE LENGTH
51   C110   30                   TSX               X POINTS TO TOP ENTRY IN STACK
52   C111   EE 04                LDX    4,X        GET SAVED PROGRAM COUNTER
53   C113   EE 02                LDX    2,X        GET REFERENCE TO TABLE LENGTH
54   C115   A6 00                LDAA   0,X        GET TABLE LENGTH
55                              * GET TABLE ADDRESS
56   C117   30                   TSX               X POINTS TO TOP ENTRY IN STACK
57   C118   EE 04                LDX    4,X        GET SAVED PROGRAM COUNTER
58   C11A   EE 00                LDX    0,X        GET REFERENCE TO TABLE
59                              * INITIALIZE ZERO COUNTER
60   C11C   5F                   CLRB
61                              *----------------------------------------------------
62                              * SEARCH TABLE FOR ZEROS
63                              *----------------------------------------------------
64                              * AT END OF TABLE?
65   C11D   4D          AGAIN    TSTA
66   C11E   27 09                BEQ    RET        BRANCH ON YES
67                              * TABLE ENTRY ZERO?
68   C120   6D 00                TST    0,X
69   C122   26 01                BNE    AHEAD      BRANCH ON NO
70                              * INCREMENT ZERO COUNTER
71   C124   5C                   INCB
72                              * DECREMENT LOOP COUNTER
73   C125   4A          AHEAD    DECA
74                              * ADVANCE TO NEXT ENTRY IN TABLE
75   C126   08                   INX
76   C127   20 F4                BRA    AGAIN
77                              *----------------------------------------------------
78                              * RETURN FROM SUBROUTINE
79                              *----------------------------------------------------
80                              * SEND RESULT TO MAIN PROGRAM
81   C129   30          RET      TSX               X POINTS TO TOP ENTRY IN STACK
82   C12A   EE 04                LDX    4,X        GET SAVED PROGRAM COUNTER
83   C12C   EE 04                LDX    4,X        GET REFERENCE TO ANSWER
84   C12E   E7 00                STAB   0,X        SEND DATA
85                              * ADJUST RETURN ADDRESS
86   C130   30                   TSX               X POINTS TO TOP ENTRY IN STACK
87   C131   EC 04                LDD    4,X        GET SAVED PROGRAM COUNTER
88   C133   C3 00 06             ADDD   #6         ADVANCE SIX BYTES
89   C136   ED 04                STD    4,X        STORE SAVED PROGRAM COUNTER
90                              * RESTORE MAIN PROGRAM REGISTERS
91   C138   38                   PULX
92   C139   33                   PULB
93   C13A   32                   PULA
94   C13B   39                   RTS
95   C13C                        END
```

Defined	Symbol Name	Value	References
65	AGAIN	C11D	76
73	AHEAD	C125	69
14	ANSWER	C000	35
12	LENGTH	0025	34
81	RET	C129	66
28	START	C100	
16	TABLE	C300	33
47	ZEROCT	C10D	32

```
      Lines Assembled :  95              Assembly Errors :  0
```

Figure 5-19   Continued.

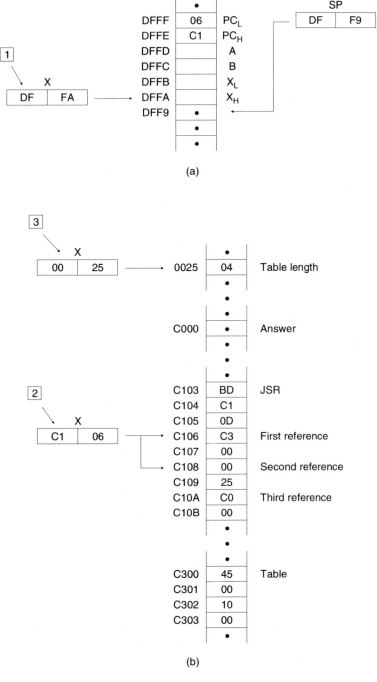

Figure 5-20   Memory diagram for the program in Figure 5-19.

executes no instructions to set up the parameters for the subroutine if the data is already in memory. Usually, the data is already in memory. Then all the work is done inside the subroutine. The setup is not reproduced in the main program each time the subroutine is called. The assembler provides the references to the data values as a list of addresses following the jump to subroutine instruction. However, the subroutine must adjust the return point so that the subroutine returns just beyond the list of addresses. You should carefully check your subroutines to be sure they do not return to an incorrect point and execute the addresses as instructions.

- *Lines 46 through 49.*   The main program registers are saved in the stack so they can be restored at the end of the subroutine. The JSR instruction has already saved the program counter in the stack. Because the main program registers are not used for passing parameters, you should not disturb them in the subroutine. Figure 5-20(a) shows a stack picture at this point in the program. The labels show where the main program registers have been saved in the stack. Also, the stack pointer is shown pointing to the next available stack location because it contains address DFF9.

- *Line 51.*   The TSX instruction points the X index register to the top of the stack—the last entry used in the stack. Carefully note that the TSX instruction transfers the contents of the stack point register *plus one* to the X index register. Remember that the location in memory of the stack is not determined until the subroutine runs. The stack may be at different locations every time the subroutine runs. Therefore, transferring the stack pointer to X allows the subroutine to access the stack contents easily. In this example, the X index register now contains DFFA, as Figure 5-20(a) shows at box 1.

- *Line 52.*   The number in the program counter was stored in the stack by the JSR instruction. The address stored is the address of the next location in memory after the JSR instruction. In this example, the stored address is the location of the first reference. The LDX instruction loads the X register with the address of the reference by offsetting four bytes from the top of the stack. The X register now contains C106 as Figure 5-20(b) shows at box 2.

- *Line 53.*   The LDX instruction loads the X register with the second reference, which is the address where the table length is stored. The LDX accesses the second reference due to the offset of 2 in the instruction. The X register now contains 0025, as Figure 5-20(b) shows at box 3.

- *Line 54.*   The A accumulator is loaded with the data value 04 from location 0025. The first parameter has now been passed into the subroutine.

- *Lines 56 through 58.*   Using the same procedure, these instructions obtain the address of the table to be searched. The X register now contains C300.

- *Lines 59 through 76.*   This part of the subroutine performs the function of counting the zeros in a table—it leaves the number of zeros in the B accumulator.

- *Lines 81 through 84.*   The subroutine sends its result to the main program in the same way that it got the data from the main program.

- *Lines 86 through 89.*   The subroutine must adjust the subroutine return address so that the subroutine returns to the correct address in the main program. In this example, the return address must be adjusted by six. This accounts for the six bytes of references that follow the JSR instruction. After line 89, the return address is C10C.

- *Lines 91 through 94.*   Removing the saved values from the stack restores the main program registers. When the RTS restores C10C to the program counter, the main program is again in control.

You should observe that this subroutine did not directly store any numbers in memory within the body of the subroutine. All numbers needed to perform the subroutine function were in the microprocessor registers.

Reference in memory—local variables in stack example

The subroutine example in Figure 5-21 demonstrates passing parameters by reference with the references in memory following the JSR instruction. The subroutine counts the leading zeros in a double-precision number in memory. A hole or working section in the stack holds all local variables. Because the parameter passing is very similar to the previous example, the following only describes details related to the local variables in the stack:

- *Lines 43 and 44.*   The two DES instructions leave two empty bytes at the top of the stack. These bytes form the hole that will hold the local variables.

- *Line 46.*   The X index register points at the hole for local variables. All instructions using the hole have indexed by X addressing. The X index register is used for no other purpose. It is left pointing to the top of the stack throughout the subroutine. Therefore, any other indexed addressed instructions use the Y index register. The X index register now contains DFF6 as the stack diagram in Figure 5-22 shows.

- *Lines 47 through 49.*   Initial values are stored in the local variables in the stack. The number 00 is stored in the zero counter and 16_{10} is stored in the loop counter.

- *Lines 51 through 53.*   The first reference is used to pass the double-precision 16-bit data value to the subroutine.

- *Lines 58, 64, 67, and 69.*   The local variables are used in the body of the subroutine. These variables are accessed with indexed addressing because the location of the stack may be different each time the subroutine is called.

- *Lines 84 and 85.*   The hole in the stack is closed, which discards the local variables now that the subroutine no longer needs them. Usually, the subroutine is easier to understand if the hole is opened and closed with the DES and INS instructions instead of the PSH and PUL instructions.

```
1                              **************************************************
2                              ** DEMONSTRATE SUBROUTINE--COUNT LEADING ZEROS IN
3                              **   DOUBLE-PRECISION NUMBER
4                              * PASS PARAMETERS BY REFERENCE IN MEMORY
5                              * USE STACK FOR LOCAL STORAGE
6                              *
7                              **************************************************
8                              *
9                              **************************************************
10                             ** DATA SECTION
11                             **************************************************
12   0025                              ORG   $25
13   0025    0678      DPNUM   FDB   $0678     DATA FOR DEMONSTRATION
14   C000                              ORG   $C000
15   C000              ANSWER  RMB   1         ANSWER PUT HERE BY SUBROUTINE
16                             *
17                             **************************************************
18                             ** MAIN PROGRAM
19                             **************************************************
20   C100                              ORG   $C100
21                             * INITIALIZE STACK
22   C100    8E DF FF   START   LDS   #$DFFF   STACK FOR MOTOROLA TRAINER
23                             * COUNT LEADING ZEROS IN DOUBLE-PRECISION NUMBER
24   C103    BD C1 0B           JSR   LEADZER
25   C106    0025               FDB   DPNUM    ADDRESS OF DATA NUMBER
26   C108    C000               FDB   ANSWER   ADDRESS OF ANSWER
27   C10A    3F                 SWI            "STOP" FOR MOTOROLA TRAINER
28                             *
29                             **************************************************
30                             ** SUBROUTINE LEADZER
31                             * COUNT LEADING ZEROS IN 16-BIT NUMBER AT PAR1,
32                             * RETURN BYTE ANSWER TO (PAR2)--MODIFIES CC
33                             **************************************************
34                             *------------------------------------------------
35                             * INITIALIZE SUBROUTINE
36                             *------------------------------------------------
37                             * SAVE MAIN PROGRAM REGISTERS
38   C10B    3C         LEADZER PSHX
39   C10C    18 3C              PSHY
40   C10E    36                 PSHA
41   C10F    37                 PSHB
42                             * MAKE HOLE IN STACK FOR LOCAL STORAGE
43   C110    34                 DES            FOR LOOP COUNTER
44   C111    34                 DES            FOR ZERO COUNTER
45                             * INITIALIZE ZERO COUNTER AND LOOP COUNTER
46   C112    30                 TSX            POINT X TO TOP ENTRY IN STACK
47   C113    4F                 CLRA
48   C114    C6 10              LDAB  #16
49   C116    ED 00              STD   0,X
50                             * GET NUMBER TO TEST
51   C118    1A EE 08           LDY   8,X      GET SAVED PROGRAM COUNTER
52   C11B    18 EE 00           LDY   0,Y      GET REFERENCE
53   C11E    18 EC 00           LDD   0,Y      GET DATA
54                             *------------------------------------------------
55                             * COUNT NUMBER OF LEADING ZEROS
56                             *------------------------------------------------
57                             * 16 BITS TESTED?
58   C121    6D 01      AGAIN   TST   1,X
59   C123    27 0D              BEQ   RET      BRANCH ON YES
60                             * NEXT BIT A ZERO?
61   C125    05                 ASLD
62   C126    24 04              BCC   AHEAD    BRANCH ON YES
63                             * TERMINATE ON A ONE
```

Figure 5-21   Call-by-reference in memory subroutine with local variables in the stack.

```
64   C128   6F 01                    CLR    1,X
65   C12A   20 04                    BRA    ENDLOOP
66                            * INCREMENT ZERO COUNTER
67   C12C   6C 00            AHEAD   INC    0,X
68                            * DECREMENT LOOP COUNTER
69   C12E   6A 01                    DEC    1,X
70   C130   20 EF            ENDLOOP BRA    AGAIN
71                           *-------------------------------------------------
72                           * RETURN FROM SUBROUTINE
73                           *-------------------------------------------------
74                           * SEND RESULT TO MAIN PROGRAM
75   C132   A6 00            RET     LDAA   0,X       GET DATA
76   C134   1A EE 08                 LDY    8,X       GET SAVED PROGRAM COUNTER
77   C137   18 EE 02                 LDY    2,Y       GET REFERENCE
78   C13A   18 A7 00                 STAA   0,Y       SEND DATA
79                           * ADJUST RETURN ADDRESS
80   C13D   EC 08                    LDD    8,X
81   C13F   C3 00 04                 ADDD   #4        SKIP TWO REFERENCES
82   C142   ED 08                    STD    8,X
83                           * CLOSE HOLE IN STACK
84   C144   31                       INS
85   C145   31                       INS
86                           * RESTORE MAIN PROGRAM REGISTERS
87   C146   33                       PULB
88   C147   32                       PULA
89   C148   18 38                    PULY
90   C14A   38                       PULX
91   C14B   39                       RTS
92   C14C                            END
```

```
   Defined            Symbol Name        Value            References

       58     AGAIN                       C121              70
       67     AHEAD                       C12C              62
       15     ANSWER                      C000              26
       13     DPNUM                       0025              25
       70     ENDLOOP                     C130              65
       38     LEADZER                     C10B              24
       75     RET                         C132              59
       22     START                       C100

          Lines Assembled :  92           Assembly Errors :   0
```

Figure 5-21   Continued.

Local variables stored in the stack no longer exist when the subroutine returns to the main program. Sometimes these variables are called *dynamic variables* or *automatic variables*. Dynamic variables come into existence as the subroutine is entered, and they are discarded before it exits or returns to the main program.

5.7   RECURSIVE SUBROUTINES

A *recursive subroutine* calls or jumps to itself. Such a subroutine must contain a decision that ultimately chooses not to call the subroutine again, but instead returns from subroutine—otherwise, the result is an infinite loop. Certain mathematical functions, such as factorial, are easy to calculate using recursion.

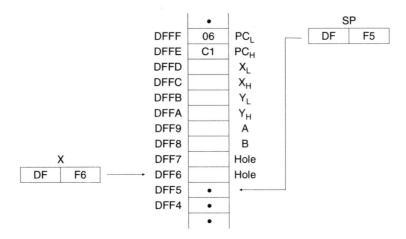

Figure 5-22   Stack diagram for the program in Figure 5-21.

Writing a subroutine that can call itself and function correctly is easy if you understand the problem. The problem is with local variables. Consider a subroutine that uses a fixed location in memory for a local variable. When it calls itself, it will overwrite the number in that fixed location and the subroutine fails. Such a fixed location usually is accessed using direct or extended addressing in the 68HC11. However, if the subroutine puts its local variables into the stack, it will create a new hole in the stack each time it is entered. Using a hole in the stack for local variables makes writing recursive subroutines easy. Although it does not use recursion, the example in Figure 5-21 demonstrates the technique necessary for recursive subroutines.

Recursive subroutines are usually clever programs that use few instructions. However, they are often difficult to understand. Also, they usually require a large stack space to hold a new copy of the local variables and a return address for each call of the subroutine. Because a nonrecursive subroutine can do the function of any recursive subroutine, most people avoid recursive subroutines when using small computers.

5.8   REVIEW

Most of the 68HC11 instructions have now been introduced. Only a few instructions remain to be covered, and they pertain to input/output hardware. Many instructions in this chapter were designed for specific applications rather than general programming. For example, the stack controlling instructions should be used only for manipulating the stack. To use them for other purposes is to invite programming errors and confusion for anyone reading the listing. In a later chapter, you will find that the stack has further uses that will prevent you from using stack instructions for other purposes.

Subroutines are very important in practical programming. This chapter introduces many parameter-passing techniques. Some of these techniques are difficult to implement in the 68HC11. Nevertheless, the techniques are useful. More complex microprocessors have additional addressing modes to make subroutines easier to program. Always, a programmer must make a wise decision about which parameter-passing technique to use for each application.

5.9 EXERCISES

5-1. Write a two-instruction program module to rotate the D accumulator left one place without including the carry bit; that is, rotate left circular D. The resultant condition codes can be anything.

5-2. Write a program module to rotate the A accumulator right one place without including the carry bit. This operation is called *rotate circular*.

5-3. Does execution of a LDY instruction with indexed by X addressing change the contents of the X register?

5-4. The ASL instruction multiplies any 8-bit unsigned number or any 8-bit two's complement number by two. (true/false/unknown)_____

5-5. Does the instruction sequence NEGA, NEGB, SBCA #0 correctly perform the operation NEGD?

5-6. How many different prebyte codes are used in the 68HC11 instruction set?

5-7. The ORAB #00 instruction will set the Z bit only if the number in the B accumulator was
_____.

5-8. If the Y index register contains C31F, a LDAA instruction with indexed by Y addressing can load data from addresses _____ through and including _____.

5-9. To logical shift the number in the A accumulator right six places using the MUL instruction, the number _____ must first be put in the B accumulator and the result is found in the (A, B, neither) _____ accumulator.

5-10. Executing a PSHX instruction results in the number in the stack pointer being (higher, lower, the same) _____ as it was before the execution.

5-11. If the Y index register contains C205 immediately after the TSY instruction is executed, then the stack pointer contains _____.

5-12. A JSR instruction with indexed by X addressing is located at address E510. When this instruction is executed, it puts the number _____ into the stack.

5-13. Write a program module that copies bits 0 through 3 of the B accumulator to bits 0 through 3, respectively, of the A accumulator without changing bits 4 through 7 of the A accumulator. The number in the B accumulator may be modified.

5-14. Write a program module that interprets the 3-bit number in bits 0 through 2 of the B accumulator as a bit number. The program sets only the bit of the A accumulator numbered in the B accumulator. It must not change any other bits in the A accumulator. Bits 3 through 7 of the B accumulator may be nonzero.

5-15.   Refer to Figure 5-17 for the following questions:

(a)   Immediately after the instruction on line 40 is executed, the number in the A accumulator is _____ and the number in the B accumulator is _____.

(b)   The lowest number in the stack pointer as this subroutine runs is _____.

(c)   The number in the A accumulator after the subroutine finishes is _____.

(d)   The (single-byte) number in memory location DFFF when the subroutine executes the instruction on line 49 is _____.

(e)   If (only) line 25 (now a comment) were changed to ORG $D100, identify all program listing lines that would have different binary codes.

(f)   If (only) line 11 were changed to ORG $D800, identify all subroutine program listing lines that would have different binary codes.

(g)   The number that results from executing the instruction on line 55 is _____.

5-16.   Supply the missing items in Figure 5-23 with good assembly language by reading the listing. The program puts 0 in bit 0 of location RESULT if either bit 0 and bit 1 of location DATAX are both 0 or if bit 7 of location DATAY is 0; otherwise, it puts 1 in bit 0 of location RESULT. It does not change any other bits in location RESULT.

5-17.   Write a new program that does the same function as the program in Figure 5-23, but uses the BSET, BCLR, BRSET, and BRCLR instructions instead of ANDA and ORAA.

5-18.   Find the complete hexadecimal code for a single branch instruction to be placed at location D002 that will test both bits 6 and 4 of memory location 00F4 and branch to an instruction at location D025 only if they are both zeros.

5-19.   A single branch instruction that branches to itself if both bits 3 and 4 of memory location 0020 are zero has a hexadecimal code of _____.

5-20.   The single instruction that can put FF in location 00F0 is _____.

```
* BIT MANIPULATION PROGRAM
MASK1     EQU    $03
MASK2     EQU    $_____   (a)
*
          ORG $0000
FILTER    FCB    %_____   (b)
DATAX     RMB    1
DATAY     RMB    1
RESULT    RMB    1
          RELATIVE
          ORG $D100
START     LDAA   DATAX
          ANDA   #_____   (c)
          _____  (d) NEXT
          LDAA   DATAY
          ANDA   FILTER
          BEQ    NEXT
          LDAA   _____  (e)
          _____  (f) #MASK2
          STAA   RESULT
          BRA    LAST
NEXT      LDAA   RESULT
          ANDA   #%_____  (g)
          STAA   RESULT
LAST      SWI            "STOP" FOR TRAINER
          END
```

Figure 5-23   Exercises 5-16 and 5-17.

5-21. If, after part of a program runs, the number in the stack pointer is the same as the original number that was loaded into it, the stack contains _____.

5-22. Immediately following the execution of a TSX instruction, the X index register points to the last byte that was stored in the stack. (true/false/unknown)_____

5-23. Write a subroutine that exchanges the numbers in the A and B accumulators. The subroutine must not use any permanently allocated read/write memory locations.

5-24. Refer to Figure 5-24, which is a listing of a main program and a subroutine with little documentation. The subroutine compares two numbers and returns the more positive of the two (two's complement) numbers by placing it in the A accumulator. Zero is returned in the B accumulator if the number passed to the subroutine in the A accumulator is unchanged, and one is returned in B if A is changed. Complete the missing parts (a) through (f) of the figure, then answer the following:

(g) Immediately after the instruction on line 24 is executed, the number in the X index register is _____.

(h) The number the RTS instruction removes from the stack is _____.

(i) The lowest number in the stack pointer when this program is run is _____.

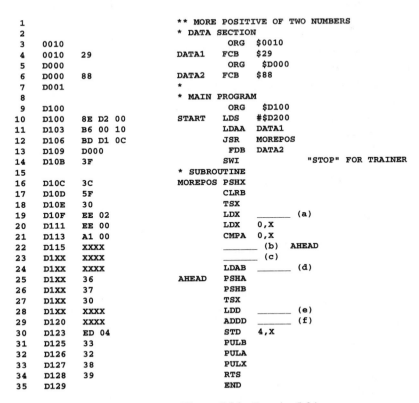

```
 1                                    ** MORE POSITIVE OF TWO NUMBERS
 2                                    * DATA SECTION
 3      0010                                  ORG   $0010
 4      0010    29              DATA1  FCB   $29
 5      D000                                  ORG   $D000
 6      D000    88              DATA2  FCB   $88
 7      D001                          *
 8                                    * MAIN PROGRAM
 9      D100                                  ORG   $D100
10      D100    8E D2 00        START  LDS   #$D200
11      D103    B6 00 10               LDAA  DATA1
12      D106    BD D1 0C               JSR   MOREPOS
13      D109    D000                   FDB   DATA2
14      D10B    3F                     SWI            "STOP" FOR TRAINER
15                                    * SUBROUTINE
16      D10C    3C              MOREPOS PSHX
17      D10D    5F                     CLRB
18      D10E    30                     TSX
19      D10F    EE 02                  LDX   _____ (a)
20      D111    EE 00                  LDX   0,X
21      D113    A1 00                  CMPA  0,X
22      D115    XXXX                   _____ (b)  AHEAD
23      D1XX    XXXX                   _____ (c)
24      D1XX    XXXX                   LDAB  _____ (d)
25      D1XX    36              AHEAD  PSHA
26      D1XX    37                     PSHB
27      D1XX    30                     TSX
28      D1XX    XXXX                   LDD   _____ (e)
29      D120    XXXX                   ADDD  _____ (f)
30      D123    ED 04                  STD   4,X
31      D125    33                     PULB
32      D126    32                     PULA
33      D127    38                     PULX
34      D128    39                     RTS
35      D129                           END
```

Figure 5-24 Exercise 5-24.

5-25. Write a subroutine that converts a two-digit BCD number in the A accumulator to an 8-bit binary number in the A accumulator. The subroutine must not use any permanently allocated read/write memory locations.

5-26. Write a subroutine that converts an 8-bit binary number in the A accumulator to a two-digit BCD number in the A accumulator. The subroutine must not use any permanently allocated read/write memory locations.

5-27. Write a subroutine that converts an 8-bit two's complement number in the A accumulator to a 16-bit two's complement number in the D accumulator.

5-28. Write a subroutine that multiplies the two two's complement signed 8-bit numbers in the A and B accumulators and returns the 16-bit product in the D accumulator. Your subroutine must use the MUL instruction to do the multiplication. The subroutine must not use any permanently allocated read/write memory locations.

5-29. Write a subroutine that forms the absolute value of the two's complement number in the D accumulator and puts the result in the D accumulator. The subroutine must not use any permanently allocated read/write memory locations.

5-30. Write a subroutine that compares the 8-bit two's complement numbers in two tables. It will determine whether all the numbers in the first table are equal to or greater than the corresponding numbers in the second table. The subroutine will return 00 in the A accumulator if the condition is false, and FF if it is true. The references to the tables will be passed in the X and Y registers and the table length will be in the A accumulator.

5-31. Write a subroutine named FILL that fills a block of bytes in memory with a number. Pass the number for the fill pattern in the A accumulator, the address of the first byte in the X register, and the 16-bit length of the block in the Y register.

5-32. Write a subroutine that uses bits 0 through 2 of the A accumulator as a bit number, bit 7 of the A accumulator as a bit value, and the address in the X index register as a reference to a memory register. The subroutine will put the value into the specified bit in the referenced memory register without changing any other bits. Bits 3 through 6 of the A accumulator are unspecified.

5-33. Write a subroutine that finds the largest 16-bit unsigned number in a table. The main program will pass the number of 16-bit numbers in the table in the A accumulator and the first address of the table as a reference in memory. The subroutine will return the number found in the D accumulator.

5-34. Change the subroutine in Figure 5-21 so that the references are passed between the main program and the subroutine using the stack. Be careful to have the stack pointer always ready to store new information into the stack. This technique of parameter passing is difficult in the 68HC11. Discuss the advantages and disadvantages of this technique of parameter passing.

5-35. Change the program in Figure 5-19 by inserting a BRA instruction in the main program between the JSR instruction and the references. The BRA must branch to the first instruction after the list of references. Change the subroutine so that it returns to the BRA instruction. Discuss the advantages and disadvantages of this technique of parameter passing.

Chapter 6

Hardware

Computer hardware is difficult to discuss in a textbook. Usually, specific integrated circuits must be covered. Using a technique that covers broad principles and applications is especially difficult. However, the Motorola 68HC11 is easier to describe than many products. The hardware that most practical applications require is already in the chip. Memory, input/output hardware, and even a selection of common input/output devices are inside the integrated circuit package.

The 68HC11 is a *single-chip microcomputer* or *microcontroller* because it contains memory and input/output hardware. Many applications require no additional external hardware. Because the hardware devices and the interconnections are all within the chip, you cannot see separate parts and interconnecting wires. Trying to use laboratory instrumentation to observe the internal hardware operation is very difficult. You have no way to connect the instrumentation! Consequently, you may find studying and using the 68HC11 as a single-chip computer quite abstract.

Your goal may be to learn to construct microcomputer hardware from a collection of integrated circuits. This goal is possible when the 68HC11 operates in the expanded mode. In expanded mode, the internal signals necessary for expansion are available at the integrated circuit pins. When using expanded mode, you must design, construct, and test the hardware. In applying the 68HC11, you would construct your own hardware if the devices available within the package were not appropriate.

This chapter does not discuss the hardware construction. The descriptions avoid the electronic circuit details and concentrate on the logical operation of the hardware and the software necessary to use it. Some examples refer to specific 68HC11 internal input/output hardware. Chapter 7 adds more hardware details.

6.1   HARDWARE/SOFTWARE SYNERGY

People frequently divide computer system development into hardware development and software development. However, computer hardware and software do not operate independently. The operation of a computer in a practical application depends on both. The word *synergy* describes the relationship between hardware and software. The word *synergy* means that each of two things helps to improve the other so that the whole is greater than the sum of the parts. This chapter shows that hardware and software are each designed to work with the other to make powerful and useful microcomputers.

If you have used large data processing or scientific computers, learning hardware operation and design may seem unnecessary. Such computers usually isolate the average user from the details of hardware. In many applications of microcomputers, however, the computer is part of the electronic control circuitry. In such applications, users cannot remain isolated from the hardware operation.

In most microcomputer applications, you make a tradeoff between software and hardware. Often you can do the same job either in software or with additional hardware. Many considerations come to bear on your decision to use software or hardware. These include cost, complexity of hardware, reliability, size, electrical power consumption, memory space available, and execution speed. Only broad knowledge and experience will lead to the best solution.

6.2   THE HARDWARE BUILDING BLOCKS

You must interconnect several hardware devices to construct a microcomputer. Wires may make the connections between separate integrated circuits. Alternatively, the connections may be within an integrated circuit chip. Where possible, the following discussion implies the first approach with distinct interconnecting wires. However, in principle, no difference exists when the hardware is inside a chip.

Microcomputer

A *microcomputer* is most often a collection of integrated circuits on a circuit board that make a complete functional computer. Typical designs usually include a microprocessor IC, a clock, several memory ICs, and several input/output ICs. When most of these components are in a single integrated circuit, the computer is called a *single-chip microcomputer*.

Memory

Let's review the signals used by the read/write memory integrated circuit described in Chapter 1. Figure 6-1 is a copy of Figure 1-10.

The figure implies that this integrated circuit contains 64_{10} kilobytes of memory because it has 16_{10} address lines. However, most applications that require 64_{10} kilobytes will need several ICs that each contains fewer than 64_{10} kilobytes of memory. These memory ICs will have fewer address pins. For example, if Figure 6-1 represented a memory IC with 14_{10} address pins, then it would contain $16,384_{10}$ 8-bit registers because 2^{14} is $16,384_{10}$.

Usually the address pins on memory integrated circuits are labeled A_0 through A_{15}, and the data lines are labeled D_0 through D_7. The subscripts are the bit numbers that identify the bits in microcomputer registers. In the figure, the IC has eight data pins corresponding to the eight bits in the registers.

The read/write or R/W control line tells the IC to read when the wire is high and to write when the wire is low. When the memory chip writes, it stores the number applied to the data pins in the addressed memory register. When the memory reads, it puts the number in the addressed memory register on the data pins. That is, when the memory is reading, it controls the logic levels on the pins. The R/W signal controls the direction of the data pins. The data pins are called *bidirectional* data pins.

The chip select or CS control line enables the IC to respond to the other signals. When the chip select pin is deasserted, the IC effectively disconnects the pins from the internal hardware. The data pins must have *tri-state* output hardware with three possible conditions—high, low, and disconnected. Tri-state hardware allows another IC to control the voltage levels on the output pins of the IC. That is, while the pins on the memory IC are internally disconnected, another IC can apply either the high or low voltage level to the pins without any interaction with the internal memory hardware.

Additional pins not shown in the figure are necessary for the electrical power and ground connections. The power pin is usually labeled V_{CC} or V_{DD}, and the ground pin is labeled V_{SS}.

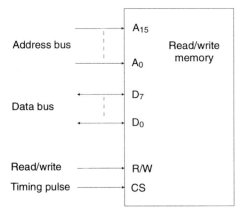

Figure 6-1   Memory signals.

Microprocessor

A *microprocessor* is a single integrated circuit that contains the control unit and processor parts of a microcomputer. Some practical ICs may vary from this definition because they include a few additional hardware items in the chip. The point is that the microprocessor doesn't contain memory or input/output hardware. Therefore, the microprocessor IC must communicate with separate memory and input/output ICs. In most microcomputers, a collection of binary signal wires connects the various ICs.

Buses

A *bus* is a collection of binary signal wires that together carry a binary number. Several electronic devices, often ICs, may connect to the bus to send or receive numbers. Usually a given bus has a particular purpose and has a name indicating that purpose.

Most microprocessors, including those manufactured by Motorola, use three buses to communicate between the microprocessor and other ICs in the microcomputer. First, the *address bus* carries address numbers from the microprocessor to other integrated circuits. Furthermore, the address bus never carries any other kind of number. Second, the *data bus* carries several kinds of data numbers between the microprocessor and other ICs in both directions. Finally, the *control bus* carries control signals between the microprocessor and other ICs. Some control signals go in both directions and some in only one direction.

The 68HC11 has a 16-bit address bus and an 8-bit data bus. The number of control signals depends on whether the design of the computer uses all the signals. Some signals are necessary only for advanced designs.

I/O Integrated Circuits

Input/output ICs transfer data numbers and control signals between the microprocessor and input/output devices. The input/output devices are external to the computer. The input/output ICs connect to both the microprocessor buses and the I/O devices.

Input/output ICs operate at logic signal power levels. Applications requiring higher power levels need additional external hardware. For example, power transistors are necessary to drive devices such as solenoids.

6.3   MEMORY CHARACTERISTICS

Memory technology falls into two broad categories called *read/write memory* and *read only memory*. Read/write memory allows the microprocessor to both write numbers into registers and read back those numbers later. Read only memories hold numbers in registers the microprocessor cannot alter—the numbers are permanent. Although the technologies in these two types of memory are different, the general principles of operation are similar.

Integrated circuit companies manufacture both read/write and read only memories using several different technologies. Memory ICs are available in many configurations and many packages with widely differing properties. Consequently, many technical terms describe these ICs.

The following section introduces the principles of integrated circuit memory technology and the related terms. The descriptions of the electronic circuits are simplified and, therefore, include only the necessary details.

Memory Terminology

Some very general terms describe the characteristics of memory devices. The terms are independent of the technology used to construct the memory. However, people commonly associate certain technologies with the general terms. Sometimes people use the technology name as a replacement for the general term. You may find this use of terms confusing.

Here are some general terms that describe memory:

- *Volatile memory.* A *volatile memory* loses the information stored in it when electrical power is removed from the memory. The information or numbers in a volatile memory evaporate when the power is turned off. Generally, losing information is a disadvantage. However, losing information is unimportant when the volatile memory holds only data numbers a running program uses. In other applications, volatile memory may hold the program instruction numbers if they can be restored easily from another device.

- *Nonvolatile memory.* A *nonvolatile memory* uses a storage technology that does not require electrical power to retain information. Most nonvolatile memories use a storage technique that makes the stored information permanent and unchangeable. Others use magnetic materials to store information, but few of these memories are used in practice.

- *Read/write memory.* A *read/write memory* uses a technology that allows the microprocessor to both store and retrieve numbers electronically at the full speed of the microprocessor. Some technologies are relatively slow, requiring many milliseconds to store a number. Therefore, these technologies are not read/write memories. Most read/write memories are also volatile because an electronic circuit must be active to retain the information. All microcomputers require some read/write memory to store data that changes as the program runs and to make a stack.

- *Read only memory.* A *read only memory* contains permanent information that cannot be changed by the microprocessor at the full speed of the microprocessor. Some technologies allow the microprocessor to change numbers in the read only memory, but the required time is usually many milliseconds. Most read only memories are also nonvolatile memories. Usually, read only memory contains the program instructions in a microcomputer used for a control application. Note that the name *read only* is an oxymoron because something must write to the memory or no information would be there to read! The writing is done either during the

manufacture of the memory or by some process that alters the physical characteristics of the memory.

- *Sequential access memory.* A *sequential access memory* reads and writes its
  registers in a sequential order. Magnetic tape is an example of sequential access
  memory because the tape must move to access the numbers. Some registers in a
  sequential access memory have longer access times than others. For example, the
  relative tape position affects the access time. A sequential access memory is not
  practical as the program storage memory of a microcomputer.

- *Random access memory.* Accessing a register in a random access memory takes
  the same time and effort as accessing any other randomly selected register. *Random
  access memories* are electronic memories. They may be either read only memory
  or read/write memory. Only random access memory is practical for program and
  data storage in a microcomputer.

The construction of the hardware determines the characteristics of these memories.
When you understand the operation of the electronic circuits, the characteristics of the various memories and their uses in practical applications will be clear.

Memory Principles

The operation of the electronic circuits in the integrated circuit determines the characteristics
of the memory. Understanding the memory characteristics is easy when you understand the
circuit inside. The construction and operation of a simple IC memory can be explained using
a diode matrix memory. The diode matrix memory then forms the basis for understanding
more complex hardware.

Diode matrix memory

The *diode matrix memory* is a nonvolatile read only memory. It is constructed from a
decoder, a grid of wires, and some diodes. Figure 6-2 shows a simple version of the circuit.
The numbers stored in a diode matrix memory depend on the wiring configuration of the
circuit. The following describes each component.

Decoder. The decoder in Figure 6-2 has two inputs that represent a 2-bit binary
number. This decoder has four outputs such that one output is high and the others are low.
The input number selects which output has the high level. This decoder is a one-out-of-four
decoder. Other decoders have more outputs, but the number of outputs is always a power
of two.

Diode. The diode in the diode matrix memory is an electronic device that passes
electrical current in one direction but not the other direction. A simple but effective model
for it is a short circuit of zero resistance when passing current, and an open circuit of nearly
infinite resistance when it is blocking current. The arrow head in the diode symbol in the
figure indicates the direction of low resistance.

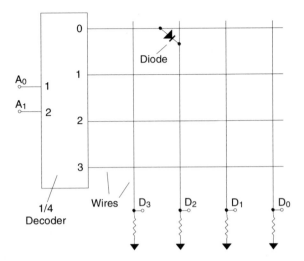

Figure 6-2   Diode matrix memory construction.

Wire grid.   The horizontal wires in the figure connect to the outputs of the decoder; they are not connected to the vertical wires. The vertical wires connect to the ground or common with a resistor to make a complete electrical circuit. The output terminals of the memory connect to the resistors. The terminals are labeled D_0 through D_3 because they provide the output data from the memory.

Operation.   Each horizontal wire in Figure 6-2 forms one register in the memory. The address that selects a register is applied to the two inputs of the decoder. These inputs are labeled A_0 and A_1 because they provide the address to the memory.

As an example, suppose that the input number or address is binary 00 because the wires are low. Then horizontal wire number 0 is high, and wires 1, 2, and 3 are low. Assume that high is +5 V_{DC} and low is 0 volts for this example. In Figure 6-2, terminals D_0, D_1, and D_3 are low because these terminals do not have a source of voltage. However, terminal D_2 is high because the diode connected to this vertical wire is passing electric current. Therefore, the terminals D_3 through D_0 are low, high, low, low, which can represent the binary number 0100. The placement of one diode on row 0 or address 0 made a single binary 1 in the output data number. The other horizontal rows did not affect the output.

Next, look at Figure 6-3. Additional diodes are connected to the other rows, but they do not affect the operation of row 0. For the next example, suppose that the input address is binary 01, so horizontal wire 1 is now high and the other horizontal wires are low. The diodes connected between horizontal wire 1 and output wires D_0 and D_1 pass current, so terminals D_0 and D_1 are high. The output data number is binary 0011.

Notice that the D_1 vertical wire has a second diode connected to it and then to horizontal wire 2. Making this connection with a wire instead of the diode would allow current to pass from horizontal wire 1 through horizontal wire 2 and make output terminal D_3 high. The diode prevents this because it passes current in only one direction. The diode prevents horizontal wire 1 from affecting terminal D_3. The diodes electrically isolate each horizontal wire from the others so that they do not interact.

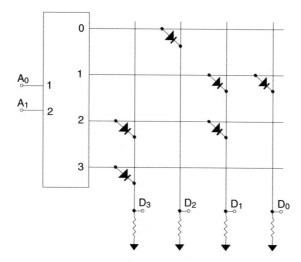

Figure 6-3 Diode matrix memory operation.

Each row or register in the diode matrix memory is programmed with a binary number by connecting diodes at the correct intersections. Figure 6-4 is a memory model using hexadecimal numbers for the memory shown in Figure 6-3. Adding more rows and columns makes a memory that holds more and larger numbers.

Characteristics.   The diode matrix memory is a read only memory. If it is used with a microprocessor, the microprocessor cannot change the pattern of diodes connected to the wire grid.

The diode matrix memory is also a nonvolatile memory. The information or numbers in the memory are stored by the presence or absence of diodes. If the electrical power is turned off to the memory, the electronic circuits stop working. However, the pattern of diodes is unchanged, so the information is not lost.

Using a pattern of diodes is a very reliable way to store information. Information is lost only through mechanical failure of the electronic components. Environmental factors such as electric or magnetic fields do not affect the pattern of diodes.

Flip flop memory

The simplest read/write memory uses flip flops for bit storage. The read/write memory uses a matrix to access the flip flops as the diode matrix memory accesses diodes. The electronics in the read/write memory is much more complicated than in the read only memory. In read/write memory, the matrix must route control signals to the flip flops. However, the diode matrix and read/write memories use the same principles of operation.

0	4
1	3
2	A
3	8

Figure 6-4 Memory model for diode matrix memory in Figure 6-3.

Read Only Memory ICs

The acronym for read only memory is ROM. Normally, people use the acronym *ROM* only to name an integrated circuit read only memory. Read only memory usually holds the program and constant data numbers in a microcomputer. Most control applications of microcomputers need the nonvolatile character of the ROM. Such applications need a nonvolatile memory, but people usually call the memory ROM because the ROMs are nonvolatile. They substitute *ROM* for *nonvolatile*.

The diode matrix memory is the basis for integrated circuit read only memories. Many variations of this design, including some with devices other than diodes, provide characteristics useful for a wide range of applications. This section discusses the common ROM integrated circuits including PROMs, EPROMs, and EEPROMs.

Masked ROM

The first practical diode matrix memories were constructed on printed circuit boards. The development of integrated circuit technology quickly replaced the printed circuits with ICs. A *masked ROM* is a custom integrated circuit containing a diode matrix memory. The ROM IC is manufactured with the necessary pattern of diodes in place. This electronic part is very reliable because it is a simple circuit in a highly reliable form; namely, the integrated circuit. Sometimes the memory electronics is incorporated into larger chips such as a single-chip microcomputer.

Usually, a large number of identical ICs are purchased from an integrated circuit manufacturer. The purchase order must document the required numbers. The IC manufacturer must design and manufacture the masks required as part of the manufacturing process. Making the masks is a costly and time-consuming task. After the masks are made, the production of large numbers of ICs can proceed. If many thousands of ICs are manufactured, the cost of the masks and the cost of putting the IC into production are inconsequential. However, if the number is small, the cost of making the masks is prohibitive.

A typical application for a masked ROM is a high-volume product containing a microcomputer. Examples include microwave ovens, cassette tape decks, and other consumer products.

Fusible link PROM

When only a few ROMs are required for laboratory development or for low-volume products, the masked ROM is impractical. A ROM memory that can be field programmed and used immediately is preferred. Such an IC is known by the acronym *PROM*, from field programmable read only memory. The user purchases the PROM as a standard IC and then programs it. If the required number of parts is relatively small, the cost of the ICs and the programming cost compare favorably with the cost of a custom-masked ROM.

Programming.   The first integrated circuit PROM was the fusible link PROM. It is a diode matrix memory with fusible links in series with the diodes. The matrix is completely filled with diodes and links when the integrated circuit is manufactured. The programming process disconnects selected diodes by removing their links.

A suitably large electric current will destroy a link. The links have resistance which converts electrical energy to heat. Therefore, the programming current causes the selected links within the chip to overheat and burn out. Consequently, programming these PROMs is called *burning*. An electronic PROM programmer can selectively burn the correct links when given the hexadecimal numbers for the registers. The chip has circuits to allow access to selected links by the programming device.

Once the PROM is programmed, the numbers are permanent. Certainly the microprocessor will not change the numbers in the PROM. To correct programming errors, you will usually discard the IC and program a new one. If an error leaves links in place, you can program the IC again to remove the incorrect links. You cannot restore a burned link.

EPROM

One disadvantage of the fusible link PROM is that changes in the stored numbers cannot be made. An incorrect IC must be discarded and replaced with a new IC.

The numbers in some ROM memory ICs can be changed. Usually, you must erase all the numbers in the chip, and then program all the numbers again. Such an erasable programmable read only memory is known by the acronym *EPROM*.

Storage mechanism.   The EPROM depends on the charge storage capability of a capacitor. In principle, an FET, or field effect transistor, replaces the diode in the diode matrix memory. The transistor acts either as the diode or as an open circuit. The transistor does one or the other depending on whether its gate is electrically charged. The gate is effectively a capacitor plate encased in a very good electrical insulator similar to glass. Once charged properly, the capacitor will retain its charge for many years.

Programming.   An electronic PROM programmer and the electronics in the chip electrically pulse the selected capacitor to charge it. The pulsing limits the heating of the chip to prevent damage. Usually, a microcomputer in the programmer optimizes the charging of each capacitor. The characteristics of the capacitors vary greatly depending on their location on the chip. The PROM programmer is given the hexadecimal numbers needed, and it automatically programs each capacitor. Usually, programming an EPROM IC requires a few minutes. Programming the IC is called *burning*, although the programming process is nondestructive.

Erasing.   Discharging the capacitors erases the IC. Shining ultraviolet light on the surface of the chip causes the charge to leak off the capacitors. The photons of light provide enough energy to allow the charge to pass through the insulator and leak off. Erasing usually takes considerable time, often as long as an hour of exposure to intense ultraviolet light. Incidentally, the photons of white or infrared light do not have enough energy to erase the IC, so ambient light does not affect EPROMs. Sunlight does not affect EPROMs, because the light intensity is too low.

Most EPROMs can be erased and programmed 100 times or more before they fail, but wide variations occur between practical parts. The programming process causes thermal stresses that can lead to eventual failure.

Packaging. The need to shine light on the chip means that the IC package must have a window to allow light to reach the chip. This window, usually made of quartz, adds to the cost of the IC. In some applications where the EPROM is part of a high-volume product, it is desirable to program the memory during the assembly of the product. Such a product will not change after the customer purchases it, so erasing is unnecessary. Then the EPROM package is made without a window to reduce cost.

The EPROM has an important disadvantage—it must be removed from the microcomputer for erasing or programming. The physical removal and replacement are inconvenient. Furthermore, to allow removal, the EPROM is mounted into a socket. Sockets are costly and less reliable than soldering. Finally, the PROM programmer device required to program the EPROM is costly.

EEPROM

Many applications require the microcomputer to store information in a nonvolatile memory. The device containing the microcomputer needs the information after it is turned off and then on again. For example, a device may measure the total amount of time it has operated so that it can recommend timely maintenance. The device must save the accumulated time when the power is turned off. Some devices use a volatile memory with a battery to maintain power to the memory. However, the battery and its associated hardware are costly and unreliable. An alternative is the electrically erasable programmable read only memory IC called the *EEPROM*. Some companies name these integrated circuits FLASH memory. The microprocessor electrically programs and erases the EEPROM, yet it is a nonvolatile memory.

Storage mechanism. The storage mechanism in the EEPROM is the same as the EPROM—electrical charge on a capacitor. However, the electronics to program the chip by charging the capacitors must now reside in the memory chip.

Programming. The microprocessor controls the programming electronics inside the EEPROM IC through software. In addition, programming some EEPROMs requires a separate voltage source. Usually, the required programming voltage is higher than the usual +5 VDC power supply. In some applications, the lack of a higher voltage source ensures that the EEPROM is never changed.

Erasing. The erasing circuit is inside the EEPROM integrated circuit. The microprocessor controls the erasing hardware through software. Some ICs have a complex erasing procedure, so accidental erasure is unlikely. Some EEPROMs erase individual bytes; others must erase the entire IC.

Packaging. The usual integrated circuit package holds the EEPROM. The only distinguishing characteristic is the part number. The control signals are similar to other memory integrated circuits. However, some EEPROMs have a pin devoted to a programming voltage supply. Other ICs create the programming voltage internally.

The EEPROM is not a read/write memory. The EEPROM erasing and programming procedures require times of a few milliseconds—much longer than the response time of flip flops. Furthermore, the lifetime, measured in erase and programming cycles, is limited. The typical lifetime of an EEPROM is many thousands of these cycles. Normal reading of the EEPROM does not affect its lifetime significantly.

Read/Write Memory ICs

Most read/write memory integrated circuits, called *RAMs*, are volatile memories. They usually hold data numbers that change as the program runs. In some applications, RAMs also may hold the program. Programming techniques such as a stack require the use of read/write memory.

RAM integrated circuits store information using two different technologies. The two types are called *static RAM* and *dynamic RAM*. Usually, static RAMs are faster and more expensive than equivalent-size dynamic RAMs.

Static RAM

The *static RAM* uses flip flops to hold the information. The flip flops require electrical power to operate and thus retain information. The flip flop circuit requires many electronic components, so the flip flops require a relatively large area on a chip.

Dynamic RAM

The *dynamic RAM*, also called *DRAM*, uses a tiny capacitor to hold charge that represents the stored information. The capacitor is part of an electronic circuit that both controls the stored charge and reads the voltage on the capacitor. Because the chip must hold thousands of capacitors, each capacitor is very tiny. Therefore, the stored charge is small and is quickly lost through the connected circuitry. Usually the charge becomes unreadable in a matter of a few milliseconds. Therefore, the dynamic RAM requires additional circuitry to *refresh* the capacitors to their original charge levels periodically. The refresh electronics adds complexity to the memory. However, a given area of a chip can hold many more dynamic RAM bits than static RAM bits. This extra capacity makes the dynamic RAM very attractive. For large memories, the reduced cost of fewer ICs easily offsets the cost of the refresh electronics.

Matching Software to Memory

Most microcomputer designs require parts of the memory to have different characteristics. First, most applications require nonvolatile memory to hold the program instructions and constant data values. Then the program is always available when the power is on. This memory is usually read only memory because it is nonvolatile. However, the program must then have pure procedure code—self-modifying programs are impossible. Furthermore, the selection of the type of ROM may depend on the particular characteristics of the ROM.

Second, most applications require some read/write memory for data that the program changes as it runs. The read/write memory usually is a volatile RAM memory. The program must not depend on volatile information at power up. Therefore, most microcomputers have both ROM and RAM memory.

6.4 MICROPROCESSOR BUSES

The microprocessor discussed in this chapter is the 68HC11 operating in a mode that allows access to the buses. Some details were omitted for clarity. Chapter 7 more fully describes the 68HC11 modes and further hardware details.

Bus Characteristics

A *bus* is a collection of binary signal wires used together for a specific purpose. Sometimes a bus carries a binary number from one place to another. Sometimes a bus carries a group of control signals that work together to control a device.

Most microprocessors use three buses to connect the microprocessor integrated circuit to the other integrated circuits in the microcomputer. In particular, the other ICs are mostly memory and input/output ICs. A single-chip microcomputer uses the same buses, but they are inside the chip.

Figure 6-5 illustrates the three buses connecting the parts of the microcomputer. Usually the clock, not shown in the figure, connects directly to the microprocessor. The figure shows only the wires at the extreme left and right of the buses instead of every wire.

Address bus

The purpose of the address bus is to carry addresses from the microprocessor to the memory and input/output ICs. The microprocessor never has an address sent into it. Instead, it always forms the addresses that are on the address bus. The other ICs only use the addresses on the address bus. The figure implies this use of the address bus because arrow heads point from the microprocessor to other ICs.

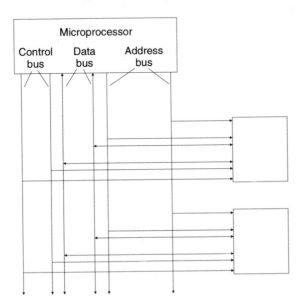

Figure 6-5 Microprocessor buses.

A 16-bit address bus can carry $65,536_{10}$, or 64K, different addresses. That is, it can directly control memory with addresses from 0000 to FFFF. If a design requires more memory than 64K, a switching technique enables memory in blocks. On the other hand, installing less than 64K of memory to reduce cost is common. Then some addresses can't be used. Some designs also omit some address wires to further reduce cost.

Data bus

The purpose of the *data bus* is to carry numbers between the microprocessor and the memory and input/output ICs. The meaning of the numbers can be almost anything, so they are called *data numbers*. For example, when the microprocessor gets numbers from the memory during the fetch phase of an instruction, the numbers are parts of instructions. When the microprocessor executes a STAA instruction, it sends the number from the accumulator to the memory on the data bus.

The data bus is a *bidirectional* bus because the same wires carry numbers in two different directions. The microprocessor sends a data number on the data bus at one time, and then it receives a number on the data bus at another time. The data bus is *time multiplexed* because its job is different at different times.

Control bus

The *control bus* contains several signal wires primarily for the microprocessor to control the memory and input/output ICs. However, some ICs send signals to the microprocessor on some control bus lines. The buses for various microprocessors from different companies are similar, but the greatest differences between manufacturers are in the character of these control signals.

Microprocessor Bus Connections

Consider now connecting the microprocessor buses to a read/write memory integrated circuit. As an example, assume that the memory IC contains four kilobytes of static RAM registers. The memory IC has 12_{10} address pins and 8 data pins.

Address bus connections

Figure 6-6 illustrates the connection of the microprocessor to the address pins of the memory IC. The address line notation is the same as the common bit numbering notation. The memory IC has address lines A_0 through A_{11}. The microprocessor controls these lines to select a particular register within the memory IC.

Data bus connections

Figure 6-7 illustrates the address and data buses connected to a memory IC. The data bus pins on the microprocessor connect to the data bus pins on the memory IC. When the microprocessor writes to a memory register, it turns the data bus direction to outgoing from the microprocessor. When the microprocessor reads from the memory, it turns the data bus

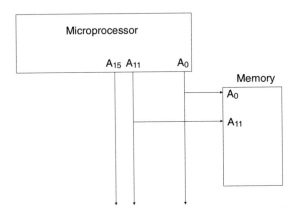

Figure 6-6   Address bus connections.

direction to ingoing to the microprocessor. Notice that the communication to the memory is always eight bits at a time. Again, the figure shows only the left and right bus wires instead of every bus wire.

Control bus connections

The control bus has several individual control wires. First, Figure 6-8 illustrates the connection of the single read/write, or R/W, control signal from the microprocessor to the memory IC. The microprocessor controls the R/W line to tell the memory whether to read or write. Usually the high level means read and the low level means write. The level on the R/W line corresponds to the direction of the data bus. When the microprocessor signals a read operation, the data bus direction is ingoing to the microprocessor. When the microprocessor signals a write operation, the data bus is outgoing from the microprocessor.

Second, Figure 6-8 illustrates the connection of the E-clock signal from the microprocessor to the memory IC. All operations of the buses must be coordinated in time. The microprocessor sends a high-going pulse called the *E-clock*. The E-clock signal enables the memory IC at the correct time using its chip enable or CE pin. Therefore, the memory IC

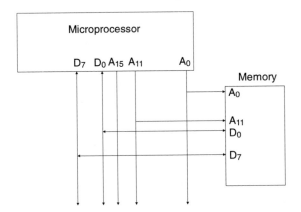

Figure 6-7   Data bus connections are added.

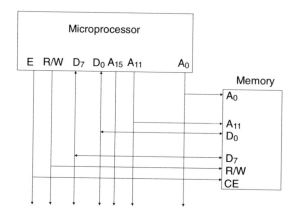

Figure 6-8   Read/write and E-clock control signals are added.

responds only when the signals on the buses are stable and unchanging. Of course, if the memory IC responds while the buses are changing, it gets erroneous information.

Bus Operation Example

Let's follow the signals in the computer as it fetches and executes an instruction. Chapter 2 uses this exercise as a block diagram to explain the operation of the microprocessor. This section adds much more detail showing the flow of numbers on the buses. The following accurately describes the operation of the computer.

The instruction

Figure 6-9 shows the read/write memory containing an extended addressed STAA instruction and the number 27 in the A accumulator. The instruction op code is B7, and the extended address is 04FE. This instruction will store the 27 from the A accumulator register into memory register 04FE. The memory register now contains the number 56.

Fetch phase

The numbers in Figure 6-9 are the register contents during the first clock cycle as the microprocessor fetches the instruction. On the first tick, the microprocessor sends the number in the program counter down the address bus. That is, it connects the outputs of the flip flops in the program counter to the microprocessor address bus pins. Because the memory IC has a 12-bit address, only the number on address lines A_0 through A_{11} tell the memory which register to access. So, it accesses address 412. Simultaneously, the microprocessor turns its data bus pins inward so that it can read a data number supplied by the memory IC. The microprocessor also sends the high or read level on the R/W line to tell the memory IC to do a read operation. When these signals are stable and unchanging, the microprocessor sends a pulse on the E-clock line to tell the memory to respond.

The memory responds by controlling the data bus lines and putting the number B7 on the data bus. The microprocessor routes the number that comes to it on the data bus to the

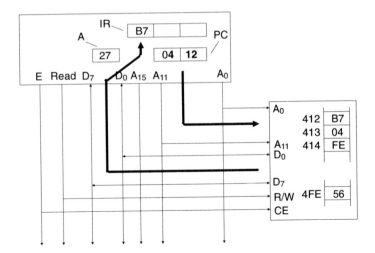

Figure 6-9 Fetching an instruction op code.

instruction register. Then the microprocessor increments the number in the program counter to 0413.

On the next tick of the clock, the same procedure brings the number 04 from address 0413 to the instruction register. Then, on the final tick of the fetch phase, the same procedure brings the number FE from address 0414 to the instruction register.

At the completion of the fetch phase, the three bytes of the instruction code have been transferred over the data bus to the microprocessor. The program counter is pointing to the next instruction because it has been incremented during the fetch phase.

Execute phase

Figure 6-10 illustrates the register contents during the execute phase of the instruction operation. The execute phase of this instruction requires only a single clock cycle.

On the next tick of the clock, the microprocessor examines the B7 op code in the instruction register and learns that it must perform a store operation with the A accumulator. The address of the memory register where the data is to be stored is in the instruction register. The microprocessor connects the 16_{10} instruction register bits that contain the address of the data to the address bus pins. The memory IC receives the 12-bit address 4FE on the address bus.

Simultaneously, the microprocessor turns its data bus outward and connects the outputs of the A accumulator flip flops to the data bus pins. The microprocessor then sends the number 27 on the data bus. The microprocessor also sets the R/W line to the low level to tell the memory IC to perform a write operation.

When the signals are stable and unchanging, the microprocessor sends a pulse on the E-clock line. The E-clock pulse tells the memory to respond to the signals from the microprocessor. When the memory responds, it stores the number 27 in register 4FE.

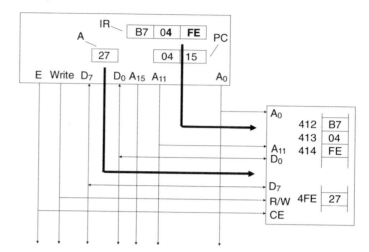

Figure 6-10   Executing a store accumulator instruction.

Computer Failures

Certain failures of the computer hardware may be anticipated. This section simulates some failures that may occur as the microprocessor fetches and executes the instruction discussed above. Of course, many more and very complex failures can occur. We will consider only very simple failures.

Grounded address line

Let's assume that the failure is a short to ground of the address line A_0 as Figure 6-11 shows. The ground simulates a low level or a logic 0 on that line. This failure could occur because some foreign material fell onto the circuit, making the undesired connection. Similarly, a failure of the electronics inside the memory chip connected to A_0 could act the same.

Further assume that this failure causes no further failures in the integrated circuits; they continue to operate normally. Return to Figure 6-9 to observe the fetch phase of the instruction.

On the first clock tick, the address sent on the address bus is 0412. The number 0412 has a 0 at bit 0. Therefore, the result with the grounded line is the same as if no failure occurred. The microprocessor correctly fetches the instruction op code B7.

On the second clock tick, the address sent on the address bus is 0413, but the ground failure causes the address to be 0412 again. The microprocessor incorrectly fetches the next byte of the instruction from address 0412 and gets B7 again.

On the third clock tick, the address is 0414, which the ground failure does not affect. The microprocessor correctly fetches the number FE. The resulting instruction code in the instruction register is B7B7FE, which is clearly incorrect.

On the fourth tick, the execute phase of the instruction sends the address B7FE on the address bus. The grounded line does not affect this address. Therefore, the microprocessor

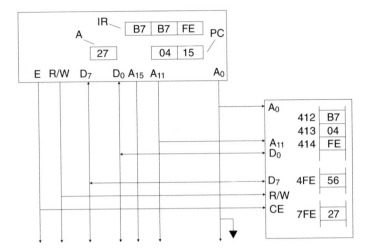

Figure 6-11   Effect of instruction fetched and executed with grounded A_0 line.

stores 27 from the A accumulator at address B7FE instead of at address 04FE. Figure 6-11
shows the final numbers after fetching and executing the instruction are complete.

Clearly, any program running in this computer is going to do strange and bizarre
things!

Grounded data bus line

Consider another failure similar to the grounded address line; only now let's ground
data bus line D_0. Figure 6-12 shows this condition. When the microprocessor fetches the
instruction at address 0412, the op code is B6 instead of B7. The complete instruction code
fetched is B604FE. This instruction is a LDAA instruction instead of the programmed STAA
instruction. Besides doing an incorrect operation, this erroneous instruction alters the number in the A accumulator by loading 3C into it. Figure 6-12 shows the numbers after this
instruction is fetched and executed.

As with the grounded address line, a ground on a data bus line can cause both incorrect
instructions and data numbers. Again, this computer will do strange and bizarre things that
are completely unpredictable.

Other failures

The failures considered here are very simple indeed. Much more complex failures can
and do occur. However, if you consider only this instruction, moving the ground to other
address or data lines will completely change the effects of the failures. When intermittent
failures are also possible, the problem becomes greater yet. This example illustrates why
computers can do things that are almost unexplainable. Such failures are a big problem when
a computer controls equipment dangerous to people and other equipment.

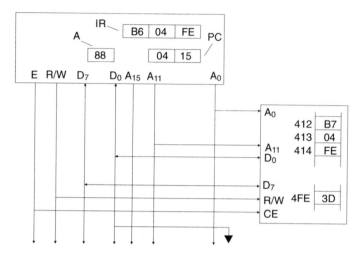

Figure 6-12   Effect of instruction fetched and executed with grounded D_0 line.

Furthermore, software in the computer can never check all the computer hardware for failures when the hardware is unpredictable. Safety will require an external hardware solution. Chapter 7 introduces some hardware to check for safe operation of the computer.

Memory Expansion

A single memory IC usually is not enough for a practical microcomputer. Additional ICs provide more memory and different types of memory. However, hardware signals must control each individual memory IC.

Memory chips in parallel

Look at the two RAM memory ICs in Figure 6-13. Each IC has the same connections to the microprocessor. If the two memory ICs contain the same numbers, they will operate in parallel correctly, but the second IC is useless. If the two ICs contain different numbers, conflicts will occur, and the hardware will not work correctly. Some failures in computer hardware make multiple ICs operate in parallel, so this exercise is useful.

Both ICs in Figure 6-13 respond to addresses in the range from 0000 to 0FFF. They would be much more useful if the top IC responded to these addresses and the second IC responded to addresses 1000 through 1FFF. That is, we want the first 4K of memory in the first IC and the second 4K in the second IC.

Address decoding

To make the ICs respond to correct addresses, a switch must enable the top IC in the address range 0000 through 0FFF while disabling the second IC in this range. Likewise, the switch must disable the first IC for the range 1000 to 1FFF while enabling the second IC.

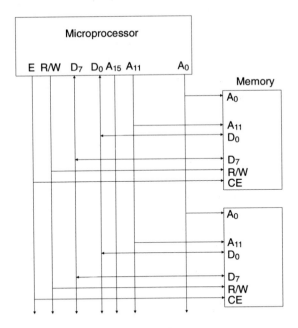

Figure 6-13   Two 4-kilobyte RAM chips in parallel.

Examination of the two address ranges shows that address line A_{12} must control the switch. Bit 12 of the address is the only bit that is different in the two address ranges.

Because the E-clock pulse makes the memory IC respond, the switch must transfer the E pulse to the memory IC only in the correct range of addresses. Figure 6-14 shows a circuit using two AND gates to switch between the two ICs. These gates are called *address decoders* because they determine the range of addresses that can enable the memory IC.

The upper AND gate in the figure passes the E-clock signal to the memory IC if line A_{12} is low. At the same time, the lower AND gate passes the E-clock to the bottom memory IC only when the A_{12} line is high. Therefore, the memory ICs respond only in the correct address ranges.

Adding more memory ICs to the computer in Figure 6-14 requires address decoder gates with many more inputs. A common alternative uses a one-out-of-sixteen decoder IC to divide the memory into 16_{10} ranges. The decoder connects to address lines A_{12} through A_{15}. The selected output of the decoder is a control input to the address decoder gate. A later section on input/output hardware illustrates this technique.

6.5   PARALLEL I/O PRINCIPLES

When hardware transfers all bits of a number at the same time, the hardware does a *parallel data transfer*. The microprocessor inherently does parallel transfers because the data bus transfers all eight bits at once. The following section describes hardware that transfers data to I/O devices in parallel.

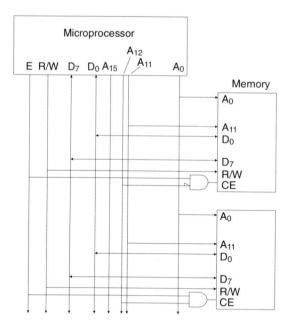

Figure 6-14   Address decoder gates to enable memory chips.

I/O Programming Model

Input/output involves two major operations; namely, *data transfer* and *timing*. Data transfer is the sending of data numbers to output devices or receiving data numbers from input devices. The input and output devices are external to the computer hardware. Timing refers to synchronizing the data transfer to the I/O device with the program running in the computer. The program must cause a data transfer only at the time the I/O device is ready for a data transfer. At other times, the I/O device cannot respond correctly to data transfers.

Data transfer

Figure 6-15 shows a programming model for the I/O section of the computer. Two register boxes represent the data transfer mechanism in the I/O section. The word *port* refers to a place where information enters or leaves the computer. In particular, numbers move between the input and output devices through these 8-bit ports.

Input port.   A transfer of data from the input pins to the microprocessor register occurs at the time an instruction is executed. A LDAA instruction transfers the data to the A accumulator on the E-clock pulse during the execute phase. The input port does not contain a register to hold input data. The data is transferred instantaneously at the time of the E-clock pulse.

Output port or register.   The output port contains a register that controls the pins on the output port hardware. The data transfer is done by an instruction such as STAA. The register is necessary because the data only exists on the data bus during the instruction execution. The data is transferred to the register at the time of the E-clock pulse. The register

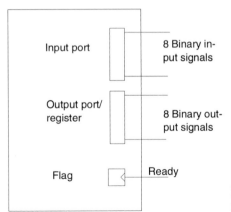

Figure 6-15   Input/output programming model.

then holds the signal voltages at the pins so that the output device can respond. The register holds the signals until another output operation is executed.

Timing or synchronization

Two methods of synchronizing I/O devices to the computer program are used, but both use the flag device shown in Figure 6-15.

Flag.   The *flag* is a single flip flop that is set to 1 by a ready signal from the I/O device. Setting the flag tells the program that the I/O device is ready for a data transfer.

Polling.   The flowchart in Figure 6-16 illustrates the *polling* technique of input/output device synchronization. The program repeatedly tests or polls the I/O flag to determine when the I/O device is ready for a data transfer. When the program detects that the flag is set, the program clears the flag and controls the I/O hardware to do the data transfer.

The main advantage of the polling technique is its simplicity when used with simple problems. When a sophisticated system using many I/O devices is designed, the polling technique becomes very complex and cumbersome. Polling also may degrade the performance of the I/O devices due to the polling overhead. A more sophisticated technique called *interrupt* is then used. A later section discusses interrupt.

6.6   PARALLEL I/O HARDWARE

The Motorola 68HC11 uses the address, data, and control buses to control input/output hardware as if it were memory. The technique of controlling I/O hardware the same as memory is called *memory-mapped I/O*. The input and output ports have memory addresses just as memory registers do. Memory-mapped I/O hardware connects to the microprocessor buses the same as does memory.

The same instructions used with memory control the input/output operations. The 68HC11 does not have any instructions specifically for input/output. However, do not assume

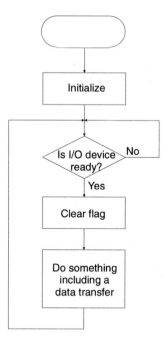

Figure 6-16 Example using polling for I/O synchronization.

that the I/O registers act like memory. Remember that an input port reads binary signals on pins of an IC—it does not read back a number previously stored in a register! For emphasis, certainly a program that fetches instructions from I/O ports will fail.

The example in Figure 6-17 illustrates some consequences of memory-mapped I/O. The memory map shows the placement of various kinds of memory and I/O registers in the memory space. It further illustrates that hardware may not be installed for all memory locations, so some addresses are unused.

Figure 6-17 Example of a practical memory map.

I/O Circuit Construction

Parallel input/output hardware is built from several different types of integrated circuits. The approach used depends upon the particular application. Sometimes using the least hardware possible reduces cost. The least cost design includes hardware for only the necessary signals. Other applications may require more flexibility because the application is not well known or may change. Flexibility is often gained by adding some redundant hardware.

Output port design

Let's design an 8-bit output port for use with a Motorola 8-bit microprocessor. Assign address 4003 to this port. Additionally, assume that address lines A_2 through A_{11} are not decoded. Further assume that no other writable device in the address range from 4000 through 4FFF has binary 11 for bits 0 and 1 of its address.

Figure 6-18 shows the circuit. Not all the address lines are decoded—some are not connected—to reduce cost and complexity. Therefore, more than one address can make the port respond because the unconnected wires make no difference. A bit value of 0 is commonly assigned to the unconnected lines.

Output register.   An 8-bit register is the principal component of the output port. The input bus to the register connects to the microprocessor data bus. The register must be clocked when the correct address appears on the address bus, the R/W signal indicates write, and a pulse occurs on the E-clock line. Clocking the register at this time transfers the correct data from the data bus into the output port register. The output register then controls the output pins sending the data to the output device outside the computer.

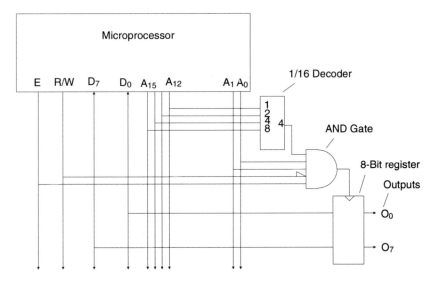

Figure 6-18   Output port circuit.

Address range decoder.   The purpose of the decoder is to indicate when the address on the address bus is within a certain range of addresses. Further decoding is usually necessary so that only the correct address makes the port respond.

In Figure 6-18, the four inputs to the one-out-of-sixteen decoder connect to address bus lines A_{12} through A_{15}. Each output of the decoder asserts only when the address is within a certain range. Output 4 asserts when the address is in the range 4000 through 4FFF. Of course, the other decoder outputs correspond to other address ranges. For example, output 15 corresponds to the range F000 to FFFF.

Final decoder.   Output 4 of the decoder is ANDed with the other necessary signals to make the output port register respond correctly. Because bits A_0 and A_1 must both be 1s when the port responds, these address lines are connected to the AND gate. To assure that the register only responds on a write signal, the R/W line is connected to a low-asserted input of the AND gate. The E-clock pulse line makes a high-going pulse when the other signals are valid. Therefore, the output of the AND gate forms a pulse to clock the register.

In this example, not all address lines are decoded; address lines A_2 through A_{11} are not connected. Not decoding all addresses reduces cost and complexity. However, the port responds to more than one address. People usually assign 0s to unconnected address lines; then, the port has only one address.

Read back.   The output port in Figure 6-18 cannot be read by the program. The program can store into the register, but it cannot read the register. Therefore, the program must remember what was last output by saving a copy in memory. Many practical output ports have this limitation. Additional hardware is necessary if the program must read the output port.

Design defect.   A serious defect in this design is the lack of a hardware signal to initialize the output register at power-up. When the electronics is turned on, the flip flops in the register have unknown states. Because the output port may control some dangerous equipment, unknown states are unacceptable in practice. Though software may initialize the output port soon after power-up, depending on the software for initial safety is unacceptable.

Input port design

Now let's design an 8-bit input port for a Motorola 8-bit microprocessor. The design will be similar to the design of the output port. Assign address 4002 to this input port and leave address lines A_2 through A_{11} unconnected. Assume that no other readable device will have an address in the range 4000 through 4FFF that also has binary 10 for bits 0 and 1. Figure 6-19 shows the circuit.

Input transfer gates.   The principal component of the input port is a set of eight transfer gates with tri-state outputs. A tri-state output will be high, low, or disconnected. When the device transfers, the output level is the same as the input level. When not transferring, it disconnects the output. The tri-state device is necessary because more than one controlling device connects to the data bus, but only one can control the bus at a time. Those devices not controlling the bus must disconnect from it.

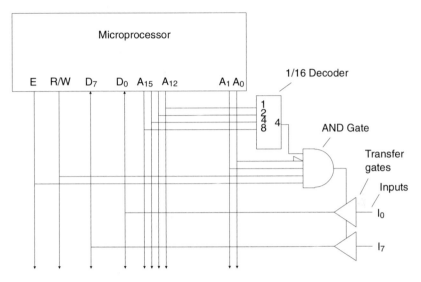

Figure 6-19   Input port circuit.

The transfer gates must transfer the input signals to the data bus at the correct time. The transfer must occur when the correct address appears on the address bus, the R/W signal says read, and the E-clock pulse occurs.

Address range decoder.   The address decoding for the input port example is the same as the output port example. Output 4 asserts when the address on the address bus is in the range 4000 through 4FFF.

Final decoder.   The principle of the final decoder is the same as the decoder for the output example. However, the address it decodes is different because the least-significant two bits must be 10. Therefore, a low-asserted input of the AND gate connects to A_0.

Programming example

Let's use the input and output ports in Figures 6-18 and 6-19. Figure 6-20 shows a high-asserting push button switch connected to input bit 6, and a low-asserting switch connected to bit 2. A high-asserting indicator light is connected to output bit 5. The example program asserts the light only if both push buttons are pushed. The program runs in an infinite loop continually controlling the light.

Figure 6-21 is the listing of the program. Here is a description of it.

- *Line 20.*   The memory register labeled IMAGE holds a copy of the last number sent to the output port. Sometimes the program needs to change only a few of the output bits, but the hardware controls all bits at once. Therefore, the program must know all the output bits. Because the program cannot read the output port, it must keep a copy of the output bits in memory. The copy provides the unchanged bits. The

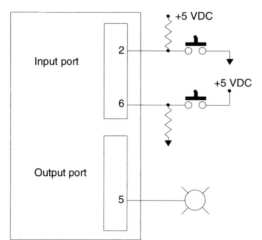

Figure 6-20 Input port connected to switches and output port connected to light.

location IMAGE is in the direct addressing range so that the bit instructions can easily access it.

- *Lines 27 through 29.* Set the output bits to 1s in both the memory register labeled IMAGE and in the output hardware. First, setting these bits shows that the output port should be set to some known initial condition. The output port used here has no hardware to initialize it. Second, setting these outputs to 1s implies that we want them to stay 1s. The rest of the program should only modify the output bit for the light. Carefully avoid changing output bits that should remain the same.

- *Line 31.* Point the index register at the input port. The branch-on-bit-condition instructions can only reach the input port with indexed addressing.

- *Lines 32 and 33.* Test the two input bits corresponding to the switches. A branch occurs if a switch is released. Because the switch at bit 6 is high-asserted, the branch occurs if the input bit is 0. The switch at bit 2 is low-asserted, so the branch occurs when the input bit is 1.

- *Line 35.* If a branch did not occur at either line 32 or line 33, then both switches are pushed and control comes to this line. The bit set instruction updates the IMAGE register to assert the light.

- *Line 38.* If a branch occurs on either line 32 or line 33, at least one switch is released. The bit clear instruction updates the IMAGE register to deassert the light.

- *Lines 40 and 41.* Transfer the output bit pattern in the IMAGE register to the output port hardware to control the light. Additions to the program that change other IMAGE bits cause no problems, because the program updates all output bits at once.

- *Line 42.* The program loops back to run in an infinite loop. Most control programs effectively run in infinite loops.

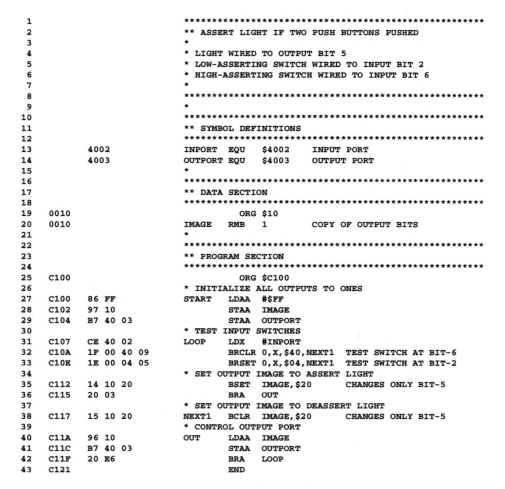

```
 1                              ********************************************************
 2                              ** ASSERT LIGHT IF TWO PUSH BUTTONS PUSHED
 3                              *
 4                              * LIGHT WIRED TO OUTPUT BIT 5
 5                              * LOW-ASSERTING SWITCH WIRED TO INPUT BIT 2
 6                              * HIGH-ASSERTING SWITCH WIRED TO INPUT BIT 6
 7                              *
 8                              ********************************************************
 9                              *
10                              ********************************************************
11                              ** SYMBOL DEFINITIONS
12                              ********************************************************
13        4002                  INPORT  EQU    $4002      INPUT PORT
14        4003                  OUTPORT EQU    $4003      OUTPUT PORT
15                              *
16                              ********************************************************
17                              ** DATA SECTION
18                              ********************************************************
19   0010                                ORG    $10
20   0010                       IMAGE   RMB    1          COPY OF OUTPUT BITS
21                              *
22                              ********************************************************
23                              ** PROGRAM SECTION
24                              ********************************************************
25   C100                                ORG    $C100
26                              * INITIALIZE ALL OUTPUTS TO ONES
27   C100   86 FF               START   LDAA   #$FF
28   C102   97 10                        STAA   IMAGE
29   C104   B7 40 03                     STAA   OUTPORT
30                              * TEST INPUT SWITCHES
31   C107   CE 40 02            LOOP    LDX    #INPORT
32   C10A   1F 00 40 09                  BRCLR  0,X,$40,NEXT1   TEST SWITCH AT BIT-6
33   C10E   1E 00 04 05                  BRSET  0,X,$04,NEXT1   TEST SWITCH AT BIT-2
34                              * SET OUTPUT IMAGE TO ASSERT LIGHT
35   C112   14 10 20                     BSET   IMAGE,$20       CHANGES ONLY BIT-5
36   C115   20 03                        BRA    OUT
37                              * SET OUTPUT IMAGE TO DEASSERT LIGHT
38   C117   15 10 20            NEXT1   BCLR   IMAGE,$20       CHANGES ONLY BIT-5
39                              * CONTROL OUTPUT PORT
40   C11A   96 10               OUT     LDAA   IMAGE
41   C11C   B7 40 03                     STAA   OUTPORT
42   C11F   20 E6                        BRA    LOOP
43   C121                                END
```

Figure 6-21   Program using an input port and an output port.

Flag design

The I/O flag may consist of a single D flip flop as shown in Figure 6-22. The previous figures show the microprocessor to the left of the I/O hardware. Therefore, Figure 6-22 shows the flip flop output signal on the left and the ready signal on the right.

The D lead is always high, so clocking the flip flop always sets it. The ready signal from the input/output device clocks the flip flop to set the flag. The computer tests the output of the flip flop to determine whether the I/O device is ready for a data transfer. The computer hardware treats this flag signal as an input bit. The computer controls the direct clear lead on the flip flop to clear the flag. However, common input ports and output ports generate a clear signal for the flag when the program reads or writes the port. Then a separate clear-the-flag operation is unnecessary.

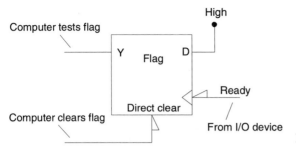

Figure 6-22   Input/output flag.

Programmable I/O Hardware

Input/output circuits are built from integrated circuits. Usually, an I/C contains some input ports, output ports, and flags. Whatever the selection of hardware, some customers will find the selection ideal and others will find it unsatisfactory. Some customers will always need a different hardware configuration. The configuration customers need usually changes as a design project proceeds. One fixed design is not adequate for all customers.

The major limitation on integrated circuit design is often the number of pins on the package. The package will hold more electronics than there are pins to use. Therefore, for each pin, the manufacturer may put both input electronics and output electronics in the chip. Then internal electronic switches can connect the correct hardware to the pins. Bits in a control register control the switches. The purchaser of the IC programs it to use the required inputs and outputs by placing numbers in control registers. While using the same IC, each customer can have the unique hardware required for their application. Furthermore, they can adapt to new requirements by changing the numbers in the control registers.

The extra complexity of the programmable I/O hardware does not greatly increase the cost. The cost of ICs decreases as the number sold increases. Because many different customers can use the same programmable hardware, the increased sales volume offsets any extra cost due to its programmability. However, the programmer must learn to program the I/O hardware configuration before using it. The programmability adds a little extra complexity to the program.

6.7 68HC11 PARALLEL I/O HARDWARE

The 68HC11 has five 8-bit I/O ports named *PORTA* through *PORTE*. These ports correspond to collections of pins on the 68HC11 IC that carry I/O signals.

The five ports and the buses are all inside the 68HC11 integrated circuit package. The registers related to the I/O hardware have addresses determined by the internal connections. The I/O registers use a block of addresses beginning at address 1000.

One of the internal ports, *PORTC*, is a programmable parallel I/O port. In contrast, *PORTB* is an output-only port. The other ports are dedicated to I/O devices in the 68HC11 chip. However, the other ports also can be parallel I/O ports. The following discusses only the PORTB and PORTC ports and the associated control registers called *DDRC* and *PIOC*. Chapter 7 discusses all the parallel I/O ports.

The PORTB Register

The 68HC11 port PORTB is an 8-bit output-only port. It is not programmable. The PORTB register is at address 1004. To do output with this port, a program need only store into the PORTB register to control the high-asserted output pins. Usually the store is done with an STAA instruction. If the program reads PORTB, it gets the number last stored into PORTB.

When the computer is powered-up or a hardware reset occurs, the PORTB register contains zeros. Figure 6-23 illustrates the PORTB register with the reset bit values shown below the figure.

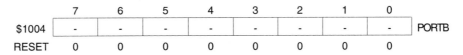

Figure 6-23   PORTB, a parallel output-only register.

The PORTC Register

The 68HC11 I/O port called *PORTC* is a programmable 8-bit I/O port. The hardware allows each bit in the port to be either an input bit or an output bit. Figure 6-24 illustrates the PORTC register. The figure also shows the values of the bits immediately after a hardware reset of the 68HC11.

After the PORTC register is programmed, some of the bits may be parallel inputs and some may be parallel outputs. The program need only store into PORTC with a STAA instruction to do output, or read PORTC with LDAA to do input. The input operation results in an instantaneous read from the IC pins. The output operation writes into a register that holds the binary signals on the output pins. All pins are high-asserted for both input and output. So, the high level represents logic 1, and the low level represents logic 0.

When the program stores to bits programmed as inputs, the hardware ignores the control signals. When the program reads bits programmed as outputs, it gets the last number stored in PORTC. The same number is read even if the output pins are shorted to ground!

Figure 6-24   PORTC, a programmable parallel I/O register.

The DDRC Register

Any 68HC11 register called a *data direction register* determines whether the bits of a programmable I/O register are inputs or outputs. The data direction register for PORTC is labeled DDRC. Figure 6-25 illustrates the DDRC register, which is at address 1007. The program can read the DDRC register as it would a memory register, but writing into DDRC connects internal hardware in a particular configuration.

The bits in DDRC correspond bit-by-bit with the bits in PORTC. When a bit in DDRC is 0, it programs the corresponding bit in PORTC as an input bit. A 1 in a DDRC bit makes the corresponding PORTC bit an output. At power-up reset, all PORTC bits are inputs because reset zeros the DDRC register.

	7	6	5	4	3	2	1	0	
$1007	-	-	-	-	-	-	-	-	DDRC
RESET	0	0	0	0	0	0	0	0	

Figure 6-25   DDRC, the PORTC data direction register.

In most programs, the initialization section stores a number in the DDRC, and then the program never again changes DDRC. However, in some practical applications, the PORTC bits change between inputs and outputs as the program runs.

The PIOC and PORTCL Registers

The 68HC11 register called *PIOC* is the parallel I/O control register. The PIOC register is at address 1002. The PIOC register controls several hardware options, but it also contains a flag bit. The following discusses only the flag.

The flag bit in PIOC is the output of the flag. A flag is not a memory bit! Therefore, storing into this register bit does not affect the flag.

The flag in PIOC has a ready signal at the 68HC11 pin named *STRA* for strobe A. The name of the flag is therefore *STAF*. The program tests the STAF flag by testing bit 7 of PIOC. Figure 6-26 shows the STAF flag in the PIOC register.

The ready signal for the STAF flag, STRA, can be active on either a low-to-high or high-to-low transition. Figure 6-26 shows the EGA bit, bit 1 of PIOC, which is the edge-for-STRA bit. Making EGA a 1 programs the ready signal to respond to low-to-high transitions; making EGA a 0 programs the ready signal for high-to-low transitions. The initialization section of the program must store the needed value into EGA.

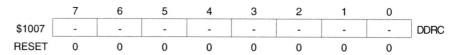

	7	6	5	4	3	2	1	0	
$1002	STAF	STAI	CWOM	HNDS	OIN	PLS	EGA	INVB	PIOC
RESET	0	0	0	0	0	U	1	1	

Figure 6-26   PIOC, the parallel I/O control register.

The STRA pin also does a second function. Besides setting the flag, it clocks or strobes a latch register called *PORTCL*. In some applications, holding an input number in a register or latch at the instant the ready signal occurs is useful. The PORTCL register latches a copy of the number in PORTC at the instant the STRA signal sets the flag. The latching occurs even if the flag was already set. The PORTCL register is at address 1005 as Figure 6-27 illustrates.

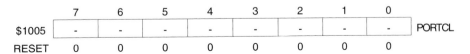

Figure 6-27   PORTCL, the PORTC input latch register.

Clearing the STAF flag requires the execution of two instructions. The first must read the PIOC register, and the second must read the PORTCL register. Use LDAA instructions for both reads.

This process may seem a strange way to clear the flag, but remember the flag is not a memory bit. It is a hardware flip flop that is set by a signal on a pin. If you are using input bits, reading the PORTCL register is probably useful anyway.

6.8   PARALLEL I/O EXAMPLE USING POLLING

The 68HC11 has several input/output ports for parallel I/O and several flags. The example that follows uses only the programmable port PORTC and the STAF flag. The example demonstrates the principles of parallel input/output with timing using this hardware and associated software. Look at Figure 6-16 to review the polling technique.

Problem Description

The computer reads a number from a thumbwheel switch, and then displays the number on a display unit. The computer reads the thumbwheel switch only when a person pushes the push button switch. Usually, a person sets a number on the thumbwheel switch and then presses the push button. The program updates the display at the time the button is pushed. The thumbwheel switch generates a 4-bit number and the display, which has an internal decoder, displays a 4-bit number.

I/O Hardware

Figure 6-28 illustrates the I/O hardware connections to the 68HC11 input/output ports. Both the thumbwheel switch and the display unit use high-asserted signals. The thumbwheel switch is connected to bits 4 through 7 of PORTC. The display unit is connected to bits 0

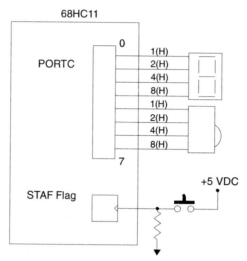

Figure 6-28   Input/output hardware example.

through 3 of PORTC. Therefore, four of the PORTC bits must be inputs, and the other four bits must be outputs.

The push button switch provides the ready signal for the STAF flag. Assume that the switch has no switch contact bounce, or that it makes no difference to the application.

Polling Software

The program in Figure 6-29 illustrates the polling technique of I/O synchronization. The program reads the thumbwheel switch and controls the display only when the push button is pushed. Here is a description of the program.

- *Lines 21 through 24.*   Symbols are defined for the addresses of the input/output control registers.

- *Line 36.*   The PORTC register is set to zero, though it may already contain zeros put there by a hardware reset. The CLR PORTC instruction affects only those PORTC pins that were previously programmed as outputs. If a hardware reset precedes this program, the PORTC pins will be all inputs.

- *Lines 38 and 39.*   The data direction register is set to make four input pins and four output pins. The output pins now respond to the bits in PORTC.

- *Lines 41 through 42.*   The PIOC register, illustrated in Figure 6-26, is initialized. The 1 put into EGA, bit 1 of PIOC, makes the STAF flag respond to a low-to-high transition on the STRA pin. The 0s put into the other bits of PIOC cause no problems. The store to bit 7 does nothing in the hardware.

- *Lines 46 and 47.*   Test the STAF flag and loop until it is set by the ready signal from the push button switch. The STAF flag condition is in bit 7 of the PIOC register.

```
1                        ****************************************************
2                        ** COPY THUMBWHEEL SWITCH TO DISPLAY WHEN
3                        ** PUSHBUTTON IS PUSHED
4                        *
5                        * USE POLLING TECHNIQUE FOR TIMING
6                        *
7                        * FOUR-BIT THUMBWHEEL SWITCH WIRED TO
8                        * ..BITS 4-7 OF PORT C
9                        *
10                       * FOUR-BIT DISPLAY WIRED TO BITS 0-3 OF PORT C
11                       *
12                       * PUSHBUTTON SWITCH WIRED TO STRA SO
13                       * ..LOW TO HIGH ON PUSH
14                       *
15                       ****************************************************
16                       *
17                       ****************************************************
18                       ** SYMBOL DEFINITIONS
19                       ****************************************************
20                       * 68HC11 REGISTERS
21          1002         PIOC    EQU   $1002    PARALLEL I/O CONTROL REGISTER
22          1003         PORTC   EQU   $1003    I/O PORT C REGISTER
23          1005         PORTCL  EQU   $1005    PORT C LATCH REGISTER
24          1007         DDRC    EQU   $1007    DATA DIRECTION REGISTER C
25          000F         IOPAT   EQU   $0F      I/O PATTERN, 0=IN 1=OUT
26                       *
27                       ****************************************************
28                       ** PROGRAM SECTION
29                       ****************************************************
30   C100                        ORG   $C100
31                       *------------------------------------------------
32                       * INITIALIZATION
33                       *------------------------------------------------
34                       * INITIALIZE PORT C
35                       *    INITIALIZE OUTPUTS TO ZEROS
36   C100  7F 10 03              CLR   PORTC    0=LOW
37                       *    SET UP INS AND OUTS
38   C103  86 0F                 LDAA  #IOPAT
39   C105  B7 10 07              STAA  DDRC     0=IN, 1=OUT
40                       * SET UP PIOC
41   C108  86 02                 LDAA  #$02     STAF RESPONDS ON LOW TO HIGH
42   C10A  B7 10 02              STAA  PIOC     ..TRANSITION ON STRA
43                       *------------------------------------------------
44                       * PUSHBUTTON PUSHED?
45                       *------------------------------------------------
46   C10D  7D 10 02      PBTST   TST   PIOC     TEST STAF FLAG
47   C110  2A FB                 BPL   PBTST    TRICK! FLAG AT SIGN BIT
48                       *------------------------------------------------
49                       * COPY THUMBWHEEL SWITCH TO DISPLAY
50                       *------------------------------------------------
51                       * READ THUMBWHEEL SWITCH AND CLEAR I/O FLAG
52   C112  B6 10 02              LDAA  PIOC     TWO INSTRUCTIONS TO CLEAR STAF
53   C115  B6 10 05              LDAA  PORTCL   ..FLAG AND INPUT SSSS----
54                       * POSITION DATA FOR OUTPUT
55   C118  46                    RORA           -SSSS---
56   C119  46                    RORA           --SSSS--
57   C11A  46                    RORA           ---SSSS-
58   C11B  46                    RORA           ----SSSS
59                       * CONTROL DISPLAY
60   C11C  B7 10 03              STAA  PORTC    OUTPUT TO PORT C
61   C11F  20 EC                 BRA   PBTST
62   C121                        END
```

Figure 6-29   Polling program.

- *Lines 52 and 53.* Both the PIOC and PORTCL registers must be read to clear the STAF flag. However, reading the PORTCL register with LDAA inputs the condition of the thumbwheel switch to the accumulator. The bit pattern in PORTCL represents the input bits latched at the instant the STAF flag was set.

- *Lines 55 through 58.* The bits representing the thumbwheel switch are positioned so that they can control the display unit.

- *Line 60.* Storing the accumulator into PORTC controls the output pins that drive the display unit.

This example program runs in an infinite loop that has complete control of the computer. Most programs for control computers are similar infinite loops.

6.9   INTERRUPT CONCEPTS

The *interrupt* technique of timing or input/output synchronization is more sophisticated than the polling technique. The performance of the computer is usually better when it uses interrupt rather than polling. The software is also easier to organize when complex I/O programs are necessary.

Figure 6-30 uses flowcharts to illustrate the fundamental concept of interrupt. A main program runs in an infinite loop possibly doing useful work. Sometimes the main program does nothing but run an empty loop. When the input/output device is ready for a data transfer, it sets the I/O flag. The I/O device sets the flag exactly as it did in the polling technique. The hardware interrupt system in the microprocessor responds to the flag by stopping the main program and sending control to the interrupt service routine. The transfer of control is done entirely by hardware. The computer executes no instructions to accomplish the transfer.

The *interrupt service routine*, or ISR, services the input/output device. The ISR clears the I/O flag, does the data transfer, and completes any other work required. When the I/O device service is complete, the ISR returns control to the main program. The main program continues execution as if nothing had happened. In many practical cases, the main program is completely unaware that the interrupt service routine has run.

The transfer to the ISR occurs only when the I/O device is ready for a data transfer, thus synchronizing the program to the I/O device. Therefore, interrupt is a timing technique.

The ISR is similar to a subroutine, but the ISR is not a subroutine. The ISR runs in response to a hardware signal. In contrast, a subroutine runs because an instruction in a program transfers control to it. No instructions in the main program cause the interrupt.

6.10   THE 68HC11 INTERRUPT SYSTEM

The 68HC11 has an interrupt system that can respond to signals from a variety of I/O devices. External I/O devices connected to the microcomputer IC can generate interrupts. In addition,

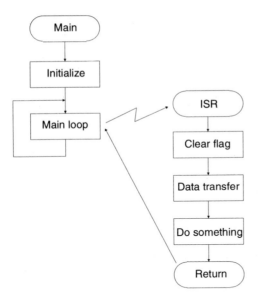

Figure 6-30   Main program and interrupt service routine.

several I/O devices inside the 68HC11 IC also can cause interrupts. The following only discusses using an external device to cause an interrupt.

Interrupt Signal Path

Figure 6-31 is a block diagram of the hardware involved in using interrupt. The figure shows only those signals specifically related to interrupt. The figure omits the address, data, and control bus signals.

The 68HC11 interrupt system hardware receives external interrupt signals through the IRQ interrupt request pin. Driving this pin low requests an interrupt. The I/O flag controls the IRQ line. Therefore, when the ready signal sets the flag, the flag sends a request for an interrupt to the interrupt system.

However, two levels of enabling are necessary for the signal to reach the interrupt system. First, the flag must have control of the IRQ wire. Usually, an electronic switch in the I/O section connects the flag to the IRQ wire. Programs use this switch to choose whether they will use interrupt.

Second, the IRQ line must have control of the interrupt system. An electronic switch in the microprocessor connects the IRQ pin to the interrupt system. The I bit in the condition code register controls this switch. When the I bit is 0, the switch closes enabling interrupts. When I is 1, the switch opens and masks or disables the interrupt signals. Therefore, the CLI instruction, which puts 0 in the I bit, enables the interrupt system.

Don't be confused by the I bit values that may seem backward. When I is 0, the interrupt system is enabled; when I is 1, the interrupt system is disabled.

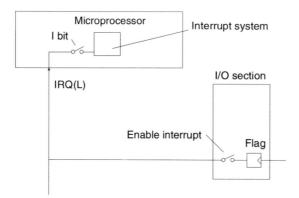

Figure 6-31 Block diagram of interrupt signal path.

Interrupt System Operation

Let's begin this discussion of the interrupt system by assuming that the main program initialization enabled the flag to cause an interrupt. Also assume that the interrupt system was enabled with the CLI instruction. Now the main program loop is running, doing some function. The main program ignores the I/O device connected to the flag.

At some time unknown to the computer, the I/O device gets ready for a data transfer and sets the flag. The flag signals the interrupt system to interrupt by driving the IRQ wire low. The interrupt system hardware then transfers program control to an interrupt service routine program. It does this transfer by placing a new address in the program counter.

The ISR now takes care of the I/O device. In particular, it clears the flag that was the source of the interrupt. Clearing the flag acknowledges that the computer has responded to the interrupt request from the I/O device. Clearing the flag is necessary so that the I/O device can set the flag again in the future to request service again. Remember, interrupt is a timing technique.

Next, the ISR must do the data transfers necessary for the I/O device and run any other necessary software. When the ISR finishes, the ISR returns control to the main program. The main program continues at the point of the interrupt. Usually, the main program doesn't know the interrupt occurred.

Saving status in the stack

When the ISR runs, it will certainly use the microprocessor registers. The ISR usually modifies the accumulators, index registers, and the condition code register. Transferring control to the ISR also modifies the program counter.

When control returns to the main program, the program can continue execution only if all the microprocessor registers are preserved. The information in these registers at any time is called the *status* of the computer.

To preserve the status, the interrupt system hardware saves the microprocessor registers in the stack. The main program has no instructions to save the status—the interrupt system hardware pushes the registers onto the stack.

The interrupt system in the 68HC11 stores nine bytes of information in the stack. At the end of the ISR, all these bytes must be restored. When they are, the main program continues as if nothing happened.

The return from interrupt instruction, RTI, restores the status. The RTI instruction pulls nine bytes from the stack into the microprocessor registers. All interrupt service routines must use the RTI instruction to return to the main program. Returning to the main program any other way makes unstructured spaghetti programs.

Stack use.   A stack is necessary to use interrupt because the interrupt system hardware needs it. Uses of the stack now include storing program data, storing subroutine return addresses, and storing the microprocessor status for the interrupt system.

The interrupt system uses the stack when an I/O device signals that it is ready. Therefore, the computer doesn't know when the interrupt system will use the stack. The main program may be using the stack when the interrupt occurs. Consequently, the main program must always use the stack correctly and orderly. Otherwise, the program crashes when an interrupt occurs.

The status dilemma.   The discussion of the computer status omitted an important point for simplicity—the microprocessor may be fetching an instruction when the interrupt signal arrives at the interrupt system. During the fetch operation, the microprocessor puts the instruction code into the instruction register. However, the interrupt system does not save the instruction register in the stack. If the interrupt system immediately transfers control to the ISR, any bytes already in the instruction register are overwritten and lost. It will not then be possible to continue with the main program execution.

The 68HC11 avoids this dilemma because the interrupt system checks the IRQ signal only at the beginning of each fetch operation. At that point, the previous instruction is completed and nothing of value is in the instruction register. Therefore, saving the instruction register is unnecessary. However, waiting until the next fetch operation delays the response to the interrupt signal by a few E-clock cycles.

The infinite interrupt dilemma.   The way the IRQ line works leads to another dilemma. The I/O flag triggers the interrupt by driving the IRQ wire low. The wire stays low until the flag is cleared in the ISR. However, the interrupt system checks the IRQ wire as it begins to fetch an instruction. Therefore, another interrupt should occur as the microprocessor fetches the first instruction in the ISR! An infinite loop of interrupts is the result.

To prevent this problem, the interrupt system automatically disables itself when it responds to an interrupt. That is, the interrupt system sets the I bit to 1 when it responds to an interrupt request, which allows the ISR to run undisturbed by further interrupt requests. Consequently, the interrupt system ignores the low level that stays on the IRQ line until the I/O flag is cleared. However, the interrupt system must be enabled again when control returns to the main program. Otherwise, the interrupt system would never honor another interrupt request.

The ISR can use the CLI instruction to enable the interrupt system again. However, reenabling is usually unnecessary because the interrupt system saves the status before disabling itself. The status saved in the stack includes the 0 in the I bit. When the RTI instruction restores the status, the I bit is automatically set back to 0. Therefore, the RTI instruction automatically enables the interrupt system at the end of the ISR.

Because the interrupt system disables and enables itself automatically during interrupt service, programs usually contain only one CLI instruction. It enables the interrupt system during the program initialization. Furthermore, programs seldom use the SEI instruction that disables the interrupt system.

Transferring control to the ISR

The interrupt system hardware must know the address of the beginning of the ISR. Remember, instructions do not send control to the ISR. Therefore, hardware must provide the address. In particular, the 68HC11 uses a block of memory locations at the top of memory to hold addresses for interrupts. For the IRQ interrupt, memory locations FFF2 and FFF3 must contain the address of the interrupt service routine. These locations are called an *interrupt vector* because they contain a pointer or vector to the ISR. When the interrupt system responds to the IRQ wire, it goes to the interrupt vector to get the address of the ISR.

The 68HC11 has many internal devices that can cause interrupts. Each I/O device has a separate interrupt vector. Therefore, each device has its own ISR. Each internal device also has a flag. Each flag has a switch to enable or disable an interrupt from it. The I bit also enables or disables interrupts from these devices.

6.11 IRQ INTERRUPT EXAMPLES

The details of using the IRQ interrupt vary somewhat depending on whether a single I/O device or multiple I/O devices are using the same interrupt hardware. This discussion looks at the IRQ interrupt facilities of the 68HC11 in each circumstance. If only a single device causes interrupts, a flag internal to the 68HC11 will likely be used. If more devices must cause interrupts, additional hardware external to the 68HC11 must be used.

Single Interrupting Device

Although other approaches are possible, the STAF flag is often used if an IRQ interrupt must be generated by a single I/O device. The previous discussion on polling I/O used the STAF flag for polling; however, the STAF flag may also cause interrupts.

Using the STAF flag to cause interrupts

The output side of the STAF flag in the PIOC register has a signal path to the IRQ interrupt line within the 68HC11 chip. The signal path includes an electronic switch as illustrated by Figure 6-31. The switch is controlled by the STAI bit in PIOC as shown by Figure 6-26.

Usually, the initialization section of the program will put the correct value into STAI so that the flag can cause an interrupt when the proper transition on the STRA pin sets it. A

1 in STAI enables interrupts by closing the switch shown in Figure 6-31. A 0 in STAI prevents the flag from causing an interrupt.

Some people describe the interrupt as *masked* when the enable bit is 0. That is, if an interrupt cannot occur, it is masked.

Because the reset condition of STAI is 0, and a program must change STAI before the STAF flag can cause interrupts, using the STAF flag in the polling example did not cause any unwanted interrupts. The computer is powered-up in a reset condition with the interrupts masked, so a polling program is not affected by the interrupt system.

Interrupt program example

The example here is a program that has the same function as the polling example in Section 6.8, and the hardware connections are the same as Figure 6-28. That is, the example uses parallel I/O to a thumbwheel switch and a display unit. Pushing a push button sets the I/O flag which, in this example, causes an interrupt to signal that the data transfer needs to be done immediately. The program must read the number from the thumbwheel switch and send it to the display.

The following discusses the operation of the program in Figure 6-32. You also should refer to the polling program in Figure 6-29, because much of this program is the same.

- *Lines 21 through 25.*   Symbols are defined for the addresses of the input/output control registers. All I/O registers in the 68HC11 are in a block of addresses beginning at address 1000. The symbol REG defines the beginning of this block of I/O addresses.

- *Line 34.*   The interrupt vector is set to the address of the IRQ interrupt service routine. Loading the program into memory sets this vector.

- *Line 44.*   The stack is initialized for use by the interrupt system.

- *Lines 47, 49, and 50.*   The PORTC initialization is the same as in the polling example. Here the DDRC programs four bits as inputs and four as outputs.

- *Lines 52 and 53.*   The PIOC register is initialized. Storing a 1 into bit 6, STAI, of PIOC enables IRQ interrupts from the STAF flag. Storing 1 into bit 1, EGA, makes the STAF flag set on a low-to-high transition of the STRA pin. The 0s put into the other bits of PIOC will cause no problems.

- *Line 55.*   The interrupt system is enabled after all I/O hardware is initialized. Initializing the interrupt system before this may let the interrupt service routine control uninitialized hardware.

- *Line 59.*   The main program loop does nothing in this example. It only provides an operating program while the computer waits for interrupts.

- *Lines 67 and 68.*   The STAF input/output flag is tested as a safety measure. A hardware failure could erroneously cause an interrupt. If an erroneous interrupt

```
 1                                    ******************************************************
 2                                    ** COPY THUMBWHEEL SWITCH TO DISPLAY WHEN
 3                                    ** PUSHBUTTON IS PUSHED
 4                                    *
 5                                    * USE INTERRUPT TECHNIQUE FOR TIMING
 6                                    *
 7                                    * FOUR-BIT THUMBWHEEL SWITCH WIRED TO
 8                                    * ..BITS 4-7 OF PORT C
 9                                    *
10                                    * FOUR-BIT DISPLAY WIRED TO BITS 0-3 OF PORT C
11                                    *
12                                    * PUSHBUTTON SWITCH WIRED TO STRA SO
13                                    * ..LOW TO HIGH ON PUSH
14                                    *
15                                    ******************************************************
16                                    *
17                                    ******************************************************
18                                    ** SYMBOL DEFINITIONS
19                                    ******************************************************
20                                    * 68HC11 REGISTERS
21             1000                   REG      EQU    $1000      BASE ADDRESS OF REGISTERS
22             1002                   PIOC     EQU    $1002      PARALLEL I/O CONTROL REGISTER
23             1003                   PORTC    EQU    $1003      I/O PORT C REGISTER
24             1005                   PORTCL   EQU    $1005      PORT C LATCH REGISTER
25             1007                   DDRC     EQU    $1007      DATA DIRECTION REGISTER C
26             000F                   IOPAT    EQU    $0F        I/O PATTERN, 0=IN 1=OUT
27                                    * MASKS
28             0080                   BIT7     EQU    %10000000
29                                    *
30                                    ******************************************************
31                                    ** DATA SECTION
32                                    ******************************************************
33   FFF2                                      ORG    $FFF2
34   FFF2      C113                            FDB    IRQISR    IRQ INTERRUPT VECTOR
35                                    *
36                                    ******************************************************
37                                    ** MAIN PROGRAM
38                                    ******************************************************
39   C100                                      ORG    $C100
40                                    *--------------------------------------------------------
41                                    * INITIALIZATION
42                                    *--------------------------------------------------------
43                                    * INITIALIZE STACK
44   C100      8E DF FF                        LDS    #$DFFF
45                                    * INITIALIZE PORT C
46                                    *     INITIALIZE OUTPUTS TO ZEROS
47   C103      7F 10 03                        CLR    PORTC     0=LOW
48                                    *     SET UP INS AND OUTS
49   C106      86 0F                           LDAA   #IOPAT
50   C108      B7 10 07                        STAA   DDRC      0=IN, 1=OUT
51                                    * SET UP PIOC
52   C10B      86 42                           LDAA   #$42      ENABLE STAF INTERRUPT
53   C10D      B7 10 02                        STAA   PIOC      ..LOW TO HIGH ON STRA
54                                    * TURN ON INTERRUPT SYSTEM
55   C110      0E                              CLI
56                                    *--------------------------------------------------------
57                                    * WAIT FOR INTERRUPTS
58                                    *--------------------------------------------------------
59   C111      20 FE                  HERE     BRA    HERE      DO NOTHING!
60                                    *
61                                    ******************************************************
62                                    ** PUSHBUTTON SWITCH INTERRUPT SERVICE ROUTINE
63                                    ******************************************************
```

Figure 6-32   IRQ interrupt program.

```
64                               *----------------------------------------------------------
65                               * VALID STAF INTERRUPT?
66                               *----------------------------------------------------------
67    C113    CE 10 00           IRQISR  LDX     #REG
68    C116    1F 02 80 0D                BRCLR   PIOC-REG,X,BIT7,RTIRQ    BRANCH ON NO
69                               *----------------------------------------------------------
70                               * COPY THUMBWHEEL SWITCH TO DISPLAY
71                               *----------------------------------------------------------
72                               * READ THUMBWHEEL SWITCH AND CLEAR I/O FLAG
73    C11A    B6 10 02                   LDAA    PIOC     TWO INSTRUCTIONS TO CLEAR STAF
74    C11D    B6 10 05                   LDAA    PORTCL   ..FLAG AND INPUT SSSS----
75                               * POSITION DATA FOR OUTPUT
76    C120    46                         RORA             -SSSS---
77    C121    46                         RORA             --SSSS--
78    C122    46                         RORA             ---SSSS-
79    C123    46                         RORA             ----SSSS
80                               * CONTROL DISPLAY
81    C124    B7 10 03                   STAA    PORTC    OUTPUT TO PORT C
82                               *----------------------------------------------------------
83                               * RETURN TO MAIN PROGRAM
84                               *----------------------------------------------------------
85    C127    3B                 RTIRQ   RTI
86    C128                               END
```

Figure 6-32   Continued.

occurs, the ISR in this example simply returns to the main program. In dangerous situations, other responses would be necessary.

- *Lines 73 through 81.* These lines of the program clear the STAF flag, read the thumbwheel switch, and control the display. They are the same instructions as in the polling example.

- *Line 85.* The RTI instruction returns program control to the main program. The RTI restores the microprocessor registers with main program data from the stack. The RTI removes nine bytes from the stack. Restoring the registers enables the interrupt system so that future interrupts can occur. Execution continues at line 59 in the main program.

The work done by the interrupt service routine is the same as that done by a polling program—only the timing technique is different. Carefully compare lines 73 through 81 of this program to lines 52 through 60 of Figure 6-29.

Multiple Interrupting Devices

The IRQ interrupt request line inside the 68HC11 also connects to a pin so that external devices (flags) can connect to the interrupt line and thus cause interrupts. Only the STAF flag connects to the IRQ line inside the 68HC11 chip.

The IRQ interrupt is the principal external interrupt line in the 68HC11. It has a single interrupt vector to send control to a single interrupt service routine. The IRQ interrupt in the 68HC11 is nearly the same as the IRQ interrupt in other Motorola 8-bit microprocessors.

Required hardware and software

Multiple external I/O devices may each have a flag connected to the IRQ pin. The flags connect with open-collector or open-drain hardware, which allows the various flags to be wire-ORed. This connection allows any device connected to the IRQ pin to pull the voltage to the low level without interference from the others.

The multiple I/O devices that control the IRQ line all send program control to the same interrupt service routine. This ISR must then determine which flag caused the interrupt by examining each flag to find which is set.

Figure 6-33 shows the flowchart for an interrupt service routine that polls two flags to determine which caused the interrupt. The string of decisions is called a *polling chain* because each flag is questioned or polled to see if it is set.

The overhead to poll the flags is much smaller than the overhead of the polling I/O technique. With interrupt, the polling occurs only once each time a device sets a flag. In the polling technique, the polling must happen continually so that the computer can respond quickly when a flag gets set.

Interrupt program example

This example using interrupt is an expansion of the interrupt example with a single device in the previous section. There, the computer reads a thumbwheel switch and displays the number when a push button is pushed. In the following example, a second device also can interrupt, so the interrupt service routine follows the flowchart in Figure 6-33.

Desired operation.   A push button switch sends a ready signal to the STAF flag to cause an interrupt. In responding to the IRQ interrupt request, the program reads a number from a thumbwheel switch and controls a display to show the number. In addition, an external free-running oscillator causes periodic interrupts. At each oscillator interrupt, the program

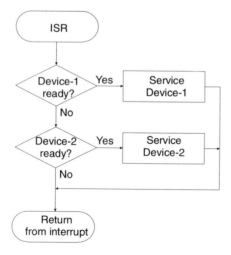

Figure 6-33   Flowchart of ISR for multiple interrupting devices.

reverses the state of a light to make a flasher. Both the display control and the light flasher operate concurrently.

Hardware.   Figure 6-34 illustrates the input/output hardware. The push button switch connects to the STRA pin to control the STAF flag. The thumbwheel switch connects to PORTC bits 4 through 7. The display unit connects to PORTB bits 4 through 7. PORTB is an output only port. This part of the hardware is similar to the connections in the previous example; however, the display unit now connects to PORTB instead of PORTC.

In addition, an external flag is made from a D flip flop. Other available flags inside the 68HC11 could be used, but this example uses an external flag. The D lead of the flip flop is connected to the high level. A transition of the flip flop clock lead by the oscillator sets the flag. Bit 0 of PORTB makes a clear signal for the flag. Driving bit 0 low clears the flag. The output of the flag drives the IRQ line through an open drain driver. The IRQ pin must have an external pull-up resistor as shown. The external flag and the internal STAF flag are therefore wire-ORed on the IRQ line. That is, if either flag is set, it will pull the IRQ line to the low level.

Program operation.   Figure 6-35 shows the program for the IRQ example. Here is a description of it:

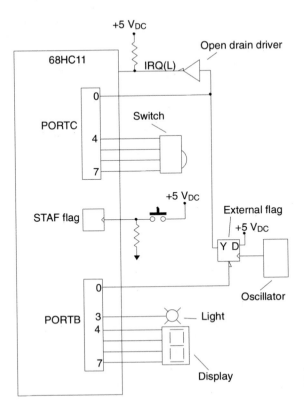

Figure 6-34   Hardware for multiple IRQ interrupt example.

```
 1                              *********************************************************
 2                              ** COPY THUMBWHEEL SWITCH TO DISPLAY WHEN
 3                              ** PUSHBUTTON IS PUSHED AND FLASH A LIGHT
 4                              *
 5                              * USE INTERRUPT TECHNIQUE FOR TIMING
 6                              *
 7                              * FOUR BIT THUMBWHEEL SWITCH WIRED TO
 8                              * ..BITS 4-7 OF PORT C
 9                              *
10                              * FOUR BIT DISPLAY WIRED TO BITS 4-7 OF PORT B
11                              *
12                              * PUSHBUTTON SWITCH WIRED TO STRA SO
13                              * ..LOW TO HIGH ON PUSH
14                              *
15                              * EXTERNAL FLAG WITH OSCILLATOR CONNECTED TO IRQ
16                              *
17                              * LIGHT WIRED TO BIT 3 OF PORT B
18                              *
19                              * FLAG DIRECT CLEAR WIRED TO BIT 0 OF PORT B
20                              *
21                              *********************************************************
22                              *
23                              *********************************************************
24                              ** SYMBOL DEFINITIONS
25                              *********************************************************
26                              * 68HC11 REGISTERS
27          1000                REG      EQU    $1000     BASE ADDRESS OF REGISTERS
28          1002                PIOC     EQU    $1002     PARALLEL I/O CONTROL REGISTER
29          1003                PORTC    EQU    $1003     I/O PORT C REGISTER
30          1004                PORTB    EQU    $1004     OUTPUT PORT B REGISTER
31          1005                PORTCL   EQU    $1005     PORT C LATCH REGISTER
32          1007                DDRC     EQU    $1007     DATA DIRECTION REGISTER C
33          00F1                IOPAT    EQU    $F1       I/O PATTERN, 0=IN 1=OUT
34                              * MASKS
35          0001                BIT0     EQU    %00000001
36          0008                BIT3     EQU    %00001000
37          0080                BIT7     EQU    %10000000
38                              *
39                              *********************************************************
40                              ** DATA SECTION
41                              *********************************************************
42   FFF2                                ORG    $FFF2
43   FFF2   C118                         FDB    IRQISR    IRQ INTERRUPT VECTOR
44                              *
45                              *********************************************************
46                              ** MAIN PROGRAM
47                              *********************************************************
48   C100                                ORG    $C100
49                              *-------------------------------------------------------
50                              * INITIALIZATION
51                              *-------------------------------------------------------
52                              * INITIALIZE STACK
53   C100   8E CF FF                     LDS    #$CFFF
54                              * INITIALIZE PORT B
55   C103   86 01                        LDAA   #BIT0     DON'T CLEAR EXTERNAL FLAG, AND
56   C105   B7 10 04                     STAA   PORTB     ..ZEROS FOR LIGHT AND DISPLAY
57                              * INITIALIZE PORT C
58                              *    INITIALIZE OUTPUTS TO ZEROS
59   C108   7F 10 03                     CLR    PORTC     0=LOW
60                              *    SET UP PORTC INS AND OUTS
61   C10B   86 F1                        LDAA   #IOPAT
62   C10D   B7 10 07                     STAA   DDRC      0=IN, 1=OUT
```

Figure 6-35 IRQ Interrupt program.

```
63                            * SET UP PIOC
64    C110    86 42                   LDAA    #$42        ENABLE STAF INTERRUPT
65    C112    B7 10 02                STAA    PIOC        ..LOW TO HIGH ON STRA
66                            * TURN ON INTERRUPT SYSTEM
67    C115    0E                      CLI
68                            *---------------------------------------------------
69                            * WAIT FOR INTERRUPTS
70                            *---------------------------------------------------
71    C116    20 FE           HERE    BRA     HERE        DO NOTHING!
72                            *
73                            ****************************************************
74                            ** INTERRUPT SERVICE ROUTINE
75                            ****************************************************
76                            *---------------------------------------------------
77                            * INTERRUPT POLLING CHAIN
78                            *---------------------------------------------------
79                            * INTERRUPT FROM PUSH BUTTON?
80    C118    CE 10 00        IRQISR  LDX     #REG
81    C11B    1E 02 80 06             BRSET   PIOC-REG,X,BIT7,COPY     BRANCH ON YES
82                            * INTERRUPT FROM OSCILLATOR?
83    C11F    1E 03 01 12             BRSET   PORTC-REG,X,BIT0,FLASH   BRANCH ON YES
84                            * ILLEGAL INTERRUPT
85    C123    20 1E                   BRA     RTIRQ       IGNORE ILLEGAL INTERRUPT
86                            *---------------------------------------------------
87                            * COPY THUMBWHEEL SWITCH TO DISPLAY
88                            *---------------------------------------------------
89                            * READ THUMBWHEEL SWITCH AND CLEAR I/O FLAG
90    C125    B6 10 05        COPY    LDAA    PORTCL      CLEAR STAF FLAG, INPUT SSSS----
91                            * FORM OUTPUT VALUE
92    C128    84 F0                   ANDA    #$F0        KEEP SWITCH BITS
93    C12A    F6 10 04                LDAB    PORTB       GET LAST OUTPUT TO PORT B
94    C12D    C4 0F                   ANDB    #$0F        REMOVE DISPLAY BITS
95    C12F    1B                      ABA                 MERGE OUTPUT BITS
96                            * CONTROL DISPLAY
97    C130    B7 10 04                STAA    PORTB       OUTPUT TO PORT B
98                            * RETURN TO MAIN PROGRAM
99    C133    20 0E                   BRA     RTIRQ
100                           *---------------------------------------------------
101                           * FLASH LIGHT
102                           *---------------------------------------------------
103                           * CLEAR OSCILLATOR FLAG
104   C135    1D 04 01        FLASH   BCLR    PORTB-REG,X,BIT0         DRIVE PIN LOW
105   C138    1C 04 01                BSET    PORTB-REG,X,BIT0         DRIVE PIN HIGH
106                           * TOGGLE LIGHT BIT
107   C13B    B6 10 04                LDAA    PORTB
108   C13E    88 08                   EORA    #BIT3
109   C140    B7 10 04                STAA    PORTB
110                           * RETURN TO MAIN PROGRAM
111   C143    3B              RTIRQ   RTI
112   C144                            END
```

Figure 6-35 Continued.

- *Line 43.* Loading the program into memory initializes the IRQ interrupt vector.

- *Lines 53 through 67.* The program initializes the computer hardware and enables the interrupt system.

- *Line 71.* The main program loop does nothing but wait for interrupts.

- *Lines 80 and 81.* Using indexed addressing, the BRSET instruction tests the STAF flag. If the flag is set because the push button was pushed, control is sent to line 90.

- *Line 83.* The external flag is connected to bit 0 of PORTC so that the flag can be tested. The BRSET instruction tests the flag. If the flag is set because the oscillator time has expired, control is sent to line 104.

- *Line 85.* If neither flag was set, an erroneous interrupt was detected. The program returns to the main program ignoring the error.

- *Line 90.* Reading the PORTCL latch register clears the STAF flag. Remember that both the PIOC and PORTCL registers must be read to clear the STAF flag. However, the BRSET instruction on line 81 already read PIOC. Reading PORTCL also inputs the number from the thumbwheel switch.

- *Lines 92 through 95.* The bits read from PORTC are masked so that only bits 4 through 7, the data from the thumbwheel switch, are retained. The bits last output to PORTB are read, and the display bits are masked. The remaining bits from the two ports are merged.

- *Lines 97 and 99.* The new output bits are stored in PORTB to update the display. Bits 0 through 3 remain the same. Then control returns to the main program.

- *Lines 104 and 105.* The direct clear lead of the external flag flip flop is first driven low and then driven high. This action clears the flip flop and then allows it to respond to the normal clock lead.

- *Lines 107 through 111.* Toggle bit 3 of PORTB to reverse the state of the flashing light. Control then returns to the main program.

You should observe that the interrupt system is disabled for the entire time that the ISR runs. Therefore, if both input/output devices interrupt at the same time, the display is updated first and then the oscillator interrupts to control the light.

Hardware failures

Flags and input/output ports connect to external I/O devices. Therefore, they are more susceptible to damage than other components within the computer. The connections may be made with lengthy cables and to devices with different power supply voltages. An improper voltage on the cable could damage some hardware.

Considering the consequences of common failures is not only good engineering, it is also a means to better understand the operation of the computer. Let's look at the effects of common hardware failures on the program in Figure 6-35. Notice that this example is not the enhanced example which would act somewhat differently in the presence of failures.

The following describes the effects of some failures:

- *STAF flag fails set.* When the flag fails, the IRQ wire goes low and requests an interrupt so that control goes to the interrupt service routine. The ISR tries to clear the flag; however, the flag does not clear, because it failed and the IRQ wire will remain low. The ISR does not enable the interrupt system with a CLI instruction, so the ISR cannot be interrupted and will continue running. The ISR copies the thumbwheel switch to the display and continues running until it executes the RTI instruction.

The RTI instruction restores the main program registers and enables the interrupt system. Remember, the IRQ line is still low because the failed flag is still set. As the microprocessor starts to fetch an instruction from the main program, the interrupt system causes another interrupt. Control goes to the ISR again and the same sequence repeats. The result is an infinite loop because control always returns to the ISR immediately after the RTI instruction.

The effects of this failure can now be determined. The main program will never run again—not even a single instruction. On the other hand, the ISR runs repeatedly as fast as the computer can go.

The effect of the ISR quickly and repeatedly servicing an I/O device depends on the kind of device it is. In the example, the computer continually reads the thumbwheel switch and updates the display without anyone pushing the push button. If someone changes the thumbwheel to a new position, the display updates as the input number changes.

Next consider the effect on the flashing light. When the oscillator sets the second flag, the flag has no effect on the IRQ line because it is already low. Furthermore, the polling chain does not detect that the second flag is set, so the light stops flashing and it remains at the state it was in when the STAF flag failed.

Such behavior from a computer can be baffling. One hardware device continues working and another stops working. It may not seem possible that a computer could do such strange things.

- *The external flag fails set.*   If the external flag is always set, the IRQ line will always be low. The oscillator has no effect on the IRQ line. The ISR is in an infinite loop servicing the flashing light repeatedly. The light flashes as fast as the ISR can toggle the output bit. A person will see only a dim light.

The STAF flag continues to work normally, and the polling chain responds to it. Therefore, the display is updated only when the push button is pushed. The main program never runs again.

- *The STAF flag fails clear.*   If the STAF flag fails clear, the push button can never cause an interrupt. The computer ignores the thumbwheel switch and the display. If the thumbwheel switch is changed, the display remains the same.

The polling chain detects the external flag, so the oscillator continues to interrupt and the light flashes normally. The main program runs normally.

- *The external flag fails clear.*   If the external flag stays clear, the oscillator can never cause an interrupt. The ISR will completely ignore the light, so the light stays in the state it was in when the flag failed.

The polling chain detects the STAF flag, so the thumbwheel switch and display work normally. The main program continues to operate normally.

- *Both flags fail set.*   If both flags fail set, the IRQ line is always low and the main program never runs again. The polling chain tests the STAF flag first, so it sends control to the thumbwheel switch and display routine. When the ISR finishes updating the display, the ISR returns control to the main program. Therefore, the interrupt service routine does not test the external flag, and the computer ignores the oscillator.

  Immediately after the RTI instruction, another interrupt occurs and the ISR sends control to the thumbwheel switch and display routine again. So, as when the STAF failed alone, the display is updated continually and the light does not flash.

- *The IRQ wire is grounded.*   The IRQ wire is always low, so the main program will never run, because interrupts will occur continually. However, the polling chain will detect when the flags set, so the display and flashing light work normally.

- *Noise pulse on IRQ wire.*   If electrical noise causes a very short duration low-going pulse on the IRQ wire, it may trigger an interrupt. If none of the flags are set, the polling chain in this example simply returns control to the main program. The computer ignores the false interrupt signal. If a flag is set at the time of the pulse, the polling chain will respond normally to the flag. The main program is not affected.

  This failure could occur in a computer controlling something that is very dangerous. If the polling chain does not detect a set flag, the ISR should then respond to the failure instead of returning control to the main program. For example, the program may somehow force a hardware reset. Alternately, it may signal a warning to a person.

Many other failures are possible. The consequences depend both on the design of the hardware and the software. These examples should help you to better understand the interrupt system.

6.12   THE CONCURRENCY PROBLEM

Programs that use interrupt are susceptible to errors caused by different parts of the program accessing common resources. The cause of these errors is called the *concurrency problem*.

The word *concurrency* implies that multiple actions are taking place simultaneously. In the case of software, two or more program sections appear to be running simultaneously, which is the case when interrupts are used. The main program and several interrupt service routines all appear to be running simultaneously, although that is not the case. The microprocessor can only execute one instruction at a time, but the switching between the main program and the interrupt service routines gives the appearance of simultaneous operation.

People who work with computers prefer to use the word *concurrent* rather than *simultaneous* to describe this action.

The resources involved in the concurrency problem include all the hardware in the computer system. In particular, memory registers and I/O devices are susceptible. The emphasis in this chapter is on how the concurrency problem affects I/O devices that use interrupt. Be careful though because the problem is just as troublesome when it affects memory.

How It Happens

The concurrency problem is caused by the fact that the computer cannot always perform actions on resources with indivisible program elements (instructions). That is, almost all operations on resources require the computer to execute multiple instructions. While the computer is in the midst of performing an operation on a resource, an interrupt can stop that operation and allow another program section access to the same resource. The second access changes the resource in some way. When the second program section has completed and control returns to the first program section so that it can continue, the state of the resource has changed. The first program section then causes problems as it continues execution.

Generally, only small portions of a total program perform operations that have the concurrency problem. The parts of a program that are susceptible to causing problems are called *critical sections*.

Preventing Problems

The solution to the concurrency problem is simply to prevent multiple program sections from concurrently accessing common resources.

Use a single instruction

If the operation to be performed on the resource is very simple, it may be possible to perform with a single instruction! The operation in this situation is then said to be an *atomic* operation because an interrupt cannot occur during the operation.

For example, to increment a single-byte number in memory, use the increment-memory instruction, INC, rather than the sequence LDAA, ADDA #1, STAA. The INC instruction cannot be interrupted, therefore, no concurrency problem occurs when altering the memory byte. However, only a few situations will allow this simple approach to solve the problem.

Prevent interrupts

In small computers with fairly simple instruction sets, such as the 68HC11, often a practical solution is to disable the interrupt system while the critical program section is executed. This technique is a "brute force" technique because it degrades the response of the computer to interrupt requests from the I/O devices. However, the program overhead to solve the concurrency problem is quite small, and it degrades the execution speed of the program very little. This approach effectively makes the sequence of instructions that operate on the resource an atomic operation.

Caution is recommended if the critical section is within a subroutine. Calling the subroutine while the interrupt system is disabled results in the subroutine enabling the interrupt system. This inadvertent enabling of the interrupt system can cause unexpected new problems.

Use locks

Various techniques can provide a lock for each resource. While a resource is locked, the resource is not available for use. A locked resource is currently owned by some program section. The owner program section may or may not be actively using the resource. When a resource is not locked, it is available for any program section to take ownership of it. Such a locking mechanism prevents concurrent accesses to the resource and thus prevents the concurrency problem. An advantage of locks is that the interrupt system can remain on so that the response to interrupts is good.

Program strategy. Using locks to prevent concurrency problems adds complexity to programs. Before a program section uses a resource, the program must check that resource's lock to see if the resource is available for use. If the resource is locked so that it is not available, the current program section cannot use the resource. Probably, it will return later and try again, but another attempt to access the resource depends on the program logic. When the current owner program section finishes with the resource, it must unlock the resource. In the future, when the first program section checks the lock again and finds it unlocked, it will take ownership of the resource by locking it, and then proceed to use the resource.

Using a memory bit. A memory bit can provide the lock for a resource. Let's call the lock bit an *availability bit*. Then a 1 in the bit indicates that the resource is available and 0 that it is not available. When a program wants to use the resource, it copies the availability bit and then makes the bit 0. If the copy of the availability bit says the bit previously was 1, the program proceeds to use the resource which is now locked. If the copy says the bit was 0, the resource was previously locked and still is; the program must return later to see if the owner of the resource has unlocked it. When a critical section completes its operation on a resource, it sets the availability bit for that resource to 1 which unlocks the resource. This availability bit is effectively a simple software flag.

Making it work. The approach described above has one small problem: the concurrency problem also applies to the availability bit! If an interrupt occurs between copying the bit and clearing the bit, another program section could "beat this section to the resource" and the program malfunctions. This problem can happen because the second access (in response to the interrupt) sees the resource as available, takes ownership, and begins using the resource. If somehow control returns to the first program before the second one relinquishes ownership, the first program still has an indication that the resource is available and will proceed to use it.

The technique described here works properly only if the copying and clearing of the availability bit can be an atomic operation. Some computers, but not the 68HC11, have special instructions, usually called test-and-set-bit, for this very purpose. Similarly, the program can disable the interrupt system while working on the availability bit.

Locks and the 68HC11.   Existing 68HC11 instructions can be used to provide most of the advantages of a special test-and-set-bit instruction. The logical-shift-right instruction is convenient.

Consider using a memory byte as the lock. Let's define a lock byte value of 00 to mean the resource is not available (locked) and 01 to mean it is available. The availability bit is bit 0 of this byte. To test if the resource is available and to lock the resource in an atomic operation, execute the LSR instruction on this byte. The LSR copies the availability bit into the C bit while storing 00 into the byte; the 00 locks the resource. Now the program can test the C bit to see if the resource is available, and if it is, proceed to operate on it. When the operation is complete, the program releases the lock making the resource available by storing 01 into the lock byte.

Other approaches.   More sophisticated techniques for solving the concurrency problem are available. A disadvantage of the lock described here is that a blocked program section must repeatedly try to access the resource and, therefore, increases the execution overhead. Other techniques involve a queue to track blocked program sections and allow them to run as the resource becomes available.

Example Program with Concurrency Problem

The concurrency problem frequently occurs when I/O devices, such as output ports, have some bits used for one purpose and other bits used for another purpose. This situation is the basis for the example considered here. Let's begin with the previous example and enhance its capability, and then look at how concurrency problems occur.

Enhanced interrupt program example

Consider modifying the program in Figure 6-35 so that the output signals from the computer in response to setting the external flag are updated as quickly as possible. This example is unrealistic and contrived because the program only flashes a light for a person to look at, but many practical applications require quick response to interrupts.

The response problem.   The example program has varying response times depending on the computer's activity when the external flag gets set. If the main program was running at the time the flag gets set, updating the light will occur quickly. If the instruction on line 81 just sent control to the push button switch service routine when the external flag gets set, it will be some time before the computer will update the output to the light. The variation, in addition to the duration, of the response time may be troublesome in some applications.

Solving the problem.   The response to the external flag can be made less variable and generally faster if the program can interrupt out of the ISR for the push button switch. Control can go immediately to the ISR for the external flag, and then return later to finish servicing the STAF flag.

To interrupt out of an ISR, the interrupt system must be reenabled within that ISR. Recall that the interrupt system is disabled automatically when an interrupt occurs, so the interrupt system is usually disabled while an ISR runs. Reenabling the interrupt system within an ISR will allow a second device to interrupt the service of the first device. Figure 6-36 illustrates this change in the program.

Clearing the interrupt request.   Reenabling the interrupt system within an ISR must be done carefully. In particular, the ISR must clear the flag that caused the interrupt before reenabling the interrupt system. Otherwise, the flag will immediately cause another interrupt. The interrupt examples in this book always test the flag at the beginning of the ISR because testing at the beginning is necessary if the interrupt system is to be reenabled.

Concurrency problems in example

The enhanced example program shown in Figures 6-35 and 6-36 has two sections of code that both modify the contents of PORTB. Therefore, this enhanced example now has concurrency problems. The original example did not.

In particular, the routine in Figure 6-36 that reads the thumbwheel switch and controls the display can be interrupted. If the interrupt from the oscillator should come after the program reads PORTB, but before storing the new result into PORTB, errors will occur. Once again, if the interrupt comes after the LDAA PORTB, ANDB #$0F, or ABA instructions but before the STAA PORTB instruction, concurrency problems will occur.

A specific situation.   Consider a specific example. Suppose that PORTB currently contains 79. The display shows the number 7, and the flashing light is now on. Now, with the thumbwheel switch set to 4, the push button is pushed. An interrupt occurs and the ISR routine for the push button switch runs to the LDAA instruction, which reads 79 from PORTB.

Next, suppose that immediately after the LDAA instruction executes, an interrupt from the external oscillator occurs. The interrupt system stores the microprocessor registers in the stack including the number 79 from accumulator A. When control gets to the flasher routine, it reads the same number 79 from the PORTB register, modifies the number to 71, and stores 71 in PORTB. The flashing light is now off. As this routine finishes, the RTI instruction

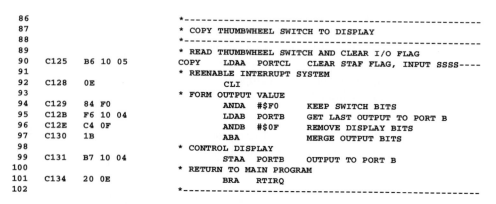

```
86              *-------------------------------------------------------------
87              * COPY THUMBWHEEL SWITCH TO DISPLAY
88              *-------------------------------------------------------------
89              * READ THUMBWHEEL SWITCH AND CLEAR I/O FLAG
90    C125  B6 10 05     COPY    LDAA   PORTCL    CLEAR STAF FLAG, INPUT SSSS----
91              * REENABLE INTERRUPT SYSTEM
92    C128  0E                   CLI
93              * FORM OUTPUT VALUE
94    C129  84 F0                ANDA   #$F0      KEEP SWITCH BITS
95    C12B  F6 10 04             LDAB   PORTB     GET LAST OUTPUT TO PORT B
96    C12E  C4 0F                ANDB   #$0F      REMOVE DISPLAY BITS
97    C130  1B                   ABA              MERGE OUTPUT BITS
98              * CONTROL DISPLAY
99    C131  B7 10 04             STAA   PORTB     OUTPUT TO PORT B
100             * RETURN TO MAIN PROGRAM
101   C134  20 0E                BRA    RTIRQ
102             *-------------------------------------------------------------
```

Figure 6-36   A portion of example program with interrupt system reenabled.

restores the microprocessor registers thus returning control to the interrupted switch routine. The A accumulator once again contains 79.

Now the switch routine modifies the A accumulator updating the number from the thumbwheel switch so that the accumulator contains 49. The routine stores 49 in PORTB, which makes the display correctly show 4, but it also erroneously turns the flashing light back on! The light should remain off at this point, but it has been changed due to the concurrency problem.

The critical section.   The critical section of the thumbwheel switch routine that leads to the concurrency problem includes the LDAA, ANDA, ABA, and STAA instructions. The critical section can be protected from interrupts as a means to eliminate the concurrency problem with PORTB.

Consider a new enhanced program that disables the interrupt system while the critical section runs. Figure 6-37 shows only the modified part of the original program from Figure 6-35. The modified version allows the switch routine to be interrupted because the CLI instruction on line 92 reenables the interrupt system. The SEI on line 95 and the CLI on line 101 prevent concurrency problems with PORTB.

This example is quite simple. Reenabling the interrupt system provided little gain in the response time for the flashing light; however, the principles are correct. In practical examples, the critical section generally is very small compared to the total routine. Therefore, much improved response from reenabling the interrupt system is possible.

6.13   THE REENTRANCY PROBLEM

Subroutines may not function correctly when a program uses interrupt. A problem may occur when a subroutine is interrupted if the interrupt causes control to go to the interrupted subroutine. In this case, the subroutine must run again from its beginning, though it has not

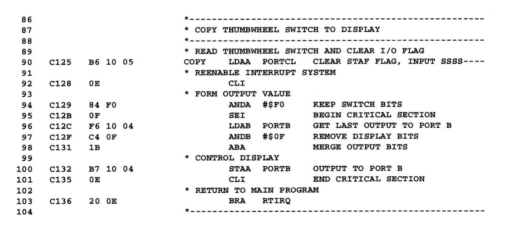

```
 86                                *---------------------------------------------------------
 87                                * COPY THUMBWHEEL SWITCH TO DISPLAY
 88                                *---------------------------------------------------------
 89                                * READ THUMBWHEEL SWITCH AND CLEAR I/O FLAG
 90    C125   B6 10 05             COPY    LDAA   PORTCL   CLEAR STAF FLAG, INPUT SSSS----
 91                                * REENABLE INTERRUPT SYSTEM
 92    C128   0E                           CLI
 93                                * FORM OUTPUT VALUE
 94    C129   84 F0                        ANDA   #$F0     KEEP SWITCH BITS
 95    C12B   0F                           SEI             BEGIN CRITICAL SECTION
 96    C12C   F6 10 04                     LDAB   PORTB    GET LAST OUTPUT TO PORT B
 97    C12F   C4 0F                        ANDB   #$0F     REMOVE DISPLAY BITS
 98    C131   1B                           ABA             MERGE OUTPUT BITS
 99                                * CONTROL DISPLAY
100    C132   B7 10 04                     STAA   PORTB    OUTPUT TO PORT B
101    C135   0E                           CLI             END CRITICAL SECTION
102                                * RETURN TO MAIN PROGRAM
103    C136   20 0E                        BRA    RTIRQ
104                                *---------------------------------------------------------
```

Figure 6-37   A portion of example program with concurrency control.

finished the first run. It must be *reentered*. Reentering a subroutine may cause problems. A subroutine that is *reentrant* always works correctly, while one that is not reentrant sometimes corrupts the data it is using.

You need to consider reentrancy only if your program uses interrupt. However, most modern applications of microcomputers use interrupt, so reentrancy is important.

Some similarity exists between the concurrency problem and the reentrancy problem. They both involve interrupts, and they both result in corruption of information.

The Data Corruption Problem

Let's first look at a good way to prevent a subroutine from being reentrant. To be nonreentrant, a subroutine must store a number in a specific memory location and then retrieve the number later in the subroutine. Suppose that an interrupt occurs between the point of saving the number and the point of retrieving the number. When the interrupt service routine uses the subroutine, the data in the specific memory location is overwritten. When execution of the subroutine continues from the point of interrupt, the subroutine retrieves erroneous data.

The data corruption problem applies only to data stored in memory by the subroutine. Data in the microprocessor registers is never lost due to an interrupt. The interrupt system automatically saves the microprocessor registers in the stack.

Nonreentrant subroutine example

The program in Figure 6-38 includes a nonreentrant subroutine. It is an example program contrived to show the problems of nonreentrant subroutines. The main program alters a data value that gets corrupted by the nonreentrant subroutine when the interrupt service routine uses it. The subroutine is very simple—it adds two to the A accumulator. The main program tests the altered data value to determine if the subroutine corrupted the data. If so, the program stops.

The example program in Figure 6-38 is almost the same as the program in Figure 6-32. The interrupt service routine reads the thumbwheel switch to get a number for the display. In this example however, the program adds two to the number before displaying it. In addition, the main program loop continually updates a value by adding two to it. This calculation is an example only—it is not a practical program.

Here is a description of the program and the problem the nonreentrant subroutine causes:

- *Lines 50 through 61.* The initialization is as before except that it initializes memory location VALUE to 00.

- *Lines 68 through 70.* The main program loop gets the number in location VALUE into the A accumulator. Let's assume that the data number is 22. At line 68, control goes to the nonreentrant subroutine.

- *Line 116.* The subroutine stores the 22 passed in the A accumulator in memory location SAVE. This number is the data number the subroutine corrupts.

```
 1                                    **************************************************************
 2                                    ** COPY THUMBWHEEL SWITCH PLUS TWO TO DISPLAY WHEN
 3                                    ** PUSHBUTTON IS PUSHED
 4                                    *
 5                                    * USE NONREENTRANT SUBROUTINE
 6                                    *
 7                                    * USE INTERRUPT TECHNIQUE FOR TIMING
 8                                    *
 9                                    * FOUR-BIT THUMBWHEEL SWITCH WIRED TO
10                                    * ..BITS 4-7 OF PORT C
11                                    *
12                                    * FOUR-BIT DISPLAY WIRED TO BITS 0-3 OF PORT C
13                                    *
14                                    * PUSHBUTTON SWITCH WIRED TO STRA SO
15                                    * ..LOW TO HIGH ON PUSH
16                                    *
17                                    **************************************************************
18                                    *
19                                    **************************************************************
20                                    ** SYMBOL DEFINITIONS
21                                    **************************************************************
22                                    * 68HC11 REGISTERS
23            1000                    REG      EQU    $1000     BASE ADDRESS OF REGISTERS
24            1002                    PIOC     EQU    $1002     PARALLEL I/O CONTROL REGISTER
25            1003                    PORTC    EQU    $1003     I/O PORT C REGISTER
26            1005                    PORTCL   EQU    $1005     PORT C LATCH REGISTER
27            1007                    DDRC     EQU    $1007     DATA DIRECTION REGISTER C
28            000F                    IOPAT    EQU    $0F       I/O PATTERN, 0=IN 1=OUT
29                                    * MASKS
30            0080                    BIT7     EQU    %10000000
31                                    *
32                                    **************************************************************
33                                    ** DATA SECTION
34                                    **************************************************************
35    0010                                     ORG    $0010
36    0010                            VALUE    RMB    1         MAIN PROGRAM DATA
37    0011                            SAVE     RMB    1         SUBROUTINE SAVE LOCATION
38                                    *
39    FFF2                                     ORG    $FFF2
40    FFF2  C125                               FDB    IRQISR    IRQ INTERRUPT VECTOR
41                                    *
42                                    **************************************************************
43                                    ** MAIN PROGRAM
44                                    **************************************************************
45    C100                                     ORG    $C100
46                                    *----------------------------------------------------------
47                                    * INITIALIZATION
48                                    *----------------------------------------------------------
49                                    * INITIALIZE STACK
50    C100  8E DF FF                           LDS    #$DFFF
51                                    * INITIALIZE MAIN PROGRAM DATA
52    C103  7F 00 10                           CLR    VALUE
53                                    * INITIALIZE PORT C
54                                    *    INITIALIZE OUTPUTS TO ZEROS
55    C106  7F 10 03                           CLR    PORTC     0=LOW
56                                    *    SET UP INS AND OUTS
57    C109  86 0F                              LDAA   #IOPAT
58    C10B  B7 10 07                           STAA   DDRC      0=IN, 1=OUT
59                                    * SET UP PIOC
60    C10E  86 42                              LDAA   #$42      ENABLE STRA INTERRUPT
61    C110  B7 10 02                           STAA   PIOC      ..LOW TO HIGH
62                                    * TURN ON INTERRUPT SYSTEM
```

Figure 6-38   Example program with nonreentrant subroutine.

```
 63     C113    0E                              CLI
 64                              *------------------------------------------------------
 65                              * WAIT FOR INTERRUPTS
 66                              *------------------------------------------------------
 67                              * GET VALUE FOR EXAMPLE
 68     C114    96 10           HERE     LDAA   VALUE
 69                              * ADD TWO TO VALUE WITH NONREENTRANT SUBROUTINE
 70     C116    BD C2 00                 JSR    ADDTWO
 71                              * TEST IF VALUE WAS CORRUPTED
 72     C119    D6 10                    LDAB   VALUE
 73     C11B    CB 02                    ADDB   #2
 74     C11D    11                       CBA
 75     C11E    26 04                    BNE    ERROR    BRANCH ON YES
 76                              * SAVE VALUE FOR EXAMPLE
 77     C120    97 10                    STAA   VALUE
 78     C122    20 F0                    BRA    HERE
 79                              * STOP IF DATA GETS CORRUPTED
 80     C124    3F              ERROR    SWI                "STOP" FOR TRAINER
 81                              *
 82                              ****************************************************
 83                              ** PUSHBUTTON SWITCH INTERRUPT SERVICE ROUTINE
 84                              ****************************************************
 85                              *------------------------------------------------------
 86                              * VALID STAF INTERRUPT?
 87                              *------------------------------------------------------
 88     C125    CE 10 00        IRQISR   LDX    #REG
 89     C128    1F 02 80 10              BRCLR  PIOC-REG,X,BIT7,RTIRQ   BRANCH ON NO
 90                              *------------------------------------------------------
 91                              * COPY THUMBWHEEL SWITCH TO DISPLAY
 92                              *------------------------------------------------------
 93                              * READ THUMBWHEEL SWITCH AND CLEAR I/O FLAG
 94     C12C    B6 10 02                 LDAA   PIOC     TWO INSTRUCTIONS TO CLEAR
 95     C12F    B6 10 05                 LDAA   PORTCL   ..FLAG AND INPUT SSSS----
 96                              * POSITION DATA FOR OUTPUT
 97     C132    46                       RORA            -SSSS---
 98     C133    46                       RORA            --SSSS--
 99     C134    46                       RORA            ---SSSS-
100     C135    46                       RORA            ----SSSS
101                              * ADD TWO TO DISPLAY NUMBER
102     C136    BD C2 00                 JSR    ADDTWO   MAY CORRUPT DATA IN MAIN!
103                              * CONTROL DISPLAY
104     C139    B7 10 03                 STAA   PORTC    OUTPUT TO PORT C
105                              *------------------------------------------------------
106                              * RETURN TO MAIN PROGRAM
107                              *------------------------------------------------------
108     C13C    3B              RTIRQ    RTI
109                              *
110                              ****************************************************
111                              ** NONREENTRANT SUBROUTINE ADDTWO
112                              * ADD 2 TO A ACCUMULATOR--MODIFIES CC
113                              ****************************************************
114     C200                             ORG    $C200
115                              * INCORRECTLY STORE DATA IN FIXED MEMORY LOCATION
116     C200    97 11           ADDTWO   STAA   SAVE
117                              * GET NUMBER TWO
118     C202    86 02                    LDAA   #2
119                              * RETRIEVE INCORRECTLY STORED DATA
120     C204    9B 11                    ADDA   SAVE
121                              * RETURN FROM SUBROUTINE
122     C206    39                       RTS
123     C207                             END
```

Figure 6-38 Continued.

- *Line 118.*   Assume that an interrupt occurs at the completion of the instruction on line 116 or line 118. The interrupt must occur at these points or the problem will not occur. Therefore, the chances of the interrupt occurring at this exact point are small. Control next goes to the instruction on line 88.

- *Lines 88 through 100.*   When an interrupt occurs, the interrupt system hardware transfers control here. This part of the interrupt service routine clears the flag, reads the thumbwheel switch, and positions the data.

- *Line 102.*   The interrupt service routine calls the subroutine to add two to the number from the thumbwheel switch. Assume we set the thumbwheel switch to 5.

- *Line 116.*   The subroutine stores the number 05 in memory location SAVE. This destroys the number 22 that the subroutine stored at SAVE on the last run. This problem makes the subroutine nonreentrant!

- *Lines 118 through 122.*   The subroutine correctly adds two and returns the correct result 07.

- *Lines 104 through 108.*   The interrupt service routine finishes correctly and returns control to the main program loop.

- *Line 118.*   The subroutine continues the first run. However, the ADDA instruction adds two to the incorrect number 07 in location SAVE. The subroutine returns the incorrect result 09 to the main program.

- *Lines 72 through 75.*   In this example program, the main program checks for a reentrancy error. If it finds one, it branches to line 80 and stops.

- *Lines 77 and 78.*   If the interrupt did not occur at a point that causes a reentrancy error, the program stores the updated number in location VALUE. It then sends control to the beginning of the main program loop.

Reentering a subroutine is only a problem if the interrupt occurs at the correct time when the subroutine is at the correct point. Even if the chances of this interrupt are small, it will eventually happen and the program will fail.

The nonreentrant subroutine solution

The nonreentrant subroutine problem has only one solution—don't write subroutines that allow it to happen! Write only good programs that don't have problems. Testing your program cannot prove that it works correctly.

Data Corruption Solutions

The best way to make a subroutine reentrant is to avoid storing data numbers so that the subroutine overwrites them upon reentry. The following discusses several techniques that avoid the problem of data corruption.

Use the stack

Saving data in the stack is the best solution to the reentrancy problem! Then, when the subroutine is reentered, it uses new locations in the stack and does not destroy any saved data. Figure 6-39 shows a new reentrant version of the nonreentrant subroutine from Figure 6-38. The subroutine makes a working location in the stack. The stack location replaces location SAVE in the original program.

The subroutine in Figure 6-39 looks very difficult when it does such a simple job, but its purpose is to illustrate the programming technique. Part of the illustration is to store data in memory in a way that allows reentrancy.

The Motorola 68HC11 instruction set makes reentrant subroutines possible. The instruction set has instructions to save the registers A, B, X, and Y in the stack. Therefore, the subroutine can store temporary data in the stack. Furthermore, all data manipulation instructions, by using indexed addressing, can reach the data values inside the stack. For example, a subroutine that uses a loop counter can put the counter in the stack.

Some Motorola 8-bit microprocessors do not have instructions to save the index register in the stack or to transfer the index register to another microprocessor register. Therefore, when using such a microprocessor, any subroutine that needs to save the index register and then restore it later cannot be reentrant. The subroutine must store the index register in a dedicated memory location.

Put no data in memory

The example subroutine can be modified to use only data values in microprocessor registers as a solution to the reentrancy problem. Storing no data in memory registers avoids the data corruption problem. Remember the microprocessor registers are saved in the stack by the interrupt system hardware. Figure 6-40 shows a variation of the subroutine from Figure 6-38 that uses only the microprocessor registers.

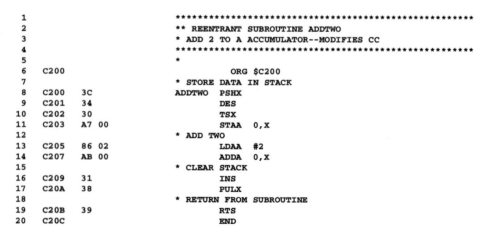

```
 1                        ******************************************************
 2                        ** REENTRANT SUBROUTINE ADDTWO
 3                        * ADD 2 TO A ACCUMULATOR--MODIFIES CC
 4                        ******************************************************
 5                        *
 6    C200                            ORG  $C200
 7                        * STORE DATA IN STACK
 8    C200  3C            ADDTWO   PSHX
 9    C201  34                     DES
10    C202  30                     TSX
11    C203  A7 00                  STAA  0,X
12                        * ADD TWO
13    C205  86 02                  LDAA  #2
14    C207  AB 00                  ADDA  0,X
15                        * CLEAR STACK
16    C209  31                     INS
17    C20A  38                     PULX
18                        * RETURN FROM SUBROUTINE
19    C20B  39                     RTS
20    C20C                         END
```

Figure 6-39 Reentrant subroutine that uses the stack.

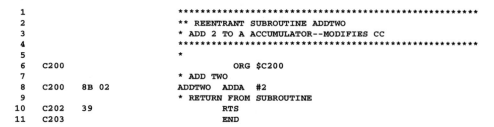

```
 1                                    ********************************************************
 2                                    **  REENTRANT  SUBROUTINE  ADDTWO
 3                                    *  ADD  2  TO  A  ACCUMULATOR--MODIFIES  CC
 4                                    ********************************************************
 5                                    *
 6    C200                                        ORG  $C200
 7                                    *  ADD  TWO
 8    C200      8B  02                ADDTWO   ADDA   #2
 9                                    *  RETURN  FROM  SUBROUTINE
10    C202      39                             RTS
11    C203                                     END
```

Figure 6-40   Reentrant subroutine that uses microprocessor registers.

This subroutine is now quite simple. However, complex subroutines usually need more storage space than the microprocessor registers can provide. When possible, using only the microprocessor registers avoids many problems.

Prevent interrupts

Preventing interrupts while the subroutine runs is another solution to the reentrancy problem. For example, the main program may disable the interrupt system using the SEI instruction before jumping to the subroutine. When control returns to the main program, it enables the interrupt system again.

Preventing interrupts to solve the reentrancy problem has a serious disadvantage. The computer will not service the I/O devices while the subroutine is running. Therefore, the I/O devices operate more slowly because they must wait for the subroutine to finish before getting service. The I/O device cannot interrupt until the subroutine returns and the main program enables the interrupt system again.

Alternatively, the subroutine can enable and disable the interrupt system. Then the subroutine can disable the interrupt system for the minimum amount of time. That is, the subroutine disables the interrupt system only during the part of the subroutine that may cause a problem. However, use this technique with caution. The subroutine may enable the interrupt system when it should not. For example, if the interrupt service routine runs with the interrupt system disabled, the subroutine would enable it prematurely.

Don't reenter

Using multiple copies of the subroutine that each store data in different memory locations avoids the reentrancy problem. Each program that can enter the subroutine uses a different copy of the subroutine. Of course, this solution defeats much of the advantage of using a subroutine.

More Examples of Reentrant Subroutines

The subroutine examples in Figures 5-11, 5-17, 5-18, 5-19, and 5-21 are good reentrant subroutines. Use those subroutines as examples of good technique. They keep temporary subroutine data in microprocessor registers or in the stack. Also look at the discussion of recursive subroutines in Chapter 5. A recursive subroutine also is reentrant.

6.14   TROUBLESHOOTING CONCURRENCY AND REENTRANCY PROBLEMS

Determining that a computer system has concurrency problems or reentrancy problems through observation of the running computer is very difficult if not impossible. Many circumstances must converge to make a program fail.

Generally, the interrupts that trigger these problems come from I/O devices that are outside the computer. The I/O devices are in no way synchronized with the operation of the computer. Effectively, the interrupts occur at random places within the program. As a result, simulating a situation that will cause a problem is almost impossible. To a large extent, luck is involved. To be more formal, these problems are not *deterministic*.

What You Are Up Against

Consider the nonreentrant subroutine example in Figure 6-37. The subroutine corrupts data because it stores a number in memory location **SAVE**. Now consider the coincidence of events that must happen for the subroutine to cause a problem.

First, the main program must have jumped to the subroutine that is nonreentrant. Next, an I/O device must cause an interrupt after the **STAA SAVE** instruction is executed but before the **LDAA SAVE** instruction executes. Remember, the I/O device is external to the computer. You have no way of knowing when it will set the I/O flag. Finally, the interrupt service routine must jump to this subroutine. All these events must occur to make the program fail.

Don't Be Misled

The chance of every one of the events necessary to cause concurrency or reentrancy problems happening is small. Therefore, problems occur infrequently and at random times. If you were testing such a program, you may run the computer for hours and hours without a problem. But, eventually, the series of events will occur and the program will malfunction. Often the failure is very surprising. You may have run the program for months without a problem.

Such behavior frequently baffles people. They sometimes dismiss the failure as a one-time occurrence due to electrical noise. Sometimes they load the program into memory again assuming that something changed the program code. They may replace memory ICs believing that an intermittent failure occurs in an IC. How else could a program that works perfectly for hours to months suddenly fail?

You should now understand that such failures can occur due to incorrectly designed and written software.

Do It Right the First Time

The best solution to these problems is to write good programs that avoid the problems. This involves extra effort to understand the problems, to remain watchful for their appearance, and to use good programming techniques at all times. Troubleshooting systems with problems will require careful scrutiny of all the program code in a system and rewriting of problem routines.

6.15   REVIEW

Chapter 6 introduced the basic hardware in the microcomputer. The microprocessor connects to the memory and input/output hardware with the address, data, and control buses. Memory-mapped I/O treats the I/O parts the same as memory using the same bus signals. Therefore, identical wiring connects the microprocessor to the memory and the I/O hardware. However, the I/O hardware does not function the same as memory. In practical applications, load and store instructions transfer data to and from I/O devices. Timing or synchronizing those transfers to the program requires an I/O flag and a means of testing that flag. The I/O device sets the flag to indicate that it is ready for a data transfer. The computer clears the flag to indicate that it is responding to the request for a data transfer. The computer detects the flag condition using either polling or interrupt. Interrupt is the most common and effective timing technique. When using interrupt, be careful to write reentrant subroutines and to prevent concurrency problems.

6.16   EXERCISES

6-1. What is a volatile memory?

6-2. If a microcomputer is embedded within a product, what types of memory ICs would be appropriate to hold the program code?

6-3. Why do embedded microcomputers usually have nonvolatile memory at the highest memory addresses and RAM memory at the lowest addresses?

6-4. Why do people usually put a paper sticker over the window of EPROMs?

6-5. Is the lifetime of an EEPROM measured in erase/program cycles approximately equal to the lifetime of a RAM measured in read/write cycles?

6-6. A RAM memory IC has 12_{10} address pins. How many registers does this IC contain?

6-7. During the execute phase of a **LDAA** instruction, is the direction of the data bus into or out of the microprocessor? What is it during the fetch phase?

6-8. Under what circumstance would thinking of the R/W line as an input/output line be appropriate?

6-9. During the execution of a **STAA** instruction, is the R/W line low during the fetch phase or during the execute phase?

6-10. How many bytes are in the largest memory that the 68HC11 can directly address?

6-11. When does the data bus carry a number into the microprocessor? Out of the microprocessor?

6-12. Does the execution of a INC memory instruction use the data bus to both read a data number from memory and to store a data number into memory?

6-13. During the fetch phase of a ROL instruction with extended addressing, the direction of the data bus is (into, out of, both directions at different times, unknown) _____ the microprocessor.

6-14. Why does the instruction in Figure 6-12 load 3C into the A accumulator?

6-15. In Figure 6-14, the lowest possible address that will access a register in the bottom memory chip is _____.

6-16. In Figure 6-14, the highest possible address that will access a register in the bottom memory chip is _____.

6-17. What is the highest possible address and the lowest possible address that will make the output port in Figure 6-18 respond?

6-18. In Figure 6-18, if the AND gate were connected to output 9 instead of output 4 of the decoder, what range of addresses would make the output port respond?

6-19. In Figure 6-18, if the inputs to the AND gate now connected to A_1 and A_0 were instead connected to A_3 and A_2, what range of addresses would make the output port respond?

6-20. Can an input port and an output port both have the same address in the same computer?

6-21. Change the circuit in Figure 6-18 so that the output register can be read by the microprocessor with the same address used to write it.

6-22. Using the fewest number of logic devices, change the circuit in Figure 6-18 by adding a second output port that responds to address 4007.

6-23. Can a single STD instruction be used to output to two 8-bit ports with the addresses 4010 and 4011?

6-24. What is the effect of changing the instruction on line 38 of Figure 6-21 to CLR IMAGE?

6-25. Discuss the pros and cons of input and output ports with addresses in the direct addressing range.

6-26. Design a circuit with both a RAM and a ROM memory at the same addresses with hardware to switch between them under program control.

6-27. Can the BCLR instruction be used to clear the STAF flag in the PIOC register?

6-28. PORTC has been programmed so that the odd-numbered bits are inputs and the even-numbered bits are outputs. Discuss the results of using the DEC instruction on PORTC as it is now programmed.

6-29. PORTB drives two display units, like those in Figure 6-34, connected to bits 0 through 3 and bits 4 through 7. Discuss the effect of trying to increment the displayed two-digit number by using the INC instruction on PORTB.

6-30. Write a polling program to read two 4-bit thumbwheel switches and display the results on two display units each time a push button is pushed. Use the PORTB and PORTC registers and the STAF flag.

6-31. Refer to the program in Figure 6-29. Describe what will be seen on the display unit if the push button is pushed and held while the thumbwheel is changed?

6-32. Refer to the program in Figure 6-32. Assume that the flag is always set because the clearing hardware failed. Describe how the program will perform including what the thumbwheel switch and the display do.

6-33. Refer to the program in Figure 6-32. Assume that the flag is always cleared because the flag hardware failed. Describe how the program will perform including what the thumbwheel switch and the display do.

6-34. Refer to the program in Figure 6-32. Describe how the program performs if a CLI instruction is inserted at line 69 and the existing lines 69 and higher are moved down one line.

6-35. Refer to Figure 6-41 which is the end of Figure 6-32 with a small change. In Figure 6-32, PORTC was read into accumulator A, accumulator A was rotated, and then accumulator A was stored back into PORTC. In Figure 6-41, the rotate instruction works directly on PORTC. Unfortunately, after this change, the program no longer works correctly. Explain why it no longer works, and exactly what will be output on PORTC.

6-36. Suppose that the 68HC11 instruction set included a test-and-set-bit instruction. This type of instruction tests a memory bit leaving a result in a condition code register bit, and then sets the memory bit. Both operations are carried out by the execution of this single instruction. Explain how this instruction and a memory bit dedicated to a critical section of memory could be used to lock a resource to prevent the concurrency problem.

6-37. Explain why the program in Figure 6-35 does not have concurrency problems.

6-38. Refer to the program listing in Figure 6-35. If the programmer erroneously entered a NOP instruction instead of the CLI instruction, what will the program do? Explain what each output does.

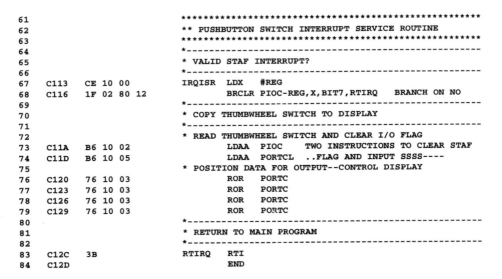

```
61              ***********************************************************
62              ** PUSHBUTTON SWITCH INTERRUPT SERVICE ROUTINE
63              ***********************************************************
64              *---------------------------------------------------------
65              * VALID STAF INTERRUPT?
66              *---------------------------------------------------------
67   C113  CE 10 00     IRQISR  LDX   #REG
68   C116  1F 02 80 12          BRCLR PIOC-REG,X,BIT7,RTIRQ   BRANCH ON NO
69              *---------------------------------------------------------
70              * COPY THUMBWHEEL SWITCH TO DISPLAY
71              *---------------------------------------------------------
72              * READ THUMBWHEEL SWITCH AND CLEAR I/O FLAG
73   C11A  B6 10 02            LDAA  PIOC    TWO INSTRUCTIONS TO CLEAR STAF
74   C11D  B6 10 05            LDAA  PORTCL  ..FLAG AND INPUT SSSS----
75              * POSITION DATA FOR OUTPUT--CONTROL DISPLAY
76   C120  76 10 03            ROR   PORTC
77   C123  76 10 03            ROR   PORTC
78   C126  76 10 03            ROR   PORTC
79   C129  76 10 03            ROR   PORTC
80              *---------------------------------------------------------
81              * RETURN TO MAIN PROGRAM
82              *---------------------------------------------------------
83   C12C  3B          RTIRQ   RTI
84   C12D                      END
```

Figure 6-41 Exercise 6-35.

6-39. Refer to the program listing in Figure 6-35. If the programmer erroneously omitted the LDS instruction, what will the program do?

6-40. Refer to the program listing in Figure 6-35. What is the effect of putting the instructions on lines 104 and 105 in the opposite order?

6-41. Using the hardware in Figure 6-34, change the program in Figure 6-35 to make the BCD display count up by one periodically. Use the oscillator interrupts as the time base. Read the thumbwheel switch to determine the rate at which counting should take place. The number on the switch multiplied by five is the number of oscillator interrupts between updates of the display. Also make the light toggle each time the push button is pushed.

Chapter 7

Advanced 68HC11 Hardware

The 68HC11 integrated circuit contains many input/output devices and several types of memory. The I/O devices are those frequently required for control applications. The 68HC11 can be expanded by adding external input/output devices and memory. If a change in an application requires additional hardware, expanding the hardware is easy.

This chapter discusses the hardware devices inside the 68HC11 chip. Example programs illustrate their operation. Furthermore, this chapter discusses 68HC11 expansion using the microprocessor buses and the SPI bus.

This chapter also introduces techniques of event timing, waveform generation, serial communications, and fail-safe operation. All these techniques are important to control applications of microcomputers. The 68HC11 internal devices are used in examples of these techniques.

7.1 THE HARDWARE CONFIGURATION

The 68HC11 hardware is programmable. Both the microcomputer hardware configuration and the input/output devices are programmable. Numbers in registers control how each hardware device operates. The design of the 68HC11 allows a wide range of applications.

Operating Modes

The 68HC11 operates in one of two modes called the *single-chip microcomputer mode* and the *expanded mode*. Hardware signals decide the operating mode as the microcomputer is powered-up. Software can never change the operating mode, because doing so would be dangerous.

In the single-chip mode, the 68HC11 uses only the resources inside the chip. These resources are primarily memory and input/output devices. The address, data, and control buses are not available for expanding the computer. However, considerable expansion is possible using the high-speed SPI serial bus with ICs designed to communicate serially.

In the expanded mode, the 68HC11 connects the internal address, data, and control buses to pins. Consequently, in expanded mode, some internal I/O devices cannot be used. The pins those devices use in the single-chip mode are used for bus signals in the expanded mode. The devices lost are two parallel I/O ports and a flag. However, having the buses available outside the chip makes expanding the microcomputer with external memory and I/O chips possible. Later sections of this chapter discuss 68HC11 expansion.

The operating mode of the microcomputer has little effect on the programmer. Only the character of the hardware affects the program.

The 68HC11 Chip Versions

The 68HC11 chip is available in several versions. Most versions differ principally in the amount and kind of memory inside the chip. Some versions omit part of the memory and some I/O devices. Others, such as the 68HC11A1, are the same part as another, but they are shipped with internal programming to disable the internal ROM.

The 68HC11A8 and 68HC11E9 versions accurately represent the entire 68HC11 family. The A8 version has eight kilobytes of masked ROM and 256_{10} bytes of RAM. The E9 version has 12_{10} kilobytes of masked ROM and 512_{10} bytes of RAM. An additional version, the 68HC711, substitutes EPROM for the masked ROM in the E9 version. The A8 and E9 versions have a full complement of I/O devices.

Each variety of the 68HC11 is functionally the same—the differences are mainly in the amount of memory in the chip. However, some versions have larger hardware differences. For example, some can operate only in the expanded mode.

The examples in this book are based on the assumption that the A8 or E9 versions are being used. However, all examples apply to other versions that have functionally equivalent hardware.

Block Diagram of the 68HC11

Figure 7-1 is the block diagram of the 68HC11 operating in the single-chip mode. All input/output devices built into the chip are available in this mode. Figure 7-2 is the block diagram for the expanded mode.

The block diagrams show the pins on the 68HC11 package and some of the pin functions. The pins connect the microcomputer to other devices, so they are most visible to the

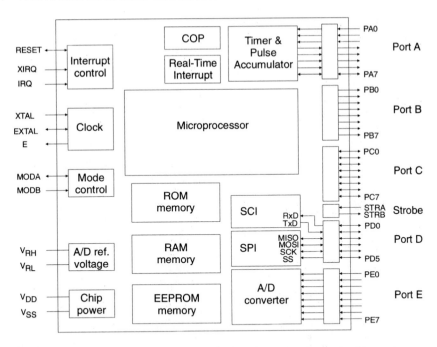

Figure 7-1   Block diagram of the 68HC11A8 hardware while in single-chip mode.

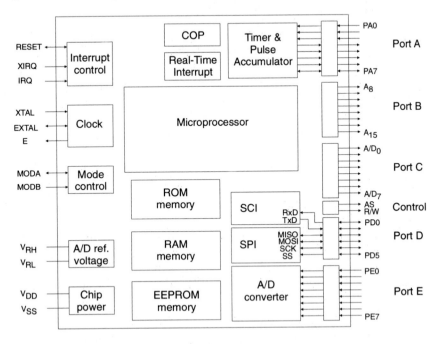

Figure 7-2   Block diagram of the 68HC11A8 hardware while in expanded mode.

user of the computer. The collections of pins called *ports A through E* are input/output, address bus, and data bus signals.

The block diagrams also show the internal organization of the chip. Here is a description of the hardware in the IC package:

- *The 68HC11 microprocessor.* The 68HC11 microprocessor executes a total of 307_{10} instruction codes.

- *ROM memory.* The A8 chip has eight kilobytes of masked ROM, and the E9 chip has 12_{10} kilobytes of masked ROM. The ROM addressing places it at the high end of memory, so the ROM includes address FFFF. A programmable control register can disable the internal ROM so that other memory can use these addresses.

- *RAM memory.* The A8 chip has 128_{10} bytes of static RAM, and the E9 chip has 512_{10} bytes of static RAM. The RAM addressing places it at the low end of the memory space, including address 0000. The RAM is thus in the direct addressing range. A programmable control register can move the memory to other addresses.

- *EEPROM memory.* Both the A8 and E9 versions contain 512_{10} bytes of EEPROM memory. The E9 chip has a control register that prevents erroneous changing of the EEPROM.

- *Internal clock oscillator.* Only a crystal and two capacitors are necessary to build the clock for the 68HC11.

- *SCI serial communications interface.* The SCI asynchronous serial device allows communication to remote serial I/O devices.

- *SPI serial peripheral interface.* The SPI interface allows both control of other ICs in the microcomputer and communications to another nearby circuit board.

- *Parallel input/output ports.* Several parallel I/O ports, including some programmable ports, allow communication to external I/O devices.

- *Programmable timer.* The timer device allows measurement of timing intervals and timed control of outputs. Quantities such as RPM are easily measured. Accurate output waveforms, such as pulse-width modulated waveforms, are easily generated.

- *Analog-to-digital converter.* Eight channels of the A/D converter measure DC voltages. Including analog hardware in the chip allows sensors of many kinds to interface easily to the 68HC11.

- *Pulse accumulator.* The pulse accumulator can both count pulses and measure the duration of pulses from I/O devices outside the computer.

- *Computer operating properly, clock monitor, and illegal instruction devices.* Each of these devices forces a safe hardware response when certain computer hardware or software has failed.

A block of registers, usually mapped to address 1000, controls all internal I/O devices. Appendix B contains a summary of all these control registers. These registers are described throughout this chapter.

68HC11 Pin Connections

Look at the pins in Figures 7-1 and 7-2. Here is a brief description of the pin functions:

- V_{DD} and V_{SS}.   The power and ground pins are called V_{DD} and V_{SS}. The 68HC11 operates with V_{DD} at +5 V_{DC}.

- *XTAL and EXTAL.*   The crystal for the clock connects to the pins named XTAL and EXTAL. The crystal usually operates near 8 MHz.

- *MODA and MODB.*   The MODA and MODB pins determine the operating mode of the 68HC11 chip.

- *RESET.*   The RESET pin initializes the microcomputer hardware.

- *IRQ and XIRQ.*   The interrupt system has two external interrupt lines called *IRQ* and *XIRQ*. The internal STAF flag connects to the IRQ interrupt line.

- *Ports A through E.*   In single-chip mode, all ports illustrated in Figure 7-1 support input/output. In the expanded mode, ports B and C, as illustrated in Figure 7-2, carry address and data bus signals while the other ports support input/output.

Some 68HC11 pins have multiple functions. For example, the MODB pin also provides standby power for the internal RAM memory. It can power the RAM while the rest of the chip is powered down.

7.2   MEMORY SPACE

Figure 7-3 illustrates the memory space in the 68HC11. The figure shows the internal memory addresses after the 68HC11 has been powered-up in the single-chip microcomputer mode. The figure shows the differences between the A8 and E9 versions of the integrated circuit. Any unlabeled area of memory can be external memory when the 68HC11 is in expanded mode. In single-chip microcomputer mode, the unlabeled areas are unused.

Internal ROM

The internal masked ROM is always located so that the high end includes address FFFF. The ROM therefore contains the interrupt vectors. Usually, the program and constant data values are stored in this ROM.

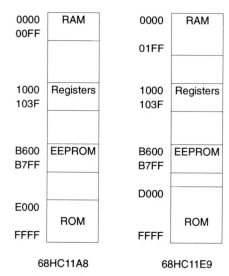

Figure 7-3   Internal memory of the 68HC11A8 and 68HC11E9 ICs.

Internal RAM

The internal static RAM is usually located so that its lowest address is 0000. Therefore, the RAM is in the direct addressing range. Because the ROM memory is usually much smaller than 64_{10} kilobytes to reduce cost, using the shorter direct addressed instructions to reach data is important. Normally, having RAM in the direct addressing range is desirable.

A separate pin on the 68HC11 powers the internal RAM. Therefore, the rest of the IC can be off while the RAM memory retains its information. The RAM usually gets its power from a backup battery while the rest of the IC is off. Retaining information during power down is very valuable when the computer stores diagnostic information in the RAM. The standby voltage supply is connected to the MODB/V_{STBY} pin.

When the RAM has a standby voltage supply, the RAM is called *keep-alive* memory. The battery does not make the RAM a nonvolatile memory, but the battery does enable the RAM to retain information while the computer is off. Regardless, the RAM with a battery is commonly called a *non-volatile RAM*.

Internal EEPROM

The internal EEPROM is always at address B600. The microprocessor reads EEPROM registers the same as any other memory registers. However, programming the EEPROM requires a setup procedure that reduces the chance of accidental changes to the EEPROM. The programming is done one byte at a time.

Programming the EEPROM changes 1s to 0s. Therefore, programming a register with 1s if 0s are already programmed requires erasing the registers first. The erased state is all 1s. The erasing hardware erases either the entire EEPROM, groups of registers, or individual registers. Erasing larger groups requires less time than erasing multiple individual bytes.

Registers

A block of memory addresses is reserved for the hardware control registers. These are all the registers for control of memory and the input/output devices. Appendix B lists all the hardware control registers in the 68HC11. Look at the table in Appendix B to get an overview of all the control registers. The purpose of these registers becomes clear as various hardware devices are introduced. Generally, you need only think about one or two of these registers at a time. Don't be discouraged by the many registers and the many control bits.

Memory configuration and mapping

The internal RAM and hardware control registers can be moved to different addresses by software. Similarly, software can disable the internal memory.

The INIT register.   Figure 7-4 shows the *INIT register*, which determines the addresses of the internal RAM and hardware control registers block. The INIT register specifies the most-significant four bits of the addresses. The reset condition makes the RAM start at address 0000 and the registers start at address 1000. For example, if software changed the INIT register to 20, the RAM would begin at address 2000 and the registers at address 0000.

If moved from the reset locations, the RAM or control registers can conflict with the ROM or with each other. If so, the RAM has precedence over the ROM, and the registers have precedence over the RAM.

	7	6	5	4	3	2	1	0	
$103D	RAM3	RAM2	RAM1	RAM0	REG3	REG2	REG1	REG0	INIT
RESET	0	0	0	0	0	0	0	1	

Figure 7-4   RAM and I/O mapping register.

The CONFIG register.   The functions of the *CONFIG register* include enabling and disabling the internal ROM and EEPROM memories. Other bits in the CONFIG register control other options.

Figure 7-5 shows the bits in the CONFIG register. The *ROMON bit* enables the ROM memory when it is 1 and disables it when it is 0. Read the name ROMON as *ROM on*. However, the ROM can never be disabled in the single-chip mode, because the computer must have the ROM to operate. If the ROM is disabled, its memory space is available on the external buses.

	7	6	5	4	3	2	1	0	
$103F	0	0	0	0	NOSEC	NOCOP	ROMON	EEON	CONFIG
RESET				CONFIG register bits are not affected by Reset.					

Figure 7-5   System configuration register.

The EEON bit in the CONFIG register enables the EEPROM memory when it is 1, and disables it when it is 0. If the EEPROM is disabled, its memory space is externally available in the expanded mode. Read the name EEON as *EE on*.

The CONFIG register is an EEPROM register. It is programmed as the EEPROM memory is programmed. Because the CONFIG register is a nonvolatile register, it determines the configuration of the microcomputer hardware when the 68HC11 is powered-up.

7.3   HARDWARE RESET

The reset procedure determines the way much of the hardware in the 68HC11 operates. Usually the RESET pin is used when the power is turned on to the computer, so it affects all future operations of the computer. For example, the MODA and MODB pins are read during reset to determine whether the 68HC11 operates in expanded mode or single-chip mode.

Reset Operation

The block diagrams in Figures 7-1 and 7-2 show the RESET pin. The RESET signal forces a hardware initialization of all hardware devices in the microcomputer—including the 68HC11. Most, but not all, control bits in the internal 68HC11 registers are forced to 0 by RESET. A low signal voltage on the RESET pin resets the hardware.

Reset sources

Many hardware devices can drive the RESET pin low. Some of these are inside the 68HC11 IC and others are outside it. Figure 7-1 shows the RESET pin as bidirectional to indicate that devices inside or outside the 68HC11 can cause the reset. Some common reasons for reset are turning on the computer or detecting a hardware error.

Software effects

Normally, the RESET pin is used at program start time to ensure that the electronic circuits in the computer start at a known condition. First, the reset initializes the electronics in the microprocessor and the rest of the microcomputer. Then the microprocessor gets an address from the reset vector and puts it into the program counter. The computer starts running at this address. The reset vector for all Motorola 8-bit microprocessors is at addresses FFFE and FFFF.

Reset is not an interrupt. The reset vector is not an interrupt vector. The reset vector only determines the starting address of the program. If something forces a reset while a program is running, the computer stops running that program and begins again at the address specified in the reset vector. The status of the microprocessor is not saved in the stack, so nothing is restored.

Power-on reset

The DC power supply for the 68HC11 requires a little time to stabilize the voltage after it is turned on. A circuit outside the 68HC11 keeps RESET low while the voltage stabilizes. After the voltage is stable, the circuit changes the level on the RESET pin to high. The microcomputer responds by loading the program counter with the address stored at the reset vector and starts running.

The 68HC11 also has an internal power-on timer. The internal circuit senses when the supply voltage V_{DD} changes from low to high. The power-on circuit forces an internal reset for the next 4064_{10} E-clock cycles, and then responds to the RESET pin. In typical applications, this causes a delay of about two milliseconds even if the power supply stabilizes faster. Therefore, the 68HC11 will always require a small time after power-on before it starts operating.

Hardware Programmable Options

The operating mode of the 68HC11, either single-chip mode or expanded mode, is selected during a hardware reset. When the RESET pin is low, the MODA and MODB pins select one of four modes. Two of these are test modes—not normal operating modes.

Table 7-1 lists the modes and the pin values. The pins use high-asserted signals. For example, if both MODA and MODB are connected to V_{DD} with pull-up resistors, a reset puts the 68HC11 into the expanded mode.

After reset, software cannot change the mode of the 68HC11. If software could change the operating mode, an errant program could accidentally change the mode. This change could lead to dangerous operation.

Timed-Write-Once Programmable Options

Many hardware options in the 68HC11 chip are programmable. Numbers in control registers determine the configuration of the programmable hardware. Reset initially configures the programmable hardware. Following reset, the program usually changes the configuration once in the initialization. Only rarely does a program change the configuration many times as the program runs.

Erroneous changes to certain programmable hardware could cause unsafe operation. To avoid errors, some 68HC11 programming bits only respond to one write after a reset.

TABLE 7-1   MODE SELECTION

MODB Pin	MODA Pin	Mode Selected
1	0	Single-chip
1	1	Expanded
0	1	Special test
0	0	Special bootstrap

Thereafter, these special bits are read-only bits; further writes cannot change them. Call these special control bits *timed-write-once bits*.

Besides the write-once feature, the timed-write-once bits also will not change after 64_{10} E-clock cycles after the RESET pin goes high. Therefore, if the reset condition of the timed-write-once bits is incorrect, the program must initialize them immediately after the RESET. Thereafter, the program cannot change them.

The timed-write-once bits add a measure of safety to the computer. They reduce the chance of an incorrect hardware configuration. The effects of hardware failures that cause errant programs are reduced if the hardware configuration cannot change.

7.4 INTERRUPT SYSTEM

Chapter 6 introduced the operation of the interrupt system. Chapter 7 adds many more details. The interrupt system in the 68HC11 supports several types of interrupts from many input/output devices. The 68HC11 also has a hardware reset function that is similar to an interrupt. Hardware signals trigger both interrupts and resets. The hardware signals may originate both within and outside the chip. Before examining the many input/output devices in the 68HC11, let's look at the details of the interrupt system.

Most interrupts in the 68HC11 have separate interrupt vectors that send control to separate interrupt service routines. The interrupt vectors contain the addresses of the interrupt service routines for each device.

The following discussion mentions only the interrupt vectors not associated with particular input/output devices. Later sections devoted to specific I/O devices discuss the interrupts for those devices.

Refer to Table 7-2. It lists the many interrupt vectors in the 68HC11. The interrupts from the various I/O devices all operate similarly. Looking at the table gives an overview of the details to follow.

IRQ Interrupt

The IRQ interrupt is named from the phrase "interrupt request." The IRQ interrupt is the interrupt discussed in Chapter 6. The examples there use both the STAF flag inside the 68HC11 and external flags connected to the IRQ pin to cause interrupts. Only the STAF flag connects to the IRQ line inside the 68HC11 chip.

The IRQ interrupt is the principal external interrupt line in the 68HC11. It has a single interrupt vector to send control to a single interrupt service routine. The IRQ interrupt in the 68HC11 is nearly the same as the IRQ interrupt in other Motorola 8-bit microprocessors.

Level or edge triggering for IRQ

The IRQ interrupt pin can be programmed to interrupt at either the low level or on a high-to-low transition. The IRQE bit determines which condition triggers the interrupt. When IRQE is 0, the IRQ pin is programmed for low-level assertion.

TABLE 7-2 INTERRUPT VECTOR ADDRESSES AND MASK BITS

Vector Address	Interrupt Device	Condition Code Mask Bit	Local Device Mask Bit
FFC0, C1 through FFD4, FFD5	Reserved	None	None
FFD6, FFD7	SCI Receive Data Register Full	I	RIE
	SCI Receiver Overun	I	RIE
	SCI Idle Line Detect	I	ILIE
	SCI Transmit Data Register Empty	I	TIE
	SCI Transmit Complete	I	TCIE
FFD8, FFD9	SPI Serial Transfer Complete	I	SPIE
FFDA, FFDB	Pulse Accumulator Input Edge	I	PAII
FFDC, FFDD	Pulse Accumulator Overflow	I	PAOVI
FFDE, FFDF	Timer Overflow	I	TOI
FFE0, FFE1	Timer Output Compare 5	I	OC5I
FFE2, FFE3	Timer Output Compare 4	I	OC4I
FFE4, FFE5	Timer Output Compare 3	I	OC3I
FFE6, FFE7	Timer Output Compare 2	I	OC2I
FFE8, FFE9	Timer Output Compare 1	I	OC1I
FFEA, FFEB	Timer Input Capture 3	I	IC3I
FFEC, FFED	Timer Input Capture 2	I	IC2I
FFEE, FFEF	Timer Input Capture 1	I	IC1I
FFF0, FFF1	Real Time Interrupt	I	RTII
FFF2, FFF3	IRQ (External Pin or Parallel I/O)	I	None
	IRQ Parallel I/O	I	STAI
FFF4, FFF5	XIRQ Pin	X	None
FFF6, FFF7	SWI	None	None
FFF8, FFF9	Illegal Opcode Trap	None	None
FFFA, FFFB	COP Timeout	None	NOCOP
FFFC, FFFD	Clock Monitor Timeout	None	CME
FFFE, FFFF	RESET	None	None

Normally, the IRQ pin is used in a wired-OR circuit using low-level assertion. The external flags use open-collector or open-drain hardware. These devices pull the IRQ line to the low level to cause an interrupt.

Figure 7-6 shows the OPTION control register. The IRQE bit is bit 5 of the OPTION register. When IRQE is 1, the IRQ pin is programmed for the high-to-low transition.

When the 68HC11 is reset, IRQE is set to 0, so IRQ is level-sensitive if the program does not change it. The IRQE bit is a timed-write-once bit.

Level-sensitive interrupt is most commonly used. It is also the only choice in other Motorola 8-bit microprocessors. Therefore, the level-sensitive interrupt is used for all example programs in this book.

	7	6	5	4	3	2	1	0	
$1039	ADPU	CSEL	IRQE	DLY	CME	0	CR1	CR0	OPTION
RESET	0	0	0	1	0	0	0	0	

Figure 7-6 Configuration options register.

Software Interrupt

An instruction can cause an interrupt. The SWI or *software interrupt* instruction triggers an interrupt. The software interrupt acts like an IRQ interrupt, except that it is triggered by an instruction and there is a separate interrupt vector. Furthermore, the software interrupt cannot be disabled; if the instruction is fetched, the interrupt occurs. However, the SWI instruction disables the interrupt system with the I bit as other interrupts do. When the interrupt service routine returns control to the main program, it continues at the next instruction after SWI.

The SWI instruction may be used when the program detects a fatal error. Instead of branching to an error routine, resetting the computer quickly for safety may be necessary. The software interrupt can vector to a routine that forces a hardware reset through additional hardware. A later section shows an example of this use along with the COP timer.

XIRQ Interrupt

The 68HC11 has an external nonmaskable interrupt; that is, an interrupt that cannot be disabled. The I bit does not disable the XIRQ interrupt. Usually, the application requiring a nonmaskable interrupt involves safety.

An application for XIRQ

The most common application of the XIRQ interrupt requires an interrupt when "the world is coming to an end!" That is, if it doesn't matter what the computer is currently doing, interrupting it and responding to the nonmaskable interrupt is best.

The classic example is a microcomputer powered from the power line. When the power fails, what the computer is doing makes no difference. Soon, the computer will stop running because the power is off—an interrupt can't hurt anything. Generally, the power supply for a microcomputer will continue to work for a few milliseconds after a power failure. The microcomputer can use this time to save important information in a nonvolatile memory. A circuit in the power supply must detect the power failure and trigger the nonmaskable interrupt.

XIRQ interrupt system operation

The XIRQ interrupt line connects only to a pin. Therefore, only input/output devices external to the 68HC11 can cause nonmaskable interrupts. An XIRQ signal causes an interrupt even if another interrupt service routine is executing. The interrupt vector for XIRQ is at addresses FFF4 and FFF5.

The X bit. The XIRQ interrupt has an interrupt-enable bit, called *X*, in the condition code register. The X bit is very different from the I bit because XIRQ is nonmaskable.

The 68HC11 disables the XIRQ interrupt at reset by forcing the X bit to 1. The program, at the correct point in the initialization, puts 0 in the X bit to enable the XIRQ interrupt. However, the X bit is a *sticky bit* that the program cannot change back to 1. Only an interrupt signal at the XIRQ pin or a hardware reset can set X to 1; only hardware can set X to 1. Therefore, once the nonmaskable interrupt is enabled, it remains enabled and is nonmaskable for the rest of the time the program runs.

The TAP instruction. Only the TAP and RTI instructions can make the X bit 0. The TAP enables XIRQ interrupts by transferring a number from the A accumulator to the condition code register. The TAP instruction usually enables both the XIRQ interrupt and interrupts under control of the I bit.

The TAP instruction cannot change X to 1. Look at the TAP instruction in the instruction set table in Appendix A. The effect on the X bit is shown with a downward arrow. This symbol means that X can be changed only from 1 to 0.

Be careful because the TAP instruction controls the other condition code bits while it zeros X and I. Generally, use the TAP instruction only in the initialization part of an interrupt-driven program. Don't use it in an interrupt service routine.

XIRQ interrupt operation. When an XIRQ interrupt occurs, the XIRQ interrupt system hardware sets both the X bit and the I bit to 1 disabling further XIRQ and IRQ interrupts. The RTI instruction returns both the X bit and the I bit to their previous states. So, return from the interrupt service routine enables the XIRQ again. In most ways, the XIRQ interrupt is identical to the IRQ interrupt.

XIRQ interrupt signal. The XIRQ pin on the 68HC11 is a level-sensitive pin, so connect hardware to the XIRQ pin as you would to the IRQ pin. The XIRQ pin is not programmable; it is only level-sensitive. If both the IRQ and XIRQ pins request interrupts simultaneously, the 68HC11 responds to the XIRQ interrupt request. That is, XIRQ has priority over IRQ. The XIRQ also can interrupt an IRQ interrupt service routine that is in progress.

Other microprocessors. Some other Motorola 8-bit microprocessors have a similar interrupt named NMI for *nonmaskable interrupt*. The NMI interrupt in some processors can truly never be disabled, and therefore must use a transition-sensitive signal. However, if an interrupt occurs soon after a program starts and before the initialization is completed, the program could crash. The nonmaskable interrupt in the 68HC11 is more sophisticated.

Illegal Instruction Interrupt

An illegal instruction op code is a number that does not represent an instruction implemented in the microprocessor. Illegal instruction codes represent either programming errors or hardware failures. The 68HC11 microprocessor detects if an illegal instruction code is fetched.

When the 68HC11 detects an illegal instruction code, it generates an internal interrupt. This interrupt causes the microprocessor to vector through addresses FFF8 and FFF9 to an interrupt service routine. Usually the illegal instruction interrupt service routine somehow forces a reset on the computer.

The illegal instruction interrupt is a nonmaskable interrupt that has priority over the IRQ interrupt because it sets the I bit to 1. The illegal instruction interrupt does not set the X bit, but if the XIRQ interrupt service routine fetches an illegal instruction, it will be interrupted. An illegal instruction interrupt is more important than the XIRQ because the microprocessor is not executing valid instructions.

If the illegal instruction results from a memory failure, the illegal instruction interrupt service routine also may have illegal instructions. This may cause an infinite loop of illegal instruction interrupts. Other safety measures should detect this problem, but the stack probably will overflow and may destroy important data in RAM. Therefore, the illegal instruction interrupt may reset the stack pointer to avoid the stack overrun. Of course, the rest of the program must then account for this.

Most of the possible 2-byte op codes and a few of the possible 1-byte op codes are not valid instructions. Therefore, the microprocessor probably can detect failures. For example, if a memory chip fails, an incorrect instruction will likely be fetched. The computer can respond to this failure if the hardware detects the incorrect instruction.

Many other Motorola 8-bit microprocessors use only single-byte op codes. Most of their possible op code numbers are valid, so detecting an illegal instruction code has little value. Those processors do not have illegal instruction detection. When they fetch and execute an illegal code, the operation of the microprocessor is unpredictable.

Interrupt Priority

If two or more devices all request an interrupt simultaneously, the interrupt system hardware must choose one to respond to first. The 68HC11 uses a strict priority system—the device with the highest priority is serviced first. In the 68HC11, the device with an interrupt vector at the highest address is also at the highest priority. When the interrupt service routine returns to the main program, the device with the highest priority at that time is serviced next.

Interrupt priority is important only when two or more devices interrupt at the same time. In most applications, interrupt priority is of little importance. However, it can be important when the program must know which of two related devices is serviced first.

7.5 PARALLEL I/O PORTS

The block diagrams in Figures 7-1 and 7-2 show ports A through E and two associated pins that are not part of a port. The ports have different functions depending on the operating mode of the 68HC11.

The ports are collections of signal wires or pins that the 68HC11 uses to communicate with devices outside the chip. The port names are arbitrary because each port has several functions.

Software can program the 68HC11 hardware to connect ports A, D, and E to input/output devices, such as an analog-to-digital converter, inside the 68HC11. People often associate the ports with those hardware devices, though the ports also do other functions. Later sections of this book discuss the special associations of the ports with particular I/O devices.

Let's begin by describing the parallel input/output capability of the ports. All the I/O port pins have either digital input or digital output capability. However, the parallel I/O capability is different for the single-chip mode and the expanded mode. For example, the ports B and C used in examples in this book have an additional function.

Hardware Initialization

A hardware reset initializes the input/output hardware in the 68HC11. The reset also puts the chip into the correct operating mode—either single-chip mode or expanded mode.

A hardware reset disables all input/output devices inside the 68HC11. The effect is to program all available I/O ports as parallel I/O ports. The reset clears all input and output registers. Also, reset clears the data direction register bits making any programmable port bits inputs. All data direction register bits make an input when they are 0 and make an output when they are 1.

Parallel I/O

Most of the parallel input/output pins function the same in both the single-chip mode and the expanded mode. Table 7-3 lists the port pins and their parallel input/output functions for both modes. It also lists the data direction register control bits for the programmable I/O pins.

The functions of the various bits of a port may seem disorganized. For example, port A has input only, output only, and programmable I/O pins. The variety results because these pins can connect to an internal I/O device that requires this configuration of input and outputs. Only ports B and C have the same functions regardless of internal programming.

Table 7-3 shows the reset condition of the ports before any software control of them. The table does not show that the electrical characteristics of the various pins differ somewhat. Some ports can be programmed for different electrical characteristics. Consult a Motorola manual for details of the electronics.

Single-chip mode parallel I/O

When a reset puts the 68HC11 into the single-chip mode, ports B and C become parallel I/O ports. In particular, port B is an output only port, so it does not have a data direction register. Port C is a programmable input/output port with a data direction register named *DDRC*.

Two pins are associated with the ports that are not part of any port. In single-chip mode, these pins are named *STRA* and *STRB*. The pin STRA is the *strobe A* or clock or ready signal for the STAF flag. Both STRA and STRB, *strobe B*, have additional programmable functions.

TABLE 7-3 PARALLEL INPUT/OUTPUT PIN FUNCTIONS

Port	Single-chip Mode			Expanded Mode		
Bit Name	Pin Name	Pin Function	DDR Bit	Pin Name	Pin Function	DDR Bit
A-0	PA0	In only		PA0	In only	
A-1	PA1	In only		PA1	In only	
A-2	PA2	In only		PA2	In only	
A-3	PA3	Out only		PA3	Out only	
A-4	PA4	Out only		PA4	Out only	
A-5	PA5	Out only		PA5	Out only	
A-6	PA6	Out only		PA6	Out only	
A-7	PA7	In/Out	DDRA7	PA7	In/Out	DDRA7
B-0	PB0	Out only		A8	Address bus	
B-1	PB1	Out only		A9	Address bus	
B-2	PB2	Out only		A10	Address bus	
B-3	PB3	Out only		A11	Address bus	
B-4	PB4	Out only		A12	Address bus	
B-5	PB5	Out only		A13	Address bus	
B-6	PB6	Out only		A14	Address bus	
B-7	PB7	Out only		A15	Address bus	
C-0	PC0	In/Out	DDRC0	A0/D0	Address/Data bus	
C-1	PC1	In/Out	DDRC1	A1/D1	Address/Data bus	
C-2	PC2	In/Out	DDRC2	A2/D2	Address/Data bus	
C-3	PC3	In/Out	DDRC3	A3/D3	Address/Data bus	
C-4	PC4	In/Out	DDRC4	A4/D4	Address/Data bus	
C-5	PC5	In/Out	DDRC5	A5/D5	Address/Data bus	
C-6	PC6	In/Out	DDRC6	A6/D6	Address/Data bus	
C-7	PC7	In/Out	DDRC7	A7/D7	Address/Data bus	
D-0	PD0	In/Out	DDRD0	PD0	In/Out	DDRD0
D-1	PD1	In/Out	DDRD1	PD1	In/Out	DDRD1
D-2	PD2	In/Out	DDRD2	PD2	In/Out	DDRD2
D-3	PD3	In/Out	DDRD3	PD3	In/Out	DDRD3
D-4	PD4	In/Out	DDRD4	PD4	In/Out	DDRD4
D-5	PD5	In/Out	DDRD5	PD5	In/Out	DDRD5
None	STRA	Strobe A		AS	Address strobe	
None	STRB	Strobe B		R/W	Read/Write	
E-0	PE0	In only		PE0	In only	
E-1	PE1	In only		PE1	In only	
E-2	PE2	In only		PE2	In only	
E-3	PE3	In only		PE3	In only	
E-4	PE4	In only		PE4	In only	
E-5	PE5	In only		PE5	In only	
E-6	PE6	In only		PE6	In only	
E-7	PE7	In only		PE7	In only	

Expanded mode parallel I/O

When a reset puts the 68HC11 into expanded mode, ports B and C become address and data bus pins. Therefore, the port B and port C input/output functions are lost. The STRA signal for the STAF flag is also lost. Table 7-3 lists the pin functions for expanded mode.

Port C is multiplexed with both address and data bus signals, so an external latch IC is necessary. All address, data, and control bus signals are available when the port B, port C, AS, and R/W pins are used.

7.6   INTERNAL FLAGS

Most I/O devices inside the 68HC11 have flags. An I/O device sets a flag to indicate completion of an operation; the flag usually causes an interrupt. The program, in responding to the flag, must clear the flag. However, not all flags in the 68HC11 clear the same way.

The STAF Flag

The STAF flag discussed earlier is cleared in an indirect way. In particular, instructions must read two registers to clear the STAF flag that is in a third register. A few other flags are cleared in a similar indirect way, but most other flags in the 68HC11 are cleared in a different and direct way.

Direct-Clearing Flags

All direct-clearing flags in the 68HC11 are cleared when an instruction writes a 1 into the flag bit! Writing a 1 to something that is in the 1-state to make it go to the 0-state may seem strange, but that is how the hardware works. The reason is that several flags are grouped together in a single register. Writing 0s into flag bits does nothing. Therefore, only those bits containing 1s in the number written clear a flag.

BSET instruction problems

A BSET instruction may seem the perfect instruction to clear one of these direct-clearing flags, but it won't clear the flags correctly! Here is an example of the problem. Suppose that a memory register is all flags with bit 7 and bit 5 now set. If a BSET instruction is to clear the bit 7 bit only, the mask in the instruction is 80. The BSET instruction ORs the mask to the memory register and writes the result back to the memory register. So, when it reads memory, it gets A0 because two flags are set. Then it ORs 80 and gets A0, which it writes back to memory. When it does, it clears both flags! Of course, this operation is incorrect. The BSET instruction clears all flags that are set, not just the one selected.

BCLR for clearing flags

The BCLR instruction can correctly clear flags; however, the complement of the normal mask is used. The BCLR instruction ANDs the complement of the mask to the memory register, and writes the result back to the memory register. To clear the flag at bit 7 in a flags register, use a mask of 7F. If the flags at bit 7 and at bit 5 are set, the AND instruction reads A0 from the flags register, ANDs the complement of 7F getting 80, and writes the 80 back to the flags register. Writing the 1 at bit 7 clears the flag at bit 7 correctly, and the 0 at bit 5 does not affect the bit 5 flag. Therefore, the BCLR instruction with a complemented mask will correctly clear flags. Be careful to use the BCLR instruction, not the BSET instruction.

7.7   REAL-TIME CLOCK

You may occasionally want a computer to perform certain I/O operations at specified times. One way for a computer to track time is to use an I/O device that causes interrupts periodically. An oscillator sets a flag at the end of each period. The interrupt service routine then can count interrupts to track time.

The RTI Device

When a computer is used as a control device, periodic timing is usually necessary. The example in the previous section included a flashing light. The time base for the timing was an external oscillator and a flag.

Because the need for periodic timing is common, the 68HC11 has an internal I/O device called the *real-time interrupt device*, or the *RTI*. The real-time interrupt device is simply an oscillator with a flag that can cause periodic interrupts. The interrupt service routine that responds to these interrupts can do data transfers at precise times determined by the interrupts.

The frequency of the oscillator that sets the flag is dependent on the E-clock rate of the 68HC11. A programmable counter driven at the E-clock rate provides four interrupt rates. The E-clock runs at one-fourth the rate of the crystal.

Figure 7-7 shows a block diagram of the RTI hardware. The control bits for the timer are in various 68HC11 control registers.

RTI flag

The real-time interrupt device causes an interrupt by setting a flag named *RTIF* for *real-time interrupt flag*. A program can examine this flag by testing bit 6 of the TFLG2 register at address 1025. Figure 7-8 shows the TFLG2 register.

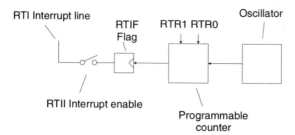

Figure 7-7   Block diagram of real-time interrupt hardware.

The RTIF flag is set periodically by the real-time interrupt oscillator. A program clears the RTIF flag by writing a 1 to bit 6 of the TFLG2 register. Writing a 0 to the RTIF bit does nothing to it. Similarly, writing 0s to the other bits of this register will not affect any other I/O hardware. Normally, you will use the STAA instruction to write the 1 to clear the flag. Section 7.6 discusses using bit instructions with flags.

If you choose not to use interrupt, the RTI device flag can be polled. Most applications use interrupt, so the device is named accordingly.

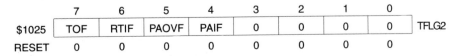

	7	6	5	4	3	2	1	0	
$1025	TOF	RTIF	PAOVF	PAIF	0	0	0	0	TFLG2
RESET	0	0	0	0	0	0	0	0	

Figure 7-8   Timer flag register 2.

RTI control bits

Bit 1 and bit 0 of the PACTL register at address 1026 program the rate of interrupts. Figure 7-9 illustrates the PACTL register. The name of this register does not relate to the real-time interrupt device.

The control bits are named *RTR1* and *RTR0*, respectively. The two control bits determine which one of four divide factors is applied to the E-clock in generating the interrupts. That is, they select one of four interrupt rates. Table 7-4 shows the interrupt rates for several different E-clock rates.

The real-time interrupt is enabled or disabled by a switch controlled by bit 6 of the TMSK2 register at address 1024. This interrupt-enable bit is named *RTII*. Figure 7-10 shows the TMSK2 register and the RTII bit. A 0 in the RTII bit disables the RTI interrupt and a 1 enables the RTI interrupt.

When using only the real-time interrupt device, avoid problems caused by other bits in the PACTL, TMSK2, and TFLG2 registers by only storing 0s to those bits.

The real-time interrupt has its own interrupt vector at address FFF0. The real-time interrupt is under control of the I bit.

RTI Programming Example

The following example uses the real-time interrupt as a time base to toggle the bits of PORTC every 8.19 milliseconds. Each time the interrupt service routine runs, all output bits in the

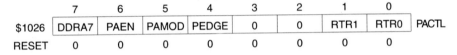

	7	6	5	4	3	2	1	0	
$1026	DDRA7	PAEN	PAMOD	PEDGE	0	0	RTR1	RTR0	PACTL
RESET	0	0	0	0	0	0	0	0	

Figure 7-9   Pulse accumulator control register.

TABLE 7-4   REAL-TIME INTERRUPT RATE

RTI Control Bits		Rate is E Divided by	Crystal Frequency		
RTR1	RTR0		8.3886 MHz	8.0 MHz	4.9152 MHz
			Interrupt Period (milliseconds)		
0	0	2^{13}	3.91	4.10	6.67
0	1	2^{14}	7.81	8.19	13.33
1	0	2^{15}	15.62	16.38	26.67
1	1	2^{16}	31.25	32.77	53.33
		E =	2.0971 MHz	2.0000 MHz	1.2288 MHz

PORTC are complemented. Figure 7-11 shows the listing of the program. Here is a description of the program:

- *Line 27.*   The RTI vector at address FFF0 sends control to the interrupt service routine at address C118. No other device shares this interrupt vector.

- *Line 37.*   The program initializes the stack because the interrupt system will use it to store the microprocessor status when an interrupt occurs.

- *Lines 40 through 43.*   The program initializes PORTC for all outputs with the outputs all cleared.

- *Lines 46 and 47.*   The interrupt rate of the real-time interrupt oscillator is set. The rate after reset is the fastest rate because reset clears the rate control bits. Here the RTR1 and RTR0 bits are set to 01 to choose the 8.19-millisecond interrupt rate.

- *Lines 49 and 50.*   The real-time interrupt is enabled to cause interrupts. Be careful because other bits in the TMSK2 register can enable other interrupts.

- *Lines 65 and 66.*   The interrupt vector sent control here because the RTI device interrupted. At this point, the interrupt system is disabled because a 1 is in the I bit. The RTIF flag is checked to see if it is set. If not, an error caused an invalid interrupt. Checking for invalid interrupts is good practice in practical programs.

- *Lines 71 and 72.*   Storing a 1 to the RTIF flag bit clears the real-time interrupt flag.

	7	6	5	4	3	2	1	0	
$1024	TOI	RTII	PAOVI	PAII	0	0	PR1	PR0	TMSK2
RESET	0	0	0	0	0	0	0	0	

Figure 7-10   Timer interrupt mask register 2.

```
  1                              *********************************************************
  2                              ** REVERSE ALL PORTC BITS EVERY 8.19 MILLISECONDS
  3                              *
  4                              * USE REAL TIME INTERRUPT AS TIME BASE
  5                              *
  6                              * USE 8.0-MHZ CRYSTAL FOR 68HC11 CLOCK
  7                              *
  8                              *********************************************************
  9                              *
 10                              *********************************************************
 11                              ** SYMBOL DEFINITIONS
 12                              *********************************************************
 13                              * 68HC11 REGISTERS
 14         1003                 PORTC   EQU   $1003     I/O PORT C REGISTER
 15         1007                 DDRC    EQU   $1007     DATA DIRECTION REGISTER C
 16         1024                 TMSK2   EQU   $1024     TIMER MASK REGISTER
 17         1025                 TFLG2   EQU   $1025     TIMER FLAG REGISTER
 18         1026                 PACTL   EQU   $1026     PULSE ACCUMULATOR CONTROL REG
 19         00FF                 IOPAT   EQU   $FF       I/O PATTERN, 0=IN 1=OUT
 20                              * MASKS
 21         0040                 BIT6    EQU   %01000000
 22                              *
 23                              *********************************************************
 24                              ** DATA SECTION
 25                              *********************************************************
 26  FFF0                                ORG   $FFF0     RTI INTERRUPT VECTOR
 27  FFF0     C118                       FDB   RTIISR
 28                              *
 29                              *********************************************************
 30                              ** MAIN PROGRAM
 31                              *********************************************************
 32  C100                                ORG   $C100
 33                              *---------------------------------------------------
 34                              * INITIALIZATION
 35                              *---------------------------------------------------
 36                              * INITIALIZE STACK
 37  C100     8E CF FF          START   LDS   #$CFFF
 38                              * INITIALIZE PORT C
 39                              *    INITIALIZE OUTPUTS TO ZEROS
 40  C103     7F 10 03                  CLR   PORTC     0=LOW
 41                              *    SET UP PORTC INS AND OUTS
 42  C106     86 FF                     LDAA  #IOPAT
 43  C108     B7 10 07                  STAA  DDRC      ALL OUTPUTS
 44                              * INITIALIZE REAL TIME INTERRUPT
 45                              *    SET INTERRUPT RATE TO 8.19 MS
 46  C10B     86 01                     LDAA  #1        ZEROS IN OTHER BITS
 47  C10D     B7 10 26                  STAA  PACTL     ..CAUSE NO PROBLEMS
 48                              *    ENABLE RTI INTERRUPT
 49  C110     86 40                     LDAA  #BIT6     ZEROS IN OTHER BITS
 50  C112     B7 10 24                  STAA  TMSK2     ..CAUSE NO PROBLEMS
 51                              * TURN ON INTERRUPT SYSTEM
 52  C115     0E                        CLI
 53                              *---------------------------------------------------
 54                              * WAIT FOR INTERRUPTS
 55                              *---------------------------------------------------
 56  C116     20 FE             HERE    BRA   HERE      DO NOTHING!
 57                              *
 58                              *********************************************************
 59                              ** INTERRUPT SERVICE ROUTINE
 60                              *********************************************************
 61                              *---------------------------------------------------
 62                              * INTERRUPT POLLING CHAIN
 63                              *---------------------------------------------------
```

Figure 7-11   Interrupt program for real-time interrupt.

```
64                              * INTERRUPT FROM REAL TIME INTERRUPT DEVICE?
65    C118    CE 10 25          RTIISR  LDX    #TFLG2
66    C11B    1F 00 40 08               BRCLR  0,X,BIT6,RTRTI   IGNORE ILLEGAL INTERRUPT
67                              *-------------------------------------------------------
68                              * CONTROL PORT C OUTPUTS
69                              *-------------------------------------------------------
70                              * CLEAR REAL TIME INTERRUPT FLAG
71    C11F    86 40                     LDAA   #BIT6      STORE 1 TO CLEAR FLAG!
72    C121    B7 10 25                  STAA   TFLG2    ..ZEROS DO NOTHING
73                              * COMPLEMENT PORT C OUTPUTS
74    C124    73 10 03                  COM    PORTC
75                              * RETURN FROM INTERRUPT
76    C127    3B               RTRTI    RTI
77    C128                              END
```

Figure 7-11   Continued.

- *Line 74.*   The PORTC outputs are complemented. This happens every 8.19 milliseconds because this interrupt service routine runs every 8.19 milliseconds. That is, it runs every time the RTI flag is set.

- *Line 76.*   Control returns to the main program, and the interrupt system is enabled again.

To change this program for longer times, first change the RTI rate control bits. If even longer times are necessary, the interrupt service routine can count multiple interrupts before performing an action.

7.8   THE PROGRAMMABLE TIMER

Control applications of computers usually require measurements of time. The performance of control systems is usually time-dependent. The 68HC11 contains a hardware timer. The timer can measure time for both inputs and outputs. The timing schemes are different for inputs and outputs.

The timer in the 68HC11 is very flexible. It will accurately measure times from microseconds to centuries! The characteristics of the timer are programmable with software. The timer uses the interrupt system, and it has several interrupt vectors. However, the accuracy of the timer does not depend on the software.

Timer Principles

The timer consists of three logical parts: a free-running counter, input-capture hardware, and output-compare hardware. Because many programmable options are available, this hardware is very complex. However, the fundamental principles of its operation are quite simple. Let's begin with an overview of the operation of the timer.

The free-running counter

An up-counter is a special register that counts up by one each time it is clocked. The free-running counter in the 68HC11 is a 16-bit up-counter driven or clocked by an accurate oscillator. The free-running counter counts up by one on each cycle of the oscillator. The counter continually counts whenever the 68HC11 is powered-up—thus the name free-running counter. The free-running counter is the time base for all timer functions.

The free-running counter counts from 0000 to FFFF, and then rolls over to 0000 again. Rolling over to zero is called an *overflow*. The number in the counter and the overflow can be used for timing. For example, if the oscillator operates at 1.0 MHz, the counter rolls over every $65,536_{10}$ microseconds, or about every 65_{10} milliseconds. By checking the number in the counter and the rolling over of the counter, timing with a resolution of 1.0 microsecond can be done. The accuracy of the timing depends on the accuracy of the oscillator only.

Input capture for timing events

Figure 7-12 shows the free-running counter connected to another register. A parallel transfer from the counter to the register occurs when the register is clocked. Clocking the register captures a copy of the number in the counter.

For example, suppose that the capture register is clocked, and it contains the number 0100. A little later, it is clocked again, and then it contains 0200. Assuming that the counter did not overflow, we conclude that 100_{16} microseconds elapsed between the captures. Therefore, we have determined that the time between captures was 256_{10} microseconds.

By using input signals to clock the capture register, and reading the capture register and noting overflows, any event outside the computer can be accurately timed. The times between events can range from microseconds to years!

Output compare for controlling events

Figure 7-13 shows the free-running counter, an output-compare register, and a comparison circuit. The comparator looks at the numbers in the counter and the compare register, and signals whenever the two numbers are the same.

As a simple example, if the number in the compare register doesn't change, the numbers will match every $65,536_{10}$ counts. If the oscillator runs at 1.0 MHz, the comparator indicates a match every 65.536_{10} milliseconds.

A more practical example requires the program to change the number in the compare register. For example, suppose that the compare register contains 0100 and a match was just

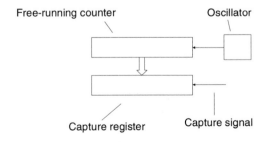

Figure 7-12 Free-running counter and capture register.

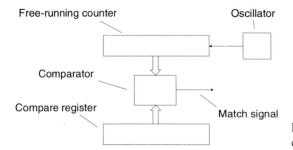

Figure 7-13   Free-running counter, compare register, and compare circuit.

indicated. The program responds by reading the compare register, adding 0100, and storing the result 0200 into the compare register. The comparator indicates the next match exactly 256_{10} microseconds after the first.

If the match signal complements an output bit, the bit is changed every 256_{10} microseconds. The timing is very accurate. The time the software takes to change the compare register does not affect the accuracy of the timing.

Free-Running Counter

The free-running counter in the 68HC11 is a 16-bit up-counter. The time base for the counter is the crystal in the 68HC11 clock. The free-running counter is the basis for all the timing functions of the programmable timer system.

The counter hardware

Figure 7-14 shows the *free-running counter's register*, called *TCNT*, and its clock. The oscillator runs at the E-clock rate and is driven by the crystal. The *prescaler* is a programmable counter that divides the E-clock rate by one of four factors; namely, 1, 4, 8, or 16_{10}.

Prescaler control.   The prescaler control bits are shown in Table 7-5 with typical clock rates. The control bits called *PR1* and *PR0* are in the *timer mask register* called *TMSK2*. Figure 7-10 shows the TMSK2 register.

The prescaler control bits are timed-write-once bits. Therefore, they must be set by the program within 64_{10} E-clock cycles after reset.

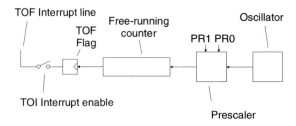

Figure 7-14   Block diagram of free-running counter hardware.

TABLE 7-5   FREE-RUNNING COUNTER RATE AND RANGE

Prescaler Control Bits		Divide Factor	Crystal Frequency		
			8.3886 MHz	8.0 MHz	4.9152 MHz
PR1	PR0		One Count(μs) / **Overflow Range**(ms)		
0	0	1	0.477 / 31.25	0.5 / 32.77	.814 / 53.33
0	1	4	1.91 / 125	2 / 131.1	3.26 / 213.3
1	0	8	3.81 / 250	4 / 262.1	6.51 / 426.7
1	1	16	7.63 / 500	8 / 524.3	13.02 / 853.7
		E =	2.0971 MHz	2.0000 MHz	1.2288 MHz

Timer-overflow flag.   When the free-running counter overflows from FFFF to 0000, it sets the *timer-overflow flag* called *TOF*. Writing a 1 into the TOF bit clears the TOF flag. The TOF flag is in the *timer flag register* called *TFLG2*. Figure 7-10 shows the TFLG2 register.

The timer-overflow flag can cause an interrupt. The enable bit for this interrupt is the TOI bit in the TMSK2 register. The timer-overflow interrupt vector is at address FFDE. This interrupt is under control of the I bit.

Reset condition.   Reset clears the free-running counter. After that, it is a read only register that is unaffected by instructions. Reads are buffered so that an instruction that reads a double-byte number gets the correct number. The least-significant byte is not out of date.

Example program

Figure 7-15 is the listing for a program that uses the timer-overflow interrupt. The program is similar to the example in Figure 7-11. The program does the same function, but it uses the timer overflow as the time base.

The example uses an 8.0-MHz crystal and programs the prescaler for a divide-by-eight rate. The free-running counter then overflows every 262.1_{10} milliseconds. Remember that the prescaler is programmed by timed-write-once bits.

This example is a simple use of the timer-overflow interrupt. More often, timer overflow is used with other timing functions to extend their time range.

Input Capture

Input capture can measure the time between events external to the computer. A signal sent to the computer must make a proper transition at the time of the event to trigger the capturing hardware. An input-capture flag is set when a capture takes place; the flag may cause an interrupt.

If the input signal is periodic, the time between identical events is the period of the waveform. When the period is known, the frequency of the input signal can be calculated. Many kinds of applications that require timing can use input capture.

```
1                                    ********************************************************
2                                    ** REVERSE ALL PORTC BITS EVERY 262.1 MILLISECONDS
3                                    *
4                                    * USE TIMER OVERFLOW AS TIME BASE
5                                    *
6                                    * USE 8.0-MHZ CRYSTAL FOR 68HC11 CLOCK
7                                    *
8                                    ********************************************************
9                                    *
10                                   ********************************************************
11                                   ** SYMBOL DEFINITIONS
12                                   ********************************************************
13                                   * 68HC11 REGISTERS
14           1003                    PORTC   EQU    $1003     I/O PORT C REGISTER
15           1007                    DDRC    EQU    $1007     DATA DIRECTION REGISTER C
16           1024                    TMSK2   EQU    $1024     TIMER MASK REGISTER
17           1025                    TFLG2   EQU    $1025     TIMER FLAG REGISTER
18           00FF                    IOPAT   EQU    $FF       I/O PATTERN, 0=IN 1=OUT
19                                   * MASKS
20           0040                    BIT7    EQU    %01000000
21                                   *
22                                   ********************************************************
23                                   ** DATA SECTION
24                                   ********************************************************
25   FFDE                                    ORG    $FFDE     TOF INTERRUPT VECTOR
26   FFDE    C113                            FDB    TOFISR
27                                   *
28                                   ********************************************************
29                                   ** MAIN PROGRAM
30                                   ********************************************************
31   C100                                    ORG    $C100
32                                   *--------------------------------------------------------
33                                   * INITIALIZATION
34                                   *--------------------------------------------------------
35                                   * INITIALIZE STACK
36   C100    8E CF FF                START   LDS    #$CFFF
37                                   * INITIALIZE PORT C
38                                   *     INITIALIZE OUTPUTS TO ZEROS
39   C103    7F 10 03                        CLR    PORTC     0=LOW
40                                   *     SET UP PORTC INS AND OUTS
41   C106    86 FF                           LDAA   #IOPAT
42   C108    B7 10 07                        STAA   DDRC      ALL OUTPUTS
43                                   * INITIALIZE TIMER OVERFLOW INTERRUPT
44   C10B    86 82                           LDAA   #$82      PR1:PR0=10 FOR 262.1 MS
45   C10D    B7 10 24                        STAA   TMSK2     ..ENABLE TOF INTERRUPT
46                                   * TURN ON INTERRUPT SYSTEM
47   C110    0E                              CLI
48                                   *--------------------------------------------------------
49                                   * WAIT FOR INTERRUPTS
50                                   *--------------------------------------------------------
51   C111    20 FE                   HERE    BRA    HERE      DO NOTHING!
52                                   *
53                                   ********************************************************
54                                   ** INTERRUPT SERVICE ROUTINE
55                                   ********************************************************
56                                   *--------------------------------------------------------
57                                   * INTERRUPT POLLING CHAIN
58                                   *--------------------------------------------------------
59                                   * INTERRUPT FROM TIMER OVERFLOW?
60   C113    CE 10 25                TOFISR  LDX    #TFLG2
61   C116    1F 00 40 08                     BRCLR  0,X,BIT7,RTTOF  IGNORE ILLEGAL INTERRUPT
62                                   *--------------------------------------------------------
63                                   * CONTROL PORT C OUTPUTS
64                                   *--------------------------------------------------------
```

Figure 7-15 Interrupt program for timer-overflow interrupt.

```
65                              * CLEAR TIMER OVERFLOW FLAG
66    C11A    86 40                     LDAA    #BIT7      STORE 1 TO CLEAR FLAG!
67    C11C    B7 10 25                  STAA    TFLG2      ..ZEROS DO NOTHING
68                              * COMPLEMENT PORT C OUTPUTS
69    C11F    73 10 03                  COM     PORTC      OUTPUT TO PORT C
70                              * RETURN FROM INTERRUPT
71    C122    3B              RTTOF     RTI
72    C123                              END
```

Figure 7-15   Continued.

The hardware

The 68HC11A8 IC has three sets of input-capture hardware, and the 68HC11E9 has four sets. The E9 version shares a pin and other hardware between an optional input-capture device and an optional output-compare device. The input-capture signals are part of the port A pins.

In all details, each input-capture functions the same. Therefore, the following discussion describes the hardware for all input captures.

Figure 7-16 shows the parts of the input-capture hardware. The edge control hardware determines the active edge or transition of the pin.

Input-capture registers.   The *timer input-capture registers* are each 16-bit registers that can hold numbers from the 16-bit free-running counter. The three registers are named *TIC1*, *TIC2*, and *TIC3* and are at addresses 1010, 1012, and 1014 respectively. A program can only read these registers—writing to them has no effect.

The E9 version of the 68HC11 has a fourth input capture that shares a pin and other hardware with an output-compare device. Then the corresponding input-capture register is called *TI4O5* and is at address 101E. Storing 0s in bit 2 and bit 3 of the PACTL register programs TI4O5 for input capture, and 1s program it for output compare.

Port signals.   The input-capture pins are part of port A. The pins are called *input captures 1, 2, and 3*. IC1 connects to pin *PA2*, IC2 connects to pin *PA1*, and IC3 connects to pin *PA0*. When IC4 exists, it connects to pin PA3. Refer to Figures 7-1 and 7-2 to clarify these connections.

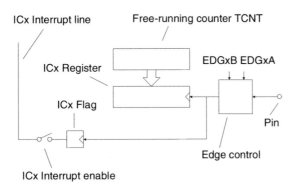

Figure 7-16   Block diagram of input-capture hardware.

Edge control.   Table 7-6 shows the *edge control bits* for the input-capture pins. The control bits called *EDGxB* and *EDGxA* are in the timer-control register called *TCTL2*. Figure 7-17 shows this register. The figure shows the register for the A8 version of the 68HC11, which has three input captures. The program may change the edge control bits at any time.

TABLE 7-6   INPUT CAPTURE EDGE

EDGxB Bit	EDGxA Bit	Active Pin Transition
0	0	Capture disabled
0	1	Capture on low-to-high
1	0	Capture on high-to-low
1	1	Capture on both transitions

Table 7-6 shows that an input capture may be disabled. While an input capture is disabled, the corresponding pin is available as an input-only pin. Refer to Table 7-3 for the parallel I/O functions of the ports.

If the pin is enabled for input captures, the capture register and the flag are clocked on a low-to-high transition, a high-to-low transition, or both transitions. These transitions are also called *rising edge*, *falling edge*, and *any edge*, respectively.

Flags.   Figure 7-18 shows the *input-capture flags* in the *timer-flag register* called *TFLG1*. The flags are named *IC1F*, *IC2F*, and *IC3F* to correspond to the input captures named *IC1*, *IC2*, and *IC3*, respectively. The input-capture flags are set by active transitions on the associated input-capture pins. Writing a 1 to an input-capture flag clears it; writing a 0 does not affect it.

In the E9 version of the 68HC11, bit 3 of TFLG1 is called *I4/O5F*. A 0 in bit 2 of the PACTL register makes I4/O5F an input-capture flag.

Interrupts.   Each input-capture flag can trigger an interrupt. The *interrupt-enable bits* for IC1, IC2, and IC3, respectively, are called *IC1I*, *IC2I*, and *IC3I*. They are in the *timer*

	7	6	5	4	3	2	1	0	
$1021	0	0	EDG1B	EDG1A	EDG2B	EDG2A	EDG3B	EDG3A	TCTL2
RESET	0	0	0	0	0	0	0	0	

Figure 7-17   Timer control register 2.

	7	6	5	4	3	2	1	0	
$1023	OC1F	OC2F	OC3F	OC4F	OC5F	IC1F	IC2F	IC3F	TFLG1
RESET	0	0	0	0	0	0	0	0	

Figure 7-18   Timer flag register 1.

mask register called *TMSK1*. A 1 in an interrupt-enable bit enables the flag to cause an interrupt, and a 0 disables it. Figure 7-19 shows the TMSK1 register. The reset condition shows that a hardware reset disables all input-capture interrupts.

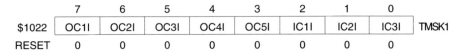

	7	6	5	4	3	2	1	0	
$1022	OC1I	OC2I	OC3I	OC4I	OC5I	IC1I	IC2I	IC3I	TMSK1
RESET	0	0	0	0	0	0	0	0	

Figure 7-19   Timer interrupt mask register 1.

In the E9 version of the 68HC11, the interrupt from the fourth input capture is enabled by bit 3 of TMSK1 called *I4O5I*. Zeros in bit 2 and bit 3 of the PACTL register enables the input-capture function.

Measuring short elapsed times

A *short time* is any time less than the time between overflows of the free-running counter. If the time is longer, then overflows must be counted and the complexity is greater. If the E-clock runs at 2.0 MHz and the timer prescaler is set for the fastest free-running counter, the overflow time is 32.77_{10} milliseconds with a resolution of 0.5 microsecond.

Figure 7-20 is an example program that illustrates the use of input capture. Assume that the input signal connects to port A at pin PA0. This pin corresponds to input capture 3. Further assume that the elapsed time is measured between low-to-high transitions of the signal. The program repeatedly measures the elapsed time between these transitions. It puts 1 in bit 0 of PORTC only if the time is at or below a limit, and it puts 1 in bit 1 of PORTC only if the time is over the limit; otherwise, these bits are 0s. If the input signal stops changing, the time will certainly be over the limit. Here is a description of the program:

- *Line 35.*   The time limit is the data number the program compares to the elapsed time to determine if the time is out of limits. The limit is a double-byte number of free-running counter periods.

- *Line 37.*   The elapsed time between input captures in free-running counter periods.

- *Lines 58 through 62.*   Input capture 3 is programmed to respond to low-to-high transitions on the PA0 pin and to cause an interrupt.

- *Lines 75 through 85.*   The main program loop continually checks the elapsed time and updates the two indicator bits in PORTC. If the elapsed time is higher than the time limit, bit 1 is set to a 1; otherwise, bit 0 is set to a 1. When an interrupt occurs, the interrupt service routine updates the elapsed time.

- *Lines 100 and 101.*   The external signal that is being timed set the flag, which caused the interrupt. The flag is now cleared by storing a 1 into the flag bit.

```
 1                              **********************************************************
 2                              ** MEASURE ELAPSED TIME BETWEEN EVENTS AND INDICATE
 3                              ** IF OUT OF ACCEPTABLE RANGE
 4                              *
 5                              * USE INPUT CAPTURE 3 TO MEASURE TIME
 6                              *
 7                              * USE 8.0-MHZ CRYSTAL FOR 68HC11 CLOCK AND RUN
 8                              * ..FREE-RUNNING COUNTER AT MAXIMUM RATE
 9                              *
10                              * MAXIMUM ELAPSED TIME MUST BE LESS THAN FREE-RUNNING
11                              * ..COUNTER OVERFLOW TIME OF 32.77 MILLISECONDS
12                              *
13                              **********************************************************
14                              *
15                              **********************************************************
16                              ** SYMBOL DEFINITIONS
17                              **********************************************************
18                              * 68HC11 REGISTERS
19          1003                PORTC   EQU   $1003     I/O PORT C REGISTER
20          1007                DDRC    EQU   $1007     DATA DIRECTION REGISTER C
21          1014                TIC3    EQU   $1014     TIMER INPUT CAPTURE 3 REGISTER
22          1021                TCTL2   EQU   $1021     TIMER CONTROL REGISTER
23          1022                TMSK1   EQU   $1022     TIMER MASK REGISTER
24          1023                TFLG1   EQU   $1023     TIMER FLAG REGISTER
25          0003                IOPAT   EQU   $03       I/O PATTERN, 0=IN 1=OUT
26                              * MASKS
27          0001                BIT0    EQU   %00000001
28          0002                BIT1    EQU   %00000010
29          0001                IC3F    EQU   BIT0
30                              *
31                              **********************************************************
32                              ** DATA SECTION
33                              **********************************************************
34    0010                              ORG   $0010
35    0010    9000              TLIMIT  FDB   $9000     MAX ALLOWED ELAPSED TIME
36    0012                      LASTTIM RMB   2         COUNTER AT LAST IC3 CAPTURE
37    0014                      ELAPSED RMB   2         ELAPSED TIME SINCE LAST CAPTURE
38                              *
39    FFEA                              ORG   $FFEA     IC3 INTERRUPT VECTOR
40    FFEA    C135                      FDB   IC3ISR
41                              *
42                              **********************************************************
43                              ** MAIN PROGRAM
44                              **********************************************************
45    C100                              ORG   $C100
46                              *---------------------------------------------------------
47                              * INITIALIZATION
48                              *---------------------------------------------------------
49                              * INITIALIZE STACK
50    C100    8E CF FF          START   LDS   #$CFFF
51                              * INITIALIZE PORT C
52                              *     INITIALIZE OUTPUTS TO ZEROS
53    C103    7F 10 03                  CLR   PORTC     0=LOW
54                              *     SET UP PORTC INS AND OUTS
55    C106    86 03                     LDAA  #IOPAT
56    C108    B7 10 07                  STAA  DDRC      TWO OUTPUT PINS
57                              * INITIALIZE INPUT CAPTURE 3
58    C10B    86 01                     LDAA  #$01      EDG3B:EDG3A=01 FOR LOW-TO-HIGH
59    C10D    B7 10 21                  STAA  TCTL2     ..CAPTURE
60                              * ENABLE IC3 INTERRUPT
61    C110    86 01                     LDAA  #BIT0
62    C112    B7 10 22                  STAA  TMSK1
63                              * INITIALIZE ELAPSED TIME
```

Figure 7-20 Input capture 3 example.

```
64   C115   CC 00 00                LDD    #0
65   C118   DD 14                   STD    ELAPSED
66                          * INITIALIZE LAST TIME
67   C11A   FC 10 14                LDD    TIC3
68   C11D   DD 12                   STD    LASTTIM
69                          * TURN ON INTERRUPT SYSTEM
70   C11F   0E                      CLI
71                          *-----------------------------------------------------
72                          * MAIN PROGRAM LOOP
73                          *-----------------------------------------------------
74                          * ELAPSED TIME OVER LIMIT?
75   C120   DC 14           HERE    LDD    ELAPSED
76   C122   1A 93 10                CPD    TLIMIT
77   C125   22 07                   BHI    AHEAD      BRANCH ON YES
78                          * SET TIME OK INDICATOR
79   C127   86 01                   LDAA   #BIT0    1 IN BIT0 = OK
80   C129   B7 10 03                STAA   PORTC
81   C12C   20 05                   BRA    AGAIN
82                          * SET TIME OVERLIMIT INDICATOR
83   C12E   86 02           AHEAD   LDAA   #BIT1    1 IN BIT1 = OVERLIMIT
84   C130   B7 10 03                STAA   PORTC
85   C133   20 EB           AGAIN   BRA    HERE
86                          *
87                          ******************************************************
88                          ** INTERRUPT SERVICE ROUTINE
89                          ******************************************************
90                          *-----------------------------------------------------
91                          * INTERRUPT POLLING CHAIN
92                          *-----------------------------------------------------
93                          * INTERRUPT FROM IC3?
94   C135   CE 10 23        IC3ISR  LDX    #TFLG1
95   C138   1F 00 01 10             BRCLR  0,X,IC3F,RTIC3   IGNORE ILLEGAL INTERRUPT
96                          *-----------------------------------------------------
97                          * SERVICE INPUT CAPTURE 3
98                          *-----------------------------------------------------
99                          * CLEAR IC3 FLAG
100  C13C   86 01                   LDAA   #BIT0     STORE 1 TO CLEAR FLAG!
101  C13E   A7 00                   STAA   0,X       ..ZEROS DO NOTHING
102                         * CALCULATE ELAPSED TIME
103  C140   FC 10 14                LDD    TIC3
104  C143   93 12                   SUBD   LASTTIM
105  C145   DD 14                   STD    ELAPSED
106                         * SAVE CURRENT CAPTURE TIME
107  C147   FC 10 14                LDD    TIC3
108  C14A   DD 12                   STD    LASTTIM
109                         * RETURN FROM INTERRUPT
110  C14C   3B              RTIC3   RTI
111  C14D                           END
```

Figure 7-20   Continued.

- *Lines 103 through 105.* The input-capture register is read and then the last value read from the input-capture register is subtracted from it. The difference is the elapsed time in free-running counter periods, so the result is stored in the elapsed time location.

The numbers in the capture register are unsigned 16-bit numbers. The subtraction gives the correct elapsed time even if the free-running counter has overflowed. Convince yourself that this calculation is correct by trying several different numbers.

- *Lines 107 and 108.*   The current number in the input-capture register is saved in the last time register.

- *Line 110.*   Control returns to the main program loop that now uses the updated elapsed time.

You can use this program as a go/no-go frequency meter. If the frequency of the input signal is too low, the program indicates an out of limit time. If the frequency is high enough, it indicates the time is within the limit. If the input signal comes from a pulse transducer on a rotating device, the program can indicate if the RPM is above or below a limit.

Measuring long elapsed times

If the time to be measured may be longer than the overflow time for the free-running counter, the program must count overflows. Therefore, both the input capture interrupt and the timer overflow interrupt are used. Each time an overflow occurs, a separate overflow counter is incremented if timing is in progress.

The most difficult part of extending the time range is deciding whether to count an overflow. If the input capture happens close to the overflow, both the input capture flag and the timer overflow flag will be set when a timer interrupt service routine runs. However, the overflow interrupt service routine must increment the overflow counter only if the overflow happened before the input capture. Because the input capture has higher interrupt priority, its interrupt will be serviced before the overflow interrupt, which makes resolving the problem possible. Then, if the captured value has a 1 in its most-significant bit, the capture was before the overflow. If the captured value has a 0 in its most-significant bit, the capture was after the overflow.

When determining whether to increment the overflow counter, the timer interrupts must be serviced within half the time of an overflow of the free-running counter. If the interrupts are not serviced this quickly, the free-running counter may have another overflow before the relative capture time is determined.

Output Compare

The output-compare hardware normally controls the timing of changes in output bits. Usually the output signal is nearly periodic; for example, a *pulse-width modulated waveform.* However, almost any kind of timing can be achieved.

The 68HC11 has five output-compare registers. Output compare 1 has characteristics very different from those of output compares 2 through 5. Because output compares 2 through 5 share some similarity with the input captures, let's begin with them.

Output compares 2 through 5

The 68HC11A8 and 68HC11E9 ICs each has five sets of output-compare hardware. The E9 version shares a pin and other hardware between an optional input-capture device and output compare 5. The output-compare signals are part of the port A pins.

In all details, each of the output compares OC2 through OC5 functions the same. The other output compare, OC1, has additional functions that are discussed later. Therefore, the following describes the hardware for the OC2 through OC5 devices.

Output-compare registers.   Figure 7-21 shows four *output-compare registers* labeled as *TOCx*. They are each 16-bit registers, so they can be compared to the free-running counter. The four output-compare registers are named *TOC2*, *TOC3*, *TOC4*, and *TOC5* and are at addresses 1018, 101A, 101C, and 101E, respectively. A program may both read and write these registers; normally it will write to them only.

The output compare 5 register in the E9 version of the 68HC11 is called *TI4O5*. Storing 1s in bit 2 and bit 3 of the PACTL register programs TI4O5 for output-compare functions.

Port signals.   Port A includes the pins that the output-compare hardware controls. The pins are named *PA3*, *PA4*, *PA5*, *PA6*, and *PA7*. Output compare OC2 connects to PA6, OC3 to PA5, OC4 to PA4, and OC5 to PA3. Remember that pin PA3 is shared by an input-capture and an output-compare.

Comparator.   The comparator hardware looks for a match between the number in a compare register and the number in the free-running counter. When the match occurs, which it inevitably will, the comparator triggers an output action and sets a flag.

Action control.   When the comparator in Figure 7-21 detects a match between a compare register and the free-running counter, it triggers some action that takes place at an output pin. Table 7-7 lists the possible actions taken when a match occurs. Two control bits determine the actions. The control bits are called *OM* and *OL* for *output mode* and *output level*. By looking at the table, you can see that a 1 in OM makes the OL bit directly control the output bit when the match occurs. When OM is 0, the output pin is either disconnected from the timer or is toggled. The program must set OM and OL for the correct action.

Figure 7-22 shows the TCTL1 register that contains the OM and OL bits for the output-compares. The action of output compare 1 is not controlled by TCTL1.

Flags.   Figure 7-18 shows the output-compare flags in the timer-flag register called *TFLG1*. The flags are named *OC1F through OC5F* corresponding to the output compares

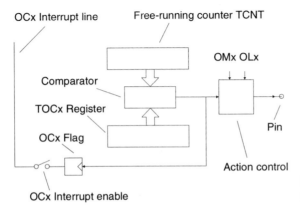

Figure 7-21   Block diagram of output-compare hardware.

TABLE 7-7   OUTPUT COMPARE ACTION

OMx Bit	OLx Bit	Successful Compare Action
0	0	Disconnect timer from output pin
0	1	Complement output pin
1	0	Set output pin low
1	1	Set output pin high

called *OC1 through OC5*. Matches between the free-running counter and the associated output-compare register set the respective flags. A flag is cleared when a 1 is written to the flag. Writing 0s to these flags does not affect them.

In the E9 version of the 68HC11, bit 3 of TFLG1, the flag for output compare 5, is called *I4/O5F*. A 1 in bit 2 of the PACTL register makes I4/O5F an output-compare flag.

Interrupts.   Each output-compare flag can cause an interrupt. The interrupt-enable bits corresponding to OC1 through OC5 are called *OC1I through OC5I*. They are in the *timer mask register* called *TMSK1*. Figure 7-19 shows this register. A 1 in an interrupt-enable bit enables the flag to cause an interrupt. The reset condition disables all output-compare interrupts.

The output compare 5 flag in the E9 version of the 68HC11 can cause an interrupt. Set bit 3 of TMSK1, called *I4O5I*, to 1 to enable the OC5 interrupt.

Instructions.   Usually, the program will change the number in the compare register to determine the time when the next output action will occur. Use a double-byte instruction to write to the 16-bit compare register or the comparison circuit may not function correctly. A match may occur at the wrong time if only one byte of the compare register is updated at a time.

Direct pin control.   While the output-compare hardware controls a pin, direct software control of the pin is not possible. Be careful because the timer may not control the output pins if the output-compare action disables the timer. When all timer output functions for a pin are disabled, the port A pins are controlled by the PORTA register. The timer does not affect the PORTA register.

Pulse-width modulation using OC2

A common digital-to-analog conversion technique is to pulse-width modulate a digital output pin. The pin is switched between low and high rapidly. Controlling the low and high times adjusts the average value of the waveform, which is used as an analog output. A

	7	6	5	4	3	2	1	0	
$1020	OM2	OL2	OM3	OL3	OM4	OL4	OM5	OL5	TCTL1
RESET	0	0	0	0	0	0	0	0	

Figure 7-22   Timer control register 1.

measure of the average value, called the *duty cycle*, is the time the pin is high as a percentage of the period.

The example in Figure 7-23 uses output compare 2 to control the duty cycle of port A pin PA6. The output pin is driven high after a time delay, and then is driven low after another time delay. Repeating this procedure generates the waveform. Here is a description of the program:

- *Lines 43 and 44.*   The OC2 is programmed to make the output low on the first compare. A hardware reset disconnects the timer from the output pin and clears port A. Therefore, after reset, the OC2 pin is also low.

  After reset, the free-running counter contains 0000 and the output-compare registers contain FFFF. Therefore, the first compare, after a reset, happens only after the full range of the free-running counter. The program may initialize the output-compare register if this is unacceptable.

- *Lines 46 through 53.*   The rest of the hardware is initialized, and the main program loop starts running waiting for interrupts.

- *Lines 62 and 63.*   The OC2 flag is tested for a valid interrupt. The X register is used as a pointer to the first hardware control register. Bit-oriented instructions with indexed addressing access the control registers.

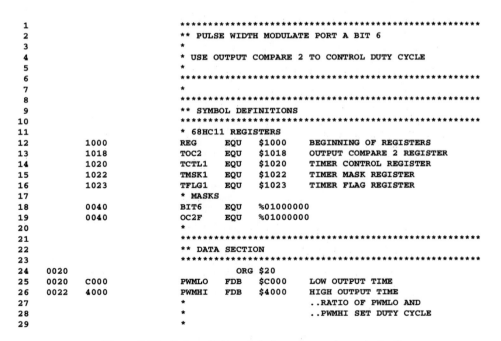

```
1                        ***********************************************************
2                        **  PULSE WIDTH MODULATE PORT A BIT 6
3                        *
4                        *  USE OUTPUT COMPARE 2 TO CONTROL DUTY CYCLE
5                        *
6                        ***********************************************************
7                        *
8                        ***********************************************************
9                        **  SYMBOL DEFINITIONS
10                       ***********************************************************
11                       *  68HC11 REGISTERS
12          1000         REG      EQU    $1000       BEGINNING OF REGISTERS
13          1018         TOC2     EQU    $1018       OUTPUT COMPARE 2 REGISTER
14          1020         TCTL1    EQU    $1020       TIMER CONTROL REGISTER
15          1022         TMSK1    EQU    $1022       TIMER MASK REGISTER
16          1023         TFLG1    EQU    $1023       TIMER FLAG REGISTER
17                       *  MASKS
18          0040         BIT6     EQU    %01000000
19          0040         OC2F     EQU    %01000000
20                       *
21                       ***********************************************************
22                       **  DATA SECTION
23                       ***********************************************************
24  0020                          ORG $20
25  0020    C000         PWMLO    FDB    $C000       LOW OUTPUT TIME
26  0022    4000         PWMHI    FDB    $4000       HIGH OUTPUT TIME
27                       *                          ..RATIO OF PWMLO AND
28                       *                          ..PWMHI SET DUTY CYCLE
29                       *
```

Figure 7-23   Pulse-width modulation using output compare 2.

```
30    FFE6                                    ORG   $FFE6
31    FFE6    C110                            FDB   OC2ISR      OC2 INTERRUPT VECTOR
32                                    *
33                                    ****************************************************
34                                    ** MAIN PROGRAM
35                                    ****************************************************
36    C100                                    ORG   $C100
37                                    *----------------------------------------------------
38                                    * INITIALIZATION
39                                    *----------------------------------------------------
40                                    * INITIALIZE STACK
41    C100    8E CF FF        START    LDS   #$CFFF
42                                    * INITIALIZE OUTPUT COMPARE OC2
43    C103    86 80                   LDAA  #$80     OM2:OL2=10 FOR SET TO LOW
44    C105    B7 10 20                STAA  TCTL1    ..ON COMPARE
45                                    * ENABLE OC2 INTERRUPT
46    C108    86 40                   LDAA  #$40
47    C10A    B7 10 22                STAA  TMSK1
48                                    * TURN ON INTERRUPT SYSTEM
49    C10D    0E                      CLI
50                                    *----------------------------------------------------
51                                    * WAIT FOR INTERRUPTS
52                                    *----------------------------------------------------
53    C10E    20 FE          HERE     BRA   HERE
54                                    *
55                                    ****************************************************
56                                    ** INTERRUPT SERVICE ROUTINE
57                                    ****************************************************
58                                    *----------------------------------------------------
59                                    * INTERRUPT POLLING CHAIN
60                                    *----------------------------------------------------
61                                    * INTERRUPT FROM OC2?
62    C110    CE 10 00       OC2ISR   LDX   #REG
63    C113    1F 23 40 1C             BRCLR TFLG1-REG,X,OC2F,RTOC2   IGNORE ILLEGAL
64                                    *                       ..INTERRUPT
65                                    *----------------------------------------------------
66                                    * SERVICE OUTPUT COMPARE 2
67                                    *----------------------------------------------------
68                                    * CLEAR OC2 FLAG
69    C117    86 40                   LDAA  #OC2F    STORE 1 TO CLEAR FLAG
70    C119    A7 23                   STAA  TFLG1-REG,X ..ZEROS DO NOTHING
71                                    * WAS LAST OUTPUT HIGH?
72    C11B    1E 20 40 0B             BRSET TCTL1-REG,X,BIT6,LASTHI  BRANCH ON YES
73                                    * PROGRAM NEXT OUTPUT TO BE HIGH
74                                    *    SET OC2 OUTPUT ACTION
75    C11F    1C 20 40                BSET  TCTL1-REG,X,BIT6 OM2:OL2=11
76                                    *    SET HIGH OUTPUT COMPARE TIME
77    C122    EC 18                   LDD   TOC2-REG,X
78    C124    D3 20                   ADDD  PWMLO
79    C126    ED 18                   STD   TOC2-REG,X
80    C128    20 09                   BRA   RTOC2
81                                    * PROGRAM NEXT OUTPUT TO BE LOW
82                                    *    SET OC2 OUTPUT ACTION
83    C12A    1D 20 40       LASTHI   BCLR  TCTL1-REG,X,BIT6  OM2:OL2=10
84                                    *    SET LOW OUTPUT COMPARE TIME
85    C12D    EC 18                   LDD   TOC2-REG,X
86    C12F    D3 22                   ADDD  PWMHI
87    C131    ED 18                   STD   TOC2-REG,X
88                                    * RETURN FROM INTERRUPT
89    C133    3B             RTOC2    RTI
90    C134                            END
```

Figure 7-23 Continued.

- *Lines 69 and 70.* Writing 1 to the flag bit clears the OC2 flag.

- *Line 72.* The program tests the OL bit to determine if the output pin was high after the last interrupt. If so, it must be changed to low on the next interrupt. Otherwise, it must be changed to high. Because this instruction can only be executing if an interrupt has occurred, this test is a sure test of the last state of the output pin.

- *Line 75.* Set the output level or OL bit so that the next compare makes the output pin high.

- *Lines 77 through 79.* The number in the output-compare register is advanced by a count that represents the low time of the output pin. Overflows from the addition cause no problems.

- *Lines 83 through 87.* These lines set up the output to go low after the proper high time.

- *Line 89.* Control returns to the main program. The low and high times must be long enough for the return and another interrupt to complete before the next output change is required.

The longest high and low time appropriate to the application should be used. Shorter times cause more frequent interrupts and greater software overhead. Very short times could cause the interrupt service overhead to be so great that nothing else could be done. The computer could spend most of its time in the interrupt service routine.

Output compare 1

Output compare 1 controls the output pin differently from the other four output compares. In addition, OC1 may control all five output compare pins at once, although four of them may also be under control of their output-compare hardware. The output-compare signals are part of the port A pins. Figure 7-24 shows the parts of the output-compare 1 hardware.

Output-compare 1 register. The output-compare 1 register that Figure 7-24 shows is a 16-bit register because it is compared to the free-running counter. The *output-compare register* is named *TOC1* and is at address 1016. A program may both read and write this register; normally the program only writes to it.

Port signals. Port A includes the pins that the output-compare hardware controls. The pins are named *PA3, PA4, PA5, PA6,* and *PA7*. Output compare OC1 may control all these output pins even if other output compares also control them.

Pin PA7 shares functions between output compare 1 and the pulse accumulator device. Pin PA7 is a programmable I/O pin. The program must make pin PA7 an output when the output-compare device uses it. Figure 7-9 shows the PACTL register and the data direction register bit DDRA7. The program puts a 1 in DDRA7 to make PA7 an output pin.

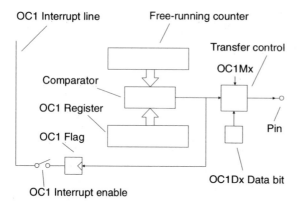

Figure 7-24 Block diagram of output compare 1 hardware.

Transfer control.   When the comparator in Figure 7-24 detects a match between the output-compare 1 register and the free-running counter, it triggers the transfer of a data bit to the output pin. A mask bit called *OC1Mx* enables or disables the transfer, and a data bit called *OC1Dx* determines the output pin level. If the program has set the mask to 1, the transfer takes place when the match occurs. The program must store the value into the data bit before the match occurs. Therefore, the timing of the transfer to the pin is independent of the software overhead.

Figure 7-25 shows the OC1 mask register called *OC1M*. Bits 3 through 7 of OC1M correspond to pins PA3 through PA7 of port A. A 1 in a mask bit enables OC1 to control the corresponding port A pin.

If a mask bit enables a transfer, the transfer occurs on the next match despite other hardware actions. For example, suppose that both OC2 and OC1 control port A pin PA6. If both try to affect the pin on the same E-clock cycle, OC1 overrides the other hardware.

Figure 7-26 shows the OC1 data register called *OC1D*. Bits 3 through 7 of OC1D correspond to pins PA3 through PA7 of port A. If the mask enables a transfer, a 1 in a data bit makes the corresponding port A pin high when the transfer occurs, and a 0 makes it low.

	7	6	5	4	3	2	1	0	
$100C	OC1M7	OC1M6	0C1M5	OC1M4	OC1M3	0	0	0	OC1M
RESET	0	0	0	0	0	0	0	0	

Figure 7-25 Output compare 1 mask register.

	7	6	5	4	3	2	1	0	
$100D	OC1D7	OC1D6	OC1D5	OC1D4	OC1D3	0	0	0	OC1D
RESET	0	0	0	0	0	0	0	0	

Figure 7-26 Output compare 1 data register.

Flag.   Figure 7-18 shows the output-compare flags in the *timer flag register* called *TFLG1*. The OC1 flag is set by matches between the free-running counter and the output-compare 1 register. Writing a 1 to an output-compare flag bit clears the flag. Writing a 0 to the flag bit does not affect it.

Interrupt.   The output compare 1 flag, OC1F, can cause an interrupt. The interrupt-enable bit OC1I corresponding to OC1 is in the *timer mask register* called *TMSK1*. Figure 7-19 shows the TMSK1 register. Writing a 1 into an interrupt-enable bit enables the flag to cause an interrupt. The reset condition disables all output-compare interrupts.

Instructions.   Always use a double-byte instruction, such as STD, to write to the 16-bit output compare 1 register or the comparison circuit may not function correctly. A match may occur at the wrong time if only one byte of the compare register is updated at a time.

Direct pin control.   When the output-compare hardware controls a pin, direct software control of the pin is not possible. Two output-compare devices must be disabled to enable direct control of a pin. First, the corresponding OC1 mask bit must be 0 to disable OC1 from controlling the pin. Second, if another output-compare corresponds to the pin, it must be disconnected. Then the port A output pins are controlled by the PORTA register. The timer does not affect the PORTA register.

Pulse-width modulation using OC1

The program in Figure 7-27 uses output compare 1 to control the duty cycle of port A pin PA7. This example that uses OC1 is very similar to the example in Figure 7-23 that uses OC2. The output pin is driven high after a time delay, then driven low after another time delay. Repeating this procedure rapidly generates the waveform with a specified duty cycle.

The example uses port A pin PA7 because this pin is more complicated than the other port A outputs. Here is a description of the program that highlights only the differences from Figure 7-23:

- *Lines 46 and 47.*   A 1 is put into the OC1 mask register to enable it to control port A pin PA7. Hardware reset clears all OC1 mask bits and disables all other output compares from controlling port A pins.

- *Line 49.*   Clearing the output data register will make the initial state of the output pin the low level.

- *Lines 51 and 52.*   Port A pin PA7 has several functions including both input and output operations, so it has a data direction bit. To use an output-compare to control this pin, the data direction must be set to output. Storing 1 in the DDRA7 bit in the PACTL register makes PA7 an output.

- *Line 80.*   Reading the last data value OC1 used indicates the state of the output pin. If the pin is low, control goes to line 83.

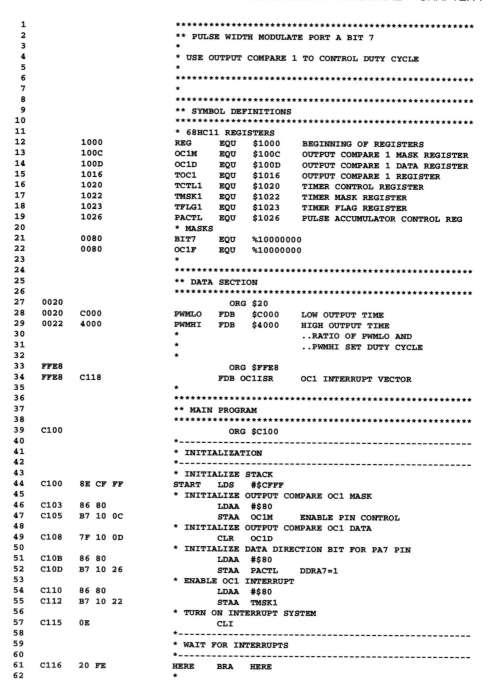

```
1                                    ******************************************************
2                                    ** PULSE WIDTH MODULATE PORT A BIT 7
3                                    *
4                                    * USE OUTPUT COMPARE 1 TO CONTROL DUTY CYCLE
5                                    *
6                                    ******************************************************
7                                    *
8                                    ******************************************************
9                                    ** SYMBOL DEFINITIONS
10                                   ******************************************************
11                                   * 68HC11 REGISTERS
12          1000                     REG      EQU    $1000     BEGINNING OF REGISTERS
13          100C                     OC1M     EQU    $100C     OUTPUT COMPARE 1 MASK REGISTER
14          100D                     OC1D     EQU    $100D     OUTPUT COMPARE 1 DATA REGISTER
15          1016                     TOC1     EQU    $1016     OUTPUT COMPARE 1 REGISTER
16          1020                     TCTL1    EQU    $1020     TIMER CONTROL REGISTER
17          1022                     TMSK1    EQU    $1022     TIMER MASK REGISTER
18          1023                     TFLG1    EQU    $1023     TIMER FLAG REGISTER
19          1026                     PACTL    EQU    $1026     PULSE ACCUMULATOR CONTROL REG
20                                   * MASKS
21          0080                     BIT7     EQU    %10000000
22          0080                     OC1F     EQU    %10000000
23                                   *
24                                   ******************************************************
25                                   ** DATA SECTION
26                                   ******************************************************
27    0020                                    ORG    $20
28    0020  C000                     PWMLO    FDB    $C000     LOW OUTPUT TIME
29    0022  4000                     PWMHI    FDB    $4000     HIGH OUTPUT TIME
30                                   *                         ..RATIO OF PWMLO AND
31                                   *                         ..PWMHI SET DUTY CYCLE
32                                   *
33    FFE8                                    ORG    $FFE8
34    FFE8  C118                              FDB    OC1ISR    OC1 INTERRUPT VECTOR
35                                   *
36                                   ******************************************************
37                                   ** MAIN PROGRAM
38                                   ******************************************************
39    C100                                    ORG    $C100
40                                   *-----------------------------------------------------
41                                   * INITIALIZATION
42                                   *-----------------------------------------------------
43                                   * INITIALIZE STACK
44    C100  8E CF FF                 START    LDS    #$CFFF
45                                   * INITIALIZE OUTPUT COMPARE OC1 MASK
46    C103  86 80                             LDAA   #$80
47    C105  B7 10 0C                          STAA   OC1M      ENABLE PIN CONTROL
48                                   * INITIALIZE OUTPUT COMPARE OC1 DATA
49    C108  7F 10 0D                          CLR    OC1D
50                                   * INITIALIZE DATA DIRECTION BIT FOR PA7 PIN
51    C10B  86 80                             LDAA   #$80
52    C10D  B7 10 26                          STAA   PACTL     DDRA7=1
53                                   * ENABLE OC1 INTERRUPT
54    C110  86 80                             LDAA   #$80
55    C112  B7 10 22                          STAA   TMSK1
56                                   * TURN ON INTERRUPT SYSTEM
57    C115  0E                                CLI
58                                   *-----------------------------------------------------
59                                   * WAIT FOR INTERRUPTS
60                                   *-----------------------------------------------------
61    C116  20 FE                    HERE     BRA    HERE
62                                   *
```

Figure 7-27   Pulse-width modulation using output compare 1.

```
63                          ***********************************************************
64                          ** INTERRUPT SERVICE ROUTINE
65                          ***********************************************************
66                          *---------------------------------------------------------
67                          * INTERRUPT POLLING CHAIN
68                          *---------------------------------------------------------
69                          * INTERRUPT FROM OC1?
70   C118   CE 10 00        OC1ISR  LDX    #REG
71   C11B   1F 23 80 1C             BRCLR  TFLG1-REG,X,OC1F,RTOC1   IGNORE ILLEGAL
72                          *                                        ..INTERRUPT
73                          *---------------------------------------------------------
74                          * SERVICE OUTPUT COMPARE 1
75                          *---------------------------------------------------------
76                          * CLEAR OC1 FLAG
77   C11F   86 80                   LDAA   #OC1F     STORE 1 TO CLEAR FLAG
78   C121   A7 23                   STAA   TFLG1-REG,X ..ZEROS DO NOTHING
79                          * WAS LAST OUTPUT HIGH?
80   C123   1E 0D 80 0B             BRSET  OC1D-REG,X,BIT7,LASTHI   BRANCH ON YES
81                          * PROGRAM NEXT OUTPUT TO BE HIGH
82                          *    SET OC1 OUTPUT DATA FOR HIGH OUTPUT
83   C127   1C 0D 80                BSET   OC1D-REG,X,BIT7
84                          *    SET HIGH OUTPUT COMPARE TIME
85   C12A   EC 16                   LDD    TOC1-REG,X
86   C12C   D3 20                   ADDD   PWMLO
87   C12E   ED 16                   STD    TOC1-REG,X
88   C130   20 09                   BRA    RTOC1
89                          * PROGRAM NEXT OUTPUT TO BE LOW
90                          *    SET OC1 OUTPUT DATA FOR LOW OUTPUT
91   C132   1D 0D 80        LASTHI  BCLR   OC1D-REG,X,BIT7
92                          *    SET LOW OUTPUT COMPARE TIME
93   C135   EC 16                   LDD    TOC1-REG,X
94   C137   D3 22                   ADDD   PWMHI
95   C139   ED 16                   STD    TOC1-REG,X
96                          * RETURN FROM INTERRUPT
97   C13B   3B              RTOC1   RTI
98   C13C                           END
```

Figure 7-27   Continued.

- *Line 83.* The output data bit is set to 1, so the pin goes high on the next match between the free-running counter and the TOC1 output-compare register.

- *Lines 85 through 87.* Because the output pin went to the low level when this interrupt occurred, the time the output must be low is added to the compare register. When the next compare occurs, the output pin will go high.

Most of this example appears similar to the example using OC2. However, the operation of the two output-compares is very different. OC2 has more functions that it can do to the output pin than does OC1.

Forcing output compares

A program can force the output-compare action to take place without a match between the free-running counter and the output-compare register. Therefore, the program can advance the time when the output action takes place.

Output-compare initialization.   Programs often force output-compares to initialize the output hardware to a known condition. Software cannot directly control an output pin while an output-compare has control of the pin. Therefore, forcing the compare hardware is the only way the program can initialize the level on the pin.

The CFORC register.   Figure 7-28 shows the *output-compare-force register* named *CFORC*. To force or trigger an output action, a program writes a 1 to a FOCx force-bit in the CFORC register. Writing 0s into force-bits does nothing. Furthermore, forcing the action does not set the flag! So forcing cannot make an interrupt occur earlier.

Each output compare can be forced by the corresponding bit in CFORC. Because the CFORC register is a hardware trigger register, reading CFORC is meaningless.

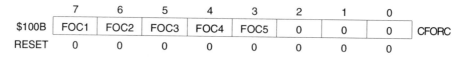

	7	6	5	4	3	2	1	0	
$100B	FOC1	FOC2	FOC3	FOC4	FOC5	0	0	0	CFORC
RESET	0	0	0	0	0	0	0	0	

Figure 7-28   Timer compare force register.

Forcing toggle actions.   Forcing the output action does not prevent the normal output action when a match occurs. Be careful using the force register if the output action is to toggle the output pin. The forced toggle of the output pin may be immediately followed by a normal toggle of the pin causing confusing results.

Using Input Capture with Output Compare

When input capture and output compare are used together, you can synchronize a timed output function to a timed input function. The synchronization is very accurate without dependence on latency time of software.

Make delayed pulse using IC1 and OC3

The example program in Figure 7-29 shows the use of both input-capture hardware and output-compare hardware together. The program uses OC3 to make a pulse on the port A pin PA5 that is delayed from the time a trigger from input capture 1 occurs. The programmable timer controls the delay time before the pulse starts and the duration of the pulse. The output pin is normally low, so the pulse is a high-going pulse.

The program is very similar to previous examples. The following describes only those parts related to the programmable timer:

- *Lines 61 through 63.*   The output compare 3 is programmed to make the pin go low on the next match between the free-running counter and the output-compare register. Then the instruction on line 63 writes to the compare-force register to initialize the pin to the low level.

```
1                                       ***********************************************************
2                                       ** MAKE DELAYED TRIGGERED PULSE
3                                       *
4                                       * USE INPUT CAPTURE 1 TO TRIGGER ON LOW-TO-HIGH
5                                       * ..TRANSITION AT PORT A BIT2
6                                       *
7                                       * USE OUTPUT COMPARE 3 TO MAKE HIGH-GOING PULSE
8                                       * ..AT PORT A BIT5
9                                       *
10                                      * USE 8.0-MHZ CRYSTAL FOR 68HC11 CLOCK AND RUN
11                                      * ..FREE-RUNNING COUNTER AT MAXIMUM RATE
12                                      *
13                                      * MAXIMUM DELAY IS 32.77 MS AND MAXIMUM DURATION
14                                      * ..IS 32.77 MS
15                                      *
16                                      ***********************************************************
17                                      *
18                                      ***********************************************************
19                                      ** SYMBOL DEFINITIONS
20                                      ***********************************************************
21                                      * 68HC11 REGISTERS
22            1000                      REG      EQU   $1000    BEGINNING OF 68HC11 REGISTERS
23            100B                      CFORC    EQU   $100B    COMPARE FORCE REGISTER
24            1010                      TIC1     EQU   $1010    INPUT CAPTURE 1 REG
25            101A                      TOC3     EQU   $101A    OUTPUT COMPARE 3 REG
26            1020                      TCTL1    EQU   $1020    TIMER CONTROL REG 1
27            1021                      TCTL2    EQU   $1021    TIMER CONTROL REG 2
28            1022                      TMSK1    EQU   $1022    TIMER INT MASK 1
29            1023                      TFLG1    EQU   $1023    TIMER INT FLAG 1
30                                      * MASKS
31            0004                      BIT2     EQU   %00000100
32            00FB                      BBIT2    EQU   %11111011
33            0020                      BIT5     EQU   %00100000
34            00DF                      BBIT5    EQU   %11011111
35                                      *
36                                      ***********************************************************
37                                      ** DATA SECTION
38                                      ***********************************************************
39   0000                                        ORG   $0000
40   0000     4000                      DELCNT   FDB   $4000    DELAY TIME IN COUNTER PULSES
41   0002     6000                      PWCNT    FDB   $6000    PULSE HIGH TIME IN PULSES
42                                      *
43   FFE4                                        ORG   $FFE4
44   FFE4     C132                               FDB   OC3ISR   OC3 INTERRUPT VECTOR
45   FFEE                                        ORG   $FFEE
46   FFEE     C118                               FDB   IC1ISR   IC1 INTERRUPT VECTOR
47                                      *
48                                      ***********************************************************
49                                      ** MAIN PROGRAM
50                                      ***********************************************************
51   C100                                        ORG   $C100
52                                      *---------------------------------------------------------
53                                      * INITIALIZATION
54                                      *---------------------------------------------------------
55                                      * INITIALIZE STACK
56   C100    8E CF FF                   START    LDS   #$CFFF
57                                      * INITIALIZE IC1 ACTIVE EDGE
58   C103    86 10                               LDAA  #$10     LTOH RISING EDGE ONLY
59   C105    B7 10 21                            STAA  TCTL2
60                                      * INITIALIZE OC3 OUTPUT TO LOW
61   C108    86 20                               LDAA  #$20     OM3:OL3=10 FOR SET TO LOW
62   C10A    B7 10 20                            STAA  TCTL1    ..ON COMPARE
63   C10D    B7 10 0B                            STAA  CFORC    FORCE OC3 ACTION
```

Figure 7-29   Delayed pulse using IC1 and OC3.

```
64                                    * ENABLE IC1 AND DISABLE OC3 INTERRUPTS
65    C110    86 04                          LDAA    #BIT2
66    C112    B7 10 22                        STAA    TMSK1
67                                    * TURN ON INTERRUPT SYSTEM
68    C115    0E                              CLI
69                                    *------------------------------------------------------------
70                                    * WAIT FOR INTERRUPTS
71                                    *------------------------------------------------------------
72    C116    20 FE           HERE    BRA     HERE
73                                    *
74                                    ************************************************************
75                                    ** INPUT CAPTURE 1 INTERRUPT SERVICE ROUTINE
76                                    ************************************************************
77                                    * CLEAR IC1 FLAG
78    C118    CE 10 00        IC1ISR  LDX     #REG
79    C11B    1D 23 FB                BCLR    TFLG1-REG,X,BBIT2   STORE 1 TO CLEAR FLAG
80                                    * SET TIME FOR OUTPUT PULSE TO GO HIGH
81    C11E    FC 10 10                LDD     TIC1
82    C121    D3 00                   ADDD    DELCNT
83    C123    FD 10 1A                STD     TOC3
84                                    * PROGRAM OC3 TO GO HIGH ON NEXT MATCH
85    C126    86 30                   LDAA    #$30       OM3:OL3=11
86    C128    B7 10 20                STAA    TCTL1
87                                    * CLEAR OC3 FLAG
88    C12B    1D 23 DF                BCLR    TFLG1-REG,X,BBIT5   STORE 1 TO CLEAR FLAG
89                                    * ENABLE OC3 INTERRUPT
90    C12E    1C 22 20                BSET    TMSK1-REG,X,BIT5
91                                    * RETURN FROM IC1 INTERRUPT
92    C131    3B                      RTI
93                                    *
94                                    ************************************************************
95                                    ** OUTPUT COMPARE 3 INTERRUPT SERVICE ROUTINE
96                                    ************************************************************
97                                    * DISABLE FURTHER OC3 INTERRUPTS
98    C132    CE 10 00        OC3ISR  LDX     #REG
99    C135    1D 22 20                BCLR    TMSK1-REG,X,BIT5   NO NEED TO CLEAR FLAG
100                                   * SET TIME FOR PULSE TO GO LOW
101   C138    FC 10 1A                LDD     TOC3
102   C13B    D3 02                   ADDD    PWCNT
103   C13D    FD 10 1A                STD     TOC3
104                                   * PROGRAM OC3 TO GO LOW ON NEXT MATCH
105   C140    86 20                   LDAA    #$20       OM3:OL3=10
106   C142    B7 10 20                STAA    TCTL1
107                                   * RETURN FROM OC3 INTERRUPT
108   C145    3B                      RTI
109   C146                            END
```

Figure 7-29   Continued.

This initialization may not be necessary. If a hardware reset precedes this program, the reset forces the pin low. Furthermore, if a reset does not occur, the pin will go low on the next match between the free-running counter and the output-compare 3 register. The match will occur sometime within the next 32.77_{10} milliseconds. Even if a reset doesn't precede the program, initializing the pin within 32.77_{10} milliseconds may be acceptable.

- *Lines 65 and 66.*   Both the IC1 and OC3 devices use interrupt, but only the IC1 interrupt is enabled here. Otherwise, OC3 will interrupt every time the output-compare register matches the free-running counter.

- *Lines 78 and 79.*   When the input signal for IC1 triggers a capture, the IC1F flag is set causing the interrupt. The IC1F flag is now cleared.

- *Lines 81 through 83.*   The time to start the output pulse with OC3 is set. First, the IC1 capture register is read to get the time when the input signal triggered an input capture. The latency time for processing the interrupt has no effect because the trigger time is in the capture register. Note that reading the free-running counter will give errors in the timing, so always read the capture register. Second, the delay time in counts of the free-running counter is added to the capture time and stored in the output-compare register. Therefore, the output pin will be affected at the time of the next match. The maximum possible delay in this program is the time between free-running counter overflows.

- *Lines 85 and 86.*   The initialization programmed OC3 to go low, so now it's programmed to go high on the upcoming match. Therefore, OC3 will generate a high-going pulse.

- *Line 88.*   While waiting for an input signal, or while processing the interrupt and running the interrupt service routine, a match for OC3 may have occurred. If a match set the OC3F flag, it must be cleared. Otherwise, an interrupt from OC3 will occur immediately instead of after the delay.

- *Line 90.*   The OC3 interrupt is enabled. After the delay, the OC3 match will make the output pin high, set the OC3F flag, and cause an interrupt to signal that the delay is complete.

- *Line 92.*   Control returns to the main program after servicing the interrupt from input capture 1. The main program and other interrupts run during the delay time.

- *Lines 98 and 99.*   The delay period is over, and OC3 caused an interrupt to send control here. Usually when a device interrupts, the interrupt service routine must clear flag so that it is ready for future interrupts. However, in this case, you do not need to know when the output pulse is complete. Furthermore, OC3 continues to set OC3F repeatedly as the free-running counter causes further matches. Therefore, OC3 must be prevented from causing further interrupts by disabling its interrupt.

- *Lines 101 through 103.*   The time at which OC3 must change the output pin is set. The high-going time is read from the compare register, the duration in counts is added, and the compare register is updated.

- *Lines 105 and 106.*   Output compare 3 is programmed to make the pin go low after the proper pulse duration. The timer continues to operate from now on independent of the program. Therefore, the pin goes low at an accurate time, independent of the program.

The time between free-running counter overflows is 32.77_{10} milliseconds in this example. Therefore, the maximum delay time and maximum pulse time are both 32.77_{10} milliseconds. Longer times require use of the timer-overflow interrupt.

In this example an input signal will not trigger a new pulse before the last pulse is complete. You should investigate the consequences of the input signal triggering a new pulse before the last one is finished. Also, consider the effects of other interrupt service routines that may delay the running of the timer interrupt service routines.

An application for a delayed pulse

Consider a gasoline engine that uses electronic fuel injection synchronized to the engine rotation. A fuel injector allows fuel flow in pulses. The pulses must occur only when the engine is at a particular point in a crankshaft revolution. A sensor on the engine makes a pulse as the crankshaft rotates to a reference position. The injector must supply fuel at some time beyond the occurrence of the sensor pulse. The program in Figure 7-29 can use the pulse from the sensor to make the control pulse for the injector.

As the engine runs at higher speeds, the delay time must be reduced. Furthermore, the duration of the injector pulse must be adjusted under different engine running conditions to deliver the correct fuel flow. A feedback control system determines the delay and pulse length. The control system program runs independently of the timer interrupt service routines. The program example here only makes the injector pulse.

Timer Flag Applications

Some applications may not need some or all input-capture functions. If not used for input-capture functions, the input-capture flags are available as general-purpose flags. Each flag has a separate interrupt vector. Ignore the input-capture register if you want to use only the flag.

Because the input-capture flags are available, the 68HC11, when using the STAF flag, has from one to five flags. Therefore, many applications will not need an external flag.

7.9 PULSE ACCUMULATOR

The pulse accumulator either counts pulses on a pin or it counts cycles of an oscillator to make a timer. The pulse accumulator operates in two modes called the *external-event-counting mode* and the *gated-time-accumulation mode.*

The primary hardware in the pulse accumulator is an 8-bit up-counter. The port A pin PA7 controls the counter. Figure 7-30 shows the *pulse-accumulator counter* called *PACNT.* The program may both read and write the PACNT register.

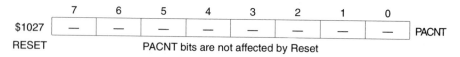

Figure 7-30 Pulse-accumulator count register.

The mode of operation of the pulse accumulator is controlled by three bits in the *pulse-accumulator-control register* called *PACTL*. Figure 7-31 is a copy of Figure 7-9 that shows the PACTL register.

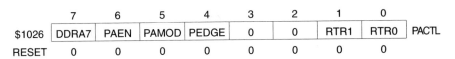

	7	6	5	4	3	2	1	0	
$1026	DDRA7	PAEN	PAMOD	PEDGE	0	0	RTR1	RTR0	PACTL
RESET	0	0	0	0	0	0	0	0	

Figure 7-31   Pulse-accumulator control register.

The *pulse-accumulator-enable bit, PAEN*, in the PACTL register enables the pulse accumulator while it is 1. The program may make the PAEN bit 0 to disable the pin from affecting the pulse accumulator. Usually, programs disable the pulse accumulator if pin PA7 is programmed for output. The DDRA7 bit in the PACTL register controls whether pin PA7 does input or output.

The *pulse-accumulator-mode bit, PAMOD*, in the PACTL register selects one of the two primary modes of operation. The program selects the external-event-counting mode by making the PAMOD bit 0, and it selects the gated-time-accumulation mode by making the PAMOD bit 1. The purpose of the *pulse-edge-bit, PEDGE*, in PACTL depends on the mode of operation.

Event-Counting Mode

The pulse accumulator counts active transitions on port A pin PA7 while the pulse accumulator is in the event-counting mode. Figure 7-32 is a block diagram of the parts of the pulse accumulator in this mode. The hardware is very flexible; it is used for many kinds of applications.

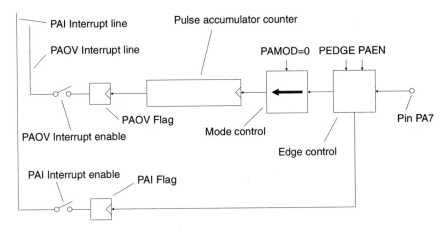

Figure 7-32   Pulse accumulator in event-counting mode.

Hardware operation

Active transitions of signals applied to the PA7 pin clock the 8-bit counter register. The counter increments by one each time it is clocked.

Flags and interrupts.   When the counter rolls over from FF to 00, it sets the *pulse-accumulator overflow flag*, *PAOVF*, in the TFLG2 register. Setting the PAOVF flag while the pulse-accumulator-overflow interrupt is enabled causes an interrupt if the interrupt system is on. The *pulse-accumulator-overflow-interrupt-enable bit*, *PAOVI*, in the TMSK2 register must be 1 to enable overflow interrupts. The overflow interrupt has an interrupt vector at address FFDC and FFDD as Table 7-2 shows.

While the pulse accumulator is in event-counting mode because PAMOD is 0, the hardware transfers the active transition from the PA7 pin to the counter. Figure 7-32 shows this transfer with a bold arrow. The active transition at pin PA7 also sets the pulse-accumu-lator input flag, PAIF. If the pulse-accumulator-input-interrupt-enable bit, PAII, is 1, setting the flag causes an interrupt. The interrupt vector is at addresses FFDA and FFDB as Table 7-2 shows.

Active transitions.   The PEDGE bit in the PACTL register determines whether the active transition at pin PA7 is high-to-low or low-to-high. While PEDGE is 0, the high-to-low transition is active. While PEDGE is 1, the low-to-high transition is active. However, if the pulse-accumulator-enable bit PAEN is 0, the input signal at the pin has no effect on any of the pulse-accumulator hardware.

Summary.   While the pulse accumulator is enabled, each active transition of the PA7 pin increments the pulse-accumulator counter and sets the pulse-accumulator-input flag. If the counter rolls over to 00, it sets the pulse-accumulator-overflow flag. If the flags are enabled to cause interrupts, setting the flags cause interrupts to unique interrupt vectors.

Programming example

Consider an application where a toothed wheel, magnetic sensor, and electronic inter-face make pulses as a mechanical device rotates. Each tooth generates a pulse as it passes the magnetic sensor. The pulses cause interrupts so that the computer responds to the rotation. However, the pulses may occur so frequently that the computer spends too much time re-sponding to the pulses. If the application does not require the computer to respond to every pulse, some pulses can be skipped. Suppose that we want the computer to respond to every fifth pulse. Rather than build an electronic circuit to generate an interrupt signal after five pulses, let's use the pulse accumulator.

For simplicity, the example program only complements port B every fifth pulse. Figure 7-33 is the listing of the example program. The following is a description of the program.

- *Lines 42 and 43.*   The program enables the pulse accumulator and puts it into the event-counting mode. It also programs pin PA7 to respond to low-to-high transitions.

The data direction bit for pin PA7, DDRA7, makes the pin an input so that the pulse accumulator responds to external pulses at the pin.

- *Lines 45 and 46.* The program enables interrupts from the pulse-accumulator-overflow flag. The counter generates an interrupt each time it overflows from FF to 00. These instructions also disable the pulse-accumulator-input-flag interrupt. This flag gets set on each input pulse, but this application doesn't need it.

- *Lines 48 and 49.* These instructions set the pulse-accumulator counter to FF, so it overflows to 00 on the very first input pulse. After the first pulse, overflows occur every fifth pulse.

- *Lines 64 and 65.* This test of the pulse-accumulator-overflow flag ensures that a valid interrupt occurred. The program ignores illegal interrupts, which is good programming practice.

- *Lines 71 and 72.* These instructions clear the pulse-accumulator-overflow flag so that it can respond to the next input pulse.

- *Lines 74 and 75.* The pulse-accumulator counter now contains 00 because an overflow occurred that sent control to this interrupt service routine. Here the

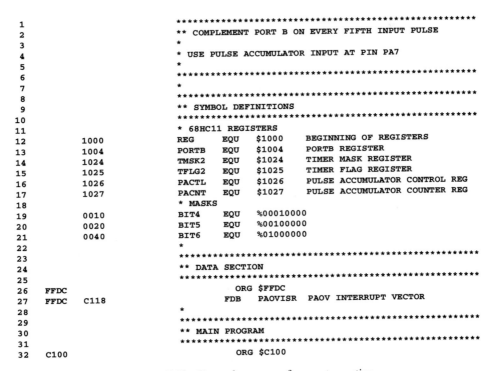

```
 1                          ***********************************************************
 2                          ** COMPLEMENT PORT B ON EVERY FIFTH INPUT PULSE
 3                          *
 4                          * USE PULSE ACCUMULATOR INPUT AT PIN PA7
 5                          *
 6                          ***********************************************************
 7                          *
 8                          ***********************************************************
 9                          ** SYMBOL DEFINITIONS
10                          ***********************************************************
11                          * 68HC11 REGISTERS
12          1000            REG      EQU    $1000       BEGINNING OF REGISTERS
13          1004            PORTB    EQU    $1004       PORTB REGISTER
14          1024            TMSK2    EQU    $1024       TIMER MASK REGISTER
15          1025            TFLG2    EQU    $1025       TIMER FLAG REGISTER
16          1026            PACTL    EQU    $1026       PULSE ACCUMULATOR CONTROL REG
17          1027            PACNT    EQU    $1027       PULSE ACCUMULATOR COUNTER REG
18                          * MASKS
19          0010            BIT4     EQU    %00010000
20          0020            BIT5     EQU    %00100000
21          0040            BIT6     EQU    %01000000
22                          *
23                          ***********************************************************
24                          ** DATA SECTION
25                          ***********************************************************
26  FFDC                             ORG $FFDC
27  FFDC    C118                     FDB    PAOVISR  PAOV INTERRUPT VECTOR
28                          *
29                          ***********************************************************
30                          ** MAIN PROGRAM
31                          ***********************************************************
32  C100                             ORG $C100
```

Figure 7-33   Example program for event counting.

```
33                          *----------------------------------------------------------
34                          * INITIALIZATION
35                          *----------------------------------------------------------
36                          * INITIALIZE STACK
37    C100    8E CF FF      START    LDS     #$CFFF
38                          * INITIALIZE PORT B
39    C103    7F 10 04               CLR     PORTB
40                          * INITIALIZE PULSE ACCUMULATOR
41                          *   EVENT COUNTING MODE
42    C106    86 50                  LDAA    #$50      DDRA7:PAEN:PAMOD:PEDGE=0101
43    C108    B7 10 26               STAA    PACTL     INPUT:ENABLE:EVENT:L-TO-H
44                          *   ENABLE PULSE ACCUMULATOR OVERFLOW INTERRUPT
45    C10B    86 20                  LDAA    #BIT5     PAOVI=1
46    C10D    B7 10 24               STAA    TMSK2
47                          *   INITIALIZE COUNTER FOR IMMEDIATE INTERRUPT
48    C110    86 FF                  LDAA    #$FF      ONE COUNT TO OVERFLOW
49    C112    B7 10 27               STAA    PACNT
50                          * TURN ON INTERRUPT SYSTEM
51    C115    0E                     CLI
52                          *----------------------------------------------------------
53                          * WAIT FOR INTERRUPTS
54                          *----------------------------------------------------------
55    C116    20 FE         HERE     BRA     HERE
56                          *
57                          ****************************************************************
58                          ** INTERRUPT SERVICE ROUTINE
59                          ****************************************************************
60                          *----------------------------------------------------------
61                          * INTERRUPT POLLING CHAIN
62                          *----------------------------------------------------------
63                          * INTERRUPT FROM PULSE ACCUMULATOR OVERFLOW?
64    C118    CE 10 00      PAOVISR  LDX     #REG
65    C11B    1F 25 20 0A            BRCLR   TFLG2-REG,X,BIT5,RTPAOV   IGNORE ILLEGAL
66                          *                              ..INTERRUPT
67                          *----------------------------------------------------------
68                          * SERVICE PULSE ACCUMULATOR OVERFLOW
69                          *----------------------------------------------------------
70                          * CLEAR PULSE ACCUMULATOR OVERFLOW FLAG
71    C11F    86 20                  LDAA    #BIT5     STORE 1 TO CLEAR FLAG
72    C121    A7 25                  STAA    TFLG2-REG,X ..ZEROS DO NOTHING
73                          * SET UP PULSE ACCUMULATOR COUNTER FOR NEXT INTERRUPT
74    C123    86 FB                  LDAA    #$FB      WILL INTERRUPT ON FIFTH PULSE
75    C125    A7 27                  STAA    PACNT-REG,X
76                          * COMPLEMENT PORTB
77    C127    63 04                  COM     PORTB-REG,X
78                          * RETURN FROM INTERRUPT
79    C129    3B            RTPAOV   RTI
80    C12A                           END
```

Figure 7-33   Continued.

counter is set to FB so that another overflow will occur after five more input pulses.

- *Line 77.*   This example program simply complements port B in response to the fifth input pulse.

Gated-Time-Accumulation Mode

The pulse accumulator, while in the gated-time-accumulation mode, counts active transitions of an oscillator while enabled by port A pin PA7. Figure 7-34 is a block diagram of the

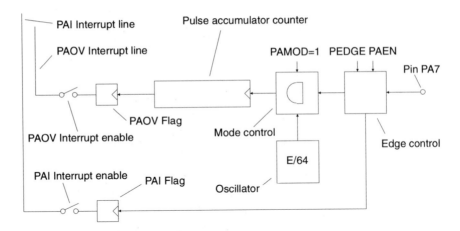

Figure 7-34   Pulse-accumulator gated-time-accumulation mode.

parts of the pulse accumulator in this mode. The counter accumulates time or counts while the signal at the pin is asserted; the PA7 pin is level-sensitive for this function.

Hardware operation

In the gated-time-accumulation mode, an oscillator based on the E-clock for the microprocessor drives the 8-bit counter. The frequency of the oscillator is always the E-clock frequency divided by 64_{10}. The signal at pin PA7 enables and disables the counter so that it responds to the oscillator at the correct times. The term *gated* means that the counter can be enabled and disabled. While the counter is enabled, it increments on each oscillator cycle. While the counter is disabled, it holds the last count. Therefore, the counter accumulates the number of oscillator cycles that occur while the counting is enabled—the pulse accumulator is a timer.

Flags and interrupts.   When the counter rolls over from FF to 00, it sets the pulse accumulator overflow flag, PAOVF, in the TFLG2 register. Setting the PAOVF flag causes an interrupt if the pulse-accumulator-overflow interrupt is enabled and the interrupt system is on. The pulse-accumulator-overflow-interrupt-enable bit, PAOVI, in the TMSK2 register must be 1 to enable overflow interrupts. The overflow interrupt has an interrupt vector at address FFDC and FFDD as Table 7-2 shows.

If the pulse accumulator is in gated-time-accumulation mode because the PAMOD bit is 1, the hardware transfers the active transitions from the oscillator to the counter while the PA7 pin is asserted. That is, the counter increments while pin PA7 is asserted. Pin PA7 is level-sensitive for this operation. Figure 7-34 shows the transfer of the oscillator signal to the counter with an AND-gate that acts as a transfer gate. However, when pin PA7 is changed to the deasserted level, this transition at pin PA7 sets the pulse accumulator-input flag, PAIF. If the pulse-accumulator-input-interrupt-enable bit, PAII, is 1, setting the PAIF flag causes an interrupt. The interrupt vector is at addresses FFDA and FFDB as Table 7-2 shows.

Active levels and transitions.   The PEDGE bit in the PACTL register determines which logic level is the active or asserted level at pin PA7. While PEDGE is 0, the high level is the asserted level. While PEDGE is 1, the low level is the asserted level. However, if the pulse-accumulator-enable bit PAEN is 0, the input signal at the pin has no effect on any of the pulse-accumulator hardware.

Remember that pin PA7 is level-sensitive while enabling the counter to respond to oscillator cycles. However, pin PA7 is transition sensitive when setting the flag.

Summary.   In the gated-time-accumulation mode, if the pulse accumulator is enabled, the counter accumulates a count of the oscillator cycles while pin PA7 is asserted. The counter holds the last count while the PA7 pin is deasserted, although the program may write a new number into the counter. A transition from asserted to deasserted at pin PA7 sets the PAIF flag. When the counter rolls over to 00, it sets the pulse-accumulator-overflow flag. If the flags are enabled to cause interrupts, setting a flag causes an interrupt to a unique interrupt vector.

Timing resolution and range

The oscillator that makes the pulse-accumulator counter increment is based on the microprocessor E-clock. The oscillator always runs at the E-clock rate divided by 64_{10}. This rate determines both the resolution of the timing and the counter overflow time.

If the crystal clock operates at the common frequency of 8.0_{10} MHz, the E-clock rate is 2.0 MHz. This rate gives a resolution of 32_{10} microseconds and an overflow time of 8.192_{10} milliseconds. Therefore, the pulse accumulator can easily measure times of approximately eight milliseconds. Longer times are measured by tracking overflows of the counter using the overflow flag, but the complexity of the program is much greater.

Programming example

Consider an application where a toothed wheel, magnetic sensor, and electronic interface make pulses as a mechanical device rotates. The pulse length in time is dependent on the physical length of the tooth and the rotation rate of the device. If the time for the tooth to pass is measured, the rotational velocity of the device can be calculated.

The example program measures the duration of a pulse in pulse-accumulator counts. If the pulse duration is long enough to allow the counter to overflow, the program indicates an error. The duration is not used for any purpose in this example.

Figure 7-35 is the listing of the example program. The example uses a pulse that is high, so the transition at the trailing edge is from high-to-low. The following is a description of the program.

- *Line 44.*   The program creates a stack because the interrupt system requires it.

- *Lines 46 and 48.*   The program variables are initialized.

- *Lines 51 and 52.*   These instructions program the pulse accumulator to operate in the gated-time-accumulation mode. They also make the high level the asserted condition of the PA7 input pin.

```
 1                                 *********************************************************
 2                                 ** MEASURE THE DURATION OF AN INPUT PULSE
 3                                 *
 4                                 * USE PULSE ACCUMULATOR INPUT AT PIN PA7 TO DETECT
 5                                 * ..HIGH PULSE
 6                                 *
 7                                 * MAXIMUM PULSE DURATION IS $FF PULSE ACCUMULATOR
 8                                 * ..COUNTS. ERROR INDICATION PROVIDED.
 9                                 *
10                                 *********************************************************
11                                 *
12                                 *********************************************************
13                                 ** SYMBOL DEFINITIONS
14                                 *********************************************************
15                                 * 68HC11 REGISTERS
16          1000                   REG      EQU    $1000     BEGINNING OF REGISTERS
17          1024                   TMSK2    EQU    $1024     TIMER MASK REGISTER
18          1025                   TFLG2    EQU    $1025     TIMER FLAG REGISTER
19          1026                   PACTL    EQU    $1026     PULSE ACCUMULATOR CONTROL REG
20          1027                   PACNT    EQU    $1027     PULSE ACCUMULATOR COUNTER REG
21                                 * MASKS
22          0010                   BIT4     EQU    %00010000
23          0020                   BIT5     EQU    %00100000
24                                 *
25                                 *********************************************************
26                                 ** DATA SECTION
27                                 *********************************************************
28   0010                                   ORG    $0010
29   0010                          PULDUR   RMB    1         PULSE DURATION IN PA COUNTS
30   0011                          ERROR    RMB    1         PULSE OVERTIME ERROR INDICATOR
31                                 *
32   FFDA                                   ORG    $FFDA
33   FFDA  C119                             FDB    PAIISR    PAI INTERRUPT VECTOR
34   FFDC  C12B                             FDB    PAOVISR   PAOV INTERRUPT VECTOR
35                                 *
36                                 *********************************************************
37                                 ** MAIN PROGRAM
38                                 *********************************************************
39   C100                                   ORG    $C100
40                                 *---------------------------------------------------------
41                                 * INITIALIZATION
42                                 *---------------------------------------------------------
43                                 * INITIALIZE STACK
44   C100  8E CF FF                START    LDS    #$CFFF
45                                 * INITIALIZE PULSE DURATION
46   C103  7F 00 10                         CLR    PULDUR
47                                 * INITIALIZE PULSE OVERTIME ERROR INDICATOR
48   C106  7F 00 11                         CLR    ERROR     00=FALSE, FF=TRUE
49                                 * INITIALIZE PULSE ACCUMULATOR
50                                 *   GATED TIME ACCUMULATION MODE
51   C109  86 60                            LDAA   #$60      DDRA7:PAEN:PAMOD:PEDGE=0110
52   C10B  B7 10 26                         STAA   PACTL     INPUT:ENABLE:TIME:H-TO-L
53                                 *   ENABLE PULSE ACC OVERFLOW AND INPUT INTERRUPTS
54   C10E  86 30                            LDAA   #$30      PAOVI:PAII=11
55   C110  B7 10 24                         STAA   TMSK2
56                                 *   INITIALIZE COUNTER FOR FIRST MEASUREMENT
57   C113  7F 10 27                         CLR    PACNT
58                                 * TURN ON INTERRUPT SYSTEM
59   C116  0E                               CLI
60                                 *---------------------------------------------------------
61                                 * WAIT FOR INTERRUPTS
62                                 *---------------------------------------------------------
63   C117  20 FE                   HERE     BRA    HERE
64                                 *
```

Figure 7-35   Example program for gated-time accumulation.

```
65                          ***********************************************************
66                          ** PULSE ACCUMULATOR INPUT INTERRUPT SERVICE ROUTINE
67                          ***********************************************************
68                          *---------------------------------------------------------
69                          * ILLEGAL INTERRUPT?
70                          *---------------------------------------------------------
71   C119   CE 10 00        PAIISR  LDX   #REG
72   C11C   1F 25 10 0A             BRCLR TFLG2-REG,X,BIT4,RTPAI    IGNORE ILLEGAL
73                          *                              ..INTERRUPT
74                          *---------------------------------------------------------
75                          * SERVICE PULSE ACCUMULATOR INPUT
76                          *---------------------------------------------------------
77                          * CLEAR PULSE ACCUMULATOR INPUT FLAG
78   C120   86 10                   LDAA  #BIT4     STORE 1 TO CLEAR FLAG
79   C122   A7 25                   STAA  TFLG2-REG,X ..ZEROS DO NOTHING
80                          * READ PULSE ACCUMULATOR COUNTER FOR PULSE DURATION
81   C124   A6 27                   LDAA  PACNT-REG,X
82   C126   97 10                   STAA  PULDUR
83                          * INITIALIZE COUNTER FOR NEXT PULSE
84   C128   6F 27                   CLR   PACNT-REG,X
85                          * RETURN FROM INTERRUPT
86   C12A   3B              RTPAI   RTI
87                          *
88                          ***********************************************************
89                          ** PULSE ACCUMULATOR OVERFLOW INTERRUPT SERVICE ROUTIN
90                          ***********************************************************
91                          *---------------------------------------------------------
92                          * ILLEGAL INTERRUPT?
93                          *---------------------------------------------------------
94   C12B   CE 10 00        PAOVISR LDX   #REG
95   C12E   1F 25 20 08             BRCLR TFLG2-REG,X,BIT5,RTPAOV  IGNORE ILLEGAL
96                          *                              ..INTERRUPT
97                          *---------------------------------------------------------
98                          * SERVICE PULSE ACCUMULATOR OVERFLOW
99                          *---------------------------------------------------------
100                         * CLEAR PULSE ACCUMULATOR OVERFLOW FLAG
101  C132   86 20                   LDAA  #BIT5     STORE 1 TO CLEAR FLAG
102  C134   A7 25                   STAA  TFLG2-REG,X  ..ZEROS DO NOTHING
103                         * INDICATE OVERTIME ERROR
104  C136   86 FF                   LDAA  #$FF
105  C138   97 11                   STAA  ERROR
106                         * RETURN FROM INTERRUPT
107  C13A   3B              RTPAOV  RTI
108  C13B                           END
```

Figure 7-35 Continued.

- *Lines 54 and 55.* Both the pulse-accumulator input and overflow interrupts are enabled. An overflow interrupt indicates an error because the interrupt occurs only if the duration of the input pulse is too long.

- *Line 57.* The pulse-accumulator counter is set to zero so that the first pulse will be accurately timed. The counter starts incrementing when the input signal enables it. In this example, the counter accumulates time while the input signal is high. When the input changes to low at the end of the pulse, the transition sets the PAIF flag, which causes an interrupt.

- *Lines 78 and 79.* At the end of the input pulse, the interrupt system sends control here. The PAIF flag is tested for an illegal interrupt as good programming practice.

● *Lines 81 and 82.*   The pulse-accumulator counter is no longer incrementing because the pulse has ended by going low. While the input is low, the counter retains the last count. The program reads the counter to determine the duration of the last pulse.

● *Line 84.*   The pulse-accumulator counter is set to zero so that the counter correctly measures the duration of the next input pulse.

● *Lines 94 and 95.*   The interrupt system sends control here if the pulse accumulator overflows. The overflow may interrupt during the pulse-accumulator-input interrupt service routine because the overflow vector has a higher priority than the input vector. The overflow flag is tested for illegal interrupts as good programming practice.

● *Lines 101 and 102.*   The program clears the overflow flag to remove the source of overflow interrupts. However, if the input pulse is long enough, multiple overflows can occur. The program provides no indication of multiple overflows.

● *Lines 104 and 105.*   The program indicates an overflow error because an overflow interrupt occurred. The pulse-accumulator hardware determined that the error occurred.

Nothing in this program example uses the pulse duration or error indication because it is an example to demonstrate technique. The error indication is permanent because nothing resets the error.

Pulse-Accumulator Flag Applications

Many applications will not need to use the pulse accumulator. If so, the pulse-accumulator-input flag can be used as a general-purpose flag. The PAIF can respond to either transition of the input signal, and it can generate an interrupt with a unique interrupt vector. The pulse-accumulator-overflow flag is less useful as a general-purpose device.

Remember that the flags associated with the programmable timer can also be used as general-purpose flags when they are not used for timing purposes. When all these programmable timer flags are used with the STAF flag, many flags are available without constructing custom hardware.

7.10   SERIAL COMMUNICATIONS INTERFACE

The serial communications interface, SCI, inside the 68HC11 makes serial transmission and reception easy. The SCI device is very flexible, so it can adapt to most applications. However, its sophistication means that the program must select many control options and control many data bits. Most programs for controlling the SCI use interrupt, so the examples in this chapter only use interrupt.

Data Transmission

Typical digital systems must send data numbers from one device to another. For example, the microprocessor must send data numbers to the memory chips. The address, data, and control buses serve this purpose. In similar applications, the distance between the two devices is small—usually a few centimeters. Therefore, using many wires as in the address, data, and control buses poses little difficulty. Transmitting the bits over many wires at the same time is called *parallel transmission*.

Other applications require data transmission over much larger distances. For example, data sent between a computer and a keyboard/display terminal, a computer and a printer, or two microcomputers may require 10 or 20 feet of cabling. Figure 7-36(a) illustrates sending 8-bit data between two such devices. The input and output ports of the computer could be port C and port B of a 68HC11 microcomputer. The figure makes clear that parallel transmission requires many wires to connect the two devices—16_{10} data wires, two timing wires, and a common wire. If the distance between the two devices exceeds 20 feet, using so many wires is likely to be impractical. Even distances of a few feet may be inconvenient and costly with so many wires.

The inconvenience of a large cable makes changing to a transmission system with fewer wires attractive. At least two wires are necessary for sending a binary signal. When using only two wires, only a single bit can be sent at once. This is the principal characteristic of a *serial transmission path*—it transmits one bit at a time. Figure 7-36(b) illustrates a common wire and two signal wires for two directions of serial transmission at once, which is equivalent to the system in Figure 7-36(a).

Some compromises are necessary to use serial transmission compared to parallel transmission. First, no signal wires are used for timing or ready signals. Therefore, the bit stream

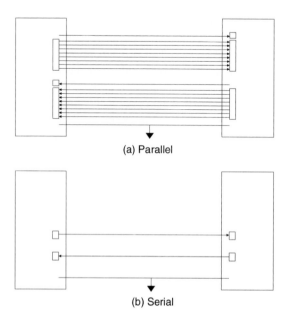

(a) Parallel

(b) Serial

Figure 7-36 Parallel and serial transmission.

must contain the timing information. Second, if equivalent hardware for transmitting 8-bit numbers is compared, the serial transmission must be at least eight times slower than the parallel transmission. Because the serial system uses only one signal path instead of eight paths, the information transmission rate must be slower. In practice, due to the timing encoding, the transmission rate is even slower. Therefore, the advantages gained from serial transmission must be greater than the disadvantage of slower speed. For example, if the transmission distance is very long, the cost of parallel transmission is prohibitive.

Communication Terminology

Several variations of the serial communication system shown in Figure 7-36(b) are practical. The communication industry developed terminology for each variation long before computer technology influenced communication. Telegraph, telephone, and radio communication preceded computer communication. Because of its origin, communication terminology is sometimes strange to people who work with computers. For example, logical terms using 1s and 0s can describe the serial signal without referring to the electrical characteristics of the hardware. However, the communication terminology calls a logical 1 a *mark* and a logical 0 a *space*.

Simplex communication

A *simplex communication system* supports communication in one direction only. The system, therefore, cannot support a response to a message. Figure 7-37(a) illustrates the hardware necessary for simplex serial communication between a computer and another device.

Simplex communication is very common—it is the typical radio and television broadcasting system. Furthermore, many computers control serial printers and cathode-ray-tube (CRT) display terminals that cannot return information to the computer. For example, the display screens at airports only display flight information.

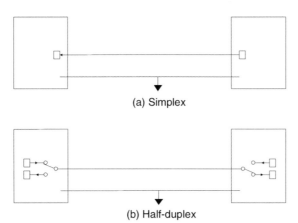

(a) Simplex

(b) Half-duplex

Figure 7-37   Serial transmission modes.

Duplex communication

A *duplex communication system* can support communication in both directions between two devices. Two variations of duplex communication are called *half-duplex* and *full-duplex*.

Half-duplex communication. A half-duplex system supports communication in two directions, but only in one direction at a time. The same hardware communicates in both directions. Figure 7-37(b) illustrates the hardware for half-duplex serial communications between two devices such as two microcomputers.

A half-duplex system can switch the signal wire between the transmitter and receiver at each end of the path. Controlling the switching between transmitter and receiver is complicated. Usually one device will transmit until both devices agree to switch roles. After switching, the second device transmits.

If something goes wrong, both devices may try to transmit at once causing a *collision* between messages. The system *protocol* or agreement between the devices must specify how to recover from such errors.

The advantage of half-duplex communication is the reduced cost of using the same wire to transmit in both directions. Often the cost of the electronic switching hardware is much less than the cost of additional wiring.

A possible disadvantage of half-duplex communication is the need for one device to wait until the other device finishes transmitting. The waiting may slow the communication process. However, in many applications, the first device interrogates the second device asking for particular information. The first device may continue only after receiving the information from the second device, so little speed is lost by using half-duplex communication.

Full-duplex communication. A full-duplex system supports communication in both directions at the same time. The hardware in Figure 7-36(b) is typical of a full-duplex serial system. It is the most general system, and it is also the most costly. However, in many applications, the cost is very small.

A typical application of full-duplex communication is the connection between a CRT terminal and a computer system. The keyboard device in the terminal transmits character codes to the computer at the same time the computer transmits to the display device. Serial communication is commonly used in this application because only a very small cable is required, and the transmission does not need to be very fast because the terminal is operated by a person.

The terminology used with commercial CRT terminals connected to mainframe computers is sometimes confusing. In this application, the term *full-duplex* usually means that the computer transmits each character it receives from the keyboard back to the display. Similarly, *half-duplex* means that the computer does not transmit a received character back to the display. The terminal must be set to operate in the mode that is compatible with the computer. Incorrect half-duplex and full-duplex settings result either in no display when keys are pressed or two copies of each key pressed. The mode settings on the terminal determine whether the terminal will internally copy the character from the keyboard to the display.

ASCII Communication Codes

This section discusses character transmission codes because most character-oriented devices use serial communication. However, many other applications besides character transmission use serial communication.

Binary codes can represent the alphabetic and numeric characters found on the typical keyboard. Most small computers use the character codes defined by the American Standards Committee for Information Interchange (ASCII). Therefore, these character codes are called the *ASCII codes*. Pronounce the acronym ASCII "ask-key."

Character codes

The ASCII code is a 7-bit code though serial systems usually transmit an 8-bit binary number. A 7-bit code can represent 128_{10} different characters.

The ASCII code is a weighted code. Therefore, the code numbers corresponding to the letters of the alphabet and the numerical characters are in numerical order. If you remember that the code 1000001 represents the capital letter A and the code 0110000 represents the digit 0, you can reproduce the codes for other letters and numbers by counting. Using hexadecimal numbers, the ASCII code for A is 41, and the code for 5 is 65.

Error detection

When a transmission system transmits character information over long distances, electrical interference may cause errors in the received codes. Interference can change ones to zeros and zeros to ones. The receiving system must detect such errors to prevent use of the erroneous information. One response to an error may be a message requesting the transmitting device to transmit the information again.

Parity bits.   Probably the simplest of many available error-detection algorithms is the *parity* scheme. When using parity, an eighth bit is added to the 7-bit ASCII code in the most-significant position. Both the transmitting device and the receiving device must know the algorithm used to select the bit. The receiving device checks received codes with the algorithm to detect erroneous codes. However, the addition of a single bit is not sufficient to detect all possible errors. Therefore, parity checking is not an adequate error-checking scheme for most applications.

Parity algorithms.   Four algorithms may be used for adding the parity bit. The two that provide the most useful error-checking ability are called *even parity* and *odd parity*. Even parity means add a 1 or a 0 so that the number of 1s in the 8-bit code is even. Odd parity means add a 1 or a 0 so that the number of 1s in the 8-bit code is odd.

Less useful are the algorithms that specify that the parity bit will always be 1 or will always be 0. Because of the communication heritage of the terminology, these algorithms are called *mark parity* and *space parity*. Table 7-8 shows some ASCII codes with parity bits added.

Even and odd parity schemes cannot detect certain multiple-bit errors. For example, suppose that the system transmits an odd-parity A with a code of 11000001. If transmission

TABLE 7-8   SELECTED ASCII CODES

Character	Even Parity	Odd Parity	Mark Parity	Space Parity
A	01000001	11000001	11000001	01000001
B	01000010	11000010	11000010	01000010
C	11000011	01000011	11000011	01000011
D	01000100	11000100	11000100	01000100
•	•	•	•	•
0	00110000	10110000	10110000	00110000
1	10110001	00110001	10110001	00110001
2	10110010	00110010	10110010	00110011
•	•	•	•	•

errors change the code to 11001101, parity checking fails to detect an error because the code still contains an odd number of 1s. Thus, parity checking has limited value for detecting errors.

Serial Communication Principles

The serial bit stream used to transmit binary numbers contains both data bits and timing information. Both the format of the bit stream and the electrical characteristics of the hardware are well established because many manufacturers' equipment must be compatible.

Serial signals

Binary signals in electronic systems usually represent the logical values 1 and 0 with two different voltage levels usually called *high* and *low*. Most serial transmission systems use voltage levels for the electronic signals; however, many serial systems instead use two different current levels. Most commercial equipment meets the requirements of a standard. The standard specifies the signals so that different devices are compatible with each other. The two most common standards are called *20-milliamp current loop* and *RS232*.

20-milliamp current loop.   Teletype machines made by Teletype Corporation use this type of logic signal. The equipment turns a source of 20-milliamps of direct current on and off to make a binary signal. A mechanism in the machine uses solenoids. The strength of the magnetic field in the solenoid is current dependent. Therefore, the amount of current in the signal is significant.

Teletype machines are no longer used with computer equipment. However, some electronic equipment still uses the signal format because it has greater immunity to interference from electrical noise than some voltage-level signal systems.

RS232.   The most common standard for serial communication signals is the Electronic Industries Association standard RS232 which is now in revision E or higher. The standard describes the characteristics of the transmitter and receiver hardware, and the connections between them. The signal levels at the receiving end of the cable must be between +3 and +25 VDC for the high level, and between −3 and −25 VDC for the low level. Using nonzero voltages for each level provides greater immunity to electrical interference. The wide range of possible voltages makes building compatible equipment easy. In practice, however, most manufacturers use +12 and −12 VDC signals.

Assertion designation.   An association between the logical value of a binary signal and the physical electrical signal must be made. For the sake of compatibility, serial transmission systems use only a single definition. In the 20-milliamp current loop system, 20-milliamps of current is a logical 1, and no current is a logical 0. In the RS232 system, the negative voltage is a logical 1, and the positive voltage is a logical 0.

Serial signal format

Whatever the meaning of the numbers transmitted, or the number of bits in the numbers transmitted, the format or arrangement of the bits is the same. Because the numbers transmitted are frequently 8-bit ASCII character codes, the following describes character transmission. Many people speak of the data number as a *character* even if the code does not represent a character. Figure 7-38 illustrates the waveform for an even parity character *M*. Remember that only one bit at a time is transmitted, so the time axis shows how the bits are spaced in time.

The format of the serial signal described here is called *non-return-to-zero*, or *NRZ*. The name means that the signal does not change within the bit time and that the signal value during the bit time determines the bit value. Other systems determine the bit values by using various kinds of transitions of the signal either during or between the bit times.

Start bit.   Before the transmission begins, the serial transmission system is idle and the signal is a binary one or mark as Figure 7-38 shows. At the beginning of the transmission of any character code, the signal changes to a binary zero or space. From this beginning, each bit is allocated a fixed amount of time until the end of the current character. The first time period or bit is called the *start bit*. The transmitter hardware automatically creates the start bit.

Data bits.   Usually eight data bits are used, so the next eight bits in Figure 7-38 represent the data for the character M. The data bits are transmitted least-significant bit first and most-significant bit last. In this example, the last data bit sent is the parity bit.

Many systems transmit five, seven, or nine data bits. For example, a system uses nine bits when an 8-bit number and a parity bit are sent.

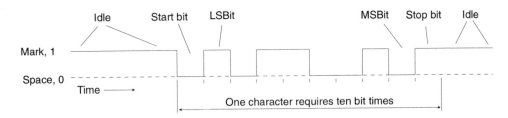

Figure 7-38   Serial signal format for an even-parity ASCII M character.

Stop bit. Following the data bits, the transmitter always sends a 1 called the *stop bit*. The stop bit usually lasts for one bit time, but some systems use two stop periods or two stop bits. The number of data bits and the time for the stop bit may vary from one system to another.

At the end of the stop period, the system is again idle. The start bit of the next character may begin immediately after the stop period of the current character. Sometimes the system may remain idle for a while.

Serial transmitter operation

A serial transmitter is a digital electronic circuit usually in an integrated circuit. The transmitter automatically generates the waveform when given the data number to transmit. It also creates the start and stop bits that are not part of the data.

A separate clock or time base enables the transmitter to control the timing of the bits. Usually, the transmitter can easily change to different bit times.

Serial receiver operation

The serial receiver is a digital electronic circuit that samples the incoming signal to extract the data from it. The receiver does not know when a character will arrive; it must determine all its information from the received waveform. The receiver must operate with the same bit time as the transmitter or extracting the data from the waveform is impossible. Furthermore, the receiver must know how many data bits were transmitted.

Figure 7-39 illustrates the principle of sampling the received waveform. The Xs show the samples that are taken at a rate much higher than the bit rate of the waveform. A typical rate is 16_{10} times faster than the bit rate—showing such a rate in the figure is difficult.

When the receiver detects the first 0 beyond the idle signal, it suspects that a character is being received. The 0 is the start bit. To detect the start bit as early as possible, the sampling rate must be much faster than the bit rate in the serial stream. However, electrical noise may make a pulse that leads to a false start. Most receivers sample the start bit at least once more to determine if a correct start bit has arrived. The simplest algorithm is to sample again after one-half bit time as the figure shows.

When a valid start bit occurs, the receiver waits one time period before sampling the waveform again. Successive samples occur every period until the stop bit is sampled. Therefore, the receiver can retrieve the data number from the bits within one character.

Most receivers use a more sophisticated algorithm than described here to detect errors due to electrical noise. For example, the sample near the middle of the bit time may consist

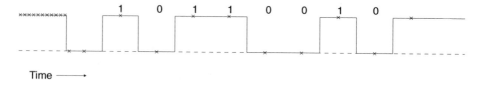

Time ⟶

Figure 7-39   Serial receiver sampling points for one character.

of three successive samples at the 16_{10} times rate. Then a voting algorithm determines the actual value of the bit.

UART hardware

Serial transmitters and receivers are available in integrated circuits. The most common IC is *universal asynchronous receiver transmitter* or *UART*. A UART contains all the electronics for both a receiver and a transmitter. The receiver and transmitter are independent so that the UART can implement a full-duplex system.

Usually a UART requires a clock signal to determine the transmission rate. It also requires additional electronics to control the relatively high positive and negative signal voltages required for many serial systems.

Serial errors

A variety of situations can cause erroneous reception of the data value. Electrical noise is a major cause of erroneous reception. However, other problems related to operating the receiver can lead to errors.

Timing errors.   Timing errors always occur because the transmitter and the receiver will not run at identical rates. They are two separate pieces of equipment. Two independent devices cannot be made to operate at the same rate. A common clock signal does not exist between the two devices. Therefore, the serial communication system described here is called *asynchronous serial communication*.

The timing error between the transmitter and receiver would always lead to data errors if the timing error accumulated from one character to the next. After enough characters were sent, a data error would occur. However, data errors do not happen, because the receiver synchronizes to each incoming character by detecting the beginning of the start bit. Because the timing error accumulates only within one character, the receiver accommodates small timing errors without data errors.

Figure 7-40 will help in determining an upper limit on the allowable timing error. In Figure 7-40(a), the receiver runs faster than the transmitter, which causes the samples to

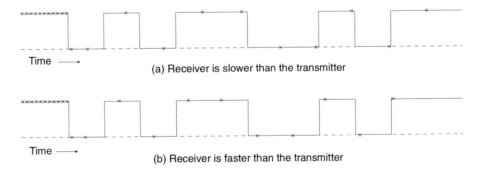

Time ⟶

(a) Receiver is slower than the transmitter

Time ⟶

(b) Receiver is faster than the transmitter

Figure 7-40   Effects of timing errors on sampling points.

occur too early. At the limit, the sample could be one-half bit time early at the end of 10_{10} bit times and still get the correct result. Similarly, Figure 7-40(b) shows the receiver running more slowly than the transmitter. Here the sample could be one-half bit time late at the end and still be correct. Therefore, the maximum allowable timing error between the transmitter and receiver is plus or minus five percent. In practice, this theoretical limit is not achieved and the allowable error is somewhat smaller. Practical electronic circuits seldom have difficulty achieving the required accuracy.

Framing errors.   Each character begins with a start bit that is 0 and ends with a stop bit that is 1. A *framing error* occurs when the received character does not end with the correct stop bit. That is, the start and stop bits do not properly frame the character.

Other than when caused by electrical noise, a framing error occurs when the receiver is out of synchronization and interprets a data bit as a start bit. Figure 7-41 illustrates sampling beginning at an incorrect point and the detection of the framing error.

If the receiver erroneously starts a character at the point where a data bit changes from 1 to 0, a 1 may not be found where the stop bit should be. Then the receiver detects the framing error. If the receiver finds a 1, it cannot detect the error. In such a case, a framing error will likely occur within the next few characters. In the example, the receiver is effectively making data numbers by using bits from two different characters.

Most serial receivers automatically check the bit stream for framing errors. The receiver sets a flag bit to indicate a framing error.

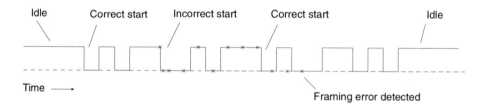

Figure 7-41   Framing error.

Overrun errors.   Immediately after the receiver has received a character, the receiver transfers the data bits to a data register for the computer to read. With a data register, the receiver can receive another character immediately even if a short delay occurs before the computer reads the previous character. Remember that the receiver cannot control when the transmitter sends a character. If the computer is too slow, the receiver may receive another character and transfer it to the data register overwriting or overrunning the last character. The last character is then lost. This problem is called an *overrun error*. The receiver hardware must detect the overrun error and indicate it.

Serial transmission rate

The time allocated for each bit determines the rate at which data bytes are transmitted. Users usually prefer the fastest possible rate of transmission. However, several factors limit the maximum rate of transmission. These include both limitations of the devices sending and receiving information and limitations of the transmission medium that carries the signals. Furthermore, the rate usually must be a standard rate so that communication with commercial equipment is possible.

Bit rate.   The rate of serial information transmission is usually specified in bits per second or in characters per second. The rate in bits per second is also called the *baud rate*. This name comes from the work of a mathematician named Baudot, who worked in the area of communication theory. Unfortunately, correct application of Baudot's work does not result in a baud rate exactly equal to the number of bits per second. However, the number is close so that people in the computer industry define baud rate as bits per second.

When determining baud rate, consider all the bits required for a character. The example in Figure 7-41 uses 10_{10} bits for each 8-bit character when you include the necessary start and stop bits. A system that can transmit a maximum of 240_{10} of these characters per second has a rate of 2400 baud.

Several standard baud rates are used by manufacturers of equipment that must communicate with other manufacturers' equipment. Table 7-9 lists the common baud rates—note that K or kilo is 1000_{10} when used in this context. The table is based on use of 8-bit characters or numbers. The table also lists the time for each bit. At the higher rates, the short bit times make responding properly difficult for the hardware.

A practical view of transmission rate.   Getting a physical feel for the standard transmission rates is valuable. Because serial communication systems often send text, let's use text as an example. A CRT terminal can display text as it receives the characters serially. Ideally, filling the screen with characters will be instantaneous.

If the transmission is at 300 baud, most people can easily read the text as the CRT prints the characters on the screen. This slow rate is unacceptable to most people unless they have no alternative. At 1200 baud, the display speed is much better. Even at 1200 baud, the screen will fill with characters so slowly that it will irritate most people. At 9600 baud, the screen will fill quickly enough to be acceptable to most people, but the significant delay is

TABLE 7-9   COMMON STANDARD BAUD RATES

Baud Rate	Number of Stop Bits	8-bit Numbers per Second	Bit Time (milliseconds)
110	2	10	9.09
150	1	15	6.67
300	1	30	3.33
1200	1	120	0.83
2400	1	240	0.42
4800	1	480	0.21
9600	1	960	0.10
19.2 K	1	1920	0.05

still noticeable. The printing of the lines of text can still be seen. At 19.2 kilobaud, the screen fills quite rapidly—most people would say that the screen is printed instantaneously.

Speeds higher than 19.2 kilobaud are much more difficult to achieve. However, for filling screens with text, the added expense of going faster would gain little. Most people would not perceive a significant improvement in speed.

Waveform distortion. Let's consider increasing the baud rate of a serial system because faster speeds are better than slow speeds. Unfortunately, physical limitations on the transmission medium make increasing the speed impractical. Figure 7-42 illustrates the problem.

When the voltage applied to the transmitter end of a cable changes abruptly, the voltage at the other end of the cable does not follow exactly. Instead, the capacitance, inductance, and resistance of the wires alter the waveshape. The top drawing in Figure 7-42 illustrates how the cable rounds the signal voltage at the receiving end of the cable.

Now consider changing the baud rate of a serial waveform. The top two drawings in Figure 7-42 show the effects of doubling the baud rate. The cable characteristics and rounding remain the same, but the effect on a pulse is to distort it a greater amount. The third drawing shows the effects of doubling the baud rate again. Now the distortion is so severe that the received waveform is likely to be useless. Doubling the baud rate again as in the bottom drawing results in almost no output at all.

The effects of different cables on the waveshape depend on the cable characteristics. Some cable types have a smaller effect of rounding than others. However, for a given cable, the rounding effect is independent of the signals sent through it. The characteristics of the transmitter and receiver electronics also affect the waveshape.

Clearly, the characteristics of the transmission medium limit the speed of serial transmission. Furthermore, rounding of the signal prevents the receiver from correctly sampling the waveform. Rounding also makes timing errors more severe. Practical systems usually require a timing error of less than plus or minus 1.5 percent. Also, if the receiver is near its limit of recognition of the bits, any noise on the signal has a greater effect.

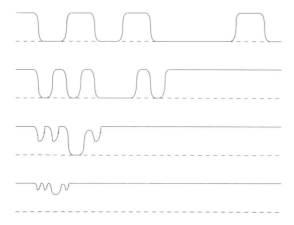

Figure 7-42 Effects of serial cable on voltage waveform as baud rate changes from a normal rate at the top to eight times faster at the bottom.

Receiver wake-up

Some receivers have special hardware features to improve the performance of multiple receiver networks. Figure 7-43 shows a typical connection of one transmitter and two receivers. Let's consider the case of the transmitter sending messages to the two receivers.

Messages.   A message simply means a block of data numbers. Each message may begin with a message identifier number. Some messages may be pertinent to both receivers, and others may be of interest to only one receiver. The receiver can decide whether the message is of interest by inspecting the message identifier.

Whenever the transmitter begins a message, both receivers respond by receiving characters. If a receiving device decides that the message is of no interest, it must still receive all the characters in the message. Usually, each incoming character causes an interrupt, and the interrupt service routine reads and discards the unwanted character. These interrupts use considerable processing time. At 9600 baud, the interrupts occur about every millisecond.

Purpose of wake-up.   If the receiver has wake-up hardware, the receiving computer can put the receiver to sleep so that it no longer responds to incoming characters. The computer will not have to waste time discarding characters. However, the receiver must be awake when a new message is transmitted.

Wake-up triggering.   Some special condition must trigger the receiver to wake up. The simplest wake-up condition is an idle transmission line. An idle condition means the transmitter has stopped transmitting. When the transmitter starts transmitting again, the receiver responds normally. If the transmitter stops for a short time at the end of each message, all receivers wake up. One character time is enough idle time to trigger wake-up, so little time is lost triggering the receivers. However, the transmitter must not hesitate during the transmission of a message.

SCI Serial I/O Hardware

The 68HC11 contains a complete serial receiver and transmitter system. This serial system is called the *SCI* for *serial communications interface*. The many control and data registers for the SCI are in the memory block beginning at address 1000. The program needs only to

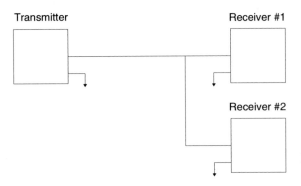

Figure 7-43   Multiple receiver serial transmission system.

use these registers to operate the SCI. One pin of the 68HC11 chip is the receiver input and another is the transmitter output. The signal levels are 0 and +5 VDC, so external hardware is necessary to make a compatible RS232 interface.

Baud rate control

The SCI transmitter and receiver can operate at a wide range of baud rates. The baud rate for the transmitter is always the same as for the receiver. The timing for the SCI is derived from the microprocessor's crystal-controlled clock. Therefore, you must choose the correct frequency of the crystal for the 68HC11 clock if a standard baud rate is necessary. The crystal also controls the E-clock frequency.

Baud rate divider chain.   The E-clock signal passes through two dividers to generate the correct clock rate for the SCI. Figure 7-44 illustrates the divider chain.

The first programmable divider circuit is called the *prescaler*. It divides the E-clock rate by one of four different factors.

The output of the prescaler drives another programmable divider called the *rate control*. This second divider reduces the prescaler rate by one of eight different factors.

The output of the rate-control divider determines the sampling rate of the serial receiver. The receiver samples at a rate 16_{10} times higher than the baud rate. Therefore, a fixed divider with a factor of 16_{10} divides the receive clock to form the transmit clock. The prescaler and rate control dividers together provide many different baud rates for a given E-clock rate.

One divide factor for the rate control divider is 1. That is, the output rate equals the input rate. Therefore, the output of the prescaler is called the *highest baud rate*. In other words, the rate control divider can only make the baud rate equal to or smaller than the rate from the prescaler.

BAUD register.   The control bits for the programmable dividers are in the BAUD register at address 102B. Figure 7-45 illustrates the BAUD register. The program must store into the *serial communication prescaler bits*, or *SCP bits*, to control the prescaler for a baud rate other than that obtained at reset. Notice that the SCP bits are set to 0s at reset. Because reset leaves the *serial communication rate bits*, *SCR bits*, undefined, the program must always store a number in the SCR bits to control the rate-control divider.

Table 7-10 shows the divide factors and the required SCP bits for the prescaler, and Table 7-11 shows the divide factors and the required SCR bits for the rate-control divider.

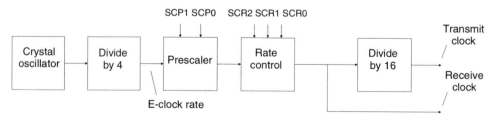

Figure 7-44   SCI baud rate divider chain.

$102B	7	6	5	4	3	2	1	0	
	TCLR	0	SCP1	SCP0	RCKB	SCR2	SCR1	SCR0	BAUD
RESET	0	0	0	0	0	U	U	U	

Figure 7-45   SCI baud rate register.

Table 7-10 also shows examples of the highest baud rate obtained at the output of the pre-scaler for three different crystal frequencies. Similarly, Table 7-11 shows examples of the final baud rate obtained at the output of the rate control for three different baud rates from the prescaler.

Consider an example using both tables. If the crystal operates at 8.0 MHz, the number 32 in the BAUD register sets the transmitter and receiver baud rates at 2400_{10} baud.

SCI data registers

The program transmits data or reads data through the serial interface by writing or reading a data register in memory. The serial hardware sets flags when it receives or transmits a character or discovers an error. The flags can trigger interrupts.

The SCI can use either 8-bit or 9-bit data numbers or characters. A control bit must be set to use 9-bit characters. The following discussion is based on 8-bit characters. A later section discusses 9-bit characters which are of interest mainly for transmitting 8-bit data numbers that include a parity bit.

Receive data register.   As a character is received, the data bits are moved into a *receiver shift register* that connects to the receive pin on the 68HC11. After all bits have arrived, the receiver automatically transfers the data bits to the *receive data register*. The data register is called *SCDR* for *serial communications data register* and is at address 102F.

Transmit data register.   The transmitter has a *transmit data register* and a *transmit shift register*. The shift register controls the transmit pin that connects to the serial line. To transmit a character, the program stores the data number into the SCDR at address 102F.

TABLE 7-10   SCI PRESCALER HIGHEST BAUD RATE

Prescaler Control Bits		Prescaler Divide Factor	Crystal Frequency		
			8.3886 MHz	8.0 MHz	4.9152 MHz
SCP1	SCP0		Highest Baud Rate (kilobaud)		
0	0	1	131.072	125.000	76.80
0	1	3	43.690	41.666	25.60
1	0	4	32.768	31.250	19.20
1	1	13	10.082	9.6000	5.907

TABLE 7-11   SELECTED SCI OUTPUT BAUD RATES

Rate-Control Control Bits			Rate-Control Divide Factor	Baud Rate from Prescaler (kilobaud)		
				131.072	76.80	9.600
SCR2	SCR1	SCR0		Output Baud Rate (kilobaud)		
0	0	0	1	131.072	76.80	9.600
0	0	1	2	65.536	38.40	4.800
0	1	0	4	32.768	19.20	2.400
0	1	1	8	16.384	9.600	1.200
1	0	0	16	8.192	4.800	0.600
1	0	1	32	4.096	2.400	0.300
1	1	0	64	2.048	1.200	0.150
1	1	1	128	1.024	0.600	0.075

The transmitter and receiver both appear to use the same data register SCDR. However, they do not; the name and the address refer to two registers. Figure 7-46 clearly shows separate receiver and transmitter data registers. The interface hardware uses the read/write signal to distinguish the receiver register from the transmitter register.

After the program stores the data number in the SCDR, the transmitter transfers the number to the transmit shift register when the shift register is empty. If the shift register is already empty, the transfer occurs immediately. This transfer scheme means that the program can easily keep the transmitter operating at full speed. While the current character is being transmitted, the program can obtain the next character and put it into the data register.

Figure 7-46 shows most of the serial communications interface hardware. The figure includes both the transmitter and receiver hardware. The receiver input is shown at the most-significant position of the shift register because the bits are received least-significant bit first.

SCI flags

All flags for the devices in the SCI are in the *serial communications status register* or *SCSR*. Figure 7-47 shows seven flags in this register. Bit 0 of SCSR is not used and always is 0. The names of the flags in SCSR do not end with the letter *F* as do most other 68HC11 flag names.

Clearing SCI flags.   All flags associated with the SCI receiver are cleared by the program reading the SCSR status register followed by reading the SCDR data register. All flags associated with the SCI transmitter are cleared by reading the SCSR followed by writing to the SCDR.

The program cannot clear a single receiver flag or a single transmitter flag. All receiver flags are cleared at once and all transmitter flags are cleared at once.

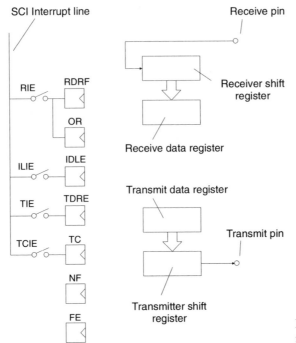

Figure 7-46   Block diagram of SCI receiver and transmitter hardware.

Only the flags set at the time the SCSR is read are cleared. When reading the SCSR to determine if a flag is set, be cautious because an additional flag could be set before the clearing is completed.

Receiver data flags.   The SCI receiver has two flags related to data reception. The receiver sets a receive flag when it transfers an incoming character from the shift register to the data register. The receiver flag is called the *receive-data-register full* or *RDRF*. The RDRF flag is in the SCSR register as Figure 7-47 shows. A program can read the SCDR to obtain the incoming character when the RDRF flag is set.

The SCI receiver also detects the incoming line becoming idle after receiving one or more characters without stopping. When the line changes to idle for one full character time, the receiver sets the IDLE flag in the SCSR register.

Receiver error flags.   The SCI receiver detects three kinds of reception errors, which it indicates by setting associated flags. The three errors are called *overrun error*, *noise error*, and *framing error*.

	7	6	5	4	3	2	1	0	
$102E	TDRE	TC	RDRF	IDLE	OR	NF	FE	0	SCSR
RESET	1	1	0	0	0	0	0	0	

Figure 7-47   Serial communications status register.

The receiver sets the *overrun-error flag*, *OR*, when a character is received and the RDRF flag is already set. Because the flag is already set, a previous character must be in the data register. This situation occurs when the program has not yet read the last character when another arrives. When an overrun occurs, the previous character remains in the data register and the incoming character is lost.

The receiver detects electrical noise on any of the received bits including the start and stop bits. It detects noise by sampling the bit at three consecutive sampling times. The receiver sets the *noise flag*, *NF*, if the three samples do not yield the same results.

The receiver also detects framing errors. If the receiver does not detect a stop bit at the end of a character, it sets the *framing-error flag* called *FE*. Because the framing error occurs at the end of a character, the FE flag is set at the same time the RDRF flag is set.

If the receiver detects both an overrun error and a framing error on the same character, it recognizes the overrun error and ignores the framing error. Furthermore, a set FE flag inhibits transfers to the SCDR.

Transmitter flags. The SCI transmitter uses two flags to indicate when a transmit operation is complete. When a program uses the transmitter, it stores into the transmit data register SCDR. The transmitter transfers the number in the data register to the transmit shift register when the transmit shift register is empty. The shift register may not be empty for a while after SCDR is written.

At the time of the transfer to the shift register, the transmitter sets the *transmit-data-register-empty flag* called *TDRE*. The actual transmission that makes the transmit pin respond begins when the transfer to the shift register occurs. After the TDRE flag is set, the program may immediately store the next character into the SCDR.

Usually, the program assumes that the character has been transmitted when TDRE is set, although the transmission will take some time. If the program must know when the transmit pin becomes idle, it must use the *transmit complete flag* called *TC*. For normal transmission, the transmitter sets TC when the transmit shift register finishes sending a character and another character has not been placed in the SCDR register.

In half-duplex applications, the serial line must be switched from transmitting to receiving at the end of a transmission. The program must know when the transmission is complete so that the switching does not cause loss of the last character. When the transmitter sets the TC flag, the switching may be done.

SCI interrupts

All SCI flags except the noise flag NF and the framing error flag FE can cause an interrupt. The five interrupting flags are all connected to a single interrupt line with an interrupt vector at address FFD6. Figure 7-46 shows the flags and the interrupt line.

Interrupt-enable bits control whether a flag can cause an interrupt. The four interrupt-enable bits for the SCI are in the second *serial communications control register* called *SCCR2*. Figure 7-48 shows this register.

The receiver data flag RDRF and the overrun flag OR share a single interrupt-enable bit called *RIE* for *receiver interrupt enable*. The *ILIE* or *idle line interrupt-enable bit* controls the idle flag interrupt. The *TIE* or *transmitter interrupt-enable bit* controls the interrupts from

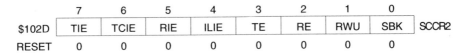

	7	6	5	4	3	2	1	0	
$102D	TIE	TCIE	RIE	ILIE	TE	RE	RWU	SBK	SCCR2
RESET	0	0	0	0	0	0	0	0	

Figure 7-48   Serial communications control register 2.

the transmitter flag. And finally, the *TCIE* or *transmit complete interrupt-enable bit* controls interrupts from the transmission complete flag TC.

Because the SCI has only a single interrupt vector, the interrupt service routine must poll the SCI flags to determine which one caused the interrupt. The overhead of polling these flags is small because the SCI is a relatively slow device. For the same reason, the SCI interrupt is at the lowest address and thus the lowest priority of all 68HC11 interrupts.

SCI character length

The SCI can use either 8-bit or 9-bit characters. The program selects the character length by controlling the M bit in the *serial communications control register* called *SCCR1*. Figure 7-49 shows the SCCR1 register. The program must put a 1 in the M bit to change to 9-bit operation. The reset condition of M is 0, so the SCI defaults to 8-bit characters in response to a power-up reset.

The serial data registers are 8-bit registers. When using 9-bit characters, the extra bit must go elsewhere. The SCI uses the bits in SCCR1 named R8 and T8 to hold bit 8 of the received or transmitted 9-bit characters, respectively.

As the program transmits characters, it needs to update the T8 bit only if a change is required. If successive characters use the same bit value for bit 8, the program need not change it. Reusing the value for bit 8 reduces the work the program must do to transmit 9-bit characters. However, using 9-bit characters is more complex than using 8-bit characters.

Receiver wake-up

Multiple-receiver applications as shown by Figure 7-43 can easily use the SCI receiver because it has wake-up hardware. The receiver can be put to sleep so that the receiving computer can ignore certain messages. While the receiver is asleep, all five SCI receiver flags are inhibited from being set. The receiver returns to normal operation after wake-up is triggered.

The SCI receiver has two ways of triggering wake-up. The WAKE bit in the SCCR1 register selects the wake-up method. Figure 7-49 shows this bit. When WAKE is 0, wake-up

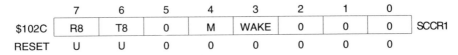

	7	6	5	4	3	2	1	0	
$102C	R8	T8	0	M	WAKE	0	0	0	SCCR1
RESET	U	U	0	0	0	0	0	0	

Figure 7-49   Serial communications control register 1.

is triggered when the receiver line is idle for one full character time. That is, the receiver must detect either 10_{10} or 11_{10} consecutive 1s—depending on the state of the M bit—to trigger wake-up. The reset condition of WAKE is 0, so the SCI defaults to idle line wake-up in response to a power-up reset.

If the program sets WAKE to 1, wake-up is triggered by reception of a character code with a 1 in its most-significant bit. This wake-up method may reduce the time lost during the idle time of the first method.

The RWU bit in the SCCR2 register is the receiver wake-up bit. The program puts the SCI receiver to sleep by storing a 1 in the RWU bit. The SCI hardware clears the RWU bit whenever wake-up is triggered. The program also may store 0 in the RWU bit, although this operation is unusual.

When using the wake-up hardware, a program probably would not use the IDLE flag and interrupt. The IDLE flag can be used to implement software wake-up.

Transmitter and receiver control

The pins associated with the serial communications interface are in port D. Pin PD0 is the receiver line and pin PD1 is the transmitter line. While the serial transmitter and receiver are disabled, these port D pins may be used for general purpose I/O. Port D has a data direction register at address 1008. When used for I/O, the program reads or writes the PORTD data register at address 1007. While the SCI transmitter and receiver are enabled, they control the associated port D pins.

The program can independently enable and disable the SCI receiver and transmitter. The receiver is enabled by storing 1 in the RE or receiver enable bit in the SCCR2 register. Figure 7-48 shows the SCCR2 register. While the receiver is enabled, bit DDRD0 is forced to a 0 to make the PD0 pin an input pin. While the receiver is disabled, the five receiver flags are inhibited.

The serial transmitter is enabled by storing 1 in the TE or transmitter enable bit in the SCCR2 register. While the transmitter is enabled, bit DDRD1 is forced to a 1 to make port D bit 1 an output.

At the time the transmitter is enabled, it automatically transmits one character time of all 1s. The initial string of 1s is called an *idle character*. Sending an idle character helps receivers accurately determine the beginning of a transmission following a period when the transmitter is disabled.

If the TE bit is changed to 0 while the shift register is transmitting a character, that character is finished before the transmitter stops controlling the output pin. Therefore, the program can disable the transmitter as soon as the TDRE flag is set for the last character.

Programming the SCI

Using the serial communications interface with practical programs is often difficult. The program must account for interactions between the serial I/O device and the computer. Usually interrupt is used with the SCI because the computer may be running other program modules between the reception and transmission of serial data bytes.

Receiver ISR

When using interrupts, the receiver interrupt service routine runs each time a character arrives. The ISR must read the character immediately in response to the interrupt so that an overrun doesn't happen.

The ISR usually stores the character in a memory buffer and then tells some other program module that a character has come in. The ISR may instead examine the received characters and only tell the other program when a particular character is received. Similarly, the number of received characters may trigger notification.

The receiver ISR must also deal with reception errors such as overruns, framing errors, and noise errors. The strategy of dealing with errors depends on the application using the serial communications.

Transmitter ISR

Usually the serial transmitter uses interrupt. The interrupt service routine does most of the transmission work. The ISR usually sends a character string that it gets from a memory buffer. Some other program module creates the character string in the buffer before transmission begins.

Interrupt response.   Consider the work the ISR must do in response to an interrupt. The ISR gets a character from the buffer, sends it to the transmitter, which clears the transmitter flag, and then returns from the interrupt. When the transmitter finishes with that character, it sets its flag causing an interrupt back to the ISR. The ISR gets the next character from the buffer and sends it to the transmitter. At each interrupt, the ISR transmits another character. The transmission is somewhat automatic as the ISR responds to each interrupt.

Stopping the transmitter.   Eventually, all characters in the buffer are transmitted. However, when the transmitter finishes the last character, the transmitter sets its flag and interrupts again. On this interrupt, the ISR has nothing more for the transmitter to do. However, the transmitter's flag is set causing the request for service. Clearing the flag is done only by transmitting a character, but no more characters are in the buffer. Returning from the interrupt immediately causes another interrupt from the transmitter.

To prevent this dilemma, the ISR must disable the transmitter interrupt or disable the transmitter when sending the last character. Usually the ISR must tell the controlling program when the transmission is finished.

Kick-starting the transmitter.   Transmission of characters is automatic once the transmitter is started. How is the transmitter started in the first place? The ISR can't start it because it only responds to interrupts after characters are transmitted. Therefore, the program module that controls the transmitter must kick-start the transmitter by sending the first character. The controlling program can then ignore the transmitter until all characters have been transmitted by the ISR.

Before beginning a new transmission, the program must enable both the transmitter and its interrupt. The sequence in which the transmitter and the interrupt are enabled must not trigger an interrupt before the first character is sent.

SCI Programming Example

The serial communications interface is often used with character-oriented devices such as a CRT terminal. A CRT terminal transmits ASCII characters serially from a keyboard, and receives and displays characters on a screen. The following example program shows using the SCI in such an application. The example hardware is common, so experimentation is easy.

The example program reads a character from the keyboard when a key is pressed, inserts the character into a message, and then transmits the message to the display. The program uses the SCI receiver and transmitter with interrupts from both devices. Testing the program is simplified by displaying the received character within the message.

Figure 7-50 is the listing of the example program. The following text describes the program.

- *Lines 4 and 5.*   The crystal for the 68HC11 clock operates at 8.0 MHz. The crystal is the time base for the SCI that controls the baud rate.

- *Line 28.*   This data location holds the count of characters that have been transmitted so far. This counter determines when the last character in the message is sent.

- *Lines 32 to 35.*   These lines define the ASCII characters in the message. The program stores the received character in location INCHAR at line 33. The carriage return and line feed characters format the displayed message in a single line.

- *Line 38.*   A single interrupt vector is set because the SCI uses only one vector though it has five flags that cause interrupts.

- *Line 48.*   A stack is created because the interrupt system requires it.

- *Line 51.*   Clearing the SCCR1 register sets the SCI mode to 8-bit characters and disables the wake-up hardware. A power-up reset also clears the SCCR1 register, so this initialization may be unnecessary.

- *Lines 53 and 54.*   The SCI receiver and transmitter baud rates are set to 1200 baud by controlling the prescaler and rate-control dividers with the BAUD register. The baud rate is unknown after reset, so the program must always initialize the BAUD register. Remember that the baud rate also depends on the crystal frequency.

- *Lines 56 and 57.*   The SCCR2 register is set to enable the receiver and its interrupt while disabling the transmitter and its interrupt. Disabling the transmitter interrupt is particularly important because the power-up reset condition of the transmitter flag is 1. At power-up, the transmitter flag correctly indicates that the transmit data register is empty. However, if the transmitter interrupt was enabled here, an erroneous transmitter interrupt would occur as soon as the interrupt system is turned on. The receiver causes no such problem. The reset condition of the receiver flag is 0, which means that a character has not been received.

- *Lines 59 and 63.*   The interrupt system is enabled, and the main program loop does nothing while waiting for an interrupt.

```
 1                                        ****************************************************
 2                                        ** TRIGGER OUTPUT MESSAGE WITH INPUT CHARACTER
 3                                        *
 4                                        * USE SCI SERIAL COMMUNICATIONS INTERFACE WITH 8.0-MHZ
 5                                        * ..CRYSTAL FOR 68HC11 CLOCK
 6                                        *
 7                                        ****************************************************
 8                                        *
 9                                        ****************************************************
10                                        ** SYMBOL DEFINITIONS
11                                        ****************************************************
12                                        * 68HC11 REGISTERS
13            1000                         REG     EQU   $1000      BEGINNING OF REGISTERS
14            102B                         BAUD    EQU   $102B      BAUD RATE REGISTER
15            102C                         SCCR1   EQU   $102C      CONTROL REGISTER 1
16            102D                         SCCR2   EQU   $102D      CONTROL REGISTER 2
17            102E                         SCSR    EQU   $102E      STATUS REGISTER
18            102F                         SCDR    EQU   $102F      DATA REGISTER
19                                        * MASKS
20            0008                         BIT3    EQU   %00001000
21            0020                         BIT5    EQU   %00100000
22            0080                         BIT7    EQU   %10000000
23                                        *
24                                        ****************************************************
25                                        ** DATA SECTION
26                                        ****************************************************
27   0010                                          ORG   $0010
28   0010                                 CHARCNT RMB   1          NUMBER OF CHARS TRANSMITTED
29                                        *
30   C500                                          ORG   $C500
31                                        * MESSAGE TABLE
32   C500  54 48 45 20 49                 BEGMSG  FCC   'THE INPUT CHARACTER WAS '
     C505  4E 50 55 54 20
     C50A  43 48 41 52 41
     C50F  43 54 45 52 20
     C514  57 41 53 20
33   C518                                 INCHAR  RMB   1          INPUT CHARACTER INSERTED IN MSG
34   C519  0D                                     FCB   $0D        CARRIAGE RETURN
35   C51A  0A                             ENDMSG  FCB   $0A        LINE FEED
36                                        *
37   FFD6                                          ORG   $FFD6
38   FFD6                                          RMB   SCIISR    SCI INTERRUPT VECTOR
39                                        *
40                                        ****************************************************
41                                        ** MAIN PROGRAM
42                                        ****************************************************
43   C100                                          ORG   $C100
44                                        *---------------------------------------------------
45                                        * INITIALIZATION
46                                        *---------------------------------------------------
47                                        * INITIALIZE STACK
48   C100  8E CF FF                       START   LDS   #$CFFF
49                                        * INITIALIZE SCI SERIAL COMMUNICATIONS INTERFACE
50                                        *    INITIALIZE SCI MODE
51   C103  7F 10 2C                               CLR   SCCR1      8-BIT CHARACTERS, NO WAKE UP
52                                        *    INITIALIZE SCI BAUD RATE TO 1200 BAUD
53   C106  86 33                                  LDAA  #$33       SCP1:SCP0=11 SCR2:SCR1:SCR0=011
54   C108  B7 10 2B                               STAA  BAUD
55                                        *    ENABLE SCI RECEIVER AND ITS INTERRUPT
56   C10B  86 24                                  LDAA  #$24       DISABLES SCI TRANSMITTER AND
57   C10D  B7 10 2D                               STAA  SCCR2      ..ITS INTERRUPT
58                                        * TURN ON INTERRUPT SYSTEM
59   C110  0E                                     CLI
```

Figure 7-50   Example program for serial communications interface.

```
60                          *-----------------------------------------------------
61                          * WAIT FOR INTERRUPTS
62                          *-----------------------------------------------------
63   C111   20 FE           HERE     BRA    HERE
64                          *
65                          ******************************************************
66                          ** SCI INTERRUPT SERVICE ROUTINE
67                          ******************************************************
68                          *-----------------------------------------------------
69                          * INTERRUPT POLLING CHAIN
70                          *-----------------------------------------------------
71                          * INTERRUPT FROM SCI RECEIVER?
72   C113   CE 10 00        SCIISR   LDX    #REG
73   C116   1E 2E 20 06              BRSET  SCSR-REG,X,BIT5,SCIRCV
74                          * INTERRUPT FROM SCI TRANSMITTER?
75   C11A   1E 2E 80 10             BRSET  SCSR-REG,X,BIT7,SCITX
76                          * ILLEGAL INTERRUPT
77   C11E   20 27                   BRA    RTSCI    IGNORE
78                          *-----------------------------------------------------
79                          * SERVICE SCI RECEIVER
80                          *-----------------------------------------------------
81                          * READ INPUT CHARACTER
82   C120   A6 2F           SCIRCV   LDAA   SCDR-REG,X  FINISH CLEARING FLAG
83                          * PUT INPUT CHARACTER INTO OUTPUT MESSAGE
84   C122   B7 C5 18                 STAA   INCHAR
85                          * INITIALIZE MESSAGE BYTE COUNT
86   C125   7F 00 10                 CLR    CHARCNT
87                          * START MESSAGE - ENABLE SCI XMTR AND ITS INTERRUPT
88   C128   1C 2D 88                 BSET   SCCR2-REG,X,BIT7+BIT3  SENDS IDLE CHAR
89                          * RETURN TO MAIN PROGRAM
90   C12B   7E C1 47                 JMP    RTSCI
91                          *-----------------------------------------------------
92                          * SERVICE SCI TRANSMITTER
93                          *-----------------------------------------------------
94                          * AT END OF MESSAGE?
95   C12E   86 1B           SCITX    LDAA   #ENDMSG-BEGMSG+1  MESSAGE LENGTH
96   C130   91 10                    CMPA   CHARCNT
97   C132   22 05                    BHI    MORE       BRANCH ON NO
98                          * TERMINATE MESSAGE - DISABLE XMTR AND ITS INTERRUPT
99   C134   1D 2D 88                 BCLR   SCCR2-REG,X,BIT7+BIT3
100  C137   20 0E                    BRA    RTSCI
101                         * TRANSMIT NEXT CHARACTER
102                         *    GET CHARACTER FROM TABLE
103                         *    POINT TO MESSAGE TABLE
104  C139   CE C5 00        MORE     LDX    #BEGMSG
105                         *    ADJUST POINTER TO NEXT CHARACTER
106  C13C   D6 10                    LDAB   CHARCNT
107  C13E   3A                       ABX
108                         *    GET NEXT CHARACTER ASCII CODE
109  C13F   A6 00                    LDAA   0,X
110                         *  OUTPUT NEXT CHARACTER
111  C141   B7 10 2F                 STAA   SCDR    FINISH CLEARING XMTR FLAG
112                         *  INCREMENT COUNT OF CHARACTERS SENT
113  C144   7C 00 10                 INC    CHARCNT
114                         *-----------------------------------------------------
115                         * RETURN FROM SCI INTERRUPT
116                         *-----------------------------------------------------
117  C147   3B              RTSCI    RTI
118  C148                            END
```

Figure 7-50 Continued.

- *Line 72.*   The interrupt system sends control here whenever any interrupt from the SCI occurs. This line is the beginning of the interrupt service routine. The index register is set to point to the 68HC11 registers so that bit instructions with indexed addressing can control the SCI hardware.

- *Lines 73 to 77.*   The SCI device has only a single interrupt vector, so software must determine which SCI device caused the interrupt. The BRSET instruction on line 73 sends control to the receiver service routine if the receiver flag is set. If the receiver flag is not set, the BRSET instruction on line 75 sends control to the transmitter service routine if the transmitter flag is set. If neither flag is set, an error occurred, and the program ignores the interrupt by returning to the main program at line 77.

- *Line 82.*   Control comes to this receiver routine before it can go to the transmitter routine because a key must be pressed to trigger an output message. The LDAA instruction reads the input character code from the receiver data register.

  This LDAA instruction also completes the clearing of the receiver flag RDRF. Clearing the receiver flag requires a sequence of two read operations while the flag is set. The first was the read of SCSR done by the BRSET instruction on line 73. The second is the read of the data register by the LDAA instruction.

- *Line 84.*   The input character from the receiver data register is stored in memory within the predetermined character string that defines the output message.

- *Line 86.*   The character counter tracks the number of characters from the message that have been transmitted. The counter is cleared so that it is ready for the start of a new message.

- *Line 88.*   This receiver service routine now enables the transmitter and its interrupt—the receiver effectively triggers the output message. Enabling the transmitter makes it control the output pin. Upon enable, the transmitter automatically sends an idle character consisting of 10_{10} consecutive 1s. This operation does not clear the TDRE flag. Therefore, a transmitter interrupt occurs immediately when the RTI instruction at the end of this ISR turns on the interrupt system. The transmitter interrupt service routine then transmits the message.

- *Lines 90 and 117.*   The interrupt service routine returns control to the main program. The interrupt service routine ran with the interrupt system off, but the RTI instruction now turns it on again. Because the transmitter flag is already set, the transmitter interrupts immediately.

- *Lines 72 to 75.*   The interrupt system sends control here. The receiver flag is clear because the receiver could not have received another character in such a short time. Therefore, control goes to the transmitter service routine because the transmitter flag is set. The transmitter service routine sends one character of the message to the serial transmitter.

- *Lines 95 to 97.* The number of characters previously transmitted is in location CHARCNT. This count is compared to the length of the message, which is 1B in this example. If more characters are to be transmitted, control goes to line 104. The first time the transmitter routine runs following an input character, the number in location CHARCNT is zero.

- *Lines 104 to 109.* The character code for the next character in the message is gotten from the message table. The character count in location CHARCNT acts as the offset from the beginning of the message table to the next character.

- *Line 111.* The character code is stored in the transmitter data register to transmit the character. This also finishes clearing the transmitter flag TDRE.

  Clearing the transmitter flag requires both a read and a write operation. The BRSET instruction on line 74 reads the SCSR register for the first part of clearing the flag. Writing to the SCDR register completes the clearing function.

  The transmitter flag remains clear until the character has effectively been transmitted by the transfer of the character from the data register to the shift register. This may take some time if the shift register is now transmitting another character. When the transfer does occur, the transmitter sets its flag causing another interrupt which gets the next character to be transmitted.

- *Line 113.* The interrupt service routine increments the character counter because the ISR has sent another character.

After the transmitter sends the last character in the message, the transmitter interrupt sends control to the transmitter service routine. Now the ISR must stop the transmitter instead of transmitting another character.

- *Lines 95 to 97.* The interrupt system, in response to the transmitter interrupt, and the polling chain, send control here. The character count is tested to determine if the last character was sent. If the transmitter has sent the complete message, control goes to line 99.

- *Line 99.* The SCI transmitter and its interrupt are both disabled. Preventing further interrupts from the transmitter effectively ends the transmission of the message, so also disabling the transmitter is appropriate.

This example ignores several important issues. The program does not use the receiver framing error and noise flags, so the program treats erroneous characters the same as any other character.

The program does not test for overrun errors, because they are not possible—the computer is fast enough to always read a character before another is received. However, receiving a new character before the message is finished starts the message again from the beginning because the character count is set to zero.

7.11   ANALOG-TO-DIGITAL CONVERTER

The *analog-to-digital* or *A/D converter* in the 68HC11 makes 8-bit unsigned numbers representing external DC voltages. The A/D converter, by using an 8-channel multiplexer, can read voltages from eight different pins on the 68HC11 package. Port E is the collection of the eight input pins for the A/D converter.

An A/D converter normally reads signals from analog sensors. Typical sensors measure temperature, pressure, or position.

Analog-to-Digital Conversion Principles

An analog-to-digital converter makes a binary number that is proportional to an unknown DC voltage. A digital-to-analog converter makes a DC voltage proportional to a binary number. Generally, an A/D converter is much more complex than a D/A converter, which is relatively simple and inexpensive.

Most A/D converters operate by comparing the unknown voltage to a voltage controlled by a D/A converter. The D/A output is varied according to some algorithm. When the two voltages are nearly the same, the number currently controlling the D/A converter becomes the digital output of the A/D converter.

Figure 7-51 shows the parts of an A/D converter. An unknown voltage and the voltage from the D/A converter are applied to an analog comparator. The comparator asserts its output while the D/A voltage is greater than the unknown voltage. The controller changes the number in the D/A register according to some algorithm until it detects a correct change in the comparator output.

Up-counting converter

The simplest algorithm for the controller starts the D/A at zero volts and then counts the register up until the D/A voltage crosses the unknown voltage. At the crossing, the counting stops and the flag is set indicating that the A/D converter has completed the conversion.

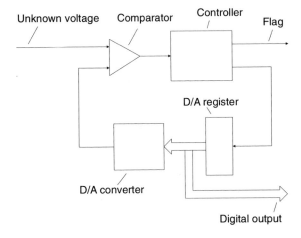

Figure 7-51   Block diagram of A/D converter hardware

The up-counting algorithm is very simple, but it has several disadvantages. The major disadvantage is that the converter takes a different amount of time to reach the result for each different voltage. If the voltage is small, the result is reached quickly, but if the voltage is large, the time is very long.

Successive-approximation converter

Another control algorithm for the A/D converter is more complex, but the performance is much better than the up-counting algorithm. Here the controller first sets the D/A register to binary 10000000 to make a voltage that is one-half of the maximum. The comparator indicates whether the unknown voltage is higher than or lower than the D/A voltage. If the unknown voltage is higher than the D/A voltage, the D/A is set to a voltage halfway between this one and the maximum. That is, the binary number is made 11000000. If now the unknown voltage is lower than the D/A voltage, the D/A is set to a value halfway between the previous two values; namely, 10100000. The converter determines all eight bits by following this procedure eight times. This algorithm is called *successive-approximation*.

One advantage of the successive-approximation algorithm is that eight trials always determine the result whatever the unknown voltage. The conversion time is always known for all voltages. Also, the average successive-approximation conversion time is many times smaller than the up-counting conversion time. However, the controller is more complex.

Multiplexer

An A/D converter is a complex device using many analog electronic components. Therefore, its cost is relatively high. If a computer must read many different analog voltage signals, using one A/D converter for each signal usually is prohibitively expensive. In such applications, compromising the speed at which the signal voltage can be read to reduce cost is often practical.

A *multiplexer* is an electronic switch that connects one of several analog signals to a single A/D converter. Typical multiplexers switch between either eight or 16_{10} different signals. A multiplexer enables the A/D converter to read many different signals, but it reduces the system performance. The computer must control the multiplexer telling it which signal to pass to the A/D converter.

The multiplexer takes a small amount of time to operate before the A/D can read the incoming voltage. Because the A/D converter can read only one voltage at a time, the rate at which all the voltages can be read is greatly reduced. In many applications, the lower cost makes the performance reduction with a multiplexer acceptable.

68HC11 A/D Converter

The 68HC11 chip contains an 8-bit successive-approximation A/D converter. No other analog input or output device is in the chip. The fact that the 68HC11 chip contains both digital and analog circuits on the same chip is noteworthy.

Converter hardware

The A/D converter in the 68HC11 chip includes a single A/D converter, a sample-and-hold circuit, and a 16-channel multiplexer. The converter can read a voltage over a 5-volt range with a maximum error of plus or minus one least-significant bit.

A/D converter.   The A/D converter is an 8-bit successive-approximation converter. The power supply for the 68HC11 must provide the A/D converter both a high and a low analog reference voltage. These voltage supplies connect to the 68HC11 pins named V_{RH} and V_{RL}. Typically +5 V_{DC} and 0 V_{DC} are used, but other positive voltages are allowable if V_{RH} is 2.5 to 5.0 volts greater than V_{RL}. Normally V_{RH} and V_{RL} also supply the voltage for the analog sensor.

A measured voltage equal to V_{RL} returns the 8-bit number 00 and a voltage equal to or greater than V_{RH} returns FF. Such a converter is called a *ratiometric A/D converter*.

Each conversion requires the time of exactly 32_{10} E-clock cycles for usual applications. With an 8.0-MHz crystal, the conversion time is 16_{10} microseconds.

Sample-and-hold.   An *analog sample-and-hold circuit* measures the unknown voltage at the beginning of a conversion. During the conversion, the sample-and-hold device holds or maintains a constant voltage for the A/D converter. Use of the sample-and-hold avoids conversion errors caused by a changing voltage during the conversion process.

Multiplexer.   The multiplexer has 16_{10} channels, although only eight channels connect to pins on the 68HC11 chip. Therefore, the multiplexer makes measuring eight different analog voltage signals possible. Of the other eight channels, four are used for internal testing and four are unused.

Certain varieties of the 68HC11 integrated circuit have fewer pins than required by the hardware in the chip. These chips can use only four channels of the multiplexer because the other four channels do not connect to pins.

A/D control

Software controls the A/D converter in ways similar to other I/O devices in the 68HC11. However, the way the A/D operates and its speed of conversion leads to some unique programming approaches.

A/D control register.   Figure 7-52 shows the *A/D control register* called *ADCTL*. As the figure shows, most of the ADCTL bits are unknown following a power-up reset.

The A/D converter starts a conversion within one E-clock cycle after the program writes a number into the ADCTL register. The number written into the control register selects one of several options on how the converter reads and stores conversion results.

	7	6	5	4	3	2	1	0	
$1030	CCF	0	SCAN	MULT	CD	CC	CB	CA	ADCTL
RESET	0	0	U	U	U	U	U	U	

Figure 7-52   Analog-to-digital control register.

A/D flag. Starting the converter clears the *conversions-complete flag*, which is called *CCF*. When a conversion sequence is complete, the A/D converter sets the flag.

If the program writes to ADCTL while a conversion is in progress, the current conversion stops and a new conversion sequence begins.

A/D interrupt. The A/D converter cannot cause an interrupt. The conversion is so fast that the overhead of servicing an interrupt makes polling attractive. However, because the exact number of E-clock cycles needed for a conversion is known, the program may execute instructions to delay 32_{10} E-clock cycles while the conversion takes place. Then the program can read the A/D immediately and it need not poll the flag.

Channel selection and scanning

When the program starts the analog-to-digital converter, the converter always does four consecutive A/D conversions as rapidly as it can. It cannot do only a single conversion. The four conversions may be either on a single channel of the multiplexer or on four different channels. Furthermore, the A/D can operate continuously by repeating the conversions without the program starting it each time.

The A/D converter sets the CCF flag only after four conversions are complete. Few programs use the flag because the A/D converter is so fast and its conversion time in E-clock cycles is known exactly.

Result registers. The A/D converter puts the numbers from the four conversions in four A/D result registers. These registers are called *ADR1*, *ADR2*, *ADR3*, and *ADR4*. The addresses of these registers are 1031, 1032, 1033, and 1034, respectively. The result of the first conversion goes to ADR1 and the last result to ADR4. The program may only read the result registers; writes to them do nothing.

The A/D puts the conversion results into the result registers as each conversion finishes. Therefore, the program may read the results as they are stored rather than waiting until all four conversions are complete. Reading the results early may be an advantage if very fast response is necessary.

Single-channel reading. When the MULT bit in the ADCTL register is 0, the A/D converter reads a single channel. The channel is selected by the number in the CD, CC, CB, and CA bits of the ADCTL register. Only the first eight channels can read external voltages, so the CD bit is set to 0 by most programs. If the program needs only a single reading of a single channel, the result is available in the ADR1 register 32_{10} E-clock cycles after the write to the control register.

Multiple-channel reading. When the MULT bit in the ADCTL register is 1, the A/D converter reads four channels in succession. Only the first eight channels of the multiplexer can read external voltages, so the CD bit is usually 0. The CC bit then selects the first group of four channels or the second group of four channels. That is, when the CD:CC bits are 00, channels 0 through 3 are read; when the CD:CC bits are 01, channels 4 through 7 are read. The CB and CA bits have no effect when MULT is 1. No other grouping of four channels is possible.

Continuous and single-channel scanning.   The A/D converter reads four times according to two different scanning methods. The program sets the SCAN bit to 1 to choose the *continuous-scan mode*. In the continuous-scan mode, the converter repeats the four readings continuously without program intervention. When four conversions are complete, the A/D converter immediately begins reading the four voltages again. Therefore, the four result registers are updated continually. The flag serves no purpose in this mode.

In the *single-scan mode*, the program starts the conversions. When the four conversions are complete, the A/D converter stops. So, the result registers are updated once each time the program starts the converter. When the four conversions are complete, the A/D sets its CCF flag.

A/D power control

The program can turn off the electrical power to the A/D converter if the converter is not needed. Power can be restored when the A/D is again needed. The power savings may be especially important if a battery powers the 68HC11. Look at the electrical specifications to find the details of the power saved.

Figure 7-53 shows the OPTION register. The OPTION register was first shown by Figure 7-6. The ADPU bit in the OPTION register controls the power to the A/D converter.

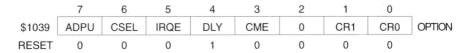

	7	6	5	4	3	2	1	0	
$1039	ADPU	CSEL	IRQE	DLY	CME	0	CR1	CR0	OPTION
RESET	0	0	0	1	0	0	0	0	

Figure 7-53   Configuration options register.

The reset condition of the A/D is off. The program must write a 1 to the ADPU bit to power-up the A/D converter. When the program turns on the A/D, it must wait at least 100_{10} microseconds before using the converter to allow the electronics to stabilize.

By changing the ADPU bit, the program can turn the A/D power on or off at any time. Some other control bits in the OPTION register are timed-write-once bits.

A/D clock options

Either one of two clocks within the 68HC11 chip may drive the A/D converter. Normally, the E-clock is the clock source for the A/D converter. A power-up reset selects the E-clock.

Another clock controlled by an internal *resistor-capacitor oscillator* is necessary when the E-clock rate is below 750_{10} KHz. Such a low E-clock rate reduces the power consumption of the 68HC11 chip. To select the R-C clock for the A/D, the program must write a 1 to the clock select bit CSEL in the OPTION register.

The R-C oscillator runs at about 1.5 MHz; it is not as accurate as the E-clock. It takes about 10 milliseconds after selecting the R-C clock before the A/D stabilizes so that the program can use it. When using the R-C clock, the program should poll the CCF flag to determine when the conversions are complete.

By changing the CSEL bit, the program can select a different clock source for the A/D at any time. Some other control bits in the OPTION register are timed-write-once bits.

The CSEL bit also enables a resistor-capacitor oscillator used in programming the EEPROM in the 68HC11 chip. Be careful to avoid problems caused by interactions due to programming the A/D clock option.

Analog Input Example

Every 8.19_{10} milliseconds, the example program reads the A/D converter and controls two digital output bits. One bit indicates if the voltage is above some specified limit. The other bit indicates if the voltage is too high for the A/D converter—it actually indicates if the A/D reading is exactly FF. The program uses the real-time interrupt device with interrupts to determine the timing. This program, shown in Figure 7-54, is based on the RTI program in Figure 7-11, so only the new parts are discussed below.

- *Lines 52 and 53.*   The reset state of the A/D converter is powered-down, so first the program powers it up and selects the E-clock to drive the converter hardware. The A/D requires about 100_{10} microseconds to stabilize after it is powered-up, so the first reading taken by this program may be erroneous.

- *Lines 81 through 88.*   The real-time interrupt service routine runs after a time delay of 8.19_{10} milliseconds.

- *Lines 90 and 91.*   A subroutine reads the A/D converter. The main program sends the channel number to the subroutine in the A accumulator. The channel number should be in the range of zero through seven.

- *Lines 111 through 113.*   Inside the subroutine, the channel number is masked to ensure that it is in a valid range for the A/D converter. The masking also sets the SCAN and MULT bits so that the A/D reads four times and stops and the four readings are all of the same channel. Storing into the ADCTL register starts the A/D conversion in the correct mode. It also clears the CCF flag, but the flag is not used in this example.

- *Lines 115 through 120.*   Because the CCF flag is not used here, the program executes a few instructions to use at least 33_{10} E-clock cycles while the A/D converter operates. After this time, the converter has completed only one A/D conversion.

- *Line 122.*   The program reads the A/D converter results from the first conversion result register. The first register must be read because only one conversion is complete at this time.

- *Line 124.*   The subroutine returns to the calling program inside the interrupt service routine.

```
 1                              **************************************************************
 2                              ** READ ANALOG VOLTAGE EVERY 8.19 MILLISECONDS AND
 3                              ** INDICATE VOLTAGE RANGE AND OVERLOAD (A/D READS FF)
 4                              *
 5                              * USE REAL TIME INTERRUPT AS TIME BASE
 6                              *
 7                              * PORTC BIT0 IS 1 FOR UPPER HALF OF VOLTAGE RANGE
 8                              * PORTC BIT1 IS 1 FOR OVERLOAD
 9                              *
10                              * USE 8.0-MHZ CRYSTAL FOR 68HC11 CLOCK
11                              *
12                              **************************************************************
13                              *
14                              **************************************************************
15                              ** SYMBOL DEFINITIONS
16                              **************************************************************
17                              * 68HC11 REGISTERS
18          1003                PORTC   EQU   $1003      I/O PORT C REGISTER
19          1007                DDRC    EQU   $1007      DATA DIRECTION REGISTER C
20          1024                TMSK2   EQU   $1024      TIMER MASK REGISTER
21          1025                TFLG2   EQU   $1025      TIMER FLAG REGISTER
22          1026                PACTL   EQU   $1026      PULSE ACCUMULATOR CONTROL REG
23          1030                ADCTL   EQU   $1030      A/D CONTROL REGISTER
24          1031                ADR1    EQU   $1031      A/D RESULT REGISTER
25          1039                OPTION  EQU   $1039      HARDWARE OPTION CONTROL REG
26                              * MASKS
27          0001                BIT0    EQU   %00000001
28          0002                BIT1    EQU   %00000010
29          0040                BIT6    EQU   %01000000
30          0080                BIT7    EQU   %10000000
31          00FF                IOPAT   EQU   $FF        I/O PATTERN, 0=IN 1=OUT
32                              * SYMBOLS
33          0003                CHNO    EQU   3          A/D CHANNEL NUMBER
34          0060                VLIMIT  EQU   $60        ANALOG VOLTAGE RANGE LIMIT
35                              *
36                              **************************************************************
37                              ** DATA SECTION
38                              **************************************************************
39   FFF0                                 ORG $FFF0      RTI INTERRUPT VECTOR
40   FFF0     C11D                         FDB  RTIISR
41   FFF2                       *
42                              **************************************************************
43                              ** MAIN PROGRAM
44                              **************************************************************
45   C100                                 ORG $C100
46                              *----------------------------------------------------------
47                              * INITIALIZATION
48                              *----------------------------------------------------------
49                              * INITIALIZE STACK
50   C100     8E CF FF          START   LDS   #$CFFF
51                              * INITIALIZE A/D CONVERTER
52   C103     86 80                      LDAA  #BIT7      POWER-UP A/D USING E-CLOCK
53   C105     B7 10 39                    STAA  OPTION    ..ADPU:CSEL=0:0
54                              * INITIALIZE PORT C
55                              *    INITIALIZE OUTPUTS TO ZEROS
56   C108     7F 10 03                    CLR   PORTC     0=LOW
57                              *    SET UP PORTC INS AND OUTS
58   C10B     86 FF                      LDAA  #IOPAT
59   C10D     B7 10 07                    STAA  DDRC      ALL OUTPUTS
60                              * INITIALIZE REAL TIME INTERRUPT
61                              *    SET INTERRUPT RATE TO 8.19 MS
62   C110     86 01                      LDAA  #BIT0      ZEROS IN OTHER BITS
63   C112     B7 10 26                    STAA  PACTL     ..CAUSE NO PROBLEMS
```

Figure 7-54   Example program for A/D converter.

```
64                              *      ENABLE RTI INTERRUPT
65    C115    86 40                    LDAA   #BIT6     ZEROS IN OTHER BITS
66    C117    B7 10 24                 STAA   TMSK2     ..CAUSE NO PROBLEMS
67                              * TURN ON INTERRUPT SYSTEM
68    C11A    0E                       CLI
69                              *----------------------------------------------------------
70                              * WAIT FOR INTERRUPTS
71                              *----------------------------------------------------------
72    C11B    20 FE            HERE     BRA    HERE      DO NOTHING!
73                              *
74                              ****************************************************************
75                              ** INTERRUPT SERVICE ROUTINE
76                              ****************************************************************
77                              *----------------------------------------------------------
78                              * INTERRUPT POLLING CHAIN
79                              *----------------------------------------------------------
80                              * INTERRUPT FROM REAL TIME INTERRUPT DEVICE?
81    C11D    CE 10 25         RTIISR   LDX    #TFLG2
82    C120    1F 00 40 1A               BRCLR  0,X,BIT6,RTRTI  IGNORE ILLEGAL INTERRUPT
83                              *----------------------------------------------------------
84                              * CONTROL PORT C OUTPUTS
85                              *----------------------------------------------------------
86                              * CLEAR REAL TIME INTERRUPT FLAG
87    C124    86 40                    LDAA   #BIT6     STORE 1 TO CLEAR FLAG!
88    C126    B7 10 25                 STAA   TFLG2     ..ZEROS DO NOTHING
89                              * READ A/D CONVERTER
90    C129    86 03                    LDAA   #CHNO
91    C12B    BD C1 3F                 JSR    ADREAD
92                              * DETERMINE VOLTAGE RANGE BIT
93    C12E    5F                       CLRB
94    C12F    81 60                    CMPA   #VLIMIT
95    C131    25 02                    BLO    NEXT
96    C133    C8 01                    EORB   #BIT0     SET BIT0
97                              * DETERMINE OVERLOAD BIT
98    C135    81 FF            NEXT     CMPA   #$FF
99    C137    26 02                    BNE    NEXT1
100   C139    C8 02                    EORB   #BIT1     SET BIT1
101                             * CONTROL OUTPUT BITS
102   C13B    F7 10 03         NEXT1    STAB   PORTC
103                             * RETURN FROM INTERRUPT
104   C13E    3B               RTRTI    RTI
105                             *
106                             ****************************************************************
107                             ** READ ANALOG-TO-DIGITAL CONVERTER SUBROUTINE
108                             * PASS CHANNEL NUMBER IN A, RETURNS READING IN A
109                             ****************************************************************
110                             * ACCEPT ONLY VALID CHANNEL NUMBERS
111   C13F    84 07            ADREAD   ANDA   #$07      ALSO SELECT SINGLE SCAN, ONE CH
112                             * START A/D CONVERSION
113   C141    B7 10 30                 STAA   ADCTL     SCAN:MULT=00
114                             * WASTE AT LEAST 33 CLOCK CYCLES WHILE A/D OPERATES
115   C144    37                       PSHB             DO NOT ALTER REGISTERS
116   C145    33                       PULB             USE MINIMUM STACK SPACE
117   C146    37                       PSHB
118   C147    3D                       MUL
119   C148    3D                       MUL
120   C149    33                       PULB
121                             * READ A/D CONVERSION RESULTS
122   C14A    B6 10 31                 LDAA   ADR1      USE FIRST RESULT REGISTER
123                             * RETURN FROM SUBROUTINE
124   C14D    39                       RTS
125   C14E                             END
```

Figure 7-54 Continued.

- *Lines 93 through 100.*   The program tests the reading from the A/D converter the subroutine returned in the A accumulator. It also puts the bit pattern in accumulator B indicating the range value and the overload bit.

- *Line 102.*   The output pattern is sent to port C indicating the range and overload condition.

- *Line 104.*   The RTI instruction returns control to the main program loop, which waits for another interrupt. The next interrupt occurs 8.19_{10} milliseconds after the previous one.

This program causes the output bit to flicker between 0 and 1 when the analog voltage is near the limit. A simple digital filter added to the program would reduce or prevent the flickering.

7.12   FAIL-SAFE OPERATION

Many microcomputers control dangerous devices. The danger may be to people or to equipment and machinery. Therefore, the computer must respond to failures that hardware and software can detect. Certain failures cannot be detected without extraordinary complication and cost.

Watchdog Timer

Many control systems, including computers, use a device called a *watchdog timer* to ensure that the system is operating. Motorola calls the timer in the 68HC11 the *computer-operating-properly timer*, or *COP*.

Principles of watchdog timer

A watchdog timer is a hardware device that can reset a system after a time period has expired. The system will restart the timer periodically if it is working correctly. If the system fails, it will likely not restart the timer. Then the timer will expire and reset the system. The reset must force everything to a safe condition. The reset may also cause the system to start operating again, and it may then continue operating correctly. If the system cannot recover from the failure, the safe condition continues.

Watchdog operation.   In a computer system, the watchdog timer is reset by a program that runs in a loop. When the program repeats the loop, it restarts the timer. If the program fails to complete the loop and restart the timer, the timer forces a system reset. The watchdog timer hardware can reset the system even when other hardware has failed.

Failures detected.   Many kinds of failures can lead to timeout of the timer. Program errors that occur only under unusual circumstances may crash the program so that it cannot restart the timer. Failure of a memory bit can change an instruction in a program leading to a crash and a timeout. Certain interrupt failures may prevent a part of the program from executing and thus restarting the timer. Therefore, a watchdog timer detects and responds to a wide range of software and hardware errors and failures that are difficult to predict.

Computer-operating-properly timer

The 68HC11 chip contains a watchdog timer called the *computer-operating-properly timer*, or *COP*. When the COP times out, it forces a reset on the 68HC11 chip and other chips connected to its reset pin. Figures 7-1 and 7-2 show the reset pin as a bidirectional signal pin. The reset signal may originate within the 68HC11 due to the COP.

COP period.   The program sets the COP timeout period to one of four times based on the E-clock rate. Figure 7-55 shows the COP divider chain and the control bits CR1 and CR0. Figure 7-53 shows the OPTION register that contains the control bits. At reset, the COP time is set to the shortest and thus safest time.

The CR1 and CR0 bits are timed-write-once bits. Therefore, the program must set the COP timeout time within 64_{10} E-clock cycles of reset. Thereafter, the program cannot change the COP time. Not allowing the program to change the COP time provides improved security that the COP will operate correctly even if failed programs try to alter it.

Table 7-12 shows typical timeout periods for several crystal frequencies. Usually the timeout period is small because the COP is used in control applications. Many control systems require sections of the program to execute periodically at a rapid rate. If the program does not execute at the correct time, damage to equipment may result. Therefore, short COP times confirm that the system is updating the control information in a timely way.

Enabling the COP.   Most control applications of the 68HC11 use the COP to provide safety. A bit in the CONFIG register enables the COP. Figure 7-5 shows the NOCOP bit in the CONFIG register. When the NOCOP bit is 0, the COP operates.

The CONFIG register is made of EEPROM cells, so it is programmed permanently as part of the control system. The program cannot affect whether the COP is active.

During software development for control system applications, the programmer usually disables the COP, which avoids resets during software debugging and testing. Likewise, applications that have no need of the COP leave it disabled.

Restarting the COP timer.   The principal purpose of the COP timer is to detect software failures. Failed software must not easily restart the timer. Consequently, restarting the timer is deliberately difficult.

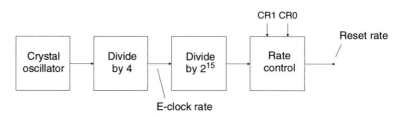

Figure 7-55   COP timer divider chain.

TABLE 7-12   COP TIMEOUT PERIOD

COP Control Bits		Rate is E Divided by	Crystal Frequency		
			8.3886 MHz	8.0 MHz	4.9152 MHz
CR1	CR0		Timeout Period (milliseconds)		
0	0	2^{15}	15.625	16.384	26.667
0	1	2^{17}	62.5	65.536	106.67
1	0	2^{19}	250	262.14	426.67
1	1	2^{21}	1000	1049	1707
		E =	2.1 MHz	2.0 MHZ	1.2288 MHz

Restarting the timer requires two distinct operations by the program. First, the program must store the number 55 into the *COP-reset register* called *COPRST*. The COPRST register is at memory address 103A. Second, the program must store the number AA into the COPRST register.

The program must complete the two stores into COPRST before the COP times out to prevent the COP from resetting the system. Each time the reset sequence is completed, a new COP timeout period begins. The program may execute any number of instructions between the two stores into COPRST. The program can obtain no useful information by reading the COPRST register. Furthermore, reading it has no affect on the COP.

The 8-bit number 55 is the complement of the number AA. Using these two numbers to reset the COP reduces the possibility of a program resetting it in error. For example, if a failure of the data bus has grounded a bus line, the COP cannot be reset.

Program strategy.   Clever programming of the COP reset can improve the safety provided by the COP. Usually, separating the two store instructions into different parts of a program is good. Then failure of one part alone prevents the COP from being reset. Similarly, if a part of the memory fails, separating the reset instructions improves the likelihood of detecting a problem.

If a certain I/O device is particularly critical to an application, part or all the COP reset may be placed in the interrupt service routine. For example, the COP can test the real-time interrupt device by using its ISR to do part of the COP reset.

COP reset.   Following a reset of the computer system by the COP, the 68HC11 continues normal operation with one exception. After the reset, the microprocessor uses a reset vector at addresses FFFA and FFFB. Therefore, the program can detect that a COP reset has occurred and take appropriate action. For example, the program may record the number of times a COP reset occurred. The number of failures may aid a service person.

Forcing a reset with COP

Programs that detect fatal errors may attempt to recover by forcing a hardware reset on the entire computer system. The microprocessor has no instructions to force reset, so the reset must be done by hardware. Usually, the program disables interrupts and then goes into an infinite loop. Soon the COP times out and forces the desired reset of all hardware in the computer system.

A typical example of forcing a reset is a response to an illegal instruction interrupt. If an illegal instruction code was fetched, the integrity of the entire computer system is in doubt. The erroneous program may have incorrectly programmed I/O hardware and destroyed information in volatile memory. Therefore, using the COP to force a reset of the system is appropriate following an illegal instruction interrupt.

Clock Failure Detection

All the many functions of the 68HC11 microprocessor and internal I/O devices depend on proper operation of the crystal clock. If the crystal clock fails completely or operates at a grossly incorrect frequency, the computer system may cause unsafe operation in certain applications.

Clock monitor

The 68HC11 contains hardware to detect if the clock is operating below a certain rate that includes a complete failure. When the clock monitor detects a clock failure, the clock monitor forces a system reset through the 68HC11 reset pin. The reset puts all devices in the computer system into the reset condition so that the system is safe.

Minimum clock rate.   The clock monitor need not be accurate, because its principal purpose is to detect failed clocks. The specifications for the 68HC11 state that clock rates above 200 KHz will not trigger a reset. Furthermore, clock rates below 10 KHz always force a reset. Special applications that use slow or changeable clock rates may have difficulty using the clock monitor.

Enabling the clock monitor.   The program enables the clock monitor by storing 1 in the *clock monitor enable bit* called *CME*. Figure 7-53 shows the OPTION register containing the CME bit. The reset condition of CME disables the clock monitor.

The program can change the CME bit at any time. Some bits in the OPTION register are timed-write-once bits.

Clock monitor reset.   If the clock is operating following a clock monitor reset, the 68HC11 operates normally except for the use of a different reset vector. The clock monitor reset vector is at addresses FFFC and FFFD. The program module that handles the reset thus knows that a clock failure occurred.

COP clock

The computer-operating-properly timer derives its time from the crystal clock for the 68HC11. If the clock fails to operate, or operates at a low frequency, the COP is nearly useless! Therefore, the COP must always be used in conjunction with the clock monitor. Then if the clock does not operate correctly, the computer system will be reset even if the COP cannot respond correctly.

7.13   I/O PORT SUMMARY

All pins of all input/output ports in the 68HC11 are digital input bits, digital output bits, or programmable digital I/O bits. All I/O pins can be used as digital pins. Overlooking the usefulness of certain pins is easy because they are usually identified with particular hardware such as the programmable timer. Some pins have some restrictions because other I/O hardware uses them.

Port A

The programmable timer and pulse accumulator use port A pins in addition to the parallel I/O functions. The PORTA register at address 1000 controls the digital input and output functions. Pins PA0, PA1, and PA2 are input-only pins. Pins PA3 through PA6 are output-only pins. Pin PA7 is bidirectional using the data direction bit DDRA7 in the PACTL register to control the direction.

In the E9 version of the 68HC11 chip, pin PA3 is also bidirectional. Then, bit 3 of the PACTL register is called *data direction bit DDRA3*.

The PORTA register may be read at any time, but information from bits corresponding to timer and pulse-accumulator functions read their information, which is not always input pin information. Writing to the PORTA register controls the corresponding output pins only if they are not being controlled by the timer. The bits written into PORTA are remembered and will control the output pins later if the timer function is disabled.

Port B

In single-chip-computer mode, port B is always an output-only digital port. The PORTB register at address 1004 controls the output pins. No other devices use the port B pins. In expanded mode, the output function of port B is lost.

Port C

In single-chip-computer mode, port C is always a programmable I/O port. The data direction register DDRC at address 1007 determines whether a pin is an input or an output. The program may either write or read the PORTC register at address 1003. No other devices use these port C pins. In expanded mode, the I/O functions of port C are lost.

In single-chip mode, the port C outputs can be either CMOS compatible or open-drain outputs. The port C wired-OR mode bit in the PIOC register controls the mode. While CWOM is 0 as it is after a reset, the port C outputs are CMOS compatible. While CWOM is 1, the outputs are open-drain outputs.

Port D

Port D is a 6-bit programmable I/O port. The PORTD register at address 1008 is used to read or write to the port. The data direction register DDRD at address 1009 controls the direction of each pin.

The outputs are affected by the DWOM bit in the SPCR register. When DWOM is 0, the outputs are CMOS compatible. When DWOM is 1, the outputs are open-drain drivers.

Two serial I/O devices also use the port D pins. When those devices are enabled, they override some of the functions specified by the DDRD and PORTD registers. Due to the multiple functions of these pins, several considerations in the electronics also come into play, so consult a Motorola manual for additional details.

Port E

The port E pins are both an input-only digital input port and the inputs to the analog-to-digital converter. The pins serve both functions simultaneously, though one function is analog and the other digital! The digital input port is read by reading the PORTE register at address 100A.

7.14 HARDWARE EXPANSION

Many applications of the 68HC11 require additional I/O hardware beyond that available within the chip. If the chip is operating in expanded mode, the microprocessor buses allow expansion, although using the buses may be too costly for some applications. Regardless, in the single-chip mode, the buses are not available! To provide for expansion in either mode, the 68HC11 chip contains a serial interface for communicating with I/O chips. An alternative application is to provide communication between two 68HC11 chips.

Serial Peripheral Interface

The 68HC11 I/O hardware can be expanded using a synchronous serial I/O bus called the *serial peripheral interface*, or the *SPI bus*. Input/output chips communicate with the 68HC11 over the SPI bus. The SPI bus does not communicate with I/O devices directly. Many commercial chips are available for serial communication over the SPI bus.

The SPI bus operates at a high baud rate, so the I/O chips must be close to the 68HC11 chip—usually both chips are on the same circuit board. Using the SPI bus leads to modest compromises in input/output speed when compared to using the microprocessor buses. This speed reduction is of little concern in many applications.

Serial communication principles

The serial peripheral interface is, in principle, a simple device based on two shift registers. Chapter 1 showed a serial communication system similar to the SPI in Figure 1-15. Figure 7-56 is a simpler version of that figure, which illustrates the principles of the SPI device.

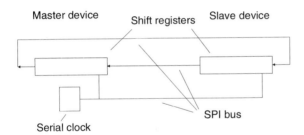

Figure 7-56 Principal hardware in SPI.

The shift registers in the two SPI devices are in a circular connection. A common clock signal clocks both shift registers. By definition, the master device contains the controlling clock. The slave device receives its clock signal from the master device.

When a clock pulses the shift leads of both devices, one bit from the master device goes to the slave device, and one bit from the slave device goes to the master device. After eight clock pulses, the numbers in the two registers have been exchanged.

The transmission and reception of bytes occur simultaneously over the SPI bus. Therefore, the SPI is a full-duplex communication system. Furthermore, the common clock makes the SPI a synchronous serial device because both devices respond in synchronism. The SCI serial device is asynchronous because it does not have a common clock. Therefore, the SCI receiver must periodically sample the input signal—the SPI does not require sampling.

SPI hardware operation

The SPI contains a single data shift register, a control register, and a status register. The SPI hardware may operate in either the master mode or the slave mode. Three pins on the 68HC11 connect the SPI bus lines to the other device. More than two devices may connect to the SPI bus. However, only one device can be the master.

Hardware connections.   Three wires form the SPI bus between a master and a single slave device that have a circuit common. One wire carries the *serial clock signal* between the pins labeled *SCK* on each device. The other two wires carry the data signals between the master and slave devices.

On each device, one pin is labeled *MOSI* for *master-out-slave-in* and the other is labeled *MISO* for *master-in-slave-out*. These labels imply that you should connect bus wires between pins with the same label on both the master and slave devices. These connections do not depend on which of the devices acts as the master.

To complete the connections, the low-asserted slave-select pin must be low on the slave device to enable it for responding to the SPI bus. A slave-select pin is not necessary on the master device, so the pin performs other functions. First, it is the input to the bus contention detection hardware. Second, it is a digital output bit.

Mode selection.   Selecting one of the two modes of operation makes the hardware operate differently. The *master-mode select bit* called *MSTR* determines the mode of operation. The MSTR bit is in the *serial peripheral control register* called *SPCR*. Figure 7-57 shows the SPCR register, which is at address 1028.

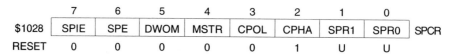

$1028	SPIE	SPE	DWOM	MSTR	CPOL	CPHA	SPR1	SPR0	SPCR
	7	6	5	4	3	2	1	0	
RESET	0	0	0	0	0	1	U	U	

Figure 7-57   Serial peripheral interface control register.

Usually, the computer chip is the master unless it is a slave to another computer chip. The master device generates the serial clock and therefore controls all data transfers over the SPI bus. The slave device cannot initiate communication. The slave receives data from the master and can transmit data to the master. However, the master must still initiate the transmission.

Baud rate.   The rate of the serial clock determines the baud rate of the SPI. The program sets the baud rate by controlling the SPR1 and SPR0 bits in the SPCR register. Figure 7-58 shows the divider chain that drives the SPI clock. Table 7-13 shows typical baud rates for the SPI.

Data register.   When the program stores into the shift register of the master, the SPI controller automatically generates eight pulses to complete the transfer of a byte between the master and slave devices. The shift register is called *SPDR* for *serial peripheral data register*. The SPDR register is at address 102A.

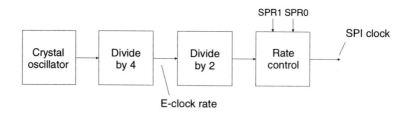

Figure 7-58   SPI clock divider chain.

Flags and interrupts.   Whether operating in the master or slave mode, when a transmission is complete, the SPI sets the *serial peripheral interface flag* called *SPIF*. The SPIF flag is in the *serial peripheral status register* called *SPSR*. Figure 7-59 shows the SPSR at address 1029.

If the 68HC11 is operating in the slave mode, the program should store into the SPI data register immediately after the flag is set. In the master mode, the program can store into the data register any time after the flag is set. In either case, if a store occurs while a transmission is in progress, a *write collision* occurs and the store has no effect.

The *write-collision bit* called *WCOL* in the SPSR register indicates whether a write collision happened. The WCOL flag is cleared by reading the SPSR register followed by reading or writing the SPDR register.

TABLE 7-13 SPI BAUD RATE

SPI Control Bits		Rate-Control Divide Factor	Crystal Frequency		
			8.3886 MHz	8.0 MHz	4.9152 MHz
SPR1	SPR0		SPI Baud Rate (kilobaud)		
0	0	1	1049	1000	614.4
0	1	2	5243	500.0	307.2
1	0	8	131.1	125.0	76.80
1	1	16	65.54	62.50	38.40
		E =	2.0971 MHz	2.0000 MHz	1.2288 MHz

The SPIF flag can cause an interrupt when it is set. The interrupt-enable bit is called *SPIE* for *serial peripheral interrupt enable*. The interrupt vector for the SPI device is at addresses FFD8 and FFD9.

If the SPI bus is operating at a high baud rate, using the flag may not be practical—especially with an interrupt. The transmission time may be short enough that the program may simply delay long enough to allow the transmission to complete. Using the interrupt may cause greater overhead rather than less.

SPI control

The *serial peripheral interface control register* called *SPCR* contains several bits that control the SPI hardware. The program enables the SPI by writing 1 to the *SPI enable bit* called *SPE* in the SPCR register. While enabled, the SPI controls the appropriate port D output pins overriding the PORTD register.

Also in the SPCR register are the *clock polarity bit* called *CPOL* and the *clock phase bit CPHA*. These bits control the timing of the serial clock. Four different timing patterns make the 68HC11 compatible with most commercially available I/O chips that use the SPI bus.

The electrical characteristics of the port D output pins are programmed by the DWOM bit in the SPCR register. The name *DWOM* means port D wired-OR mode selection bit. While DWOM is 0 as it is following a reset, the port D outputs are CMOS compatible. While DWOM is 1, the port D outputs are open-drain drivers. This electrical difference may be important when more than two devices are on a single SPI bus.

	7	6	5	4	3	2	1	0	
$1029	SPIF	WCOL	0	MODF	0	0	0	0	SPSR
RESET	0	0	0	0	0	0	0	0	

Figure 7-59 Serial peripheral interface status register.

SPI bus with multiple slaves

The SPI bus can communicate with several I/O chips or other devices with a single bus. Most applications use one of two fundamental configurations. The first configuration is a simplex system where the master device broadcasts data to multiple slave devices. This system is similar to the system described with the SCI interface.

The second configuration is a full-duplex system with multiple slave devices. In this case, only one slave device can send data to the master at a time; otherwise the signals would collide. While the slave is transmitting, all other slave devices must be disabled.

The slave-select pin labeled SS enables or disables the slave devices. Usually, digital output pins on the 68HC11 control the slave select pins of the slave devices. Then the program in the master 68HC11 controls the slave devices enabling them as needed.

Microprocessor Bus Expansion

The 68HC11 can be expanded by adding additional memory and I/O chips. The chip must be operated in expanded mode so that the microprocessor buses are available external to the chip. In the single-chip computer mode, the buses are not available externally, so the only expansion means is the SPI bus.

Mode selection

The operating mode of the 68HC11 chip is determined by high-asserted signals applied to the MODA and MODB pins during a hardware reset. Table 7-1 shows the modes selected. The operating mode is selected by hardware and not software, which prevents dangerous operation of the computer due to erroneous programs.

The single-chip computer mode and the expanded mode are the normal operating modes. Two other special modes are for testing and other purposes and are not normal operating modes. Only the normal modes are discussed here.

Figure 7-60 shows the *highest-priority register*, *HPRIO*; the bits RBOOT, SMOD, MDA, and IRV relate to the special modes. Consult a Motorola manual for further details.

Expanded mode

When the 68HC11 chip operates in the expanded mode, the pins of port B and port C become address and data bus pins. Figure 7-2 is a block diagram for the expanded mode. The port C pins carry both address and data signals at different times, which does not slow the computer's operation.

	7	6	5	4	3	2	1	0	
$103C	RBOOT	SMOD	MDA	IRV	PSEL3	PSEL2	PSEL1	PSEL0	HPRIO
RESET	—	—	—	—	0	1	0	1	

Figure 7-60 Highest-priority register.

To separate the address and data bus information, an external latch is necessary. In expanded mode, the STRA pin becomes a strobe for the latch so that it captures the address information. The STRA pin is then named *AS* for *address strobe*. Similarly, the STRB pin becomes the *read/write control line* and is named *R/W* for the expanded mode. Figure 7-61 shows a block diagram of the hardware necessary to complete the address and data buses in expanded mode.

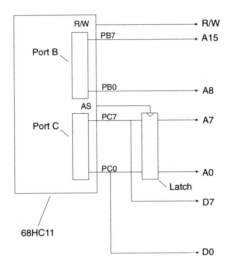

Figure 7-61   Expanded mode bus expansion using external latch.

When the 68HC11 is used with the external latch and operated in the expanded mode, the buses are compatible with the buses of other Motorola 8-bit microprocessors. Expansion with memory and I/O chips is straightforward. However, the use of port B and port C for I/O is lost. Table 7-14 shows all pin connections for the ports in each of the normal operating modes.

Port replacement unit

In expanded mode, port B and port C no longer function as I/O ports. To regain the same I/O functions that the 68HC11 has in single-chip computer mode, additional hardware is necessary.

The 68HC24 chip, known as the *Port Replacement Unit*, or *PRU*, replaces the lost ports. The PRU connects to the buses and fully implements the port B and port C input/output functions. When using the PRU chip, all single-chip functions, including the memory addresses, are the same as those of the single-chip mode.

Using the PRU during development of a system that will use single-chip mode in the final design is very helpful. During development, the internal signals can be seen on the external buses when expanded mode with a PRU is used. When all software development is complete, the same software operates the chip in single-chip mode.

TABLE 7-14   PORT SIGNAL SUMMARY FOR 68HC11A8

Pin Designation Port-Bit	Pin Function in Single-chip Mode	Pin Function in Expanded Mode
A-0	PA0/IC3	PA0/IC3
A-1	PA1/IC2	PA1/IC2
A-2	PA2/IC1	PA2/IC1
A-3	PA3/OC5/and-or OC1	PA3/OC5/and-or OC1
A-4	PA4/OC4/and-or OC1	PA4/OC4/and-or OC1
A-5	PA5/OC3/and-or OC1	PA5/OC3/and-or OC1
A-6	PA6/OC2/and-or OC1	PA6/OC2/and-or OC1
A-7	PA7/PAI/and-or OC1	PA7/PAI/and-or OC1
B-0	PB0	A8
B-1	PB1	A9
B-2	PB2	A10
B-3	PB3	A11
B-4	PB4	A12
B-5	PB5	A13
B-6	PB6	A14
B-7	PB7	A15
C-0	PC0	A0/D0
C-1	PC1	A1/D1
C-2	PC2	A2/D2
C-3	PC3	A3/D3
C-4	PC4	A4/D4
C-5	PC5	A5/D5
C-6	PC6	A6/D6
C-7	PC7	A7/D7
D-0	PD0/RxD	PD0/RxD
D-1	PD1/TxD	PD1/TxD
D-2	PD2/MISO	PD2/MISO
D-3	PD3/MOSI	PD3/MOSI
D-4	PD4/SCK	PD4/SCK
D-5	PD5/SS	PD5/SS
None	STRA	AS
None	STRB	R/W
E-0	PE0/AN0	PE0/AN0
E-1	PE1/AN1	PE1/AN1
E-2	PE2/AN2	PE2/AN2
E-3	PE3/AN3	PE3/AN3
E-4	PE4/AN4	PE4/AN4
E-5	PE5/AN5	PE5/AN5
E-6	PE6/AN6	PE6/AN6
E-7	PE7/AN7	PE7/AN7

The PRU also makes upgrading a product design practical. When changing a design that uses the 68HC11 in single-chip mode, the PRU makes adding memory or I/O devices easy. All the original design is retained when more hardware is added.

7.15   SPECIAL HARDWARE OPERATION

Several instructions and control register bits alter the normal behavior of the 68HC11. Most of these special operations involve changing the performance of the interrupt system and lowering power consumption. Some bits are for factory test functions.

Highest-Priority Interrupt

The priority of interrupts is of little concern in many applications. However, certain applications require very quick response to a particular device. Often, the device has critical timing requirements, and the delay caused by the processing of higher-priority interrupts is a problem. To overcome this difficulty, the 68HC11 can promote a single I-bit controlled interrupt to the highest priority. Promoting the priority of one interrupt does not affect the relative priorities of the remaining interrupts.

Four *priority-select bits* named *PSEL3*, *PSEL2*, *PSEL1*, and *PSEL0* choose an interrupt to promote to the highest-priority level. These priority-select bits are in the *highest-priority register* called *HPRIO*. Figure 7-60 shows the HPRIO register at address 103C. The PSEL bits choose an interrupt according to the values in Table 7-15.

The program can change the PSEL bits only while the I-bit is 1. That is, while the interrupt system is disabled, the program can select a new highest-priority interrupt. Usually, the program sets the HPRIO register during its initialization. However, an interrupt service routine also can modify the HPRIO register so that software can create a sophisticated priority interrupt system.

The reset condition of the HPRIO register puts 0101 in the PSEL bits. Therefore, normally the reserved highest-priority interrupt defaults to the IRQ interrupt. This selection is consistent with the address order of the interrupt vectors, which places IRQ at the highest address.

Wait Mode

Through program control, the microprocessor can enter a wait mode where program execution ceases. The processor exits the wait mode in response to an interrupt. The wait instruction WAI first puts the microprocessor status on the stack and then enters the wait state. During the wait, the clocks continue to operate, but the microprocessor does not execute instructions.

The wait mode serves two purposes. First, the microprocessor can respond very quickly to an interrupt. Because the microprocessor status is already on the stack at the time of an interrupt, the interrupt response is much quicker than for other interrupts. Of course,

TABLE 7-15 HIGHEST-PRIORITY INTERRUPT SELECT BITS

PSEL3	PSEL2	PSEL1	PSEL0	Highest Interrupt
0	0	0	0	Timer Overflow
0	0	0	1	Pulse-Accumulator Overflow
0	0	1	0	Pulse-Accumulator Input Edge
0	0	1	1	SPI Serial Transfer Complete
0	1	0	0	SCI Serial System
0	1	0	1	Reserved (Default to IRQ)
0	1	1	0	IRQ (External pin or Parallel I/O)
0	1	1	1	Real-Time Interrupt
1	0	0	0	Timer Input Capture 1
1	0	0	1	Timer Input Capture 2
1	0	1	0	Timer Input Capture 3
1	0	1	1	Timer Output Compare 1
1	1	0	0	Timer Output Compare 2
1	1	0	1	Timer Output Compare 3
1	1	1	0	Timer Output Compare 4
1	1	1	1	Timer IC4/OC5

the price for this response is that no instructions are executed while the microprocessor waits for the interrupt.

Second, while the microprocessor is not executing instructions, the power consumption of the chip is reduced. Power reduction may be very important for a device operated by a battery.

Stop Mode

The STOP instruction puts the microprocessor into the stop mode, which stops all clocks. The purpose of the stop mode is to reduce power consumption to the minimum possible. Recovery from the stop mode requires a hardware reset or an XIRQ or IRQ interrupt.

The response to an XIRQ interrupt request is unusual when the processor is in the STOP mode. An XIRQ interrupt forces recovery from the stop mode whatever the state of the X bit. However, if X is 0, a normal XIRQ interrupt sequence follows an XIRQ interrupt. If the X bit is 1, an XIRQ interrupt causes execution to continue with the instruction after the STOP instruction without requesting an XIRQ interrupt.

A STOP instruction is only effective if the *stop disable* or *S bit* in the condition code register is 0. Otherwise, STOP acts as a NOP instruction. Only the TAP instruction can change the S bit. Stopping the microprocessor may be dangerous in some applications—all bit values, including the I/O bits, are retained. The S bit locks out the STOP instruction to make the microcomputer less likely to execute a STOP instruction by accident.

Because the STOP instruction stops the clock oscillator, recovery from the stop mode requires time for the oscillator to stabilize. Normally, the processor will begin operation only after 4064_{10} clock cycles to allow the clock to reach normal operation. If an application uses a separate clock instead of the internal clock, this delay is unnecessary. Making the *delay bit*, *DLY*, in the OPTION register 0 disables the clock start-up delay. The DLY bit is a timed-write-once bit.

Factory Tests

The register named *TEST1* at address 103E is used only during factory testing of the 68HC11 chip. Several bits in other registers are also useful only for factory testing. These include the TCLR and RCKB bits in the BAUD register.

The TEST instruction is used only in factory testing when the 68HC11 chip is in a special mode. In normal operating modes, it is an illegal instruction and will cause an illegal instruction interrupt.

7.16   REVIEW

The 68HC11 chip is available in many variations. Most of the variations are quite similar in principle and vary mostly in details such as the amount of memory, number of input/output signals, and clock speed. The reset and interrupt systems are nearly the same as are the many input/output devices within the 68HC11 chip. The various I/O devices provide parallel and serial input/output; timing of inputs, outputs, and program execution; counting of external events; measurement of analog voltages; and detection of failures of hardware and software.

7.17   EXERCISES

7-1. How could a program bug accidentally change the operating mode of the 68HC11 from expanded mode to single-chip computer mode?

7-2. The block diagram for the 68HC11 in Figure 7-1 shows the MODA pin with a double-headed arrow. Look in a Motorola manual to find the meaning of the outward direction of this arrow.

7-3. What effect does a hardware reset have on the stack if the reset occurs while a program is running?

7-4. How much time following reset does a 68HC11 computer with an 8.0-MHz clock have to program the timed-write-once bits?

7-5. Why would you want to use the INIT register to move the internal RAM out of the direct-addressing range?

7-6. Following a hardware reset, are the IRQ and XIRQ pins level-sensitive or transition-sensitive?

7-7. Why does Figure 7-1 show the reset signal with a double-headed arrow? Does this mean that a reset signal can come out of the 68HC11?

7-8. In an application that uses the XIRQ interrupt, beyond enabling the interrupt, does the X bit serve any purpose?

7-9. Does a 1 in the I bit prevent an interrupt from the SWI instruction? Does executing the SWI instruction affect the I bit?

7-10. Refer to Figure 7-11. If the instruction on line 71 was written incorrectly so that it loaded the accumulator with zero, what is the effect on the port C bits? How frequently will the instruction on line 56 be executed?

7-11. Modify the program in Figure 7-11 so that the port C bits are toggled every 65.54_{10} milliseconds.

7-12. Complete the missing parts of Figure 7-62 by reading the comments on the listing, then:
    (q)   When light LT2 is flashing, the time the light is on is _____ milliseconds.
    (r)   After starting the program and getting to the main program loop, but before the first interrupt, LT1 is (on, off, unknown)_____ and LT2 is (on, off, unknown)_____.
    (s)   The lowest number in the stack pointer at any time while this program runs is _____.
    (t)   If light LT2 were moved from bit 5 to bit 6 of PORTB, the single line of the program that must be changed is _____.
    (u)   To make the period of the flashing three times as long as it is, the single line of the program that must be changed is _____.
    (v)   What does this program do if an invalid interrupt occurs?

7-13. Using the RTI device as a time base, write an interrupt program to pulse-width modulate four output bits. That is, change the output between low and high to make an average level that is a percentage of the high level—this percentage is called a *duty cycle*. Use PORTC for the output bits. The duty cycle for bit 0 is 20%, for bit 1 is 40%, for bit 2 is 60%, and for bit 3 is 80%. The total cycle should be 10_{10} periods of the RTI.

```
1                        ** INTERRUPT DRIVEN LIGHT FLASHER PROGRAM
2                        * NOTE: THIS PROGRAM IS NOT WELL DOCUMENTED
3                        *
4                        * THE FOLLOWING ARE ALL CORRECT COMMENTS
5                        *
6                        * ON EVERY FIFTH INTERRUPT, COMPLEMENT LIGHT LT2 IF
7                        * ..SW1 IS PUSHED
8                        *
9                        * LIGHT LT1 IS SET EQUAL TO SW2 ON EACH INTERRUPT
10                       *
11                       * TIME IS BASED ON REAL TIME INTERRUPT DEVICE AND
12                       * ..68HC11 CRYSTAL CLOCK RATE IS 8.0 MHZ
13                       *
14                       * SWITCH SW1 IS WIRED TO BIT 7 AND SW2 TO BIT 6 OF
15                       * ..PORTC SO THAT A 1 RESULTS WHEN PUSHED
16                       *
17                       * LIGHT LT1 IS WIRED TO BIT 0 OF PORTC, LIGHT LT2 TO
18                       * ..BIT 5 OF PORTB, LIGHTS ILLUMINATE WITH A 1
19                       *
20                       * OTHER DEVICES ARE WIRED TO THE OTHER PINS OF BOTH
21                       * ..PORTB AND PORTC
22                       *
23      1004             PORTB   EQU     $1004
24      XXXX             PORTC   EQU     _____  (a)
25      1007             DDRC    EQU     $1007
26      1024             TMSK2   EQU     $1024
```

Figure 7-62   Exercise 7-12.

27		1025		TFLG2	EQU	$1025	
28		1026		PACTL	EQU	$1026	
29		XXXX		IOPAT	EQU	_____	(b)
30		XXXX		MASK1	EQU	_____	(c)
31		0040		MASK2	EQU	$40	
32		0020		REV	EQU	$20	
33		XXXX		ON	EQU	_____	(d)
34		XXXX		OFF	EQU	_____	(e)
35				*			
36	0020				ORG	$20	
37	0020			COUNT	RMB	1	
38	0021	XXXX		INIT	FCB	_____	(f)
39				*			
40	XXXX				ORG	_____	(g)
41	XXXX	D121			FDB	RTIISR	
42	FFFE				ORG	$FFFE	
43	FFFE	_____	(h)		FDB	START	
44				* PROGRAM START			
45	D100				ORG	$D100	
46				* INITIALIZE STACK			
47	D100	_____	(i)	START	LDS	#$D2FF	
48				* INITIALIZE I/O HARDWARE			
49	D103	86 XX			LDAA	#IOPAT	
50	D105	B7 10 07			STAA	DDRC	
51	D108	86 33			LDAA	#$33	
52	D10A	B7 10 03			STAA	PORTC	
53	D10D	7F 10 04			CLR	PORTB	
54	D110	86 03			LDAA	#$3	
55	D112	B7 10 26			STAA	PACTL	
56	D115	86 XX			LDAA	_____	(j)
57	D117	B7 10 24			STAA	TMSK2	
58				* INITIALIZE PROGRAM			
59	D11A	96 21			LDAA	INIT	
60	D11C	97 20			STAA	COUNT	
61				* TURN ON INTERRUPT SYSTEM			
62	D11E	0E			CLI		
63	D11F	20 FE		HERE	BRA	HERE	
64				*			
65				* REAL TIME INTERRUPT SERVICE ROUTINE			
66	D121	B6 10 25		RTIISR	LDAA	TFLG2	
67	D124	84 40			ANDA	#$40	
68	D126	27 31			BEQ	RTISR	
69	D128	B7 10 25			STAA	_____	(k)
70	D12B	B6 10 03			LDAA	PORTC	
71	D12E	84 40			ANDA	#MASK2	
72	D130	26 07			BNE	NEXT	
73	D132	B6 10 03			LDAA	PORTC	
74	D135	84 FE			ANDA	#_____	(l)
75	D137	20 05			BRA	AHEAD	
76	D139	86 01		NEXT	LDAA	#_____	(m)
77	D13B	BA 10 03			ORAA	PORTC	
78	D13E	B7 10 03		AHEAD	STAA	PORTC	
79	D141	7A 00 20			DEC	COUNT	
80	D144	26 13			BNE	RTISR	
81	D146	XXXX			LDAA	_____	(n)
82	D149	85 80			BITA	#MASK1	
83	D14B	27 08			BEQ	THERE	
84	D14D	B6 10 04			LDAA	PORTB	
85	D150	88 XX			EORA	#_____	(o)
86	D152	B7 10 04			STAA	PORTB	
87	D155	96 21		THERE	LDAA	INIT	
88	D157	97 20			STAA	_____	(p)
89	D159	3B		RTISR	RTI		
90	D15A				END		

Figure 7-62 Continued.

7-14. Modify the program in Figure 7-20 to indicate OK if the length of the pulse from the input signal is less than 10_{10} milliseconds. The length here is defined as the elapsed time from the low-to-high transition until the high-to-low transition. To measure this time only, the capture edge will be changed after each capture.

7-15. Modify the program in Figure 7-23 to pulse-width modulate port A bit 5. The duty cycle will vary from 1 to 99_{10} percent in increments of 1 percent based on an 8-bit number in the range 1 to 99_{10} stored in location DUTYCYC. The full cycle of the output must always be 10_{10} milliseconds. Why does the program in Figure 7-23 not work correctly if the low or high times are nearly zero?

7-16. If you are an electrical engineer, derive an expression for the RMS value of the waveform generated by Exercise 7-15 in terms of the duty cycle.

7-17. Does generating a pulse-width modulated waveform effectively perform a digital-to-analog conversion?

7-18. Which pins can output compare OC1 control? Can the PORTA register control an output pin at the same time that OC1 is also controlling it?

7-19. Modify the program in Figure 7-23 to pulse-width modulate port A bit 7 using OC1 instead of OC2.

7-20. Write a program to read an analog voltage once a second using the output compare OC2 interrupt as a time base. Output the binary number for the lowest voltage read on port B and the number for the highest voltage read on port C. When a push button switch sets the IC1 flag, reset the lowest value to FF and the highest value to 00.

7-21. Modify the program in Figure 7-8 to use the pulse-accumulator input flag PAIF instead of the STAF flag. The push button switch will be connected to pin PA7.

7-22. Write the 8-bit ASCII codes for the letter D and the digit 8 using even, odd, mark, and space parity.

7-23. If a serial cable is much too long for your application, is coiling the excess cable around a cable spool to prevent the cable from getting tangled a good idea?

7-24. Change the SCI serial message program in Figure 7-50 so that pressing a key can only start a new message after the current message is finished. After the completion of the message, the next key pressed determines the character inserted in the next message.

7-25. Write a program for a 68HC11 to read an analog voltage and send the 8-bit number to another computer using the SCI at 9600 baud. The program must transmit approximately once a second using the real-time interrupt device as the time base. Then write a program for a second 68HC11 to receive the number using its SCI, and output the number on port B.

Chapter 8

Real-time Operating Systems

The main focus of this chapter is the quite complex software required by most embedded computers. When a microcomputer is part of a larger product, it is said to be an *embedded* computer. The 68HC11 was developed specifically for embedded applications. Such applications involve the monitoring and control of other electronics and mechanical devices. The embedded computer acquires data from the product's environment through sensors, and may control that environment with actuators. The I/O devices built into the 68HC11 chip allow it to meet the needs of a wide range of product and control applications.

The software is complex because, first, it will be interrupt-driven—the embedded computer must respond to signals generated by the sensors in a timely manner. Such a system is said to be a *real-time system*. Second, the software must handle a variety of different jobs simultaneously. Sometimes these jobs may be related, or they may be quite independent of each other. For example, an automobile engine controller will handle the firing of spark plugs, the metering of fuel, monitoring of engine temperature, monitoring of oil pressure, engine run time, and a host of other functions all at the same time.

Partitioning the software into functional sections that relate to the individual jobs or control situations is desirable. Each program section is called a *task*, and the overall system is called a *multitasking system*.

Design and development of real-time multitasking software requires a sizable programming effort. To organize and simplify this effort, the software is often written in two distinct parts. First, the software to handle interrupts, I/O devices, and the scheduling and control of the tasks is written. This software is called an *operating system*. It provides a base upon which the remainder of the software can be built. The second part of the effort is the development of the programs that carry out the required functions of the computer. Once again, these programs are called *tasks*. Developing the tasks is much easier when the operating system is in place. The operating system provides services to control and coordinate the system hardware and I/O devices.

This chapter discusses the basic principles of operating systems as used in real-time environments. A basic real-time multitasking operating system is developed in the next chapter. Then, several tasks are written to illustrate its use in a product.

In developing small computer applications, including many embedded applications, writing a custom operating system is common and practical. In larger applications, the operating system may be purchased from a company that specializes in developing operating system software.

8.1 SYSTEM CHARACTERISTICS

Most microcomputers used in embedded applications use real-time multitasking operating systems. The characteristics and requirements of such systems are important. Larger stand-alone computer systems share many of the same characteristics, but the discussion here is focused on small embedded applications.

Such an embedded system generally will not have a user interface for a human operator to communicate with the operating system. The emphasis of the operating system is on the device under its control. Furthermore, only ROM and RAM memory are available for programs and data; no consideration is given to mass storage devices because they are seldom used in small embedded applications.

Real-time Systems

The definition of a real-time computer system is not precise. Usually, a real-time computer system is part of a control system, and it must generate responses to input signals fast enough to make the control system work properly. However, the required time from input event to output signal may vary dramatically from one application to another and still meet the definition.

When a computer system uses sensors to acquire data directly from the place where the data is generated, it is called an *online system*. The real-time systems considered here are online systems.

Examples

Microwave ovens and modern electric kitchen ranges are examples of devices with embedded real-time computers. Electric ranges use digital displays and touch-sensitive push buttons for control. The computer controls the cook time, delayed start of the oven, change of oven temperature during the cooking cycle, temperature indicators, and a host of other devices. Convection ovens that have a fan to circulate the hot air tend to cook food more quickly than conventional ovens, so the computer automatically adjusts the cooking time so that recipes can be followed without error.

Virtually all modern automobiles use real-time computer systems for engine and transmission control and monitoring. The computer makes achieving low exhaust emissions and good fuel economy possible. An engine controller certainly requires quick response to fire spark plugs and fuel injectors properly as an engine is running at high speeds. The rate at which clutches in the transmission operate determines the shift feel for passengers. These are critical time-sensitive operations that can cause damage and safety problems if not carried out properly.

Many modern applications for embedded computers require response times of a few milliseconds or less. However, even less time-critical applications, such as automated-teller machines, must still provide proper response to make them practical.

Interrupt consequences

Although many computer systems sample input signals, most time-critical inputs to a real-time system cause interrupts. The use of interrupts means the programmer must give careful consideration to the reentrancy problem and the concurrency problem. The programmer must consider both of these problems throughout the design of the software in order to design a reliable system.

In addition, software that disables the interrupt system for any significant amount of time is unacceptable. If the interrupt system is disabled, the time to respond to an interrupt request is increased and real-time response may sometimes be inhibited.

What is not a real-time system

Many computer systems are capable of tracking the time-of-day. The time-of-day can be used to log when disk files are updated, letters are written, and so on. Some computer systems can cause certain actions to occur at preprogrammed times. For example, a backup of a hard disk to magnetic tape may occur at a scheduled time each night.

However, just because a system uses time does not mean that it is a real-time system. Real-time systems meet critical response time needs, which are another thing altogether. On the other hand, many real-time systems can track time-of-day and schedule events to occur at specified times.

Multitasking Systems

Computer systems that make the computer appear to be running several tasks or programs simultaneously are said to be *multitasking systems*. When the computer hardware has only a single microprocessor to execute instructions, the computer must switch control between multiple tasks in a way to make it appear that all tasks are running simultaneously. Because a computer with a single microprocessor can execute only one program at a time, it is common to speak of the multiple tasks as running *concurrently*.

A useful question is: How can a computer run multiple tasks? Won't just one task keep the computer busy? The answer is that tasks, especially in a control environment, do not run all the time. In fact, some only need to run when a particular event occurs outside the computer. Others need to run periodically at regular intervals or certain times of the day. Furthermore, even as a program is running, it may stop executing while it waits for an I/O operation to complete. Many I/O devices are relatively slow compared to the computer, so I/O operations give the computer free time to run other tasks.

Most tasks, especially in a control environment, are effectively infinite loops. A task will run to some point at which control returns to the operating system. Later, the operating system continues the task from that point. The task probably follows a loop back to the stopping point—then control returns to the operating system once again.

8.2   OPERATING SYSTEM STRUCTURE

Operating system software is built from a variety of parts including *system tables*, the *dispatcher*, and interrupt service routines. The tasks that run under control of the operating system and use its services tend to be the most visible parts of a working computer system. The tasks carry out the functions of the application.

System tables

The operating system maintains entries in several *system tables* that describe the state or condition of the tasks and the I/O devices. In particular, the *task-control-blocks*, or *TCBs*, hold information necessary to run the tasks. Similarly, *device-control-blocks*, or *DCBs*, hold the information necessary for the operating system to use the input/output devices. A device may have an additional table or buffer to hold I/O information as the I/O device operates. *Service-control-blocks*, or *SCBs*, specify the parameters for a request for the operating system to perform some function. A complete system has one TCB for each task, one DCB for each I/O device, and many SCBs that correspond to the requests the tasks make of the operating system.

Dispatcher

In a running multitasking system, control of the computer will pass from task to task based on program logic and I/O events. The *dispatcher* is a program module that determines which task will run next after something triggers the system to switch control to a new task. Different operating systems use different scheduling algorithms for determining when to switch control from one task to another.

Interrupt service routines

Each I/O device that uses interrupt has an associated interrupt service routine, or ISR, that services that particular device. The ISR knows the intimate details of the I/O device's operation. It controls the device to provide data transfer between the outside world and the inside of the operating system.

An ISR will likely run with the interrupt system disabled, so making an effort to write the ISR program code so that it executes quickly is practical. Reduced overhead in the ISR will improve the real-time response of the computer.

System services

The operating system provides *services* to tasks. For example, a task may request the operating system to schedule execution of a task after a specified delay time. The operating system then must activate the specified task at the proper time. One use of delay scheduling is to make tasks run periodically—each time a task runs, it requests to be scheduled to run again after a delay. As another example, a task may request an operating system service to output a string of characters on a serial port.

The system services are implemented as interrupt service routines. The tasks will request service by causing a software interrupt with a SWI instruction. A system service ISR will then carry out the requested service.

Using an instruction to cause an interrupt is a new idea in this book. Previously, only physical devices caused interrupts. Software interrupts are essential to operating system software.

Software strategy

The interrupt examples you saw in Chapters 6 and 7 consist of a very simple main program loop and some interrupt services routines. Most of the action in the program took place in the interrupt service routines. Indeed many practical programs are structured this very way. However, the structure of operating system software is quite different.

When you use the operating system software, you will find most of the action occurring in the main program (the tasks) with the interrupt service routines providing services to it. The tasks run with the interrupt system enabled, so interrupts will steal time from them. This is another reason it is desirable for the ISRs to run as quickly as possible.

The interrupt service routines will not, generally, immediately return program control to the point of the interrupt. Instead, the operating system decides where to send control after the interrupt service is completed.

8.3   TASKS AND THEIR CONTROL

A task is a program designed to run under control of an operating system and to use its services. In the embedded microcomputer environment, the task will effectively be a loop that runs repeatedly each time it is required.

Tasks are also known by other names. Some people speak of *processes* because the program processes information. Others speak of *application programs* because the computer is being applied to do useful work by this software. Some operating systems can further divide large tasks into smaller *threads* of execution.

Tasks used with a multitasking operating system have characteristics similar to other programs. In particular, a task must have a stack for data storage and subroutine return addresses. Each task is a separate program; therefore, each task must be assigned a stack for its use. The operating system must provide the multiple stacks required by the multiple tasks.

Task State

At various times, a task can be in one of three *states* for purposes of this discussion. Some systems have several more possible states due to enhanced features. In this case, the states will be called the *running state*, the *suspended state*, and the *ready state*. These states are defined as follows:

- *Running state.* The task that is currently running, that is, executing instructions, is in the running state. With only a single microprocessor, only one task at a time can be in the running state. While in the running state, this program has control of the computer hardware and will retain control until it gives up control or control is taken away from it through an interrupt.

- *Suspended state.* A task that doesn't want to run or is unable to run because it is waiting for something is in the suspended state. The task could be waiting for an event to occur outside the computer, for time to pass, or for a resource to become available. Such a task is described as being *blocked*. At any particular time, several tasks may be in the suspended state.

- *Ready state.* A task in the ready state wants to run, has all the resources needed to run, but is not running because the operating system has chosen another task to be the running task. At any particular time, several tasks may be in the ready state.

The state information for a task is held in the task-control-block for that task. The operating system will generate and use the information within the TCB. Generally, the task has no access to the TCB.

Task Scheduling

The operating system decides which task is to be the running task. Furthermore, it determines when a decision about finding a new task to run should be made. The algorithms the operating system uses to make these decisions profoundly affects the character of the final running system. Let's consider some possible approaches.

Multitasking schemes

A multitasking operating system runs multiple tasks concurrently. That is, it switches control from one task to another so that all tasks appear to be running alone on the computer. Consider how a task gives up control of the computer to allow other tasks to run. How this transfer of control happens is the first part of scheduling when the next task will run.

Cooperative multitasking. The running task can choose to stop running by making a service request to the operating system. If a task only gives up control of the computer by making such a service request (and hardware interrupts return control immediately to the task, not the operating system), the operating system is said to use *cooperative multitasking*. This approach is very common in practical computer systems. Cooperative multitasking depends on the running task relinquishing control so that other tasks can run.

If a task chooses not to relinquish control, other tasks could be permanently blocked from running—an unacceptable situation for many applications, especially those with safety aspects. However, even if the task gives up control at some point in its execution, it may take a while before it does and hence it may delay the running of another important task.

Cooperative multitasking, in some ways, allows the tasks to control the operating system instead of the operating system controlling the tasks.

Preemptive multitasking.   If the operating system takes control away from a task without a request from the task, the operating system is said to use *preemptive multitasking*. Certainly this preemption uses the interrupt system. A preemptive operating system can take control away from a task that "hogs" the system. Preemptive multitasking is generally desirable in real-time control systems because one task cannot prevent other tasks from running regardless of the reason.

The complexity of a preemptive operating system and the tasks that run under its control is generally greater than that of a cooperative system. However, real-time systems need preemptive multitasking so that real-time response can be attained. Only preemptive multitasking systems will be further considered here.

Scheduling algorithms

At the appropriate time, the dispatcher uses a scheduling algorithm to choose a task to be the next running task. After the dispatcher finishes, the operating system returns control to the selected task at the point control was last taken away from it, and the task continues to run.

Here are some scheduling algorithms that are in common use:

Round-robin scheduling.   The simplest scheduling algorithm is to rotate control between all tasks. That is, the dispatcher checks one task after another looking for a ready task to run. When all tasks have been checked (and possibly run), it starts over with the first task. This algorithm is known as *round-robin scheduling*—the name apparently was taken from a children's game.

The principal characteristic of round-robin scheduling is that all tasks are treated equally. No one task is considered any more important than another. Consequently, there is no control over the response time to an external event associated with a particular task. Only luck determines which tasks will run before the task associated with a particular event runs. A worst-case analysis can predict the longest delay though.

If the dispatcher does not find a ready task, it will continue going "round the robin" checking for a ready task. Eventually, an interrupt will occur that will make some task ready and the dispatcher will find it.

Priority scheduling.   A priority scheduling system assigns a priority to each task. Each time the dispatcher must choose a task, which is generally at every interrupt, it checks for ready tasks beginning at the task with the highest priority. It continues down the list of tasks in priority order. The first task it finds that's in the ready state is made the running task. This approach is called a *strict priority scheduling algorithm.*

Priority scheduling provides some control over the response time to important events in a preemptive multitasking system. An interrupt that makes the highest-priority task ready will cause it to run with the minimum delay possible. A lower-priority task may be delayed by higher-priority tasks that need to run, which is the purpose of priority. The more important—that is, response-sensitive—tasks run first. Therefore, priority scheduling is the normal algorithm used for real-time multitasking systems.

The background task.   An interesting dilemma occurs with priority scheduling if no tasks are in the ready state. Effectively, this means the computer has nothing to do! The dilemma is usually solved by creating one additional task, called the *background task*, that is always in the ready state. The background task may be a do-nothing infinite loop. Due to the background task, the dispatcher will always find some task in the ready state, and will always specify a task to run next.

In real-time systems, the background task often provides system information for maintaining and troubleshooting the system. Because these chores are not part of normal application tasks, it's reasonable to say the background task is doing nothing.

A priority-scheduled system with a background task will be running some task all the time, except when it is servicing interrupts. Sometimes deciding which part of a computer system is the operating system and which parts are the tasks is difficult. A simple approach is to say that anything that runs with interrupts enabled is part of the tasks or application programs, and any part that runs with interrupts disabled is the operating system.

Combination scheduling.   Some systems use a combination of scheduling algorithms. Each priority level of a priority system often has multiple tasks that rotate in round-robin fashion. Another approach is to rotate priorities so that tasks' priorities change over time. Such special scheduling algorithms may be required by certain applications. For example, rotating the priorities can guarantee that all tasks have an opportunity to run sooner or later.

Task Switching

A preemptive multitasking operating system must be able to take control of the computer away from the running task and give it to another. The process of making this change is called *task switching.*

Hardware context

While a task is running, it controls the computer's resources. Because tasks run with interrupts enabled, an interrupt can take control away from the task. When an interrupt occurs, the interrupt system first saves the microprocessor's status in the stack and then sends control to an ISR by placing the interrupt vector in the program counter. The microprocessor's status includes the information in the microprocessor's registers, excluding the stack pointer.

In the 68HC11, the interrupt system saves nine bytes in the stack while using the stack pointer to keep track of position within the stack.

Each task, however, must have its own stack. Therefore, to save all information about a task at an interrupt, the stack pointer must be saved also. Although the interrupt system hardware does most of the work, the stack pointer must be saved by program code in the 68HC11.

The complete hardware status of a task, that is, the content of the microprocessor's registers including the stack pointer, is called the *context* of the task. In order for the operating system to switch control from one task to another, it must save the running task's context, and then restore the context of the new task. This process is sometimes called *context switching* to emphasize that more is involved in the process of switching from one task to another than handling an interrupt.

Software state

Each task has a software state that the operating system uses to properly control that task. When the operating system switches from one task to another, it must change the software state of the task in addition to doing a hardware context switch. Figure 8-1 illustrates the basic states of a single task.

Running state.   Let's start at the running state, which means we are discussing the task that is currently running. A task will continue running until something happens to stop it. Usually the task will request the operating system to suspend it with or without scheduling it to run again sometime in the future. The task may also request the operating system to do an input/output operation, and to be suspended until the I/O operation is completed. In both of these cases, the operating system will change the task's state to suspended.

A running task may also be preempted (interrupted) by the operating system due to a variety of things. If it is preempted, the task is still able to run, so the operating system changes its state from running to ready. If this task is the highest-priority task, control will

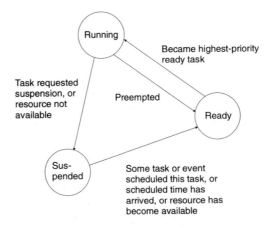

Figure 8-1   Task state diagram.

probably return to it almost immediately. If it has a lower priority, some other tasks may run before it gets started up again.

Suspended state.   A *suspended* task does not currently want to run or something is preventing it from running. Because tasks commonly run periodically, many tasks spend much of their time waiting for the next time they are to run. Generally, the run time of the task is far less than the period on which they are scheduled, so they are suspended most of the time. Similarly, a task waiting for an I/O operation to complete will be in the suspended state while it waits. Finally, a task may be waiting for an event outside the computer to occur. When that event does occur, it will cause an interrupt, which will result in the task being moved from the suspended state to the ready state.

Ready state.   A task in the *ready state* wants to run. The only reason a ready task is not running is that a higher-priority task also wants to run or is running. The higher-priority task gets use of the computer before the lower-priority task.

8.4 OPERATING SYSTEM SERVICES

An operating system may provide various services to the tasks. Generally, the services available will be those that most tasks need. Tasks request service from the operating system by executing a software interrupt. The use of an interrupt is not only simple and easy, it also makes the operating system easier to build. The operating system is interrupt-driven and thus is structured for interrupts.

System services enable tasks to use system resources such as a serial communications port. They also enable tasks to control their execution and the execution of other tasks.

Here is a list of some services that are commonly needed by embedded computer systems:

- *Schedule a task.*   A task may choose to ask the operating system to schedule it to run again at a later time. The time may be specified as a delay from now, or at a specified time-of-day. The task specified may be the requesting task or another task. The ability of a task to control the execution of both itself and other tasks is very powerful. It makes programming control systems easy because the system can be broken into logical pieces that are executed separately.

- *Suspend a task.*   A task may choose to stop running for various reasons. It may have finished its job for now, an error may have occurred, or it may want to wait until some external event starts it once again.

  A task may also request that a different task be suspended. Of course, the other task is not currently running because the running task is making the request. So the other task is probably in the ready state waiting to run, and the running task chooses to prevent it from running.

- *Start a task.*   One task may want to start another task because the other task carries out a needed function not in the running task. For example, the other task may be started by several different tasks that effectively share its abilities. For simplicity, this service may be omitted, and the schedule service used instead—the task could be scheduled immediately.

- *Transmit and receive through a serial port.*   Serial communication to other devices and computers is common and practical. The SCI device in the 68HC11 is typical. Serial devices are quite complicated because their nature is to transmit multiple characters, and to cause interrupts after every character. Furthermore, all characters sent by a task must be kept together making a complete message. Mixing characters from several tasks into one message is quite useless! The operating system service can take care of most of the details. The task simply makes a service request, and the operating system does the rest.

- *Read or set the system clock.*   In order to do time scheduling of tasks, the operating system must maintain a clock. The system clock typically is maintained by an interrupt service routine for an accurate periodic interrupt. System services to read the clock provide tasks with the ability to make things happen at desired times and to know when things happened. For example, if an external event was detected, a task could record the time at which the event took place. The set the clock service is seldom used as the system runs normally and may cause unexpected problems.

- *Communicate with another task.*   In embedded systems one task often sends data numbers to another task. This transfer is quite difficult to accomplish because of the concurrency problem. For example, one task may want to send four bytes to another task by storing them in a table in memory. Suppose that after it stores three bytes, an interrupt occurs that causes a higher-priority receiving task to run and read the four bytes. Of course, the fourth byte it receives is incorrect. By a variety of means, the operating system can provide the communication and prevent this concurrency problem.

Operating systems can provide many other services. The need for additional services often becomes apparent as an embedded system is developed. Adding more services to the operating system is usually easy to do.

8.5   REVIEW

This chapter introduces the fundamental principles and concepts of operating systems, and in particular, real-time operating systems. The principal problems encountered when developing an operating system include the reentrancy problem and the concurrency problem. All the action within a real-time system is driven by interrupts, and interrupts lead to these problems.

8.6 EXERCISES

8-1. Which of these embedded computer applications require a real-time system: windmill controller, airplane autopilot, automobile cruise control, laser printer, cellular telephone.

8-2. Could round-robin scheduling ever be practical for a real-time system?

8-3. How does the priority of a task relate to the hardware priority of the interrupt system?

8-4. Could the background task stop the computer with a WAI instruction instead of running an infinite loop?

8-5. If a task calls a subroutine, does that subroutine need to be reentrant?

8-6. Why does each task need a stack? Why not let the tasks share the same stack?

8-7. Is the information in the state entry of the TCB for the running task any different from the state entry for a ready task?

8-8. Do tasks make changes in the TCB entries?

8-9. Is a service control block or SCB part of a task or part of the operating system?

8-10. Section 8.4 discussed how the concurrency problem applied to a low-priority task sending four bytes to a higher-priority task because the higher-priority task could preempt the lower one before it finished updating the bytes. Would changing the priorities of the two tasks so that the lower one is now higher and the higher one is now lower eliminate any concurrency problems?

8-11. Why is making an embedded control system with a multitasking operating system easier than just writing a single in-line control program?

Chapter 9

Real-time System Design

Designing a real-time computer to embed within another system is often completed in two distinct phases. First, a basic yet sufficiently capable real-time operating system is developed. Second, appropriate tasks that carry out the required functions of the system are attached to the operating system. Both these efforts are preceded by carefully examining the requirements and specifications of the intended system.

This chapter examines all these aspects through the study of a complete working design of a weather station. The specifications for the weather station are first stated, then the operating system is developed. The application of the computer influences the design of the operating system. Finally, the tasks necessary to implement the weather station features are developed. Only a small amount of creativity will be necessary for you to make a working system from the code presented.

One goal of this chapter is to provide an implementation of a fairly sophisticated real-time system. Wherever possible, the software is designed to be easily expanded. Sometimes this leads to less efficient code, but the greater goal is to make modifying the software and adding new features easy. The software design follows the principles of structured programming.

You should realize the importance of flexibility in software design, and plan for changes to occur that you cannot anticipate during the initial design. Clever approaches leading to maximally efficient code often lead to inflexible and difficult-to-change systems.

A second goal of this chapter is to include a variety of I/O devices and a variety of uses for system tasks. The approaches taken are often selected simply to illustrate a particular technique in both hardware and software.

Finally, this chapter discusses assembly language code in detail. The listings are fully documented, so flow charts are not provided, which significantly reduces the number of pages needed to discuss the material—a necessary condition for this book.

9.1   WEATHER STATION DESIGN

The primary functions of the weather station developed here are to display wind speed and wind direction on both digital displays and on a CRT display. The specifications state additional functions. The digital display units, plus indicator lights and push button switches, are mounted near the 68HC11-based computer board in a decorative enclosure. The CRT display may be placed remotely from the computer.

Sensors in the form of a digital anemometer and an analog wind vane are interfaced to the computer. Note that the wind vane cannot correctly indicate wind direction as the wind speed approaches zero.

Weather Station Specifications

The following lists the required operations to be performed by the weather station:

- The station must display the current wind speed in miles-per-hour at a suitable location on the CRT screen with a message to identify the number as wind speed. The one- or two-digit numerical part of the message must be updated at least every 150 milliseconds. The message must remain fixed at a suitable position on the screen.

- The station must display the highest wind speed measured since the last reset of the highest wind speed at a suitable location on the CRT screen. A message must identify the number as the highest wind speed. The one- or two-digit numerical part of the wind speed must be updated at least every 150 milliseconds. The message must remain fixed at a suitable position on the screen.

- The station must display the wind direction as one of eight one- or two-character directions at a suitable location on the CRT screen. The display will have a message to identify the characters as wind direction. The direction part of the message must be updated at least every 110 milliseconds if the direction changes. The message must remain fixed at a suitable position on the screen.

- The computer must control the CRT display using a serial communication rate of 9600 baud.

- The current wind speed or the highest wind speed in miles-per-hour must be displayed on a two-digit 7-segment display unit and updated at least every 100_{10}

milliseconds. A push button switch will be pushed if the highest speed is desired, and released for the current speed.

- If the number displayed on the 7-segment display units is less than 10, the left display unit will be blanked.

- If the highest speed is being displayed on the 7-segment displays, a computer-controlled light will illuminate to identify it as the highest speed.

- A display unit consisting of eight lights arranged in a circle will indicate wind direction. A "no-wind" light near the wind-direction lights will be illuminated to indicate that the wind direction cannot be displayed. The direction lights must be updated at least every 110 milliseconds.

- When the wind speed is zero, the entire wind direction message on the CRT screen must be blanked. Likewise, the eight wind-direction lights must be blanked, and the "no-wind" light must be illuminated.

- A push button switch for resetting the highest wind speed must be sampled at least every 150 milliseconds, and its purpose carried out immediately upon detection.

- A header message must be displayed on the CRT display at system startup. This message must identify the CRT display as a weather station.

- A light must be flashed by complementing its state every 30 milliseconds. This light will be used to indicate an operational system and for troubleshooting.

- A system free-time output bit must be set low when the system is processing useful information, and set high when the system has free time. This output will be used for system performance monitoring.

- The system must determine whether it becomes overloaded and cannot perform the required functions within the specified time frames. An overload is a fatal system error.

- If the system detects fatal system errors, the system must be halted and suitable information maintained to facilitate determination of the malfunction's cause.

Weather Station Hardware

Several devices are assumed to be available for the construction of this system. Figure 9-1 shows all the custom hardware needed. The hardware includes the following:

- *Microcomputer.*   A prototype computer can be a trainer unit such as the Motorola 68HC11EVB evaluation board. Because this is a common computer board, the software will be written to accommodate the characteristics of this board.

- *Anemometer.*   The anemometer is a device that rotates as the wind blows over it. Its rate of rotation is determined by the wind speed. A switch inside the anemometer opens and closes once on each revolution. Therefore, the rate at which the switch opens and closes is related to the wind speed.

The relationship between the period of the rotation and the wind speed in miles-per-hour is available as a calibration table. So, if the period of the rotation can be measured, the wind speed is known. The rate of rotation of the anemometer is quite low for low wind speeds; at the lowest speeds, the period of rotation will be on the order of 2.5 seconds.

The anemometer switch contact bounce can be eliminated by using a Schmitt trigger with capacitor filtering. The output of the Schmitt trigger controls the STRA pin on the 68HC11. The STAF flag will then be set on each revolution of the anemometer.

- *Wind vane.*   The wind vane is a device that points in the direction from which the wind is blowing. The device available contains a potentiometer. When a voltage source of 5 VDC is applied across the potentiometer, the potentiometer provides an output voltage that varies between 0 and 5 VDC. When the vane points slightly to the east of north, the voltage is nearly zero, and it increases as the direction is changed toward the east. When the vane is rotated to a position just west of north, the voltage is nearly 5 VDC. So, within the accuracy of the potentiometer, the voltage jumps from 5 VDC to zero as the direction passes north. If the output voltage of the potentiometer is measured, the wind direction can be determined.

  When the wind speed approaches zero, the wind vane cannot correctly indicate wind direction. The wind speed must be at least one mile-per-hour before the direction is meaningful.

- *Wind direction lights.*   A display device made of eight light-emitting diodes (LEDs) arranged in a circle is available to indicate wind direction. The lights are labeled *N, NE, E, SE*, and so on to indicate eight directions. Each LED has a transistor driver controlled by PORTB of the 68HC11. The lights are arranged so that the north light is controlled by PB0, the northeast light by PB1, and so on.

- *Digital display units.*   Two 7-segment MAN6760 display units driven by 74LS47 chips provide wind speed displays. To conserve output pins on the 68HC11, both displays are controlled by the PORTC bits PC0 through PC3. The least significant display has a 74LS75 latch chip to hold the bit pattern for the display unit. The PORTC bit PC4 controls the latch signal to the 74LS75.

  With this hardware arrangement, the computer must update the least-significant digit and latch its information, and then update the most-significant digit. This operation causes the most-significant display to display incorrect information momentarily, so the update must be fast enough that a person cannot see the change.

- *Push button switches.*   A push button switch is connected to PORTC bit PC7 so that the pin is at the high level when the button is pushed and at the low level when it is released. This switch is the highest wind speed reset button.

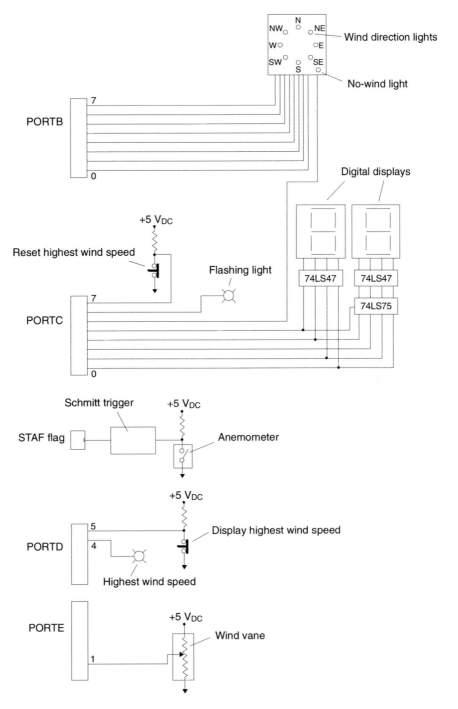

Figure 9-1   Weather station hardware.

Another similar push button switch is connected to PORTD bit PD5. When this button is pushed, the digital displays must show the highest wind speed.

- *LED lights.* An LED with transistor driver is controlled by PORTC bit PC6. When the pin is at the high level, the light illuminates. This LED is the flashing light.

  A similar LED mounted near the wind direction lights is controlled by PORTC bit PC5. This light illuminates to indicate no wind when the direction lights are blanked.

  Another similar LED is mounted near the digital displays. This LED, controlled by PORTD bit PD4, indicates that the highest wind speed is being displayed on the digital displays.

- *CRT terminal.* A cathode-ray-tube (CRT) terminal unit is available for displaying the desired messages. The display requires standard ASCII (space-parity) codes and common escape sequence control codes.

9.2 OPERATING SYSTEM DESIGN

Consider the requirements of the operating system essential to meeting the weather station specifications. Of course, many other practical applications have similar requirements.

Multitasking

Certainly the weather station software can be divided into several tasks. Because many of the weather station functions must happen periodically at a specified rate, designing appropriate tasks so that they run periodically at the required repetition rate is a straightforward solution. Therefore, the operating system must support multiple tasks and a system clock to control the times at which tasks run.

Furthermore, the weather station has several requirements that are independent of each other. Developing the software in independent tasks for each requirement that can be implemented and tested separately will aid the overall development. However, caution will be necessary if tasks must communicate data from one to another, and when they share I/O devices, to prevent concurrency problems. Subroutines shared by tasks must be reentrant.

System Clock

Many functions in the weather station depend on time. Signals must be sampled at a certain rate, or elapsed time between events must be measured. Therefore, the operating system must support a system clock. To be general, the clock must be capable of both delay timing and time-of-day timing.

The system clock can be implemented by using an oscillator to interrupt periodically. If the interrupt rate is very high, the resolution of timing will be very fine, but the overhead from the interrupts will use valuable processor time. If the rate is slow, overhead will be minimized, but timing resolution will suffer. A good compromise is to have the oscillator interrupt every 10 milliseconds. The required time periods for the weather station functions are all multiples of 10 milliseconds, so 10 milliseconds provides adequate resolution without significant processor overhead. For this reason, many commercial operating systems use 10 milliseconds for the system clock resolution.

The anemometer rotation rate, or the period of rotation, must be determined in order to find the wind speed. If the system clock is used as the time base, the period of rotation will have a resolution of 10 milliseconds. This resolution is adequate for the given anemometer. For common wind speeds, the period will be on the order of one second or longer. A resolution of one part in a hundred is considered adequate. At extreme high wind speeds, a rotation period of only one or two 10-millisecond periods corresponds to wind speeds that most likely will damage the anemometer! Therefore, wind speeds corresponding to poor resolution are not likely to be a problem.

On the other hand, because we are designing the operating system, if greater resolution is required, we can change the interrupt rate to any desired value.

I/O Support

In our multitasking system, there certainly will be multiple tasks concurrently using the input/output devices. The concurrency problem must be considered. The I/O devices necessary for the weather station include digital I/O ports, the analog-to-digital converter, and the SCI serial port.

The analog-to-digital conversion time is of little concern because the A/D in the 68HC11 operates so quickly that it is practical for the program to wait for it by executing a few instructions.

Digital I/O ports operate virtually instantly as an instruction is executed. Therefore, the only problem is to prevent concurrent access. The simplest low-overhead approach to controlling concurrency is to disable the interrupt system while the port is accessed. This approach is used here.

The SCI serial device in the 68HC11 is much more complicated than the parallel I/O ports. The nature of the CRT display is that a series of characters must be transmitted in a sequence. Furthermore, each character takes nearly 1 millisecond to transmit at 9600 baud. Multiple tasks will need to use the CRT display, each task placing its message at an appropriate position on the screen. The SCI device must be used with interrupt. When all these problems are considered, it becomes apparent that the operating system must control the SCI device, and the operating system must provide concurrency control.

9.3 OPERATING SYSTEM IMPLEMENTATION

The weather station can be built on a real-time multitasking operating system. Let's develop the basic operating system. In this chapter, all program code discussion will refer to source code instead of program listings. The source code provided here can be assembled and run on a trainer or other computer. Because the complete program code is provided, you must ignore some details at first reading that will be clarified later in the chapter.

The operating system developed here does not have a user interface because there is no need for one in making the weather station. It also has no mass storage device, only ROM and RAM. The operating system code and the tasks are assembled as one large program. More sophisticated operating systems can attach tasks at run time.

Task-Control-Blocks

The fundamental controlling mechanism for the tasks are a set of tables containing information about the tasks. These tables are commonly known as *task-control-blocks* or *TCBs*.

To get started looking at the design of the weather station software, let's assume that the operating system software will need to support eight tasks. Of course, because we have full control over the software design, we will program in a way to make rewriting the software to change the number of tasks easy.

TCB purpose

A task-control-block must hold all information that the operating system needs in order to run the task associated with the TCB. The operating system will initialize the TCBs at startup to initial conditions for each task. Then, each time a task is interrupted for any reason, the complete condition of the task, called its *context*, will be stored in the TCB. When the task is to run once again, the operating system will restore the context from the information in the TCB.

TCB design

Consider each element of the TCB design shown in Figure 9-2:

Task ID. This byte provides a way to number the task-control-blocks for easy identification by humans! Its only purpose is for ease of troubleshooting and ease of reading the listing. The operating system will search the TCBs in a specified order based on their locations in memory, so this identifier is redundant. The TCB will be identified in the code by the address label on its first entry. Figure 9-2 uses the label *TCB0* for *task zero*.

State. This byte indicates the state of a task. The tasks need only three states called *running*, *ready*, and *suspended*. Only a few bits are required to indicate the state, so they can be packed into a single byte. In this design, bit 7 will be the ready bit that says that a task would like to run. Bits 5 and 4 will be blocking bits that enable the serial SCI transmitter and receiver, respectively, to suspend the task. The running task and any ready tasks will have a 1 in bit 7 and 0s in the blocking bits of their state entries. Suspended tasks have any other combination of bits. In addition, bit 0 of the state entry will indicate whether the task is scheduled to run based on time.

Time-to-run. This 3-byte entry is a place to store the time-of-day at which the task is supposed to run the next time. The system time is the number of 10-millisecond periods

```
*  TASK CONTROL BLOCKS
***  THESE TCBS MUST REMAIN CONTIGUOUS AND IN ORDER
*  TASK0  TASK CONTROL BLOCK
TCB0     FCB   0          TASK ID
TCBST    RMB   1          STATE
TCBTM    RMB   3          TIME TO RUN
TCBSP    RMB   2          STACK POINTER
         RMB   SS         STACK SPACE
TCBCN    RMB   9          INITIAL CONTEXT
TCBLEN   EQU   *-TCB0
*      TASK CONTROL BLOCK OFFSETS
STATE    EQU   TCBST-TCB0  BIT0 TIMED
TIMEGO   EQU   TCBTM-TCB0
STKPNTR  EQU   TCBSP-TCB0
CNTXT    EQU   TCBCN-TCB0
*
```

Figure 9-2 RTOS task-control-block.

since midnight as a 24-bit number. A task can adjust this number on each of its runs to make periodic delay scheduling possible.

Stack and hardware context. The remainder of the TCB is all stack space, but it is broken into sections. The initial startup stack for the task is in the context area. The rest is stack space to store the temporary information the task needs as it executes.

Labels. The address label TCB0 is the only label used throughout the code. The other locations have labels to support the EQU statements here. The EQUs define some symbols for indexed-addressing offsets in the program code.

TCB initialization at system startup

When the operating system is first started, an initial context for each task must be created in the TCBs. The initial values will depend partly on the application of the operating system. For example, some tasks may need to be ready immediately as the system starts, while others will be suspended until started by some means at a later time.

Figure 9-3 shows a table of the required task startup values for the weather station. The programmer building an application on the operating system determines the values required in this table.

TCB initialization table. Each task only has two entries in the initialization table, the starting address of the task and its initial state. The initialization routine will copy the state entry into the TCB's state. The example in Figure 9-3 sets the states to either ready or suspended. The SCI blocking bits are cleared, and the bit-0 values are set so that none of the tasks use the system clock.

Task address. The starting address of a task is part of the initial hardware context, so the initialization routine copies it into the last two bytes of the TCB. Then, when the hardware context is restored through an RTI (return from interrupt) instruction, the starting address will be placed in the program counter.

```
* INITIAL TASK CONTROL BLOCK VALUES
INITTSK  FDB     TASK0        STARTING ADDRESS
         FCB     0            STATE: SUSPENDED
         FDB     TASK1
         FCB     BIT7         READY
         FDB     TASK2
         FCB     BIT7         READY
         FDB     TASK3
         FCB     0            SUSPENDED
         FDB     TASK4
         FCB     BIT7         READY
         FDB     TASK5
         FCB     BIT7         READY
         FDB     TASK6
         FCB     BIT7         READY
         FDB     TASK7
         FCB     BIT7         READY
*
```

Figure 9-3   Initial values for task-control-blocks.

Stack pointer.   The initialization routine also provides the initial value of the stack pointer. It will point to the location TCBCN-1 inside the TCB. The first time the task runs, this address, TCBCN-1, is in the stack pointer.

Time-to-run.   The time-to-run entry is not initialized here, although it could be. Instead, the time will be controlled entirely by tasks, so any value in time-to-run will be overwritten.

System Startup

Figure 9-4 shows the entire system initialization required for the weather station. At this point, take notice of the routine that initializes the task-control-blocks. Many other items are initialized here that will be discussed later.

The final two initializations near the end of Figure 9-4 actually start the system running. First, the current running task is set to be Task7 using the location RUNNING. Of course, any other task could be chosen as the startup task, but Task7 is the lowest-priority or background task which requires nothing special in order to run.

The second and final stage of the initialization restores the context of Task7. Because the Task7 task-control-block has been initialized with a context, the system startup mimics returning to a task from an interrupt. To do this, the stack pointer is loaded so that Task7's stack in its TCB is the current stack, and then a return from interrupt is executed. Following this execution, Task7 is running and the operating system is functional.

The Background Task

The background task, Task7 here, is the task that runs when the system has nothing else to do. Figure 9-5 shows Task7.

The background task begins by initializing several items, and then branches back in an infinite loop to repeat two of the initializations. Little productive work is done here. This

```
********************************************************
** SYSTEM INITIALIZATION
*** INITIALIZE OPERATING SYSTEM AND TASKS
********************************************************
* CREATE TEMPORARY STACK
START    LDS    #$00FF
* INITIALIZE PORT B
         CLR    PORTB     ALL PINS LOW INITIALLY
* INITIALIZE PORT C
         CLR    PORTC     ALL PINS LOW INITIALLY
         LDAA   #%01111111    ONE IN AND SEVEN OUTS
         STAA   DDRC
* INITIALIZE PORT D
         CLR    PORTD
         LDAA   #$1C      BIT2, BIT3, AND BIT4-OUTPUTS
         STAA   DDRD      BIT5-INPUT
* INITIALIZE A/D CONVERTER
         LDAA   #$90      POWER UP A/D
         STAA   OPTION
* INITIALIZE OUTPUT COMPARE 1
         CLR    OC1M      AFFECT NO OUTPUT PINS
         LDAA   #BIT7     ENABLE ONLY OC1 INTERRUPT
         STAA   TMSK1
* INITIALIZE STRA INTERRUPT
         LDAA   #BIT6+BIT1  STAI=1:EGA=LTOH
         STAA   PIOC
* INITIALIZE SERIAL COMMUNICATIONS INTERFACE
         CLR    SCCR1     8-BIT CHARS, NO WAKE UP
         LDAA   #$30      9600 BAUD
         STAA   BAUD
         LDAA   #0        DISABLE TX, TX INT, TC INT,
         STAA   SCCR2     ..RC, RC INT (RC NOT USED)
* INITIALIZE SCI AVAILABILITY BITS AND QUEUE
         BSET   DCBTX,BIT0  TX AVAILABLE
         BCLR   DCBRC,BIT0  RC UNAVAILABLE (NOT USED)
         CLR    TXDEPTH
* INITIALIZE OVERLOAD INDICATOR
         BCLR   OVERLOD,BIT0
* INITIALIZE TASK CONTROL BLOCKS FOR SYSTEM STARTUP
*    INITIALIZE LOOP
TCBINIT  LDX    #TCB0     X POINTS TO TCBS
         LDY    #INITTSK  Y POINTS TO INITIAL VALUES
         CLR    COUNT
*    MORE TCBS TO INITIALIZE?
TCBIN1   LDAA   COUNT
         CMPA   #NUMBTSK
         BHS    NEXT      BRANCH ON NO
*    SET TCB IDENTIFIER
         STAA   0,X
*    COPY STARTING ADDRESS AND STATE
         LDD    0,Y
         STD    CNTXT+7,X
         LDAA   2,Y
         STAA   STATE,X
*    SET INITIAL CCR
         LDAA   #$C0
         STAA   CNTXT,X
*    SET INITIAL STACK POINTER
         PSHY
         PSHX
         PSHX              TRANSFER X TO Y
         PULY              Y NOW POINTS TO TCB
         XGDX
         ADDD   #CNTXT-1
```

Figure 9-4 System initialization.

```
              STD     STKPNTR,Y
              PULX
              PULY
  *   NEXT TASK
              LDAB    #TCBLEN
              ABX
              LDAB    #3
              ABY
              INC     COUNT
              BRA     TCBIN1
  * INITIALIZE TASK VARIABLES
  *   TASK0
  NEXT    CLR     WSPEED
          CLR     HIGHWND
  *   TASK1
          CLR     WSECTOR
          CLR     LASTWDR
  *   TASK2
          LDD     #$1B59     POSITION CURSOR
          STD     MSGB22
          LDD     #$3151
          STD     MSGB22+2
          LDD     #$1B59     POSITION CURSOR
          STD     MSG2202
          LDD     #$3351
          STD     MSG2202+2
  *   TASK3
          CLR     BLANKED
          LDD     #$1B59     POSITION CURSOR
          STD     MSGB33
          LDD     #$2C52
          STD     MSGB33+2
  * MAKE TASK7 THE INITIAL RUNNING TASK
          LDX     #TCB7
          STX     RUNNING
  * INSTALL INITIAL CONTEXT FOR BACKGROUND TASK
          LDS     #TCB7+CNTXT-1
          RTI                SYSTEM STARTS RUNNING!
  *
```

Figure 9-4 Continued.

task simply provides something for the computer to do when it has nothing else to do! The work it does is for troubleshooting and system maintenance—it has nothing to do with the weather station.

At this point, we have an infinite loop running that has been started through an elaborate initialization process, but little more. However, notice that while the infinite loop is executing, the interrupt system is enabled. Several hardware devices have been initialized so that they can interrupt, and, of course, they will.

By looking back to Figure 9-4, you will see that the OC1 programmable timer was initialized to interrupt every 10 milliseconds. Also, the IRQ interrupt is enabled to provide interrupts from the anemometer. Any further action in this system will be triggered by interrupts from these two devices. The next step in following the operation of this system is to consider interrupts from these devices.

Notice that it is likely that both of these devices have set their flags before the program was started. In other designs, clearing these flags in the initialization may be necessary.

```
*----------------------------------------------------------
** TASK7 - BACKGROUND: MUST BE INFINITE LOOP
*** RUNS AT STARTUP AND DURING SYSTEM FREE TIME
*----------------------------------------------------------
* INITIALIZE SYSTEM CLOCK
TASK7    LDX    #SCB71
         SWI
* CLEAR OVERLOAD INDICATOR
TSK71    BCLR   OVERLOD,BIT0
* SET FREE TIME INDICATOR TO FREE
         LDX    #PORTD
         BSET   0,X,BIT2
* GO AROUND AGAIN ALWAYS
         BRA    TSK71
*
*** INITIAL SYSTEM CLOCK TIME
INITTIM FCB     0,0,0    ..MIDNIGHT
*** SERVICE CONTROL BLOCK
SCB71    FCB    SETCLK   SET SYSTEM CLOCK
         FDB    INITTIM
*
```

Figure 9-5   RTOS background task.

System Service Calls

Tasks may ask the operating system to perform certain functions. Due to the need for multi-tasking and the need to control concurrency problems, the request is made by causing a software interrupt with the SWI instruction. Requesting operating system services is the principal purpose for the SWI instruction.

Software interrupt

The SWI instruction causes control to go to an interrupt service routine (ISR) that is part of the operating system. All ISRs will be considered part of the operating system. The ISR performs the required function and returns to run a task. The operating system may return to the same task or a different task as it decides.

Because the operating system will receive requests to do a variety of functions, a way must be developed to tell the ISR which function to perform. Furthermore, some requests may need to provide a list of parameters, not just one parameter. It is thus necessary to supply one or more parameters to the operating system when the software interrupt is executed.

Service-control-block

A common approach to passing service-request parameters to the operating system is to collect them into a table or block, and then give the operating system the address of the table. We will call this block of parameters a *service-control-block*, or *SCB*.

Some computers have software interrupt instructions that contain an address to specify the location of the SCB. Because the SWI instruction in the 68HC11 does not have this provision, let's put the address in the X register before executing the SWI instruction. Then when the ISR starts executing, the address of the SCB will be in the X register, and the ISR can easily access the SCB parameters.

The set-clock service

The background task in Figure 9-5 initialized the system clock by making a SETCLK service request to the operating system. The request is initiated by loading the X register with the address SCB71 followed by executing an SWI instruction. The service-control-block at address SCB71 is so named because it is the first SCB used in Task7.

The SCB.   The SCB71 service-control-block contains two entries. The first entry is a single-byte number that specifies the desired service, that is, set the system clock to a particular time. Figure 9-5 uses the symbol SETCLK for the function number. Figure 9-6 shows the symbolic names of this function and five other needed functions.

The second entry in the SCB is the address of the time to be put into the system clock. By comparison, another function shown in Figure 9-6 is the read system clock function. The format of the SCB for READCLK is the same as for SETCLK, except that the address for the time is the location where the operating system will store the current time. The SCBs for both of these functions have the same format.

The example program in Figure 9-5 sets the startup time to zero, which we can call midnight. With additional programming, the time could be set to any desired time by a variety of means. Zero is used here for simplicity.

```
* SERVICE CONTROL BLOCK SYMBOLS FOR TASKS
SUSPEND EQU   0          OPERATING SYSTEM SERVICE
SCHED   EQU   1          ..FUNCTION NAMES
SCIRCV  EQU   2
SCITX   EQU   3
READCLK EQU   4
SETCLK  EQU   5
```

Figure 9-6   Operating system service function identifiers.

Memory considerations.   The SCBs described here contain the address of the time, not the time. Remember that an embedded computer will have its software in nonvolatile ROM memory. The SCB will be in ROM. Therefore, it contains an address of a place in read/write RAM memory. It would not be useful to store the time within the SCB in ROM.

Calling a service function

Figure 9-7 shows the interrupt service routine that processes system service calls. Ignore the code relating to the free-time indicator at this point.

The service function begins by saving the context of the task that was running when the interrupt occurred. Because the running task just caused the interrupt, location RUNNING contains the address of the TCB for this task. Figure 9-8 shows the location RUNNING and several other locations that are essential to the operating system.

Continuing in Figure 9-7, the address of the calling task's TCB is obtained from RUN-NING and put into the Y index register. The Y index register now points to the current task's TCB, and it will continue to do so throughout the ISR.

After the address of the TCB is obtained from RUNNING, the current address in the stack pointer is saved in the task's TCB. Saving the stack pointer completes the saving of the

```
*********************************************************
** SWI INTERRUPT SERVICE ROUTINE
*** OPERATING SYSTEM SERVICES REQUESTED BY TASKS
*** WARNING: X POINTS TO CALLING SCB
*********************************************************
*-------------------------------------------------------
* SET FREE-TIME INDICATOR TO BUSY
*-------------------------------------------------------
SWIISR  LDY    #PORTD
        BCLR   0,Y,BIT2
        *-----------------------------------------------
* FINISH SAVING CONTEXT OF TASK (FOLLOWING INTERRUPT)
*-------------------------------------------------------
* DETERMINE CALLING (RUNNING) TASK
        LDY    RUNNING
* SAVE STACK POINTER IN CALLING TASK'S TCB
        STS    STKPNTR,Y
*
*-------------------------------------------------------
* EXECUTE REQUESTED SERVICE FUNCTION
*** SERVICE FUNCTION IDENTIFIED BY NUMBER IN SCB
*-------------------------------------------------------
* REQUESTED FUNCTION NUMBER VALID?
        LDAA   0,X       GET FUNCTION IDENTIFIER FROM SCB
        CMPA   #NUMSERV
        BLS    SWIISR1   BRANCH ON YES
* SYSTEM ERROR
        JSR    ERROR     HALT SYSTEM
* RUN REQUESTED FUNCTION
SWIISR1 LDY    #SERVTAB
        TAB
        ASLB
        ABY
        LDY    0,Y
        JMP    0,Y       DO IT
*
*** TABLE OF OS SERVICE FUNCTION EXECUTION ADDRESSES
SERVTAB FDB    XSUSPND   0
        FDB    XSCHED    1
        FDB    XSCIRCV   2
        FDB    XSCITX    3
        FDB    XREDCLK   4
SERVEND FDB    XSETCLK   5
NUMSERV EQU    (SERVEND-SERVTAB)/2
*
```

Figure 9-7   Call system services.

hardware context of the running task in its TCB. Remember that the interrupt system saved all the other microprocessor registers.

Now the desired system function can be carried out. The next part of Figure 9-7 shows that the function number in the SCB is obtained and tested for validity. Because the address of the SCB is in the X register, this testing is easy to do.

```
* OPERATING SYSTEM VARIABLES
RUNNING RMB   2         TCB ADDRESS FOR RUNNING TASK
CLOCK   RMB   3         SYSTEM CLOCK
OVERLOD RMB   1         SYSTEM OVERLOAD INDICATOR-BIT0
ERRLOC  RMB   2         SYSTEM ERROR EXECUTION ADDRESS
COUNT   RMB   1         TASK INITIALIZATION LOOP COUNTER
*
```

Figure 9-8   Operating system variables.

If the function number is satisfactory, the address of the function routine is obtained from a table and the ISR jumps to the appropriate module to carry out the requested function.

System error routine

Before leaving Figure 9-7, consider the error checking performed on the function number. The routine checks the function number for valid numbers, and jumps to a subroutine to handle an error. The error subroutine, shown in Figure 9-9, stops the program from continuing after indicating an error occurred, and therefore is especially useful for diagnosing programming problems. If the system continued to run with erroneous data, the whole system could crash and debugging would be much more difficult.

The operating system can call the error subroutine from various places throughout the program as it detects errors. The error routine identifies where in memory an error occurred. It can do this because the JSR instruction puts the address of the next location on the stack. The error routine saves this address for troubleshooting purposes. Although the error "subroutine" will never return, the program calls it with a JSR instruction that provides the needed troubleshooting address.

System service example

Figure 9-10 shows both the read and set modules for the system clock. Each simply copies the 3-byte time from one place in memory to another. The location CLOCK, defined in Figure 9-8, holds the system time. The symbol SYSTIME has the value 1 to specify the offset of the time entry into the service-control-block.

No concurrency problems occur here because the SWI instruction disabled the interrupt system. If the interrupt system was not disabled, an interrupt from the real-time clock could change the time after the read routine had read only one or two of the time bytes. Likewise, a clock interrupt could occur when the set routine has updated only one or two of the time bytes.

After carrying out a service function, the operating system must return to a task. The clock services are very short, so control is returned immediately to the calling task to reduce system overhead. All service functions return through a common routine shown in Figure 9-10. Because the Y index register still points to the task's TCB, the task's stack pointer is

```
*----------------------------------------------------------
* SYSTEM ERROR ROUTINE
*----------------------------------------------------------
* PREVENT ANY FURTHER INTERRUPTS TO STOP SYSTEM
ERROR    SEI
* SAVE LOCATION WHERE ERROR WAS GENERATED
         PULX
         STX     ERRLOC    FOR TROUBLESHOOTING
* TURN ON SYSTEM ERROR LIGHT
         LDX     #PORTD
         BSET    0,X,BIT3
* HANG UP SYSTEM
HANG     BRA     HANG
*
```

Figure 9-9   System error routine.

```
*-----------------------------------------------------------
* OS SERVICE: READ SYSTEM CLOCK
*-----------------------------------------------------------
* COPY TIME FROM SYSTEM CLOCK TO TASK'S SCB
XREDCLK LDX     SYSTIME,X     X POINTS TO TIME BLOCK
        LDD     CLOCK+1
        STD     1,X
        LDAA    CLOCK
        STAA    0,X
* RETURN TO CALLING TASK IMMEDIATELY
        LDY     RUNNING
        BRA     RTSWI      USE COMMON SWI RETURN ROUTINE
*
*-----------------------------------------------------------
* OS SERVICE: SET SYSTEM CLOCK
*-----------------------------------------------------------
* COPY TIME FROM TIME BLOCK TO SYSTEM CLOCK
XSETCLK LDX     SYSTIME,X    X POINTS TO TIME BLOCK
        LDD     1,X
        STD     CLOCK+1
        LDAA    0,X
        STAA    CLOCK
* RETURN TO CALLING TASK IMMEDIATELY
        LDY     RUNNING
        BRA     RTSWI      USE COMMON SWI RETURN ROUTINE
*
*-----------------------------------------------------------
* COMMON RETURN FOR SWI SERVICE ROUTINES
*-----------------------------------------------------------
* RESTORE NEXT TASK'S CONTEXT
RTSWI   LDS     STKPNTR,Y
        RTI
*
```

Figure 9-10   Read and set system clock services.

restored from the TCB, and the RTI instruction restores all the other microprocessor registers. Because the registers have been restored, the interrupted task is running once again.

In this example, the background task requested the SETCLK function, so control will return to the background task. Looking again at Figure 9-5, the background task runs in an infinite loop. Nothing further will happen in the system until an interrupt occurs.

Schedule Task with an Interrupt

Not all requests to the operating system for service can be initiated by tasks. In particular, requests from devices outside the computer are required in typical control system applications. Many situations arise where critical timing requirements must be met. That is, the operating system must schedule a task to run as soon as possible following some event outside the computer. Of course, such events are not synchronized to the computer's operation in any way. Requests to the operating system can be made in response to external interrupts.

Timing external events

Measuring the elapsed time between events is a common requirement of applications based on real-time systems. In many situations, elapsed time can be found by reading the system clock on each event, and determining the time between reads. If a task runs immediately in response to the external event, adequately accurate timing is possible.

The weather station requires timing to determine the wind speed, so the operating system needs the ability to schedule a task due to an external interrupt. The anemometer is connected to the STRA pin, so it sets the STAF flag which causes an IRQ interrupt. Therefore, let's use the IRQ interrupt to schedule the highest priority task, Task0, to run immediately by setting its state to ready.

IRQ interrupt service routine

The example in Figure 9-11 uses the IRQ interrupt service routine to schedule Task0 to run immediately, or as soon as the operating system can run the task. This technique can schedule any task to run when triggered by an external interrupt.

Look at the code in Figure 9-11, and ignore the reference to a free-time indicator. Keep in mind that, in the weather station, this ISR runs because the anemometer rotated.

First, the ISR clears the STAF flag removing the source of the interrupt. Next, the ready bit for Task0 is set to 1, thus scheduling Task0 to run the next time the operating system looks for a task to run. That is, the state of Task0 is set to "ready."

Notice the use of the offset symbol STATE in the BSET instruction. Figure 9-2 defined this symbol, and some others, to make accessing entries in the TCB easy.

Next, this ISR saves the stack pointer for the task that was interrupted in that task's TCB. Remember the variable RUNNING, defined in Figure 9-8, holds the address of the TCB of the running task. The complete context of the interrupted task is now saved.

Finally, this ISR must return from the interrupt. Let's choose to have the operating system return to the highest-priority task that is now ready, which is Task0. If control returns to the interrupted task, the system will not consider higher-priority tasks until some future interrupt. Therefore, to get the most accurate timing, Task0 should run as soon as possible.

```
********************************************************
** IRQ INTERRUPT SERVICE ROUTINE
*** SCHEDULE TASK0 IMMEDIATELY
********************************************************
* SET FREE-TIME INDICATOR TO BUSY
IRQISR  LDY    #PORTD
        BCLR   0,Y,BIT2
* CLEAR STAF INTERRUPT FLAG
        LDAA   PIOC      READ TWO REGISTERS TO
        LDAA   PORTCL    ..CLEAR FLAG
* SET TASK0 READY
        LDX    #TCB0
        BSET   STATE,X,BIT7
* SAVE STACK POINTER IN INTERRUPTED TASK'S TCB
        LDY    RUNNING
        STS    STKPNTR,Y
* RUN DISPATCHER TO CHOOSE NEXT TASK
        BSR    DISPTCH
* RESTORE NEXT TASK'S CONTEXT
        LDS    STKPNTR,Y
        RTI
*
```

Figure 9-11   IRQ interrupt service routine.

Near the end of Figure 9-11, the *dispatcher* is called. In this implementation, the dispatcher is a subroutine that determines the highest-priority task that can run. The dispatcher puts the address of that task's TCB in the Y index register. At the end of this ISR, the stack pointer is thus loaded from the new task's TCB, and the RTI instruction restores the registers for that task. In this example, Task0 is now running.

The Dispatcher

The dispatcher, although a small program module, has a major impact on the characteristics of the operating system. The dispatcher decides which task will run next following an interrupt. Only the operating system, never a task, uses the dispatcher.

Operation

The dispatcher searches in order from the highest-priority task to the lowest-priority task until it finds a ready task that is not blocked. It looks only at the state entries in the TCBs. The dispatcher always finds some task that is ready because the background task is always ready. The background task runs continually whenever there are no other tasks to run.

Figure 9-12 shows the dispatcher subroutine. The dispatcher checks the ready bit, bit 7, and the blocking bits, bit 5 and bit 4, in this example. When the dispatcher finds an acceptable task, it puts that task's TCB address in location RUNNING and in the Y index register. Doing this sets up a context switch to a new task.

Scheduling algorithm

Because the dispatcher in Figure 9-12 searches through the tasks in order, it implements a strict priority scheduling algorithm. The tasks get to run in priority order. Because the dispatcher runs when an interrupt occurs, the running task is preempted. Therefore, a

```
*----------------------------------------------------------
* PRIORITY DISPATCHER SUBROUTINE
*** PUTS NEXT RUNNING TASK IN 'RUNNING' AND Y
*----------------------------------------------------------
* START AT HIGHEST-PRIORITY TASK
DISPTCH LDY   #TCB0    LAST TASK MUST BE READY
* INITIALIZE TCB LENGTH
        LDAB  #TCBLEN
* NEXT TASK READY & NOT WAITING ON TX OR RC?
DISP1   BRCLR STATE,Y,BIT7,DISP2  BRANCH ON NOT RDY
        BRSET STATE,Y,BIT5,DISP2  BRANCH ON TX BLOCK
        BRSET STATE,Y,BIT4,DISP2  BRANCH ON RC BLOCK
        BRA   DISP3    READY TO GO
* ADVANCE TO NEXT TASK
DISP2   ABY
* TRY NEXT-LOWEST-PRIORITY TASK
        BRA   DISP1
* SAVE HIGHEST-PRIORITY TASK TCB ADDRESS
DISP3   STY   RUNNING
        RTS
*
```

Figure 9-12 RTOS dispatcher.

task cannot prevent higher-priority tasks from running by using too much of the computer's time. A preemptive priority scheduling algorithm is generally required for real-time system performance.

The Suspend Service

At this point, the operating system has enough code to start the background task and to make Task0 run in response to an IRQ interrupt from the anemometer. The next required function is a way to stop Task0 from running. Task0 must stop to allow the operating system to run other lower-priority tasks. Task0 will request a system service, called *suspend*, to stop itself from running.

The operation

The suspend system service is simply a way to set a task to the "not ready" state. The suspend service does this easily by clearing the ready bit in the state entry of the TCB.

Remember that a task makes a system service call by loading the X register with the address of a SCB, and then executing a SWI instruction. The service-control-block for suspend contains two items—the first specifies what function to perform, and the second specifies which task is to be affected. The symbol TARGET is always used to specify the offset into the SCB for the location of the affected task's TCB.

For the suspend service, the service function program in Figure 9-7 sends control to the program module in Figure 9-13. This module clears the ready bit in the state entry of the specified task. Because task states have been altered, the dispatcher is called to find the highest-priority task, and that task is continued from its last run point. Regardless, the dispatcher will no longer find the suspended task ready, so that task has been stopped.

A suspend example

Figure 9-14 is a very short task to illustrate the use of the suspend service. This task is set ready by an external IRQ interrupt. Its only job is to complement an output bit each time it runs, and then suspend itself. Every time an external interrupt occurs, the output bit is complemented. This task, as are almost all tasks, is effectively an infinite loop.

Notice that concurrency problems with the output port are anticipated. Therefore, the interrupt system is disabled while PORTC is changed. Then, if some other task also uses PORTC, concurrency problems will not occur.

```
*-----------------------------------------------------------
* OS SERVICE: SUSPEND THE SPECIFIED TASK
*-----------------------------------------------------------
* CLEAR READY BIT OF SPECIFIED TASK
XSUSPND LDY    TARGET,X    X POINTS TO SCB, Y TO TCB
        BCLR   STATE,Y,BIT7
* RETURN THROUGH DISPATCHER THAT CHOOSES NEXT TASK
        JSR    DISPTCH    RETURNS TCB IN Y
        JMP    RTSWI       USE COMMON SWI RETURN ROUTINE
*
```

Figure 9-13 Suspend service.

```
*-----------------------------------------------------------
** TASK1 - RUNS ON EACH IRQ INTERRUPT
*-----------------------------------------------------------
* COMPLEMENT PORTC BIT 6
TASK1    SEI                    AVOID CONCURRENCY PROBLEM
         LDAA    PORTC
         EORA    #BIT6
         STAA    PORTC
         CLI
* SUSPEND TASK1
         LDX     #SCB11
         SWI
* GO AROUND AGAIN AT NEXT ACTIVATION
         BRA     TASK1
*
*** SERVICE CONTROL BLOCK
SCB11    FCB     SUSPEND   SUSPEND TASK1
         FDB     TCB1
*
```

Figure 9-14   Sample task to complement output bit on each IRQ interrupt.

An easier way?

A good question to ask is "To suspend a task, why not avoid the overhead of the software interrupt by using the task to clear the ready bit?" The answer is the concurrency problem. By making a proper system service call, no concurrency problems occur; furthermore, the dispatcher is also run properly.

The operating system must always handle such chores. If a task alters anything used by the operating system code, intermittent problems will occur.

The System Clock

Next, let's develop a system clock so that tasks can be scheduled to run based on time and so that tasks can measure elapsed time. The system clock consists of an accurate oscillator that causes periodic interrupts, and an interrupt service routine to track time. The oscillator chosen here is the output compare 1 (OC1) device. In addition to tracking time, the clock routine also determines whether the time has come to run time-scheduled tasks.

The real-time clock hardware

The programmable timer's output compare 1 device can be used to cause periodic interrupts. Although OC1 can also control output pins, that ability is disabled because it isn't needed for the system clock function. The initialization program in Figure 9-4 initializes OC1 as described here. OC1 will interrupt very soon after system startup because its OC1F flag will already be set at startup.

The OC1 interrupt service routine

Figure 9-15 shows the interrupt service routine for OC1. Let's follow the program code, but you must ignore the references to free-time and overload at this point.

The ISR first clears the OC1F flag, and then updates the output compare register to provide periodic interrupts. The increment added to the TOC1 register contents represents

10 milliseconds. This sets the resolution of the system clock because OC1 will interrupt every 10 milliseconds. Time will be measured in "ticks" of this 10-millisecond clock. The resolution of the system clock can be changed by modifying the number added to the TOC1 register.

The system time, in 10-millisecond ticks, is updated next by adding one to the 24-bit time number. The resultant time is checked to see if 24 hours have passed and, if so, the time is reset to zero, which is called *midnight*.

The next part of the ISR is a loop that operates on all the tasks scheduled by time. A task is scheduled by time if a 1 is in bit 0 of its state entry in its TCB. A timed task desires to run at a certain time in the future. Scheduling the task requires putting the time it should run into the task's time-to-run entry in its TCB. The clock ISR (that runs every 10 milliseconds) checks the time-to-run entries of the timed tasks to see if they match the current time. If the time matches, the ISR makes that task ready by setting its ready bit in the state entry of its TCB. If the time-to-run has not arrived, or the task is not timed, the ISR does nothing for that task. The ISR needs to avoid tasks that are not timed because the unused time-to-run in the TCB will match the current time eventually.

As in previous ISRs, this one ends by saving the context of the interrupted task, calling the dispatcher, and returning to the next task.

```
**********************************************************
** OC1 INTERRUPT SERVICE ROUTINE
**********************************************************
* SET FREE-TIME INDICATOR TO BUSY
OC1ISR  LDY   #PORTD
        BCLR  0,Y,BIT2
* SYSTEM OVERLOADED?
        BRCLR OVERLOD,BIT0,OC11   BRANCH ON NO
* SYSTEM ERROR-SYSTEM OVERLOADED
        JSR   ERROR      HALT SYSTEM
* SET OVERLOAD BIT
OC11    BSET  OVERLOD,BIT0
* CLEAR OC1 INTERRUPT FLAG
        LDAA  #BIT7      STORE 1 TO CLEAR FLAG
        STAA  TFLG1      ..ZEROS DO NOTHING
* SET UP NEXT OC1 INTERRUPT
        LDD   TOC1
        ADDD  #OC1TIME
        STD   TOC1
* UPDATE SYSTEM CLOCK
        LDD   CLOCK+1
        ADDD  #1
        STD   CLOCK+1
        BCC   OC12
        INC   CLOCK
* MIDNIGHT?
OC12    LDD   #MIDNLSB
        CPD   CLOCK+1
        BNE   OC13
        LDAA  #MIDNMSB
        CMPA  CLOCK
        BNE   OC13
* RESTART CLOCK AT ZERO
        CLR   CLOCK      IT'S MIDNIGHT
        CLR   CLOCK+1
        CLR   CLOCK+2
```

Figure 9-15   Real-time clock interrupt service routine.

```
* INITIALIZE TO FIRST TASK
OC13    LDX    #TCB0
* ALL TASK'S TIME-TO-RUN CHECKED?
OC14    CPX    #LASTTCB
        BEQ    OC16       BRANCH ON YES
* IS THIS A TIMED TASK?
        BRCLR STATE,X,BIT0,OC15  BRANCH ON NO
* TCB TIME SAME AS SYSTEM CLOCK TIME?
        LDD    TIMEGO+1,X
        CPD    CLOCK+1
        BNE    OC15       BRANCH ON NO
        LDAA   TIMEGO,X
        CMPA   CLOCK
        BNE    OC15       BRANCH ON NO
* SET TASK READY (TIME MATCHES)
        BSET   STATE,X,BIT7
* REMOVE TIMED ATTRIBUTE OF READY TASK
        BCLR   STATE,X,BIT0
* ADVANCE TO NEXT TASK
OC15    LDAB   #TCBLEN
        ABX
        BRA    OC14
* SAVE STACK POINTER IN INTERRUPTED TASK'S TCB
OC16    LDY    RUNNING
        STS    STKPNTR,Y
* RUN DISPATCHER THAT CHOOSES NEXT TASK
        JSR    DISPTCH    RETURNS TCB IN Y
* RESTORE NEXT TASK'S CONTEXT
        LDS    STKPNTR,Y
        RTI
*
```

Figure 9-15   Continued.

Because the real-time clock is interrupting every 10 milliseconds, the operating system "stirs" the tasks at least every 10 milliseconds. Stirring the tasks ensures that high-priority tasks have frequent opportunities to preempt lower-priority tasks even if no other event triggers the preemption.

The Schedule Service

The system service function called *schedule* enables a task to schedule the execution of a task using time. Specifically, the schedule service tells the operating system to start running a selected task at the requested time. The schedule service uses the system clock.

Generally, operating systems allow both time-of-day and delay scheduling. That is, a task can be scheduled to run at a specified time-of-day, or to run after a specified delay from the current time. Let's provide both, although the weather station specifications require only delay scheduling. Then it will be a simple matter to add new time-based features to the weather station later, and the extra complexity is minimal.

Tasks can be made to run periodically using delay scheduling; the task simply reschedules itself on each run. Frequently, a task will want to be suspended immediately after scheduling its next run, so let's provide automatic suspension as an option in the schedule service. Then a task can either continue running after the service request, or be suspended.

A task that wants to continue running after the scheduling is completed is likely scheduling some other task, while one that wants to be suspended is likely scheduling itself.

The service-control-block

The SCB for scheduling is quite complicated—it contains seven bytes. These include the schedule function identifier byte, the (double-byte) TCB address for the task to be scheduled, a parameter byte, and three time bytes.

The function identifier and TCB address are specified the same as for previously defined services. The parameter byte specifies the type of time scheduling and the return policy. Bit 0 of the parameter byte is 1 to indicate time-of-day scheduling, and 0 to indicate delay scheduling. Bit 1 of the parameter byte is 0 for immediate return and 1 for suspension.

Figure 9-16 shows some symbol definitions that make writing and using SCBs easier. The example in Figure 9-18 includes a typical SCB.

```
* SERVICE-CONTROL-BLOCK SYMBOLS
DEL       EQU    $00       DELAY TIMING
TOD       EQU    BIT0      TIME-OF-DAY TIMING
IMM       EQU    $00       IMMEDIATE RETURN
SUS       EQU    BIT1      SUSPEND UNTIL TIME SCHEDULED
PARAM     EQU    3         SCHED: OFFSET-RETURN PARAMETER
SCBTIME   EQU    4         SCHED: OFFSET-TIME ENTRY
CHARS     EQU    1         SCITX: OFFSET-NUMBER OF CHARS
MSGADDR   EQU    3         SCITX: OFFSET-SCI MSG ADDRESS
```

Figure 9-16   Service-control-block symbol definitions.

Schedule operation

A task requests the schedule service by loading the X register with the address of the SCB and executing an SWI instruction. The service function program in Figure 9-7 sends control to the schedule module. Let's follow the schedule module's program in Figure 9-17.

Set time-to-run.   The schedule module first makes the task specified in the SCB a timed task by setting bit 0 of its state entry. Next it determines whether time-of-day or delay scheduling was requested.

If delay scheduling was the choice, the sum of the time entry in the SCB and the current time in the system clock is put into the time-to-run entry of the selected task's TCB. If the resultant time goes past 24 hours, the time is adjusted by subtracting 24 hours from the resultant time.

If time-of-day scheduling was the choice, the time in the SCB is copied to the time-to-run entry of the TCB. For either type of scheduling, the real-time clock interrupt service routine in Figure 9-15 now checks this time-to-run every 10 milliseconds.

Return as requested.   The schedule service next determines the type of return requested by testing bit 1 of the parameter entry of the SCB. If immediate return is requested, the calling task's context is restored. If suspension is requested, the calling task is suspended, the dispatcher is called to determine the next task to run, and that task's context is restored.

A scheduling example

Figure 9-18 shows a task that uses delay scheduling to make it run periodically. The only function of this task is to complement a PORTC output bit periodically to make a flashing

```
*-------------------------------------------------------------
* OS SERVICE: SCHEDULE RUN TIME FOR SPECIFIED TASK
*-------------------------------------------------------------
* IDENTIFY SELECTED TASK CONTROL BLOCK
XSCHED  LDY    TARGET,X    X POINTS TO SCB, Y TO TCB
* MAKE SCHEDULED TASK TIMED
        BSET   STATE,Y,BIT0
* TIME OF DAY SCHEDULING?
        BRSET  PARAM,X,BIT0,SCH1  BRANCH ON YES
* ADD DELAY TIME TO CURRENT TIME
        LDD    CLOCK+1
        ADDD   SCBTIME+1,X
        STD    TIMEGO+1,Y
        LDAB   CLOCK
        ADCB   SCBTIME,X
        STAB   TIMEGO,Y
* DELAY TIME PAST MIDNIGHT?
        LDD    TIMEGO+1,Y
        SUBD   #MIDNLSB
        LDAB   TIMEGO,Y
        SBCB   #MIDNMSB
        BLO    SCH2        BRANCH ON NO
* ADJUST DELAY TIME FOR NEXT DAY
        LDD    TIMEGO+1,Y
        SUBD   #MIDNLSB
        STD    TIMEGO+1,Y
        LDAB   TIMEGO,Y
        SBCB   #MIDNMSB
        STAB   TIMEGO,Y
        BRA    SCH2
* SET TIME-TO-GO TO DESIRED TIME OF DAY
SCH1    LDD    SCBTIME÷1,X
        STD    TIMEGO+1,Y
        LDAA   SCBTIME,X
        STAA   TIMEGO,Y
* RETURN IMMEDIATELY OR AT REQUESTED TIME?
SCH2    BRCLR  PARAM,X,BIT1,SCH3  BRANCH ON IMM
* SUSPEND CALLING (RUNNING) TASK
        LDX    RUNNING
        BCLR   STATE,X,BIT7
* RETURN THROUGH DISPATCHER THAT CHOOSES NEXT TASK
        JSR    DISPTCH  RETURNS TCB IN Y
        BRA    SCH4
* RETURN IMMEDIATELY TO CALLING TASK
SCH3    LDY    RUNNING
SCH4    BRA    RTSWI   USE COMMON SWI RETURN ROUTINE
*
```

Figure 9-17   Schedule service.

light. The SCB sets the half-period of the flashing to 30 milliseconds. Notice the use of the DEL and SUS symbols from Figure 9-16 to set the parameter byte.

Once again, to avoid concurrency problems, the interrupt system is disabled as PORTC is modified. The task in Figure 9-14 could easily preempt this task and cause concurrency problems with PORTC.

This example shows that making the computer carry out functions periodically is very easy when using the operating system. Changes in the task are quite simple, and the timing is very accurate. Small tasks such as this one can perform useful functions in a larger program with little overhead or added complexity.

```
*----------------------------------------------------------
** TASK2 - FLASH LIGHT
*** RUNS PERIODICALLY EVERY 30 MILLISECONDS
*----------------------------------------------------------
* COMPLEMENT OUTPUT BIT
TASK2    SEI                 AVOID CONCURRENCY PROBLEM
         LDAA    PORTC
         EORA    #BIT5
         STAA    PORTC
         CLI
* DELAY MYSELF 30 MILLISECONDS
         LDX     #SCB21
         SWI
* GO AROUND AGAIN AT NEXT ACTIVATION
         BRA     TASK2
*
*** SERVICE CONTROL BLOCK
SCB21    FCB     SCHED     SCHEDULE TASK2
         FDB     TCB2
         FCB     DEL+SUS   ..DELAY WITH SUSPEND
         FCB     0,0,3     ..30 MILLISECONDS
*
```

Figure 9-18   Sample periodic task using the schedule system service.

The SCI Transmit/Receive Services

Most of the operating system features needed for implementing the weather station are now in place. The last major piece required is a way to transmit characters serially to a CRT terminal. The operating system will use the SCI serial receiver/transmitter hardware in the 68HC11 to provide this service.

The serial transmit SCB

Let's begin discussion of the serial transmit service with an example of a task that uses it. Figure 9-19 shows a short task that periodically displays a message on a CRT display.

Look at the SCB labeled SCB31 in Figure 9-19. The first entry is the function identifier as in all SCBs. The second entry is the (double-byte) number of characters to transmit. The last entry is the address of the message table holding the character string.

Also look at the message table labeled MSGB31. This label defines the beginning of the first message in Task3.

The message table can be in ROM or RAM because the SCB contains the address of the message table. This flexibility allows the message to be fixed if it is in ROM, or to be changed as the program runs if it is in RAM.

Serial transmit concurrency problem

The concurrency problem, as it applied to an output port, was relatively simple to solve. A useful approach is to disable the interrupt system while modifying the port. Such an approach is not useful when dealing with the SCI serial port.

First, the nature of the SCI is to output multiple bytes (characters) as a group to form a message. If two tasks concurrently accessed the transmitter, the characters from the two messages would be scrambled together! Second, the SCI device is relatively slow because

```
*---------------------------------------------------------------
** TASK3 - DISPLAY 'TEST MESSAGE' ON CRT
*** RUNS PERIODICALLY EVERY 500 MILLISECONDS
*---------------------------------------------------------------
* START SCI TRANSMITTER TO SEND MESSAGE
TASK3   LDX     #SCB31
        SWI
* DELAY MYSELF FOR 500 MILLISECONDS
        LDX     #SCB32
        SWI
* GO AROUND AGAIN AT NEXT ACTIVATION
        BRA     TASK3
*
*** TASK 3 MESSAGE 1 TABLE
MSGB31  FCB     $1B,$59,$2C,$3A   POSITION CURSOR
        FCC     'TEST MESSAGE'
MSGE31  EQU     *
*
*** SERVICE CONTROL BLOCKS
SCB31   FCB     SCITX        TRANSMIT TO CRT
        FDB     MSGE31-MSGB31
        FDB     MSGB31
SCB32   FCB     SCHED        SCHEDULE TASK3
        FDB     TCB3
        FCB     DEL+SUS   ..DELAY WITH SUSPEND
        FCB     0,0,50    ..500 MILLISECONDS
*
```

Figure 9-19 Sample task using the serial transmitter service.

bits are transmitted serially. If the interrupt system were disabled for the duration of the entire message transmission, the real-time system would no longer be real-time! The computer could not respond to other devices, could not run multiple tasks, and it could not keep track of time with the system clock while the message was transmitted.

Instead, a lock called an *availability bit* is necessary to lock the SCI transmitter for the duration of a transmission. The lock will prevent other tasks from mingling their messages with one already in progress. It is common to speak of the task currently controlling a device as *owning* the device. Only the current owner can use it—other tasks are locked out.

The lock will also allow the use of interrupt I/O with the SCI device so that the rest of the system can continue to function. To avoid concurrency problems with the availability bit, it will be controlled by the operating system—it will be changed within ISRs that run with the interrupt system disabled.

Transmitter device-control-block

The SCI transmit routine needs to maintain several pieces of information as the transmission proceeds, one of which is the availability bit. Information must be carried over between interrupts from the SCI transmitter. This information is collected together in a small table called a *device-control-block*, or *DCB*.

Each I/O device using interrupt must have one DCB that is controlled by the operating system. For example, another one is necessary for the SCI receiver.

Figure 9-20 shows the DCB for the SCI transmitter. The first entry is a status byte. Bit 0 of the status byte is the availability bit, and it is the only bit currently used. The second entry is the address of the TCB for the owning task. The third entry is the number of characters

```
* DEVICE CONTROL BLOCK FOR SCI TRANSMITTER
DCBTX    RMB    1        BIT0=1=AVAIL, 0=NOT
DCBTX1   RMB    2        CALLING TASK TCB ADDRESS
TXCNT    RMB    2        NUMBER OF CHARS TRANSMITTED
MSGLENT  RMB    2        MESSAGE LENGTH-NONZERO
MSGPT    RMB    2        POINTER TO TX MESSAGE
*
* SCI TRANSMITTER REQUEST LIST
TXDEPTH  RMB    1        POSITION OF NEXT REQUEST
TXWAIT   RMB    28       ENOUGH ROOM FOR EVERY TASK
MAXDEP   EQU    *-TXWAIT
*
```

Figure 9-20   SCI transmitter device-control-block.

already transmitted. The fourth entry is the total number of characters to be transmitted for the current service request. The final entry is the memory address of the next character to be transmitted. All this information is necessary to transmit characters.

The DCB must be initialized at system startup. In particular, Figure 9-4 shows how the availability bit in the DCB is set to 1 making the transmitter available for use.

Serial transmit operation

A task requests the serial transmit service by loading the X register with the address of the SCB and executing an SWI instruction. The service function program in Figure 9-7 sends control to the transmit module in Figure 9-21.

The operation of the SCI transmit function is much more complicated than that of the previously discussed devices. The example here is similar to the example using the SCI transmitter in Chapter 7.

The transmit module assumes that the SCI transmitter will block the owner task from running while the message is transmitted, or while it is waiting to have its message transmitted. Many tasks cannot process further information until a message transmission is completed, so continuing to run the task while the message is transmitting would not be advantageous. If blocking the task while the message is transmitted is too restrictive, a task can start another task that outputs the message. Let's follow the program code.

Block owner task.   The service module in Figure 9-21 first blocks the task requesting transmit service. The task is blocked by setting the SCI blocking bit in the task's TCB state entry. The dispatcher prevents the task from running while the blocking bit is set. Blocking the task, whether the transmitter is available or not, implies that the task will eventually be able to transmit its message.

Transmitter availability.   Continuing in Figure 9-21, the transmit service next tests the SCI transmitter's availability bit. Let's assume that the transmitter is currently available.

The routine makes the transmitter unavailable to lock out further transmit requests. Now the current owner's TCB address is stored in the DCB so that the transmitter knows which task owns it. Next, the length and address of the message is copied from the task's SCB to the transmitter's DCB. Figure 9-16 defines the symbols used here. Finally, the entry in the DCB that holds the number of characters already transmitted is set to zero.

```
*------------------------------------------------------
* OS SERVICE: SCI TRANSMIT CHARACTERS
*------------------------------------------------------
* SET TX BLOCKING BIT IN CALLING TCB
XSCITX  LDY     RUNNING   X POINTS TO SCB, Y TO TCB
        BSET    STATE,Y,BIT5
* IS TRANSMITTER AVAILABLE?
        BRCLR   DCBTX,BIT0,TX1  BRANCH ON NO
* MAKE SCI TX UNAVAILABLE
        BCLR    DCBTX,BIT0
* TELL DCB WHICH TASK OWNS TX
        STY     DCBTX1
* INITIALIZE MESSAGE USING SCB PARAMETERS
        LDD     CHARS,X
        STD     MSGLENT
        LDD     MSGADDR,X
        STD     MSGPT
* INITIALIZE MESSAGE BYTE COUNT
        CLR     TXCNT
        CLR     TXCNT+1
* ENABLE TX AND TX INTERRUPT
        LDX     #REG
        BSET    SCCR2-REG,X,BIT7+BIT3 SENDS IDLE
        BRA     TX3
* ROOM FOR TRANSMIT REQUEST ON TX WAIT LIST?
TX1     PSHY
        LDY     #TXWAIT
        LDAB    TXDEPTH
        CMPB    #MAXDEP
        BLO     TX2       BRANCH ON YES
* SYSTEM ERROR-NO ROOM ON LIST
        JSR     ERROR     HALT SYSTEM
* PUT SCB AND TCB ON LIST
TX2     ABY
        STX     0,Y       SCB ON LIST
        PULX
        STX     2,Y       TCB ON LIST
        ADDB    #4
        STAB    TXDEPTH
* RETURN THROUGH DISPATCHER THAT CHOOSES NEXT TASK
TX3     JSR     DISPTCH   RETURNS TCB IN Y
        BRA     RTSWI     USE COMMON SWI RETURN ROUTINE
*
* SCI TRANSMITTER REQUEST LIST
TXDEPTH RMB     1         POSITION OF NEXT REQUEST
TXWAIT  RMB     28        ENOUGH ROOM FOR EVERY TASK
MAXDEP  EQU     *-TXWAIT
*
```

Figure 9-21   Serial transmit service.

Transmit message.   The DCB information is now in place, so the SCI hardware can be started. The SCI transmitter is enabled and the transmitter interrupt is enabled which causes the transmitter to send an "idle" character. Momentarily, the transmitter will cause an interrupt, and the ISR will handle things from here.

Transmitter unavailable.   If the transmitter tests as unavailable, the transmit request is queued by placing it on a waiting list. The transmit request information includes the address of the SCB and the address of the requesting task's TCB; therefore, a transmit request is identified by a total of four bytes.

The waiting list is a table for storing transmit requests. The number of bytes in the table is held at the location labeled TXDEPTH, so the routine tests location TXDEPTH to be sure room is on the waiting list. If there is room on the waiting list, the TCB address and SCB address from the current transmit request are added to the list, and TXDEPTH is adjusted accordingly. If there is not room on the list, a system error has occurred because the system cannot properly handle all the transmit requests.

The transmit requests are queued on a last-in-first-out basis for simplicity. That is, requests are added at the end of the waiting list and removed later from the end of the list.

Figure 9-21 shows the waiting list, and Figure 9-4 shows how it is initialized at startup. Of course, the waiting list must be placed in RAM memory.

Return. In all cases, the transmit service module returns from the SWI interrupt by calling the dispatcher and returning to the appropriate task. Remember that the requesting task will be blocked when the dispatcher looks for ready tasks.

Serial transmit ISR

The SCI transmitter ISR is similar to the example in Chapter 7 with one major exception. When the end of the message arrives, the operating system must ensure that all the bits of the last character have been transmitted before switching to a new task. If the switching is not carefully handled, the last character could be overwritten by the first character transmitted by a different task. Use of half-duplex communication leads to the same kind of problem when the direction of the communication link is reversed at the end of a message. This situation is handled by using the transmission complete interrupt to ensure that all bits have indeed been transmitted.

Polling chain. Look at Figure 9-22. The SCI devices share a single interrupt vector, so there must be a common interrupt service routine for all SCI devices. The figure shows the common SCI interrupt service routine although the emphasis here is on the transmitter device. Ignore references to the free-time indicator at this point.

Due to the shared nature of this ISR, it stores the stack pointer for the interrupted task at the beginning. The saving of the running task's context following the interrupt from an SCI device is, therefore, completed.

Next, the polling chain examines flags to determine which SCI device caused the interrupt. Assume that the transmitter caused the interrupt and that control therefore goes to its service routine.

The transmitter service. In Figure 9-22, the transmitter service first checks the number of characters already transmitted in the DCB to determine whether the current character is the last character of the message. Assume that it is not the last character.

The transmit routine is straightforward from here. A pointer to the next character is formed with information from the DCB; the character code is obtained from the message table and sent to the transmitter hardware; and finally, the number of characters already transmitted is adjusted in the DCB.

Return from the interrupt is through a common SCI module that calls the dispatcher and restores the context of the appropriate task. Calling the dispatcher means that any task

```
********************************************************
** SCI RCVR, TX, AND TC INTERRUPT SERVICE ROUTINE
********************************************************
*------------------------------------------------------
* SET FREE-TIME INDICATOR TO BUSY
*------------------------------------------------------
SCIISR  LDY   #PORTD
        BCLR  0,Y,BIT2
*------------------------------------------------------
* SAVE STACK POINTER IN INTERRUPTED TASK'S TCB
*------------------------------------------------------
        LDY   RUNNING    SAVE CONTEXT COMMON TO ALL
        STS   STKPNTR,Y  ..SCI DEVICES
*------------------------------------------------------
* SCI INTERRUPT POLLING CHAIN
*------------------------------------------------------
* INTERRUPT FROM SCI RECEIVER?
        LDX   #REG       READ SCSR-FIRST STEP TO
        BRSET SCSR-REG,X,BIT5,SCIRCVR ..CLEAR FLAG
* IS TC TESTING REQUIRED? TDRE ALSO SET ON TC!
        BRCLR SCCR2-REG,X,BIT6,SCIC1  BRANCH ON NO
* INTERRUPT FROM SCI TRANSMISSION COMPLETE?
        BRSET SCSR-REG,X,BIT6,SCITC   BRANCH ON YES
* INTERRUPT FROM SCI TRANSMITTER?
SCIC1   BRSET SCSR-REG,X,BIT7,SCITXR  BRANCH ON YES
* SYSTEM ERROR-ILLEGAL INTERRUPT
        JSR   ERROR      HALT SYSTEM
*
*------------------------------------------------------
* INTERRUPT SERVICE FOR SCI RECEIVER
*------------------------------------------------------
* RECEIVER NOT USED
SCIRCVR BRA   RTSCI      USE COMMON SCI RETURN ROUTINE
*
*------------------------------------------------------
* INTERRUPT SERVICE FOR SCI TRANSMITTER
*------------------------------------------------------
* AT LAST CHARACTER OF MESSAGE?
SCITXR  LDD   MSGLENT    MESSAGE LENGTH
        SUBD  #1
        CPD   TXCNT
        BHI   SCITX1     BRANCH ON NO
* TERMINATE MESSAGE - DISABLE TX INTERRUPT
        BCLR  SCCR2-REG,X,BIT7  X POINTS TO REGISTERS
* ENABLE TRANSMISSION COMPLETE INTERRUPT
        BSET  SCCR2-REG,X,BIT6
* MAKE POINTER TO NEXT CHARACTER
SCITX1  LDD   MSGPT
        ADDD  TXCNT
        XGDX             X POINTS TO TABLE
* GET NEXT CHARACTER ASCII CODE
        LDAA  0,X
* TRANSMIT NEXT CHARACTER
        STAA  SCDR       FINISH CLEARING FLAG
* INCREMENT COUNT OF CHARACTERS SENT
        LDD   TXCNT
        ADDD  #1
        STD   TXCNT
* RETURN FROM INTERRUPT
        BRA   RTSCI      USE COMMON SCI RETURN ROUTINE
*
*------------------------------------------------------
* COMMON RETURN FOR SCI INTERRUPT SERVICE ROUTINES
*------------------------------------------------------
```

Figure 9-22   SCI transmitter interrupt service routine.

```
* RUN DISPATCHER TO CHOOSE NEXT TASK
RTSCI   BSR    DISPTCH  RETURNS TCB IN Y
* RESTORE NEXT TASK'S CONTEXT
        LDS    STKPNTR,Y
        RTI
*
```

Figure 9-22   Continued.

other than the transmitter-owner task now has an opportunity to run while the current character bits are being serially transmitted.

Last character.   If the current interrupt will result in transmitting the last character of the message, the transmit routine disables the transmitter interrupt and enables the transmission complete interrupt. Consequently, the next interrupt will not occur until all the bits of the last character have been sent. Otherwise, this character is treated as any other.

Transmission complete.   When the interrupt at the end of the last character occurs, the polling chain in Figure 9-22 detects an enabled transmission complete interrupt. If the transmission complete flag is also set, it sends control to the service routine in Figure 9-23.

Look at Figure 9-23. The transmission complete routine first disables interrupts from the SCI transmitter because the transmitter has finished sending all the bits of the message. Then it clears the SCI transmitter blocking bit in the owner task's TCB state entry. Remember the SCI DCB is holding the owning task's TCB address. That is, the DCB retained the identity of the owning task so that the task could be unblocked when the message was complete.

The transmission complete routine would now be finished if there was no way to queue multiple transmission requests. However, provision was made to put requests on a waiting list. So the service must now check for waiting requests by testing the TXDEPTH location.

If no requests are waiting, the SCI transmitter is made available, the dispatcher is called, and the context of the next task is restored.

If a request is waiting, the TCB and SCB addresses are removed from the list, and a new transmit request is made. The request for a new transmission is made the same way as the original request in Figure 9-22. Because requests on the waiting list are honored when the transmitter is available, eventually all requests made will be carried out as was assumed at the beginning of this section.

Serial receiver service

Although the SCI receiver is not needed for the weather station design, provision has been made for it to be added. The polling chain in Figure 9-22 tests for an interrupt from the serial receiver. Figure 9-24 is an appropriate device-control-block.

System Performance Indicators

Predicting the load on a computer when multiple tasks run under a real-time operating system is very difficult. Just the same, a measure of the load is very useful to the system developer. The following discusses making an indicator that measures the system load. Of course, this indicator cannot predict the load, but at least, it can measure the load as the system is built and after it is finished.

```
*----------------------------------------------------------
* INTERRUPT SERVICE FOR SCI TRANSMISSION COMPLETE
*----------------------------------------------------------
* TERMINATE MESSAGE - DISABLE TX AND TC INTERRUPTS
SCITC   BCLR   SCCR2-REG,X,BIT6+BIT3   X POINTS TO REG
* REMOVE SCI BLOCKING OF OWNING TASK
        LDX    DCBTX1     X POINTS TO TCB
        BCLR   STATE,X,BIT5
* ANY MORE TRANSMIT REQUESTS PENDING?
        LDAB   TXDEPTH
        BNE    SCITC1     BRANCH ON YES
* MAKE SCI TX AVAILABLE
        BSET   DCBTX,BIT0
        BRA    SCITC2
* MOVE BACK ONE REQUEST (TX STILL UNAVAILABLE)
SCITC1  SUBB   #4
        STAB   TXDEPTH
* RESTORE POINTERS
        LDY    #TXWAIT
        ABY
        LDX    0,Y
        LDY    2,Y        X POINTS TO SCB, Y TO TCB
* TELL DCB WHICH TASK OWNS TX
        STY    DCBTX1
* SET UP TX CHARACTER STRING USING SCB PARAMETERS
        LDD    CHARS,X
        STD    MSGLENT
        LDD    MSGADDR,X
        STD    MSGPT
* INITIALIZE MESSAGE BYTE COUNT
        CLR    TXCNT
        CLR    TXCNT+1
* ENABLE TX AND TX INTERRUPT
        LDX    #REG
        BSET   SCCR2-REG,X,BIT7+BIT3   SENDS IDLE
* RETURN FROM INTERRUPT
SCITC2  BRA    RTSCI      USE COMMON SCI RETURN ROUTINE
*
```

Figure 9-23   SCI transmission complete interrupt service routine.

Free-time indicator

The easiest to make free-time indicator effectively shows the percentage of time the computer is free to run other tasks. From practical experience, you will find that if the free-time is 5 to 20 percent, the computer is very busy and the response to needed processing will be delayed. If the free-time is 80 or 90 percent, the computer will appear to be doing almost nothing, and the response for each task is nearly as good as if it were the only task in the system.

```
* DEVICE CONTROL BLOCK FOR SCI RECEIVER
DCBRC    RMB   1        BIT0=1=AVAIL, 0=NOT AVAIL
DCBRC1   RMB   2        CALLING TASK TCB ADDRESS
RCVCNT   RMB   2        NUMBER OF CHARS RECEIVED
MSGLENR  RMB   2        TOTAL MESSAGE LENGTH
MSGPTRR  RMB   2        POINTER TO RC BUFFER
*
```

Figure 9-24   SCI receiver device-control-block.

Definition of free-time.   Let's define *free-time* as any time the operating system spends running the background task. If the background task, as the one shown in Figure 9-5, only runs an infinite loop when the computer has nothing else to do, this is reasonable. Then, any time not spent in the background task will be considered useful processing time.

Measuring free-time.   The operating system only leaves the background task due to interrupts. Therefore, every time an interrupt occurs, a period of useful work is beginning. In contrast to hardware interrupts occurring while the background task is running, software interrupts (due to an SWI instruction) occur when useful work was already being done inside a task.

Suppose that the program code at the beginning of each ISR clears an output bit, and program code in the background task sets this output bit. Then the voltage at the output pin will be high whenever the computer is in the background task, and low whenever it is doing work. If we observe the voltage on the output pin with an analog DC voltmeter, the voltmeter needle will average the output transitions, and give a relative indication of free-time.

The high level is 5 VDC. So, for example, if the voltmeter reads 4 VDC, the computer is free 80% of the time. As the activity level of the computer varies from moment to moment, the voltmeter needle will wiggle indicating the changes in activity.

Making the indicator.   The example program code in this chapter includes a free-time indicator. The background task in Figure 9-5 sets bit 2 of PORTD. The ISRs in Figures 9-7, 9-11, 9-15, and 9-22 each clear bit 2 of PORTD. The bit is cleared at the beginning of the ISRs for the greatest accuracy. The overhead of adding this feature to the operating system is unnoticeably small.

Overload shutdown

The computer can become overloaded if it has too much to do. In particular, external interrupts, such as those from the anemometer, can happen very frequently possibly causing a large processing burden on the system. If the system has too much to do, it is possible that the response to external interrupts will be slow and that tasks will not run at the proper times.

Predicting the system load is difficult. Also, because the load changes from instant to instant, an averaging free-time indicator cannot indicate that the system was overloaded for a short period of time. However, knowing whether a system is overloaded is important.

Definition of overload.   The operating system developed here uses a real-time clock that interrupts every 10 milliseconds. Therefore, scheduling of tasks is based on periods of 10 milliseconds. Let's define *overload* to mean the system has not completed all the tasks that needed to run within any particular 10-millisecond period. That is, all tasks that want to run should run and complete within each 10-millisecond period. In other words, the background task must run within each 10-millisecond period, or the system is overloaded.

An overload detector.   Let's take the point of view that an overloaded system is not working correctly and that it should be shut down; i.e., the program should stop running. If the system stops running upon overload, troubleshooting the problem and redesigning the offending parts of the software will be relatively easy.

Consider this strategy. Let's have the real-time clock (OC1) interrupt service routine, before it returns from interrupt, set an overload bit to indicate that an overload has occurred. The background task will clear the overload bit. At the beginning of the OC1 ISR, the ISR will test the overload bit. If the overload bit is clear, the ISR determines that an overload did not occur because the background task must have run in the last 10 milliseconds. If the overload bit is set, the background task did not run in the last 10 milliseconds, and the system is overloaded. The OC1 ISR will then shut down the system.

The background task in Figure 9-5 clears an overload bit as described. The OC1 ISR in Figure 9-15 handles the overload bit near the beginning.

Shutting down a computer system that is controlling dangerous equipment must include making the system safe. Generally, a shutdown requires that the computer set all outputs to the hardware reset state.

9.4   WEATHER STATION IMPLEMENTATION

The real-time operating system (RTOS) is now complete, so the weather station can be built using the RTOS as the foundation. Tasks can be added to the operating system to carry out the required functions listed in the specifications. The following sections outline the functions performed by each task, and then discuss the program code. The background task, Task7, has already been considered in Figure 9-5.

As with any design project, design requirements can be met in many ways. You should think about alternative approaches to those discussed here. Sometimes, the approach used here was chosen to illustrate a technique of using the operating system. The multitasking system has been exploited by using one task to control another, for example. Better approaches to meeting any particular requirement may be available.

Task Variables

The tasks needed to implement the weather station require a few associated data locations in RAM memory, and many more in ROM. Assume that the tasks' program code will be in ROM as is typical of embedded applications and that their ROM-based data locations are in memory adjacent to the program code. Figure 9-25 shows the RAM locations used in the following discussions. Figure 9-4 shows the initialization code necessary for the tasks. References will be made to these figures as needed.

Task0—Wind Speed

The weather station specifications say the wind speed must be determined and displayed on both digital displays and on a CRT display. If the wind speed drops to zero, a light on the direction display must be turned on to indicate no wind. In addition, the highest wind speed must be determined and displayed on both digital displays and on a CRT display. The highest wind must be reset to zero with a push button switch.

```
* TASK VARIABLES
*    TASK0 VARIABLES
WSPEED  RMB   1              CURRENT WIND SPEED
HIGHWND RMB   1              HIGHEST WIND SPEED
CURRTIM RMB   3              SYSTEM TIME - CURRENT INTERRUPT
LTIME0  RMB   3              TIME - LAST ANEMOMETER INTERRUPT
TICKS0  RMB   3              PERIOD OF ROTATION OF ANEMOMETER
*    TASK1 VARIABLES
WSECTOR RMB   1              WIND DIRECTION SECTOR NUMBER
LASTWDR RMB   1              LAST WIND DIRECTION
*    TASK2 VARIABLES
MSGB22  RMB   4              WIND SPEED MESSAGE
MSG2201 RMB   2              ..CURRENT
MSG2202 RMB   4
MSG2203 RMB   2              ..HIGHEST
MSGE22  EQU   *
*    TASK3 VARIABLES
BLANKED RMB   1
MSGB33  RMB   6              WIND DIRECTION MESSAGE
MSGE33  EQU   *
*    TASK4 VARIABLES
*    TASK5 VARIABLES
*    TASK6 VARIABLES
*    TASK7 VARIABLES
*
```

Figure 9-25   Task data locations in RAM memory.

Task strategy

If the computer can measure the amount of time for each revolution of the anemometer, it can determine the wind speed. The anemometer opens and closes a switch on each revolution. Figure 9-1 shows that the anemometer was wired to cause an interrupt on each revolution, so let's use the interrupt to schedule a task to run. The task can measure time and determine the period of rotation.

Immediate response to the interrupt is necessary so that the timing will be as accurate as possible. The highest-priority task will run as close to immediately after an interrupt as is possible, so let's use Task0 to determine the wind speed.

Measuring rotation time.   The time for a revolution of the anemometer is relatively large—about one second for common wind speeds. Using the programmable timer in the 68HC11 is much too difficult, and its range of time without considering overflows is much too small. So, to measure the time, let's use the RTOS system clock which has a resolution of 10 milliseconds. Because a revolution of the anemometer will take much longer than 10 milliseconds, using the system clock will give adequate results.

On each revolution of the anemometer, Task0 will run and read the system time. If the time of the last run was saved, the difference can be calculated giving the elapsed time which is the period of rotation. Task0 can then determine the wind speed and display it. However, several problems relating to this scheme must be overcome.

Timing problems.   If Task0 updates the CRT wind speed message on each anemometer interrupt, the interrupts may come faster than the CRT can be updated. The CRT display uses serial communication, so it is relatively slow. Let's just have Task0 save the wind speed in memory, and use another task (let's use Task2) to read the speed and do the display work at its own pace.

Zero wind speed problem.   Another problem is the lack of interrupts when the wind speed drops to zero. Task0 will not run, so the last wind speed will be left in memory, and a speed of zero will never be displayed. Let's overcome this problem by using another task in our multitasking system. Because zeroing the wind speed is not a time critical operation, let's use Task5 for this job. Task5 is at a relatively low priority.

Task0 can start Task5 after a delay, and Task5 can set the wind speed to zero. If Task0 runs again before the delay runs out, it can start Task5 again with delay. If the wind continues blowing, Task5 will never run. But if the wind stops, Task5 will run after a delay. If the delay corresponds to a time longer than the anemometer rotation period for a 1 mph wind, Task5 will make the wind speed zero.

The Task0 code

Figure 9-26 shows Task0. Let's look at the program code. It begins by scheduling Task5 to run after 2560_{10} milliseconds. This scheduling period is longer than one rotation period at 1 mph, and longer than the maximum rotation period the program can handle.

Next, the no-wind light on the wind direction display is turned off. If Task0 is running, wind must be rotating the anemometer, so this light should go off.

Now Task0 reads the current system time and finds the difference between it and the time of the previous interrupt. If the current time went past midnight, a correction is made. The resulting elapsed time is the period of rotation for the anemometer measured in 10-millisecond ticks. The current time is stored for use on the next interrupt. As the final part of measuring the period, Task0 tests the elapsed time to determine if it exceeds a single-byte value, and if so, saturates the time at FF ticks. Saturating the number of ticks limits the greatest elapsed time the program can handle.

To find wind speed, the program uses the elapsed time as the independent variable in a table lookup; Figure 9-26 includes a wind speed table for the anemometer. A table lookup is appropriate here because the table can be built from calibration data for the anemometer resulting in good accuracy. The program stores the wind speed obtained from the table lookup at location WSPEED so that other tasks can use it.

Figure 9-27 shows a table-lookup subroutine that uses linear interpolation between points. Notice the importance of making it a reentrant subroutine. That is, other tasks may want to use this subroutine as the system is built. It is almost certain that sometime another task, after calling this subroutine, will be interrupted and Task0 will run in response. Thus the subroutine must be reentrant. Making the subroutine reentrant is not difficult—it just uses the microprocessor registers and the stack.

Returning to Figure 9-26, the current wind speed found by the table lookup subroutine is compared to the previously stored highest wind speed. The higher of the two is stored at location HIGHWND for other tasks to use.

Finally, Task0 suspends itself so that other tasks can run until another anemometer interrupt schedules Task0 to run. When it does run again, Task0 loops back to the beginning effectively making an infinite loop of the entire task.

```
*-------------------------------------------------------
** TASK0 - DETERMINE WIND SPEED
*** RUNS ON EACH REVOLUTION OF ANEMOMETER
*-------------------------------------------------------
* START TASK5 TO ZERO WIND SPEED AFTER DELAY
TASK0   LDX    #SCB01
        SWI                RETURN IMMEDIATELY
* TURN OFF NO-WIND LIGHT
        SEI                AVOID CONCURRENCY PROBLEM
        LDAA   PORTC
        ANDA   #$FF-BIT5
        STAA   PORTC
        CLI
* READ SYSTEM CLOCK
        LDX    #SCB02
        SWI
* FIND TIME SINCE LAST TIME (PERIOD OF ROTATION)
        LDX    #CURRTIM
        LDD    1,X
        SUBD   LTIME0+1
        STD    TICKS0+1
        LDAA   0,X
        SBCA   LTIME0
        STAA   TICKS0
* SYSTEM CLOCK ROLLED OVER BEYOND ZERO?
        BCC    TSK01     BRANCH ON NO
* CORRECT FOR ROLLOVER
        LDD    #MIDNLSB
        ADDD   TICKS0+1
        STD    TICKS0+1
        LDAA   #MIDNMSB
        ADCA   TICKS0
        STAA   TICKS0
* UPDATE LAST TIME
TSK01   LDD    1,X
        STD    LTIME0+1
        LDAA   0,X
        STAA   LTIME0
* SATURATE AT MAXIMUM USABLE ROTATION PERIOD
        LDD    TICKS0    UPPER 16 BITS 0?
        BEQ    TSK02     BRANCH ON YES
        LDAA   #$FF      SATURATE AT $FF TICKS
        STAA   TICKS0+2
* UPDATE CURRENT WIND SPEED
TSK02   LDAA   TICKS0+2 8-BIT TIME
        LDX    #WINDSPD
        JSR    TABLKUP   USE CALIBRATION TABLE
        STAA   WSPEED
* CURRENT WIND SPEED HIGHER THAN HIGHEST WIND SPEED?
        CMPA   HIGHWND
        BLS    TSK03     BRANCH ON NO
* UPDATE HIGHEST WIND SPEED
        STAA   HIGHWND
* SUSPEND TASK0
TSK03   LDX    #SCB03
        SWI
* GO AROUND AGAIN AT NEXT ACTIVATION
        BRA    TASK0
*
*** ANEMOMETER WIND SPEED CALIBRATION TABLE
WINDSPD FCB    99,48,23,15,11,8,6,5,4
        FCB    4,3,3,2,2,1,1,0
```

Figure 9-26   RTOS Task0.

```
*** SERVICE CONTROL BLOCKS
SCB01   FCB   SCHED     SCHEDULE TASK 5
        FDB   TCB5
        FCB   DEL+IMM
        FCB   0,1,0     ..2560 MILLISECONDS
SCB02   FCB   READCLK   READ SYSTEM TIME
        FDB   CURRTIM
SCB03   FCB   SUSPEND   SUSPEND TASK0
        FDB   TCB0
*
```

Figure 9-26 Continued.

Task5—Zero Wind Speed

Task5 sets the wind speed to zero because Task0 cannot handle this case, so consider it next. The primary job is to zero the wind speed value and update the wind speed message. It also turns on the No-Wind light as the weather station specifications require. Task5 is very short with little complexity.

Task strategy

While there is wind, Task0 repeatedly reschedules Task5 to run in the future, and Task5 never runs. If Task0 doesn't run within 2560_{10} milliseconds, Task5 runs as scheduled. When

```
*-----------------------------------------------------
* TABLE LOOKUP SUBROUTINE
*** 8-BIT UNSIGNED NUMBERS EVERY $10, X AND B CHANGED
*-----------------------------------------------------
* DIVIDE BY $10 TO GET TABLE OFFSET AND DELTA X
TABLKUP LDAB  #$10
        MUL             A-OFFSET, B-DELTA X TIMES $10
* FORM TABLE POINTER
        PSHB
        TAB
        ABX
* POSITIVE SLOPE?
        LDAA  0,X
        SUBA  1,X       CALCULATE DELTA Y
        BLO   TAB1      BRANCH ON YES
* INTERPOLATE VALUE WITH NEGATIVE SLOPE
        PULB
        MUL             DROP B TO DIVIDE BY $100
        ADCA  #0        ROUND
        TAB
        LDAA  0,X
        SBA             SUBTRACT DELTA Y
        BRA   TAB2
* INTERPOLATE VALUE WITH POSITIVE SLOPE
TAB1    LDAA  1,X
        SUBA  0,X       CALCULATE DELTA Y
        PULB
        MUL             DROP B TO DIVIDE BY $100
        ADCA  0,X       ADD DELTA Y, ROUND
* RETURN
TAB2    RTS
*
```

Figure 9-27 Table lookup subroutine.

it runs, it zeros the wind speed and starts another task to display the new wind speed. Task5 is set ready by the initialization program in Figure 9-4 so that it runs at system startup.

The Task5 code

Look at Figure 9-28. Task5 completes its primary function in one instruction when it zeros the wind speed.

Next, the wind speed message on the CRT screen must be updated for the no-wind condition. Task5 hands this job off to another task (let's use Task3) by scheduling it with the shortest possible delay. That is, Task3 will run during the next 10-millisecond period. For simplicity, the operating system does not have a system service to start another task immediately, but a small delay in displaying a message is of no consequence to the weather station.

Task3 will display the appropriate wind direction messages on the CRT. Task5 could display the CRT message, but other tasks will display wind direction, so it would be redundant for Task5 to make wind direction messages.

Another requirement of the weather station is a no-wind light. The no-wind light indicates that the wind direction lights cannot display the correct wind direction. Task5 sets PORTC bit PC5 to turn on this light. PORTC is susceptible to concurrency problems because other tasks will use its other bits. Therefore, to prevent concurrent access to PORTC, the interrupt system is disabled while PORTC is changed. This brute force solution to the concurrency problem increases the system's minimum response time to interrupts. On the other hand, the time the interrupt system is disabled here is very small and of little consequence to the weather station. In other applications, this delay may be unacceptable.

```
*-------------------------------------------------------
** TASK5 - ZERO WIND SPEED
*** RUNS AT STARTUP AND AS SCHEDULED BY TASK0
*-------------------------------------------------------
* SET WIND SPEED TO ZERO
TASK5   CLR   WSPEED
* START TASK3 TO BLANK WIND DIRECTION MESSAGE
        LDX   #SCB51
        SWI
* TURN ON NO-WIND LIGHT
        SEI              AVOID CONCURRENCY PROBLEM
        LDAA  PORTC
        ORAA  #BIT5
        STAA  PORTC
        CLI
* SUSPEND TASK5
        LDX   #SCB52
        SWI
* GO AROUND AGAIN AT NEXT ACTIVATION
        BRA   TASK5
*
*** SERVICE CONTROL BLOCKS
SCB51   FCB   SCHED   SCHEDULE TASK3
        FDB   TCB3
        FCB   DEL+IMM  ..DON'T SUSPEND TASK5
        FCB   0,0,1    ..ALMOST IMMEDIATELY
SCB52   FCB   SUSPEND  SUSPEND TASK5
        FDB   TCB5
*
```

Figure 9-28   RTOS Task5.

Finally, Task5 suspends itself because its job is complete. It will run again in the future only when scheduled again by Task0. When it does run, it loops back to its beginning, running the entire task once again.

Task1—Wind Direction

The weather station specifications require it to display the wind direction on both the direction indicator lights and the CRT screen. The lights and the CRT message must be updated periodically every 110 milliseconds if the message changes. Furthermore, if the wind speed is zero, the direction lights must be turned off and the CRT message blanked. Let's use Task1 to meet these requirements.

Task strategy

Task1 must read the voltage from the wind vane and convert it into one of eight directions. It then must update the direction lights and the message on the CRT display accordingly.

The priority of Task1 is second highest, but it probably doesn't make too much difference when updating wind direction—no stringent requirements are on the timing. Task1 will run periodically every 110 milliseconds to meet the timing specifications. It could run more frequently, but that would just increase the system overhead with little gain.

Task1 is set ready by the initialization program in Figure 9-4, so it runs at system startup. The initialization program also zeros the wind speed, so Task1 will turn off the wind direction lights and blank any wind direction message left on the CRT prior to the system startup.

The Task1 code

Look at Figure 9-29 for Task1. It begins by determining whether the wind speed is zero. Remember the wind speed is left in a memory byte by Task0 or Task5. Because wind speed is a single-byte number, a concurrency problem does not occur as Task0, Task1, and Task5 all access it.

Wind. If the wind speed is not zero, the program branches to the section that reads the analog voltage from the wind vane. The A/D converter is started, the program delays long enough for the A/D to do its conversion, and the voltage is read.

The weather station specifications require the direction indication to specify one of eight directions. However, a given direction should be indicated if the actual direction is within 22.5 degrees either side of that direction. The easiest way to solve this problem is to divide the direction into 16 sectors of 22.5 degrees each. Then two adjacent sectors will both indicate the same direction. To do this, the program divides the voltage from the wind vane by 16. The result is one of 16 values for the direction which the program stores at WSECTOR.

Because the wind direction does not change from one 45-degree sector to another frequently, the system overhead of updating the direction displays can be reduced by

```
*-----------------------------------------------------------
** TASK1 - WIND DIRECTION
*** RUNS AT STARTUP AND PERIODICALLY EVERY 110 MSEC.
*-----------------------------------------------------------
* WIND SPEED ZERO?
TASK1   TST     WSPEED
        BNE     TSK11       BRANCH ON NO
* INDICATE WIND DIRECTION NOT POSSIBLE
        BSET    WSECTOR,$FF   MAKE INVALID WIND DIRECTION
        BSET    LASTWDR,$FF   UPDATE LAST WIND DIRECTION
* BLANK WIND DIRECTION LIGHTS
        CLR     PORTB
* START TASK3 TO BLANK CRT WIND DIRECTION DISPLAY
        LDX     #SCB11
        SWI
        BRA     TSK12
* READ ANALOG VOLTAGE FROM WIND VANE
TSK11   LDAA    #$01        SCAN=MULT=0, CHANNEL 1
        STAA    ADCTL       START A/D CONVERTER
        MUL                 WAIT FOR A/D TO OPERATE
        MUL
        MUL
* UPDATE NEW WIND DIRECTION
        LDAB    #%00010000  A LITTLE MORE TIME
        LDAA    ADR1        READ WIND VANE VOLTAGE
        MUL                 16 SECTORS/360 DEGR/8 LIGHTS
        STAA    WSECTOR
* WIND DIRECTION CHANGED?
        CMPA    LASTWDR
        BEQ     TSK12       BRANCH ON NO
* UPDATE LAST WIND DIRECTION
        STAA    LASTWDR
* UPDATE WIND DIRECTION DISPLAY UNIT
        LDX     #TABWDIR USE TABLE TO CONVERT SECTOR TO
        TAB                 ..DISPLAY PATTERN
        ABX
        LDAA    0,X
        STAA    PORTB
* START TASK3 TO DISPLAY NEW WIND DIRECTION ON CRT
        LDX     #SCB11
        SWI
* DELAY MYSELF 110 MILLISECONDS
TSK12   LDX     #SCB12
        SWI
* GO AROUND AGAIN AT NEXT ACTIVATION
        BRA     TASK1
*
*** WIND DIRECTION LIGHT TABLE
TABWDIR FCB     BIT0,BIT7,BIT7,BIT6,BIT6,BIT5,BIT5,BIT4
        FCB     BIT4,BIT3,BIT3,BIT2,BIT2,BIT1,BIT1,BIT0
*** SERVICE CONTROL BLOCKS
SCB11   FCB     SCHED       SCHEDULE TASK3
        FDB     TCB3
        FCB     DEL+IMM   ..DELAY BUT DON'T SUSPEND
        FCB     0,0,1     ..ALMOST IMMEDIATELY
SCB12   FCB     SCHED       SCHEDULE TASK1
        FDB     TCB1
        FCB     DEL+SUS   ..DELAY WITH SUSPEND
        FCB     0,0,11    ..110 MILLISECONDS
*
```

Figure 9-29   RTOS Task1.

updating the displays only when needed. So the program determines whether the wind direction has changed since the last update. If the wind direction has not changed, Task1 branches to the end and reschedules itself (with suspend) to run 110 milliseconds later. If the wind direction has changed, Task1 saves the new direction and updates the eight wind direction display lights. The program uses a basic table lookup to convert wind sector to the bit pattern for the display unit.

Look carefully at the wind-direction light table to see that the north light turns on for both the smallest and the largest voltage from the wind vane. Because direction was broken into 16 sectors, making the north light work correctly is easy.

The bit pattern from the table is put into PORTB which updates the eight wind direction display lights. All the bits of PORTB are used exclusively for these lights which only Task1 controls; therefore, a concurrency problem does not occur here.

Now, Task1 starts Task3 as soon as possible to update the CRT wind direction message. Task5 uses this same approach to update the direction message. Task3 uses the information in WSECTOR to determine the proper message to display.

Finally, Task1 reschedules itself (with suspend) to run 110 milliseconds later. When it does run, it loops back to the beginning and runs the entire task once again.

No wind.   If, at the beginning, Task1 finds that the wind speed is zero, it makes the wind direction invalid. Because only 16 sectors are possible, valid sectors have 0s in bits 4 through 7 of WSECTOR. Putting FF in WSECTOR makes the direction invalid. Task3 must blank invalid directions.

Task1 updates the direction only if it has changed; therefore, the changed flag must be set if the wind speed is zero. Otherwise, the direction will not be updated properly.

Finally, Task1 clears PORTB to turn off the direction lights, starts Task3 to update the CRT display, and reschedules itself (with suspend) to run 110 milliseconds later.

Task3—Wind Direction Message

The weather station specifications require a wind direction message on the CRT display. The message must be blanked if the wind speed drops to zero. The blanking can be done by printing space characters on the screen.

Task strategy

Task3 displays one of two wind direction messages on the CRT display. One message is a series of space characters to blank the previous wind direction message. Once the message has been blanked, there is no need to blank it again, so this task tracks whether the CRT display is already blanked to reduce the system overhead. The second message includes the wind direction and an identifying character string.

Both Task 1 and Task5 schedule Task3 when it is needed to update the wind direction message on the CRT. Task3 runs to do its job, and then suspends itself. Therefore, Task3 only runs when scheduled by another task. Task3 has no time critical requirements because it only controls the CRT display through a relatively slow serial connection; it could run at any priority without any significant change in its performance.

The Task3 code

Look at Figure 9-30 for Task3. The program begins by testing for a valid wind sector number in WSECTOR. Because there are only 16 valid values, bit 7 will be 0 for valid directions. Valid wind sector values tell Task3 to display the wind direction. Invalid values tell it to blank the wind direction message.

Valid message. Consider the case of a valid wind direction. The wind direction message consists of two parts. The first part, called the *header*, is a fixed text string such as "THE WIND DIRECTION IS ." The second part, called the *value*, is the two-character string that indicates the current wind direction.

The SCI transmit system service requires the character strings to be in memory. Because the first part of the message is fixed, it can be put into ROM; the second part is variable, so it must be put into RAM. To accommodate the memory types, the two parts of the wind direction message are transmitted with two service calls.

Continue looking at the program in Figure 9-30. After Task3 determines that a valid message is needed, it first clears the indicator that says the message is blanked. Then it converts the value in WSECTOR to a two-character string by using a basic table lookup. The

```
*-------------------------------------------------------
** TASK3 - DISPLAY WIND DIRECTION ON CRT
*** RUNS WHEN SCHEDULED BY TASK1 OR TASK5
*-------------------------------------------------------
* WIND DIRECTION VALID?
TASK3   TST    WSECTOR
        BPL    TSK31      BRANCH ON YES
* MESSAGE ALREADY BLANKED?
        TST    BLANKED
        BNE    TSK32      BRANCH ON YES
* SET MESSAGE BLANKED FLAG TRUE
        BSET   BLANKED,$FF
* BLANK WIND DIRECTION MESSAGE ON CRT
        LDX    #SCB31
        SWI
        BRA    TSK32
* SET MESSAGE BLANKED FLAG FALSE
TSK31   CLR    BLANKED
* PUT WIND DIRECTION INTO MESSAGE
        LDAB   WSECTOR
        LDX    #TABWCHR CONVERT SECTOR TO CHARACTERS
        LSLB            ..TWO CHARACTERS PER SECTOR
        ABX
        LDD    0,X
        STD    MSGB33+4
* START SCI TRANSMITTER TO SEND DIRECTION MESSAGE
        LDX    #SCB32     HEADER
        SWI
        LDX    #SCB33     VALUE
        SWI
* SUSPEND MYSELF
TSK32   LDX    #SCB34
        SWI
* GO AROUND AGAIN WHEN ACTIVATED BY TASK1 OR TASK5
        BRA    TASK3
*
```

Figure 9-30   RTOS Task3.

```
*** DIRECTION SECTOR TO CHARACTER STRING TABLE
TABWCHR FCC     ' NNWNW W WSWSW S SSESE E ENENE N'
*** TASK 3 MESSAGE 1 TABLE
MSGB31  FCB     $1B,$59,$2C,$3A  POSITION CURSOR
        FCC     '                              '
*** TASK 3 MESSAGE 2 TABLE
MSGB32  FCB     $1B,$59,$2C,$3A  POSITION CURSOR
        FCC     'WIND DIRECTION IS FROM  '
*** SERVICE CONTROL BLOCKS
SCB31   FCB     SCITX     BLANK THE MESSAGE
        FDB     MSGB32-MSGB31
        FDB     MSGB31
SCB32   FCB     SCITX     WIND DIRECTION HEADER
        FDB     SCB31-MSGB32
        FDB     MSGB32
SCB33   FCB     SCITX     WIND DIRECTION VALUE
        FDB     MSGE33-MSGB33
        FDB     MSGB33
SCB34   FCB     SUSPEND   SUSPEND TASK3
        FDB     TCB3
*
```

Figure 9-30 Continued.

FCC assembler directive makes creating the ASCII character codes in the table easy. Notice, by looking at the conversion table, that the lowest and highest wind sector values result in the character string identifying north.

Task3 stores the direction character string from the table in a RAM location (see Figure 9-25) specified by SCB33. Then, it makes two system service calls, using SCB32 and SCB33, to transmit the two message parts to the CRT display.

Finally, Task3 suspends itself so that it waits to be scheduled again in the future. When it runs again, it loops back to repeat the entire task.

Notice that the initialization program in Figure 9-4 supplied some of the message characters in RAM. For example, it provided the characters to position the cursor at desired places on the CRT screen for a neat message format.

Blank message. If the wind sector was invalid, the Task3 program first checks whether the message has already been blanked. If already blanked, the task goes to its end.

If an existing message must be blanked, Task3 sets the blanked indicator and makes a system service call using SCB31. The service call sends space characters to the CRT to remove the previous wind direction message. Following this operation, Task3 suspends itself so that it waits to be scheduled again in the future.

Task2—Wind Speed Message on CRT

The specifications for the weather station require current wind speed and highest wind speed messages on the CRT display. These messages must be updated at least every 150 milliseconds.

Another requirement of the weather station is a header message on the CRT display to identify the display as a weather station. The header message does not need to change. Let's use Task2 to meet these requirements.

Task strategy

The first job of Task2 is to provide the header message on the CRT. Displaying the message simply requires a system service call. The second job is displaying the wind speed messages every 150 milliseconds. This part of the task must run periodically using an infinite loop to display the wind speeds provided by Task0 and Task5. The first part of Task2 only needs to run once at system startup; therefore, it will be placed outside the infinite loop. The system initialization in Figure 9-4 sets this task ready at startup so that the header is displayed as soon as possible.

Effectively, the first part of Task2 does an initialization, and the loop provides periodic functions. Using part of a task to do an initialization is very straightforward and very useful. The initialization can be written as part of the task rather than putting it into the operating system code.

The Task2 code

Remember that Task2 runs at system startup. So the program in Figure 9-31 begins by displaying the header message on the CRT. Near the end of this task, it schedules itself (with suspend) to run again in the future. When it does, notice that it does not loop back to the beginning, but to an intermediate point. Therefore, the header message is transmitted only at system startup. This task has effectively carried out an initialization that will not be repeated as the system runs.

As in Task3, the wind speed message is transmitted in two parts, a fixed part and a variable part. In this task, the fixed part was combined with the CRT header message as specified by SCB21. Look at the message in the table labeled MSGB21 to see the header and two wind speed messages merged into one message.

```
*------------------------------------------------------------
** TASK2 - DISPLAY CURRENT/HIGHEST WIND SPEEDS ON CRT
*** RUNS AT STARTUP AND PERIODICALLY EVERY 150 MSEC
*------------------------------------------------------------
* START SCI TRANSMITTER TO DISPLAY HEADER MESSAGE
TASK2    LDX    #SCB21
         SWI
* PUT CURRENT WIND SPEED INTO CRT MESSAGE
TSK21    LDAA   WSPEED
         JSR    BINBCD   CONVERT BINARY TO BCD
         JSR    BCDTXT   CONVERT BCD TO ASCII
         STD    MSG2201
* PUT HIGHEST WIND SPEED INTO CRT MESSAGE
         LDAA   HIGHWND
         JSR    BINBCD   CONVERT BINARY TO BCD
         JSR    BCDTXT   CONVERT BCD TO ASCII
         STD    MSG2203
* START SCI TRANSMITTER TO SEND MESSAGE
         LDX    #SCB22
         SWI
* DELAY MYSELF 150 MILLISECONDS
         LDX    #SCB23
         SWI
* GO AROUND AGAIN AT NEXT ACTIVATION
         BRA    TSK21
*
```

Figure 9-31   RTOS Task2.

```
*** TASK 2 MESSAGE 1 TABLE
MSGB21   FCB    $0C,$1B,$59,$21,$3F   POSITION CURSOR
         FCC    'WEATHER STATION'
         FCB    $1B,$59,$23,$40       POSITION CURSOR
         FCC    'Gene H. Miller'
         FCB    $1B,$59,$31,$3A       POSITION CURSOR
         FCC    'CURRENT WIND SPEED IS  '
         FCB    $1B,$59,$33,$3A       POSITION CURSOR
         FCC    'HIGHEST WIND SPEED IS  '
MSGE21   EQU    *
*** SERVICE CONTROL BLOCKS
SCB21    FCB    SCITX      SCREEN HEADER MESSAGE
         FDB    MSGE21-MSGB21
         FDB    MSGB21
SCB22    FCB    SCITX      WIND SPEED MESSAGE
         FDB    MSGE22-MSGB22
         FDB    MSGB22
SCB23    FCB    SCHED      SCHEDULE TASK2
         FDB    TCB2
         FCB    DEL+SUS    ..DELAY TASK2 WITH SUSPEND
         FCB    0,0,15     ..150 MILLISECONDS
*
```

Figure 9-31   Continued.

Following displaying the header message, the program prepares the variable wind speed messages. The wind speed values must be converted to character strings for the CRT. The first subroutine in Figure 9-32 converts the wind-speed binary values to BCD numbers. Then, the second subroutine in Figure 9-32 converts the BCD numbers to character strings. Because these subroutines will likely be used in other tasks, they must be reentrant subroutines. These character strings are put into the RAM area designated by SCB22.

Finally, in Figure 9-31, Task2 transmits the variable part of the wind speed messages. It then schedules itself to run again in 150 milliseconds, and it is suspended until then.

```
*------------------------------------------------------
* BINARY TO BCD SUBROUTINE
*** CONVERT 8-BIT NUMBER IN A TO TWO-DIGIT BCD
*** ..NUMBER WITH MSD IN A AND LSD IN B
*------------------------------------------------------
* SAVE X REGISTER
BINBCD   PSHX
* GET LEAST SIGNIFICANT DIGIT
         TAB              CONVERT 8-BIT TO 16-BIT
         CLRA
         LDX    #10
         IDIV
         PSHB
* GET MOST SIGNIFICANT DIGIT
         XGDX
         LDX    #10
         IDIV
         TBA
* RETURN
         PULB
         PULX
         RTS
*
```

Figure 9-32   Subroutines.

```
*-----------------------------------------------------------
* BCD TO ASCII SUBROUTINE
*** CONVERT BCD DIGITS IN A AND B TO TWO ASCII
*** ..CHARACTERS IN  A AND B
*-----------------------------------------------------------
* CONVERT MOST SIGNIFICANT DIGIT TO ASCII OR SPACE
BCDTXT   TSTA
         BNE     BIN1      BRANCH ON NONZERO CHARACTER
         LDAA    #$20      SPACE CHARACTER
         BRA     BIN2
BIN1     ADDA    #$30
* CONVERT LEAST SIGNIFICANT DIGIT TO ASCII
BIN2     ADDB    #$30
* RETURN
         RTS
*
```

Figure 9-32   Continued.

Task4—Wind Speed Digital Displays

The weather station specifications require that two digital displays show the current wind speed or the highest wind speed. A push button switch is pushed for the highest speed and released for the current speed. The weather station must update the digital displays at least every 100 milliseconds. In addition, an LED indicator must illuminate when the highest speed is displayed. Let's use Task4 to meet these requirements.

Task strategy

Task4 is very straightforward because its sole job is to control the digital displays. Either Task0 or Task5 provide the wind speed value to display. The initialization program in Figure 9-4 sets the task ready so that it runs at system startup making the displays show the initial zero wind speed.

The Task4 code

Look at the program in Figure 9-33. It begins by testing PORTD bit PD5 to determine if the highest-speed push button is pushed. If the button is pushed, the program turns on the highest wind speed light using PORTD bit PD4, and then reads the highest speed. If the push button is not pushed, the program turns off the highest speed light and reads the current wind speed.

The wind speed is a binary number. However, BCD numbers are required to control the display hardware. The BINBCD subroutine in Figure 9-32 converts the binary wind speed number to a two-digit BCD number.

Notice the need for the BINBCD subroutine to be reentrant. So far, it is also used by Task2, and it could easily be interrupted and reentered. Remember, the tasks, and therefore this subroutine, run with interrupts enabled.

Following the conversion to BCD, the most significant digit is tested for zero. If it is zero, the digital displays will be easier to read if this digit is not displayed. Such a zero digit is converted to the hexadecimal digit F, which causes the display unit to blank all its lights.

```
*-------------------------------------------------------------
** TASK4 - UPDATE DIGITAL WIND SPEED DISPLAYS
*** RUNS AT STARTUP AND PERIODICALLY EVERY 100 MSEC
*-------------------------------------------------------------
* DISPLAY CURRENT OR HIGHEST SPEED?
TASK4   LDX    #PORTD
        BRCLR  0,X,BIT5,TSK41  BRANCH ON CURRENT
* HIGHEST SPEED
        BSET   0,X,BIT4    TURN ON HIGHEST SPEED LIGHT
        LDAA   HIGHWND
        BRA    TSK42
* CURRENT SPEED
TSK41   BCLR   0,X,BIT4    TURN OFF HIGHEST SPEED LIGHT
        LDAA   WSPEED
* GET WIND SPEED DIGITS
TSK42   JSR    BINBCD
* BLANK MSD DISPLAY IF MSD IS ZERO
        TSTA
        BNE    TSK43
        LDAA   #$F
* UPDATE LOW DIGIT DISPLAY (USES LATCH)
TSK43   PSHB               SAVE WIND SPEED DIGITS IN STACK
        PSHA
        TSY                Y POINTS TO DIGITS
        LDX    #PORTC      X POINTS TO PORTC
        SEI                AVOID CONCURRENCY PROBLEM
        LDAA   0,X
        ANDA   #%11100000
        ORAA   1,Y         INSERT DIGIT
        STAA   0,X
        CLI
        BSET   0,X,BIT4    LATCH LOW DISPLAY
        BCLR   0,X,BIT4
* UPDATE HIGH DIGIT DISPLAY
        SEI                AVOID CONCURRENCY PROBLEM
        LDAA   0,X
        ANDA   #%11100000
        ORAA   0,Y         INSERT DIGIT
        STAA   0,X
        CLI
        INS                CLEAN UP STACK
        INS
* DELAY MYSELF 100 MILLISECONDS
        LDX    #SCB41
        SWI
* GO AROUND AGAIN ON NEXT ACTIVATION
        BRA    TASK4
*
*** SERVICE CONTROL BLOCK
SCB41   FCB    SCHED       SCHEDULE TASK4
        FDB    TCB4
        FCB    DEL+SUS   ..DELAY TASK4 WITH SUSPEND
        FCB    0,0,10    ..100 MILLISECONDS
*
```

Figure 9-33   RTOS Task4.

Task4 now updates the digital displays. Look at Figure 9-1 that specifies the weather station hardware. PORTC bits PC0 through PC3 connect both to a latch chip and to the high-digit or most-significant display unit. The outputs of the latch chip connect to the low-digit or least-significant display unit. The program must update the latch for the least-significant digit first, and then update the most-significant digit. PORTC bit PC4 controls the latch enable.

Look at the program in Figure 9-33 once again. Observe that the program reads PORTC and masks the least-significant five bits. Then, it packs the four bits for the least-significant display unit into the number, and writes it back to PORTC which applies the 4-bit value to the latch. Because other tasks control the three most-significant PORTC bits, the interrupt system is disabled during these manipulations to prevent concurrency problems.

The next step is to update the latch. The program drives PORTC bit PC4 to the high level to enable the latch, and then to the low level so that it holds the 4-bit number. A concurrency problem will not occur while latching because the BSET and BCLR instructions cannot be interrupted.

The program next updates the most-significant or high-digit display unit. The program reads PORTC and masks the least-significant five bits again. Then it packs the four bits for the most-significant display unit into the number, and writes it back to PORTC which applies the 4-bit number to the display unit. The program once again disables the interrupt system during these manipulations to prevent concurrency problems.

After updating the displays, Task4 finally reschedules itself to run in 100 milliseconds. It is suspended until then.

Task6—Reset Highest Wind Speed and Flash Light

The weather station specifications require a push button switch that zeros the highest wind speed. Following such a reset, the highest wind speed will be updated starting with the current wind speed almost immediately. In addition, a flashing light for troubleshooting must be provided. Figure 9-1 shows the hardware connections.

If the flashing light is flashing properly, probably most of the weather station is operating correctly. Many commercial devices use this flashing-light technique to make the user confident that a system is working properly.

The flashing light can be quite useful in other ways. The human eye is very sensitive to changes in the rate of a rapidly flashing light. When you observe the light, you should see what apparently is a perfectly regular flash rate. However, in actuality, the period of the changes varies according to the tasks run by the operating system. Because Task6 is the lowest-priority task, anything that wants to run and delay this task will do so. Yet your eye should not be able to detect any changes.

If you see the light seemingly hesitate at times, you probably have concurrency problems. The light bit is being changed incorrectly as other bits in the output port are updated.

Task strategy

Task6 is very simple. It monitors a push button switch, and zeros the highest wind speed if the button is pushed. It also complements the output bit that controls the flashing light.

Task6 is initialized by the program in Figure 9-4 to run at startup. Probably the only value to this initialization is that Task6 will turn the flashing light on at startup. If an error in the system blocked lower-priority tasks from running, the error would be apparent because the light would not change immediately upon startup.

The Task6 code

Look at the program in Figure 9-34. Task6 first tests PORTC bit PC7. If it is zero, the main function of the task is finished. If it is one, the highest wind speed value is set to zero.

A PORTC input bit is only read here; therefore, no concurrency problem occurs because nothing is stored back to PORTC. Similarly, the highest wind speed can be set to zero because it is a single-byte number, and the CLR instruction cannot be interrupted. No concurrency problem occurs when a single instruction can complete an operation.

To make a flashing light, Task6 complements the output PORTC bit PC6 each time it runs. Because PORTC is changed, the program must disable the interrupt system in the critical section to avoid the concurrency problem.

Finally, Task6 schedules itself to run again after a delay so that it runs periodically every 30 milliseconds.

```
*-----------------------------------------------------------
** TASK6 - HIGHEST WIND SPEED RESET AND FLASH LIGHT
*** RUNS PERIODICALLY EVERY 30 MILLISECONDS
*-----------------------------------------------------------
* RESET BUTTON PUSHED?
TASK6   LDAA    PORTC
        ANDA    #BIT7
        BEQ     TSK61       BRANCH ON NO
* ZERO HIGHEST WIND SPEED
        CLR     HIGHWND
* REVERSE LIGHT 6
TSK61   SEI                 AVOID CONCURRENCY PROBLEM
        LDAA    PORTC
        EORA    #BIT6
        STAA    PORTC
        CLI
* DELAY MYSELF 30 MILLISECONDS
        LDX     #SCB61
        SWI
* GO AROUND AGAIN AT NEXT ACTIVATION
        BRA     TASK6
*
*** SERVICE CONTROL BLOCK
SCB61   FCB     SCHED       SCHEDULE TASK6
        FDB     TCB6
        FCB     DEL+SUS     ..DELAY WITH SUSPEND
        FCB     0,0,3       ..30 MILLISECONDS
*
```

Figure 9-34 RTOS Task6.

Final Details

The weather station is an interrupt-driven system, so the interrupt vectors must be initialized. Figure 9-35 shows the vectors for a production 68HC11 system, or a trainer that allows interrupt vectors in RAM such as the Motorola M68HC11EVM. If the Motorola M68HC11EVB trainer is used, the vectors are set up as in Figure 9-36. See Appendix C for information on a hardware reset with this trainer.

The weather station is now complete. The source code presented in this chapter can be assembled, loaded, and run. You must supply appropriate ORG statements to place the

```
**********************************************************
* INTERRUPT VECTORS FOR MOTOROLA 68HC11
**********************************************************
            ORG $FFD6
        FDB SCIISR
            ORG $FFE8
        FDB OC1ISR
            ORG $FFF2
        FDB IRQISR
            ORG $FFF6
        FDB SWIISR
            ORG $FFFE
        FDB START
    *
```

Figure 9-35   RTOS interrupt vectors for native 68HC11.

```
**********************************************************
* INTERRUPT VECTORS FOR MOTOROLA EVB
**********************************************************
* VECTORS FOR EVB
            ORG $00C4
        JMP SCIISR
            ORG $00DF
        JMP OC1ISR
            ORG $00EE
        JMP IRQISR
            ORG $00F4
        JMP SWIISR
    *
```

Figure 9-36   RTOS interrupt vectors for Motorola M68HC11EVB trainer.

code at correct memory addresses. Any special directives required by your assembler must be added. Finally, you must define all symbols, including those for the I/O registers, not explicitly defined. The programs presented here should then run without any difficulty.

9.5 REVIEW

The operating system developed in this chapter illustrates operating system principles without the complexity of commercial operating systems. It also provides a practical approach to the development of commercial embedded-computer applications. The weather station example illustrates many principles and programming techniques without undo complexity.

The weather station software could be built many other ways using more or fewer tasks. One goal here was to illustrate how a working application can be built with the functions of the application separated into tasks. When the system presented here runs on the Motorola M68HC11EVM trainer with a 2-MHz E-clock rate, the free-time indicator shows about 90% free time regardless of wind speeds from zero to 99 MPH.

You are encouraged to expand the weather station by adding more features such as temperature monitoring. As more tasks are added to the system, additional weather station features should have little impact on the existing software. The ability to easily expand the software is one of the lessons of structuring the software as an operating system and tasks.

9.6   EXERCISES

9-1.   Why do concurrency problems not occur with location RUNNING?

9-2.   When does the weather station program use the stack at address 00FF?

9-3.   Would scheduling multiple periodic tasks so that they run at the same clock tick be a better idea than running on different clock ticks?

9-4.   Do the weather station tasks ever run on the same clock tick (OC1 interrupt), or is their scheduling such that they never coincide?

9-5.   Why bother having a system service to read the system clock? Why not just have a task read the three bytes from location CLOCK?

9-6.   If a task goes into an infinite loop with no possible exit, what effect does this loop have on higher-priority tasks? On lower-priority tasks?

9-7.   Consider a system with a Task3 that is scheduled by an external interrupt. What is the effect on the operating system if the external interrupts arrive so fast that the task cannot finish executing before another interrupt arrives?

9-8.   Add a new task that frequently displays, on the CRT, the current state of each task in the weather station. Also provide an input switch to enable and disable this task from displaying its message. Explain the significance of the message displayed.

9-9.   What is the effect of scheduling a task with a time delay of zero?

9-10.   Add a system service to the operating system that will make a specified task ready immediately, and then call the dispatcher.

9-11.   Complete the SCI receiver programming, and then add the ability to type "reset" on the keyboard to zero the highest wind speed.

9-12.   Add programming to the RTOS to use the COP timer. Would putting the two parts of the COP reset in the background task and in the OC1 interrupt service routine be a good idea?

9-13.   Does the operating system look for higher-priority tasks to run after each character is sent to the SCI transmitter?

9-14.   What would be the effect on the system performance if the dispatcher was not run each time a character was transmitted by the SCI device?

9-15.   Modify the weather station program to display the wind speed in miles per hour or kilometers per hour based on the position of a switch.

9-16.   Carefully look at the initializations for the weather station. When the system is started up, which task will run first, assuming that the anemometer is not turning and has not turned since the computer was powered-up?

9-17.   Consider the free-time indicator. Is the time required for the interrupt system to save, and then to restore, the microprocessor registers, counted against free-time or against useful work time?

9-18. While the weather station is operating with a constant (frequency) wind speed signal generated in the laboratory, the wind displays flicker back and forth between two different values. Explain why this happens.

9-19. Would a concurrency problem occur with the weather station's wind speed variable if the wind speed was a double-byte number? A triple-byte number?

9-20. Suppose that the weather station software is running on a microcomputer trainer, but the flashing light and the wind speed displays don't change (they hold a single value). How would you go about troubleshooting this problem?

9-21. Run the given weather station program on a microcomputer with the I/O hardware as described. When you observe the flashing light operated by Task6, can you see any variation in its flashing rate as the operating system runs different tasks? The eye is quite sensitive to variations in the rate of flashing.

9-22. Add an analog temperature sensor to your computer hardware, then modify the weather station to display the temperature on the CRT.

9-23. Carefully look at the initializations for the weather station. If the anemometer is turning before and at a hardware reset of the computer, which task will run first?

9-24. Change the initialization of all task variables to be within the tasks.

9-25. The initialization program section in Figure 9-4 does not set the time-to-run entries in the TCBs. What effect does this have on the operating system, especially at system startup?

9-26. Can TCBs, SCBs, and DCBs be placed either into ROM or into RAM?

9-27. Devise a method of setting the system clock to a chosen time-of-day at system startup. If the anemometer causes an interrupt at the same time that OC1 causes an interrupt, will the time measured by Task0 include or not include the latest 10-millisecond period?

9-28. If the operating system is running normally, and a task uses the SETCLK function to set the clock to a later time-of-day than the current time, what effect will this have on tasks scheduled with delay scheduling? time-of-day scheduling?

9-29. Make an output driven by code in the OC1 ISR that makes a pulse on each interrupt. Then, use an oscilloscope, triggered by this pulse, to observe the free-time indicator. The triggering will be necessary for the oscilloscope to provide a steady and useful display. Make observations about the system free-time.

9-30. Watch how the messages on the CRT are displayed at system startup. Explain what you see.

Appendix A

The 68HC11 Instruction Set

The instruction set of a computer is the collection of all the codes that are recognized as valid instructions by the microprocessor. The instruction set table lists all the valid op codes and the details of the corresponding instruction. The instruction set table is the principal aid to machine language and assembly language programmers. Almost any detail needed to code a program is contained in the instruction set table.

The instruction set table lists the instructions in alphabetical order by mnemonic name. Unfortunately, alphabetical order means you must know the name of an instruction before you can find it in the table. The names of most instructions strongly imply the operation performed, so a little creative guessing will help you find the desired instruction quickly.

The reverse instruction set table lists the valid instructions in op code numerical order. The reverse instruction set is valuable for analyzing a program. For example, if you entered erroneous codes into memory and ran them as a program, strange things may have happened. To determine what did happen, trace the program by determining what each number did when it was used as an instruction.

The op code maps in this appendix illustrate the valid and invalid codes graphically. The gray areas in the maps represent invalid codes.

A.1 INSTRUCTION SET TABLE AND PROGRAMMING MODEL

Columns of instruction set table

Source form	instruction mnemonic and any operands required
Operation	word description of the function performed
Addr. mode	microprocessor register if appropriate and memory addressing mode
Op code	one or two byte op codes in hexadecimal format
Operand	format and use of the operand bytes
Bytes	number of bytes of memory occupied by the instruction
Cycles	number (decimal) of E-clock cycles required by the Motorola 68HC11 to fetch and execute the instruction
Condition codes	operation performed on the condition code bit during instruction execution

Source form operand notation

(opr)	operand—data, data address, or offset for a memory reference instruction
(msk)	mask byte—1s in the mask select operand bits
(rel)	relative offset of program relative addressed branch instruction

Operation notation

s	instruction intended for two's complement signed numbers only
u	instruction intended for unsigned numbers only

Boolean Expression

+	add
∨	or
−	subtract
∧	and
⩡	exclusive or
×	8-bit by 8-bit unsigned multiply
/	16-bit by 16-bit divide
( )	the contents of
→	make the contents of destination register the same as the contents of the source register
:	concatenate two registers
0	1-bit number zero
00	8-bit number zero
$FF	8-bit hexadecimal number FF
1	1-bit number 1
+1	add one to the contents of the register
−1	subtract one from the contents of the register

Address	address formed by the instruction
A	the 8-bit contents of accumulator A
B	the 8-bit contents of accumulator B
BCD	binary coded decimal
C	the 1-bit contents of carry/borrow bit
CCR	the 8-bit contents of the condition code register
D	the 16-bit contents of double accumulator D
D7	contents of bit 7 of the double accumulator D
I	the 1-bit contents of the interrupt mask bit
mm	8-bit mask byte—1s in the mask select operand bits
$\overline{mm}$	the complement of the 8-bit mask byte
M	the contents of the memory register addressed by the instruction
$\overline{M}$	the complement of the contents of the memory register
(M + 1)	the contents of the next higher memory register
PC	contents of program counter register
r	16-bit remainder from division
SP	the contents of the stack pointer register
Stk	the contents of the memory register at the top of the stack
V	the 1-bit contents of the two's complement overflow bit
X	the 16-bit contents of index register X
Y	the 16-bit contents of index register Y

Addressing mode

A	The instruction accesses the A accumulator.
B	The instruction accesses the B accumulator.
D	The instruction accesses the D accumulator.
DIR	direct addressing
EXT	extended addressing
IMM	immediate addressing
IND,X	indexed by X addressing
IND,Y	indexed by Y addressing
INH	inherent addressing
REL	program relative addressing
S	The instruction accesses the stack pointer register.
X	The instruction accesses the X index register.
Y	The instruction accesses the Y index register.

Instruction operand notation

ii	8-bit immediate data
dd	low byte of a direct address
hh ll	high and low bytes of an extended address
ff	unsigned 8-bit offset in indexed addressed instruction
jj kk	high and low bytes of 16-bit immediate data
mm	8-bit mask byte—1s in the mask select operand bits
rr	signed 8-bit relative offset in branch instruction

Cycles

3	requires three E-clock cycles for fetch and execute
?	depends on external hardware signals

Condition code bit notation

-	Bit is unaffected by this instruction.
0	Bit is always cleared to 0 by instruction.
1	Bit is always set to 1 by instruction.
↕	Bit is set or cleared depending on instruction.
↓	Bit can change from 1 to 0 but not 0 to 1, or can remain at either 1 or 0.

Motorola 68HC11 Programming Model

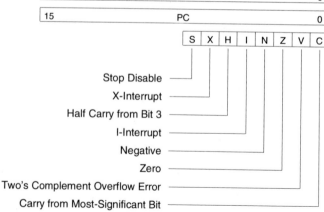

Motorola 68HC11 Instruction Set

Source Form	Operation	Boolean Expression	Addr. Mode	Op Code	Operand	Bytes	Cycles	S	X	H	I	N	Z	V	C
ABA	Add Accumulators	$A + B \rightarrow A$	INH	1B		1	2	-	-	↕	-	↕	↕	↕	↕
ABX	Add B to X	$X + 00{:}B \rightarrow X$	INH	3A		1	3	-	-	-	-	-	-	-	-
ABY	Add B to Y	$Y + 00{:}B \rightarrow Y$	INH	18 3A		2	4	-	-	-	-	-	-	-	-
ADCA (opr)	Add with Carry to A	$A + M + C \rightarrow A$	A IMM	89	ii	2	2	-	-	↕	-	↕	↕	↕	↕
			A DIR	99	dd	2	3								
			A EXT	B9	hh ll	3	4								
			A IND,X	A9	ff	2	4								
			A IND,Y	18 A9	ff	3	5								
ADCB (opr)	Add with Carry to B	$B + M + C \rightarrow B$	B IMM	C9	ii	2	2	-	-	↕	-	↕	↕	↕	↕
			B DIR	D9	dd	2	3								
			B EXT	F9	hh ll	3	4								
			B IND,X	E9	ff	2	4								
			B IND,Y	18 E9	ff	3	5								
ADDA (opr)	Add Memory to A	$A + M \rightarrow A$	A IMM	8B	ii	2	2	-	-	↕	-	↕	↕	↕	↕
			A DIR	9B	dd	2	3								
			A EXT	BB	hh ll	3	4								
			A IND,X	AB	ff	2	4								
			A IND,Y	18 AB	ff	3	5								
ADDB (opr)	Add Memory to B	$B + M \rightarrow B$	B IMM	CB	ii	2	2	-	-	↕	-	↕	↕	↕	↕
			B DIR	DB	dd	2	3								
			B EXT	FB	hh ll	3	4								
			B IND,X	EB	ff	2	4								
			B IND,Y	18 EB	ff	3	5								
ADDD (opr)	Add Memory to D	$D + M{:}(M+1) \rightarrow D$	D IMM	C3	jj kk	3	4	-	-	-	-	↕	↕	↕	↕
			D DIR	D3	dd	2	5								
			D EXT	F3	hh ll	3	6								
			D IND,X	E3	ff	2	6								
			D IND,Y	18 E3	ff	3	7								
ANDA (opr)	And Memory to A	$A \wedge M \rightarrow A$	A IMM	84	ii	2	2	-	-	-	-	↕	↕	0	-
			A DIR	94	dd	2	3								
			A EXT	B4	hh ll	3	4								
			A IND,X	A4	ff	2	4								
			A IND,Y	18 A4	ff	3	5								
ANDB (opr)	And Memory to B	$B \wedge M \rightarrow B$	B IMM	C4	ii	2	2	-	-	-	-	↕	↕	0	-
			B DIR	D4	dd	2	3								
			B EXT	F4	hh ll	3	4								
			B IND,X	E4	ff	2	4								
			B IND,Y	18 E4	ff	3	5								
ASL (opr)	Arithmetic Shift Left Memory		EXT	78	hh ll	3	6	-	-	-	-	↕	↕	↕	↕
			IND,X	68	ff	2	6								
			IND,Y	18 68	ff	3	7								
ASLA	Arithmetic Shift Left A		A INH	48		1	2	-	-	-	-	↕	↕	↕	↕
ASLB	Arithmetic Shift Left B		B INH	58		1	2	-	-	-	-	↕	↕	↕	↕
ASLD	Arithmetic Shift Left D		D INH	05		1	3	-	-	-	-	↕	↕	↕	↕

Source Form	Operation	Boolean Expression	Addr. Mode	Op Code	Operand	Bytes	Cycles	S	X	H	I	N	Z	V	C
ASR (opr)	Arithmetic Shift Right Memory[S]	(shift diagram → C)	EXT	77	hh ll	3	6	-	-	-	-	↕	↕	↕	↕
			IND,X	67	ff	2	6								
			IND,Y	18 67	ff	3	7								
ASRA	Arithmetic Shift Right A[S]		A INH	47		1	2	-	-	-	-	↕	↕	↕	↕
ASRB	Arithmetic Shift Right B[S]		B INH	57		1	2	-	-	-	-	↕	↕	↕	↕
BCC (rel)	Branch if Carry Clear	? C = 0	REL	24	rr	2	3	-	-	-	-	-	-	-	-
BCLR (opr) (msk)	Clear Memory Bit(s)	$M \wedge \overline{mm} \to M$	DIR	15	dd mm	3	6	-	-	-	-	↕	↕	0	-
			IND,X	1D	ff mm	3	7								
			IND,Y	18 1D	ff mm	4	8								
BCS (rel)	Branch if Carry Set	? C = 1	REL	25	rr	2	3	-	-	-	-	-	-	-	-
BEQ (rel)	Branch if = Zero	? Z = 1	REL	27	rr	2	3	-	-	-	-	-	-	-	-
BGE (rel)	Branch if Gr. Than or Equal[S]	? N ⊻ V = 0	REL	2C	rr	2	3	-	-	-	-	-	-	-	-
BGT (rel)	Branch if Greater Than[S]	? Z ∨ (N ⊻ V) = 0	REL	2E	rr	2	3	-	-	-	-	-	-	-	-
BHI (rel)	Branch if Higher[U]	? C ∨ Z = 0	REL	22	rr	2	3	-	-	-	-	-	-	-	-
BHS (rel)	Branch if Higher or Same[U]	? C = 0	REL	24	rr	2	3	-	-	-	-	-	-	-	-
BITA (opr)	Bit(s) Test A with Memory	A ∧ M	A IMM	85	ii	2	2	-	-	-	-	↕	↕	0	-
			A DIR	95	dd	2	3								
			A EXT	B5	hh ll	3	4								
			A IND,X	A5	ff	2	4								
			A IND,Y	18 A5	ff	3	5								
BITB (opr)	Bit(s) Test B with Memory	B ∧ M	B IMM	C5	ii	2	2	-	-	-	-	↕	↕	0	-
			B DIR	D5	dd	2	3								
			B EXT	F5	hh ll	3	4								
			B IND,X	E5	ff	2	4								
			B IND,Y	18 E5	ff	3	5								
BLE (rel)	Branch if Less Than or Equal[S]	? Z ∨ (N ⊻ V) = 1	REL	2F	rr	2	3	-	-	-	-	-	-	-	-
BLO (rel)	Branch if Lower[U]	? C = 1	REL	25	rr	2	3	-	-	-	-	-	-	-	-
BLS (rel)	Branch if Lower or Same[U]	? C ∨ Z = 1	REL	23	rr	2	3	-	-	-	-	-	-	-	-
BLT (rel)	Branch if Less Than[S]	? N ⊻ V = 1	REL	2D	rr	2	3	-	-	-	-	-	-	-	-
BMI (rel)	Branch if Minus[S]	? N = 1	REL	2B	rr	2	3	-	-	-	-	-	-	-	-
BNE (rel)	Branch if Not = Zero	? Z = 0	REL	26	rr	2	3	-	-	-	-	-	-	-	-
BPL (rel)	Branch if Plus[S]	? N = 0	REL	2A	rr	2	3	-	-	-	-	-	-	-	-
BRA (rel)	Branch Always	? 1 = 1	REL	20	rr	2	3	-	-	-	-	-	-	-	-
BRCLR(opr) (msk) (rel)	Branch if Memory Bit(s) Clear	? M ∧ mm = 00	DIR	13	dd mm rr	4	6	-	-	-	-	-	-	-	-
			IND,X	1F	ff mm rr	4	7								
			IND,Y	18 1F	ff mm rr	5	8								
BRN (rel)	Branch Never	? 1 = 0	REL	21	rr	2	3	-	-	-	-	-	-	-	-
BRSET(opr) (msk) (rel)	Branch if Memory Bit(s) Set	? $\overline{M}$ ∧ mm = 00	DIR	12	dd mm rr	4	6	-	-	-	-	-	-	-	-
			IND,X	1E	ff mm rr	4	7								
			IND,Y	18 1E	ff mm rr	5	8								
BSET (opr) (msk)	Set Memory Bit(s)	$M \vee mm \to M$	DIR	14	dd mm	3	6	-	-	-	-	↕	↕	0	-
			IND,X	1C	ff mm	3	7								
			IND,Y	18 1C	ff mm	4	8								
BSR (rel)	Branch to Subroutine	See Text	REL	8D	rr	2	6	-	-	-	-	-	-	-	-
BVC (rel)	Branch if Overflow Clear	? V = 0	REL	28	rr	2	3	-	-	-	-	-	-	-	-
BVS (rel)	Branch if Overflow Set	? V = 1	REL	29	rr	2	3	-	-	-	-	-	-	-	-

Source Form	Operation	Boolean Expression	Addr. Mode	Machine Code Op Code	Operand	Bytes	Cycles	S	X	H	I	N	Z	V	C
CBA	Compare A to B	A – B	INH	11		1	2	-	-	-	-	↕	↕	↕	↕
CLC	Clear Carry Bit	0 → C	INH	0C		1	2	-	-	-	-	-	-	-	0
CLI	Clear Interrupt Mask	0 → I	INH	0E		1	2	-	-	-	0	-	-	-	-
CLR (opr)	Clear Memory Byte	00 → M	EXT	7F	hh ll	3	6	-	-	-	-	0	1	0	0
			IND,X	6F	ff	2	6								
			IND,Y	18 6F	ff	3	7								
CLRA	Clear Accumulator A	00 → A	A INH	4F		1	2	-	-	-	-	0	1	0	0
CLRB	Clear Accumulator B	00 → B	B INH	5F		1	2	-	-	-	-	0	1	0	0
CLV	Clear Overflow Flag	0 → V	INH	0A		1	2	-	-	-	-	-	-	0	-
CMPA (opr)	Compare A to Memory	A – M	A IMM	81	ii	2	2	-	-	-	-	↕	↕	↕	↕
			A DIR	91	dd	2	3								
			A EXT	B1	hh ll	3	4								
			A IND,X	A1	ff	2	4								
			A IND,Y	18 A1	ff	3	5								
CMPB (opr)	Compare B to Memory	B – M	B IMM	C1	ii	2	2	-	-	-	-	↕	↕	↕	↕
			B DIR	D1	dd	2	3								
			B EXT	F1	hh ll	3	4								
			B IND,X	E1	ff	2	4								
			B IND,Y	18 E1	ff	3	5								
COM (opr)	Complement Memory Byte	$FF – M → M	EXT	73	hh ll	3	6	-	-	-	-	↕	↕	0	1
			IND,X	63	ff	2	6								
			IND,Y	18 63	ff	3	7								
COMA	Complement A	$FF – A → A	A INH	43		1	2	-	-	-	-	↕	↕	0	1
COMB	Complement B	$FF – B → B	B INH	53		1	2	-	-	-	-	↕	↕	0	1
CPD (opr)	Compare D to Memory	D – M:(M + 1)	D IMM	1A 83	jj kk	4	5	-	-	-	-	↕	↕	↕	↕
			D DIR	1A 93	dd	3	6								
			D EXT	1A B3	hh ll	4	7								
			D IND,X	1A A3	ff	3	7								
			D IND,Y	CD A3	ff	3	7								
CPX (opr)	Compare X to Memory	X – M:(M + 1)	X IMM	8C	jj kk	3	4	-	-	-	-	↕	↕	↕	↕
			X DIR	9C	dd	2	5								
			X EXT	BC	hh ll	3	6								
			X IND,X	AC	ff	2	6								
			X IND,Y	CD AC	ff	3	7								
CPY (opr)	Compare Y to Memory	Y – M:(M + 1)	Y IMM	18 8C	jj kk	4	5	-	-	-	-	↕	↕	↕	↕
			Y DIR	18 9C	dd	3	6								
			Y EXT	18 BC	hh ll	4	7								
			Y IND,X	1A AC	ff	3	7								
			Y IND,Y	18 AC	ff	3	7								
DAA	Decimal Adjust A	Adjust Sum to BCD	A INH	19		1	2	-	x	-	-	↕	↕	↕	↕
DEC (opr)	Decrement Memory Byte	M – 1 → M	EXT	7A	hh ll	3	6	-	-	-	-	↕	↕	↕	-
			IND,X	6A	ff	2	6								
			IND,Y	18 6A	ff	3	7								
DECA	Decrement Accumulator A	A – 1 → A	A INH	4A		1	2	-	-	-	-	↕	↕	↕	-
DECB	Decrement Accumulator B	B – 1 → B	B INH	5A		1	2	-	-	-	-	↕	↕	↕	-
DES	Decrement Stack Pointer	SP – 1 → SP	S INH	34		1	3	-	-	-	-	-	-	-	-
DEX	Decrement Index Register X	X – 1 → X	X INH	09		1	3	-	-	-	-	-	↕	-	-

Source Form	Operation	Boolean Expression	Addr. Mode	Op Code	Operand	Bytes	Cycles	S	X	H	I	N	Z	V	C
DEY	Decrement Index Register Y	Y − 1 → Y	Y INH	18 09		2	4	-	-	-	-	-	↕	-	-
EORA (opr)	Exclusive OR A with Memory	A ∨ M → A	A IMM	88	ii	2	2	-	-	-	-	↕	↕	0	-
			A DIR	98	dd	2	3								
			A EXT	B8	hh ll	3	4								
			A IND,X	A8	ff	2	4								
			A IND,Y	18 A8	ff	3	5								
EORB (opr)	Exclusive OR B with Memory	B ∨ M → B	B IMM	C8	ii	2	2	-	-	-	-	↕	↕	0	-
			B DIR	D8	dd	2	3								
			B EXT	F8	hh ll	3	4								
			B IND,X	E8	ff	2	4								
			B IND,Y	18 E8	ff	3	5								
FDIV	Fractional Divide 16 by 16[U]	D/X → X; r → D*	INH	03		1	41	-	-	-	-	-	↕	↕	↕
IDIV	Integer Divide 16 by 16[U]	D/X → X; r → D*	INH	02		1	41	-	-	-	-	-	↕	0	↕
INC (opr)	Increment Memory Byte	M + 1 → M	EXT	7C	hh ll	3	6	-	-	-	-	↕	↕	↕	-
			IND,X	6C	ff	2	6								
			IND,Y	18 6C	ff	3	7								
INCA	Increment Accumulator A	A + 1 → A	A INH	4C		1	2	-	-	-	-	↕	↕	↕	-
INCB	Increment Accumulator B	B + 1 → B	B INH	5C		1	2	-	-	-	-	↕	↕	↕	-
INS	Increment Stack Pointer	SP + 1 → SP	S INH	31		1	3	-	-	-	-	-	-	-	-
INX	Increment Index Register X	X + 1 → X	X INH	08		1	3	-	-	-	-	-	↕	-	-
INY	Increment Index Register Y	Y + 1 → Y	Y INH	18 08		2	4	-	-	-	-	-	↕	-	-
JMP (opr)	Jump	Address → PC	EXT	7E	hh ll	3	3	-	-	-	-	-	-	-	-
			IND,X	6E	ff	2	3								
			IND,Y	18 6E	ff	3	4								
JSR (opr)	Jump to Subroutine	See Text	DIR	9D	dd	2	5	-	-	-	-	-	-	-	-
			EXT	BD	hh ll	3	6								
			IND,X	AD	ff	2	6								
			IND,Y	18 AD	ff	3	7								
LDAA (opr)	Load Accumulator A	M → A	A IMM	86	ii	2	2	-	-	-	-	↕	↕	0	-
			A DIR	96	dd	2	3								
			A EXT	B6	hh ll	3	4								
			A IND,X	A6	ff	2	4								
			A IND,Y	18 A6	ff	3	5								
LDAB (opr)	Load Accumulator B	M → B	B IMM	C6	ii	2	2	-	-	-	-	↕	↕	0	-
			B DIR	D6	dd	2	3								
			B EXT	F6	hh ll	3	4								
			B IND,X	E6	ff	2	4								
			B IND,Y	18 E6	ff	3	5								
LDD (opr)	Load Accumulator D	M:(M + 1) → D	D IMM	CC	jj kk	3	3	-	-	-	-	↕	↕	0	-
			D DIR	DC	dd	2	4								
			D EXT	FC	hh ll	3	5								
			D IND,X	EC	ff	2	5								
			D IND,Y	18 EC	ff	3	6								
LDS (opr)	Load Stack Pointer	M:(M + 1) → SP	S IMM	8E	jj kk	3	3	-	-	-	-	↕	↕	0	-
			S DIR	9E	dd	2	4								
			S EXT	BE	hh ll	3	5								
			S IND,X	AE	ff	2	5								
			S IND,Y	18 AE	ff	3	6								

Source Form	Operation	Boolean Expression	Addr. Mode	Op Code	Operand	Bytes	Cycles	S	X	H	I	N	Z	V	C
LDX (opr)	Load Index Register X	$M:(M+1) \rightarrow X$	X IMM	CE	jj kk	3	3	-	-	-	-	$\updownarrow$	$\updownarrow$	0	-
			X DIR	DE	dd	2	4								
			X EXT	FE	hh ll	3	5								
			X IND,X	EE	ff	2	5								
			X IND,Y	CD EE	ff	3	6								
LDY (opr)	Load Index Register Y	$M:(M+1) \rightarrow Y$	Y IMM	18 CE	jj kk	4	4	-	-	-	-	$\updownarrow$	$\updownarrow$	0	-
			Y DIR	18 DE	dd	3	5								
			Y EXT	18 FE	hh ll	4	6								
			Y IND,X	1A EE	ff	3	6								
			Y IND,Y	18 EE	ff	3	6								
LSL (opr)	Logical Shift Left Memory Byte		EXT	78	hh ll	3	6	-	-	-	-	$\updownarrow$	$\updownarrow$	$\updownarrow$	$\updownarrow$
			IND,X	68	ff	2	6								
			IND,Y	18 68	ff	3	7								
LSLA	Logical Shift Left A		A INH	48		1	2	-	-	-	-	$\updownarrow$	$\updownarrow$	$\updownarrow$	$\updownarrow$
LSLB	Logical Shift Left B		B INH	58		1	2	-	-	-	-	$\updownarrow$	$\updownarrow$	$\updownarrow$	$\updownarrow$
LSLD	Logical Shift Left D		D INH	05		1	3	-	-	-	-	$\updownarrow$	$\updownarrow$	$\updownarrow$	$\updownarrow$
LSR (opr)	Logical Shift Right Memory Byte		EXT	74	hh ll	3	6	-	-	-	-	0	$\updownarrow$	$\updownarrow$	$\updownarrow$
			IND,X	64	ff	2	6								
			IND,Y	18 64	ff	3	7								
LSRA	Logical Shift Right A		A INH	44		1	2	-	-	-	-	0	$\updownarrow$	$\updownarrow$	$\updownarrow$
LSRB	Logical Shift Right B		B INH	54		1	2	-	-	-	-	0	$\updownarrow$	$\updownarrow$	$\updownarrow$
LSRD	Logical Shift Right D		D INH	04		1	3	-	-	-	-	0	$\updownarrow$	$\updownarrow$	$\updownarrow$
MUL	Multiply A by B[U]	$A \times B \rightarrow D, D7 \rightarrow C$	INH	3D		1	10	-	-	-	-	-	-	-	$\updownarrow$
NEG (opr)	Negate Memory Byte[S]	$00 - M \rightarrow M$	EXT	70	hh ll	3	6	-	-	-	-	$\updownarrow$	$\updownarrow$	$\updownarrow$	$\updownarrow$
			IND,X	60	ff	2	6								
			IND,Y	18 60	ff	3	7								
NEGA	Negate Accumulator A[S]	$00 - A \rightarrow A$	A INH	40		1	2	-	-	-	-	$\updownarrow$	$\updownarrow$	$\updownarrow$	$\updownarrow$
NEGB	Negate Accumulator B[S]	$00 - B \rightarrow B$	B INH	50		1	2	-	-	-	-	$\updownarrow$	$\updownarrow$	$\updownarrow$	$\updownarrow$
NOP	No Operation		INH	01		1	2	-	-	-	-	-	-	-	-
ORAA (opr)	OR A with Memory	$A \vee M \rightarrow A$	A IMM	8A	ii	2	2	-	-	-	-	$\updownarrow$	$\updownarrow$	0	-
			A DIR	9A	dd	2	3								
			A EXT	BA	hh ll	3	4								
			A IND,X	AA	ff	2	4								
			A IND,Y	18 AA	ff	3	5								
ORAB (opr)	OR B with Memory	$B \vee M \rightarrow B$	B IMM	CA	ii	2	2	-	-	-	-	$\updownarrow$	$\updownarrow$	0	-
			B DIR	DA	dd	2	3								
			B EXT	FA	hh ll	3	4								
			B IND,X	EA	ff	2	4								
			B IND,Y	18 EA	ff	3	5								
PSHA	Push A onto Stack	$A \rightarrow Stk, SP-1 \rightarrow SP$	A INH	36		1	3	-	-	-	-	-	-	-	-
PSHB	Push B onto Stack	$B \rightarrow Stk, SP-1 \rightarrow SP$	B INH	37		1	3	-	-	-	-	-	-	-	-
PSHX	Push X onto Stack	$X \rightarrow Stk, SP-2 \rightarrow SP$	X INH	3C		1	4	-	-	-	-	-	-	-	-
PSHY	Push Y onto Stack	$Y \rightarrow Stk, SP-2 \rightarrow SP$	Y INH	18 3C		2	5	-	-	-	-	-	-	-	-

Source Form	Operation	Boolean Expression	Addr. Mode	Op Code	Operand	Bytes	Cycles	S	X	H	I	N	Z	V	C
PULA	Pull A from Stack	SP+1→SP,Stk→A	A INH	32		1	4	-	-	-	-	-	-	-	-
PULB	Pull B from Stack	SP+1→SP,Stk→B	B INH	33		1	4	-	-	-	-	-	-	-	-
PULX	Pull X from Stack	SP+2→SP,Stk→X	X INH	38		1	5	-	-	-	-	-	-	-	-
PULY	Pull Y from Stack	SP+2→SP,Stk→Y	Y INH	18 38		2	6	-	-	-	-	-	-	-	-
ROL (opr)	Rotate Left Memory Byte	(rotate left diagram)	EXT	79	hh ll	3	6	-	-	-	-	↕	↕	↕	↕
			IND,X	69	ff	2	6								
			IND,Y	18 69	ff	3	7								
ROLA	Rotate Left A		A INH	49		1	2	-	-	-	-	↕	↕	↕	↕
ROLB	Rotate Left B		B INH	59		1	2	-	-	-	-	↕	↕	↕	↕
ROR (opr)	Rotate Right Memory Byte	(rotate right diagram)	EXT	76	hh ll	3	6	-	-	-	-	↕	↕	↕	↕
			IND,X	66	ff	2	6								
			IND,Y	18 66	ff	3	7								
RORA	Rotate Right A		A INH	46		1	2	-	-	-	-	↕	↕	↕	↕
RORB	Rotate Right B		B INH	56		1	2	-	-	-	-	↕	↕	↕	↕
RTI	Return from Interrupt	See Text	INH	3B		1	12	↕	↓	↕	↕	↕	↕	↕	↕
RTS	Return from Subroutine	See Text	INH	39		1	5	-	-	-	-	-	-	-	-
SBA	Subtract B from A	A − B → A	INH	10		1	2	-	-	-	-	↕	↕	↕	↕
SBCA (opr)	Subtract with Carry from A	A − M − C → A	A IMM	82	ii	2	2	-	-	-	-	↕	↕	↕	↕
			A DIR	92	dd	2	3								
			A EXT	B2	hh ll	3	4								
			A IND,X	A2	ff	2	4								
			A IND,Y	18 A2	ff	3	5								
SBCB (opr)	Subtract with Carry from B	B − M − C → B	B IMM	C2	ii	2	2	-	-	-	-	↕	↕	↕	↕
			B DIR	D2	dd	2	3								
			B EXT	F2	hh ll	3	4								
			B IND,X	E2	ff	2	4								
			B IND,Y	18 E2	ff	3	5								
SEC	Set Carry	1 → C	INH	0D		1	2	-	-	-	-	-	-	-	1
SEI	Set Interrupt Mask	1 → I	INH	0F		1	2	-	-	-	1	-	-	-	-
SEV	Set Overflow Flag	1 → V	INH	0B		1	2	-	-	-	-	-	-	1	-
STAA (opr)	Store Accumulator A	A → M	A DIR	97	dd	2	3	-	-	-	-	↕	↕	0	-
			A EXT	B7	hh ll	3	4								
			A IND,X	A7	ff	2	4								
			A IND,Y	18 A7	ff	3	5								
STAB (opr)	Store Accumulator B	B → M	B DIR	D7	dd	2	3	-	-	-	-	↕	↕	0	-
			B EXT	F7	hh ll	3	4								
			B IND,X	E7	ff	2	4								
			B IND,Y	18 E7	ff	3	5								
STD (opr)	Store Accumulator D	D → M:(M + 1)	D DIR	DD	dd	2	4	-	-	-	-	↕	↕	0	-
			D EXT	FD	hh ll	3	5								
			D IND,X	ED	ff	2	5								
			D IND,Y	18 ED	ff	3	6								
STOP	Stop Internal Clocks		INH	CF		1	2	-	-	-	-	-	-	-	-
STS (opr)	Store Stack Pointer	SP → M:(M + 1)	S DIR	9F	dd	2	4	-	-	-	-	↕	↕	0	-
			S EXT	BF	hh ll	3	5								
			S IND,X	AF	ff	2	5								
			S IND,Y	18 AF	ff	3	6								

Source Form	Operation	Boolean Expression	Addr. Mode	Op Code	Operand	Bytes	Cycles	S	X	H	I	N	Z	V	C
STX (opr)	Store Index Register X	X → M:(M + 1)	X DIR	DF	dd	2	4	-	-	-	-	↕	↕	0	-
			X EXT	FF	hh ll	3	5								
			X IND,X	EF	ff	2	5								
			X IND,Y	CD EF	ff	3	6								
STY (opr)	Store Index Register Y	Y → M:(M + 1)	Y DIR	18 DF	dd	3	5	-	-	-	-	↕	↕	0	-
			Y EXT	18 FF	hh ll	4	6								
			Y IND,X	1A EF	ff	3	6								
			Y IND,Y	18 EF	ff	3	6								
SUBA (opr)	Subtract Memory from A	A – M → A	A IMM	80	ii	2	2	-	-	-	-	↕	↕	↕	↕
			A DIR	90	dd	2	3								
			A EXT	B0	hh ll	3	4								
			A IND,X	A0	ff	2	4								
			A IND,Y	18 A0	ff	3	5								
SUBB (opr)	Subtract Memory from B	B – M → B	B IMM	C0	ii	2	2	-	-	-	-	↕	↕	↕	↕
			B DIR	D0	dd	2	3								
			B EXT	F0	hh ll	3	4								
			B IND,X	E0	ff	2	4								
			B IND,Y	18 E0	ff	3	5								
SUBD (opr)	Subtract Memory from D	D – M:(M+1) → D	D IMM	83	jj kk	3	4	-	-	-	-	↕	↕	↕	↕
			D DIR	93	dd	2	5								
			D EXT	B3	hh ll	3	6								
			D IND,X	A3	ff	2	6								
			D IND,Y	18 A3	ff	3	7								
SWI	Software Interrupt	See Text	INH	3F		1	14	-	-	-	1	-	-	-	-
TAB	Transfer A to B	A → B	INH	16		1	2	-	-	-	-	↕	↕	0	-
TAP	Transfer A to Condition Codes	A → CCR	INH	06		1	2	↕	↓	↕	↕	↕	↕	↕	↕
TBA	Transfer B to A	B → A	INH	17		1	2	-	-	-	-	↕	↕	0	-
TEST	Test (Only in Test Modes)	Addr. Bus Counts	INH	00		1	?	-	-	-	-	-	-	-	-
TPA	Transfer Condition Codes to A	CCR → A	INH	07		1	2	-	-	-	-	-	-	-	-
TST (opr)	Test Memory Byte	M – 00	EXT	7D	hh ll	3	6	-	-	-	-	↕	↕	0	0
			IND,X	6D	ff	2	6								
			IND,Y	18 6D	ff	3	7								
TSTA	Test Accumulator A	A – 00	A INH	4D		1	2	-	-	-	-	↕	↕	0	0
TSTB	Test Accumulator B	B – 00	B INH	5D		1	2	-	-	-	-	↕	↕	0	0
TSX	Transfer Stack Pointer to X	SP + 1 → X	INH	30		1	3	-	-	-	-	-	-	-	-
TSY	Transfer Stack Pointer to Y	SP + 1 → Y	INH	18 30		2	4	-	-	-	-	-	-	-	-
TXS	Transfer X to Stack Pointer	X – 1 → SP	INH	35		1	3	-	-	-	-	-	-	-	-
TYS	Transfer Y to Stack Pointer	Y – 1 → SP	INH	18 35		2	4	-	-	-	-	-	-	-	-
WAI	Wait for Interrupt	Stack Regs, Wait	INH	3E		1	?	-	-	-	-	-	-	-	-
XGDX	Exchange D with X	X → D, D → X	INH	8F		1	3	-	-	-	-	-	-	-	-
XGDY	Exchange D with Y	Y → D, D → Y	INH	18 8F		2	4	-	-	-	-	-	-	-	-

* FDIV and IDIV: Z is set if quotient is zero; C is set if denominator is zero

  FDIV: V is set if denominator is lower or same as numerator

A.2 REVERSE INSTRUCTION SET TABLE

Op Code	Operands	Mnemonic	Addr. Mode	Op Code	Operands	Mnemonic	Addr. Mode
00		TEST	INH	38		PULX	INH
01		NOP	INH	39		RTS	INH
02		IDIV	INH	3A		ABX	INH
03		FDIV	INH	3B		RTI	INH
04		LSRD	INH	3C		PSHX	INH
05		ASLD/LSLD	INH	3D		MUL	INH
06		TAP	INH	3E		WAI	INH
07		TPA	INH	3F		SWI	INH
08		INX	INH	40		NEGA	INH
09		DEX	INH	43		COMA	INH
0A		CLV	INH	44		LSRA	INH
0B		SEV	INH	46		RORA	INH
0C		CLC	INH	47		ASRA	INH
0D		SEC	INH	48		ASLA/LSLA	INH
0E		CLI	INH	49		ROLA	INH
0F		SEI	INH	4A		DECA	INH
10		SBA	INH	4C		INCA	INH
11		CBA	INH	4D		TSTA	INH
12	dd mm rr	BRSET	DIR	4F		CLRA	INH
13	dd mm rr	BRCLR	DIR	50		NEGB	INH
14	dd mm	BSET	DIR	53		COMB	INH
15	dd mm	BCLR	DIR	54		LSRB	INH
16		TAB	INH	56		RORB	INH
17		TBA	INH	57		ASRB	INH
18		Page 2 Switch		58		ASLB/LSLB	INH
19		DAA	INH	59		ROLB	INH
1A		Page 3 Switch		5A		DECB	INH
1B		ABA	INH	5C		INCB	INH
1C	ff mm	BSET	IND,X	5D		TSTB	INH
1D	ff mm	BCLR	IND,X	5F		CLRB	INH
1E	ff mm rr	BRSET	IND,X	60	ff	NEG	IND,X
1F	ff mm rr	BRCLR	IND,X	63	ff	COM	IND,X
20	rr	BRA	REL	64	ff	LSR	IND,X
21	rr	BRN	REL	66	ff	ROR	IND,X
22	rr	BHI	REL	67	ff	ASR	IND,X
23	rr	BLS	REL	68	ff	ASL/LSL	IND,X
24	rr	BCC/BHS	REL	69	ff	ROL	IND,X
25	rr	BCS/BLO	REL	6A	ff	DEC	IND,X
26	rr	BNE	REL	6C	ff	INC	IND,X
27	rr	BEQ	REL	6D	ff	TST	IND,X
28	rr	BVC	REL	6E	ff	JMP	IND,X
29	rr	BVS	REL	6F	ff	CLR	IND,X
2A	rr	BPL	REL	70	hh ll	NEG	EXT
2B	rr	BMI	REL	73	hh ll	COM	EXT
2C	rr	BGE	REL	74	hh ll	LSR	EXT
2D	rr	BLT	REL	76	hh ll	ROR	EXT
2E	rr	BGT	REL	77	hh ll	ASR	EXT
2F	rr	BLE	REL	78	hh ll	ASL/LSL	EXT
30		TSX	INH	79	hh ll	ROL	EXT
31		INS	INH	7A	hh ll	DEC	EXT
32		PULA	INH	7C	hh ll	INC	EXT
33		PULB	INH	7D	hh ll	TST	EXT
34		DES	INH	7E	hh ll	JMP	EXT
35		TXS	INH	7F	hh ll	CLR	EXT
36		PSHA	INH	80	ii	SUBA	IMM
37		PSHB	INH	81	ii	CMPA	IMM

Op Code	Operands	Mnemonic	Addr. Mode
82	ii	SBCA	IMM
83	jj kk	SUBD	IMM
84	ii	ANDA	IMM
85	ii	BITA	IMM
86	ii	LDAA	IMM
88	ii	EORA	IMM
89	ii	ADCA	IMM
8A	ii	ORAA	IMM
8B	ii	ADDA	IMM
8C	jj kk	CPX	IMM
8D	rr	BSR	REL
8E	jj kk	LDS	IMM
8F		XGDX	INH
90	dd	SUBA	DIR
91	dd	CMPA	DIR
92	dd	SBCA	DIR
93	dd	SUBD	DIR
94	dd	ANDA	DIR
95	dd	BITA	DIR
96	dd	LDAA	DIR
97	dd	STAA	DIR
98	dd	EORA	DIR
99	dd	ADCA	DIR
9A	dd	ORAA	DIR
9B	dd	ADDA	DIR
9C	dd	CPX	DIR
9D	dd	JSR	DIR
9E	dd	LDS	DIR
9F	dd	STS	DIR
A0	ff	SUBA	IND,X
A1	ff	CMPA	IND,X
A2	ff	SBCA	IND,X
A3	ff	SUBD	IND,X
A4	ff	ANDA	IND,X
A5	ff	BITA	IND,X
A6	ff	LDAA	IND,X
A7	ff	STAA	IND,X
A8	ff	EORA	IND,X
A9	ff	ADCA	IND,X
AA	ff	ORAA	IND,X
AB	ff	ADDA	IND,X
AC	ff	CPX	IND,X
AD	ff	JSR	IND,X
AE	ff	LDS	IND,X
AF	ff	STS	IND,X
B0	hh ll	SUBA	EXT
B1	hh ll	CMPA	EXT
B2	hh ll	SBCA	EXT
B3	hh ll	SUBD	EXT
B4	hh ll	ANDA	EXT
B5	hh ll	BITA	EXT
B6	hh ll	LDAA	EXT
B7	hh ll	STAA	EXT
B8	hh ll	EORA	EXT
B9	hh ll	ADCA	EXT
BA	hh ll	ORAA	EXT
BB	hh ll	ADDA	EXT
BC	hh ll	CPX	EXT
BD	hh ll	JSR	EXT
BE	hh ll	LDS	EXT

Op Code	Operands	Mnemonic	Addr. Mode
BF	hh ll	STS	EXT
C0	ii	SUBB	IMM
C1	ii	CMPB	IMM
C2	ii	SBCB	IMM
C3	jj kk	ADDD	IMM
C4	ii	ANDB	IMM
C5	ii	BITB	IMM
C6	ii	LDAB	IMM
C8	ii	EORB	IMM
C9	ii	ADCB	IMM
CA	ii	ORAB	IMM
CB	ii	ADDB	IMM
CC	jj kk	LDD	IMM
CD		Page 4 Switch	
CE	jj kk	LDX	IMM
CF		STOP	INH
D0	dd	SUBB	DIR
D1	dd	CMPB	DIR
D2	dd	SBCB	DIR
D3	dd	ADDD	DIR
D4	dd	ANDB	DIR
D5	dd	BITB	DIR
D6	dd	LDAB	DIR
D7	dd	STAB	DIR
D8	dd	EORB	DIR
D9	dd	ADCB	DIR
DA	dd	ORAB	DIR
DB	dd	ADDB	DIR
DC	dd	LDD	DIR
DD	dd	STD	DIR
DE	dd	LDX	DIR
DF	dd	STX	DIR
E0	ff	SUBB	IND,X
E1	ff	CMPB	IND,X
E2	ff	SBCB	IND,X
E3	ff	ADDD	IND,X
E4	ff	ANDB	IND,X
E5	ff	BITB	IND,X
E6	ff	LDAB	IND,X
E7	ff	STAB	IND,X
E8	ff	EORB	IND,X
E9	ff	ADCB	IND,X
EA	ff	ORAB	IND,X
EB	ff	ADDB	IND,X
EC	ff	LDD	IND,X
ED	ff	STD	IND,X
EE	ff	LDX	IND,X
EF	ff	STX	IND,X
F0	hh ll	SUBB	EXT
F1	hh ll	CMPB	EXT
F2	hh ll	SBCB	EXT
F3	hh ll	ADDD	EXT
F4	hh ll	ANDB	EXT
F5	hh ll	BITB	EXT
F6	hh ll	LDAB	EXT
F7	hh ll	STAB	EXT
F8	hh ll	EORB	EXT
F9	hh ll	ADCB	EXT
FA	hh ll	ORAB	EXT
FB	hh ll	ADDB	EXT

Op Code	Operands	Mnemonic	Addr. Mode
FC	hh ll	LDD	EXT
FD	hh ll	STD	EXT
FE	hh ll	LDX	EXT
FF	hh ll	STX	EXT
18 08		INY	INH
18 09		DEY	INH
18 1C	ff mm	BSET	IND,Y
18 1D	ff mm	BCLR	IND,Y
18 1E	ff mm rr	BRSET	IND,Y
18 1F	ff mm rr	BRCLR	IND,Y
18 30		TSY	INH
18 35		TYS	INH
18 38		PULY	INH
18 3A		ABY	INH
18 3C		PSHY	INH
18 60	ff	NEG	IND,Y
18 63	ff	COM	IND,Y
18 64	ff	LSR	IND,Y
18 66	ff	ROR	IND,Y
18 67	ff	ASR	IND,Y
18 68	ff	ASL/LSL	IND,Y
18 69	ff	ROL	IND,Y
18 6A	ff	DEC	IND,Y
18 6C	ff	INC	IND,Y
18 6D	ff	TST	IND,Y
18 6E	ff	JMP	IND,Y
18 6F	ff	CLR	IND,Y
18 8C	jj kk	CPY	IMM
18 8F		XGDY	INH
18 9C	dd	CPY	DIR
18 A0	ff	SUBA	IND,Y
18 A1	ff	CMPA	IND,Y
18 A2	ff	SBCA	IND,Y
18 A3	ff	SUBD	IND,Y
18 A4	ff	ANDA	IND,Y
18 A5	ff	BITA	IND,Y
18 A6	ff	LDAA	IND,Y
18 A7	ff	STAA	IND,Y
18 A8	ff	EORA	IND,Y
18 A9	ff	ADCA	IND,Y

Op Code	Operands	Mnemonic	Addr. Mode
18 AA	ff	ORAA	IND,Y
18 AB	ff	ADDA	IND,Y
18 AC	ff	CPY	IND,Y
18 AD	ff	JSR	IND,Y
18 AE	ff	LDS	IND,Y
18 AF	ff	STS	IND,Y
18 BC	hh ll	CPY	EXT
18 CE	jj kk	LDY	IMM
18 DE	dd	LDY	DIR
18 DF	dd	STY	DIR
18 E0	ff	SUBB	IND,Y
18 E1	ff	CMPB	IND,Y
18 E2	ff	SBCB	IND,Y
18 E3	ff	ADDD	IND,Y
18 E4	ff	ANDB	IND,Y
18 E5	ff	BITB	IND,Y
18 E6	ff	LDAB	IND,Y
18 E7	ff	STAB	IND,Y
18 E8	ff	EORB	IND,Y
18 E9	ff	ADCB	IND,Y
18 EA	ff	ORAB	IND,Y
18 EB	ff	ADDB	IND,Y
18 EC	ff	LDD	IND,Y
18 ED	ff	STD	IND,Y
18 EE	ff	LDY	IND,Y
18 EF	ff	STY	IND,Y
18 FE	hh ll	LDY	EXT
18 FF	hh ll	STY	EXT
1A 83	jj kk	CPD	IMM
1A 93	dd	CPD	DIR
1A A3	ff	CPD	IND,X
1A AC	ff	CPY	IND,X
1A B3	hh ll	CPD	EXT
1A EE	ff	LDY	IND,X
1A EF	ff	STY	IND,X
CD A3	ff	CPD	IND,Y
CD AC	ff	CPX	IND,Y
CD EE	ff	LDX	IND,Y
CD EF	ff	STX	IND,Y

A.3   THE 68HC11 OP CODE MAPS

The gray areas in the op code maps represent invalid codes. The 68HC11 recognizes 307_{10} op codes. Not many invalid single-byte codes are possible, because 235_{10} of the possible 256_{10} values for the first byte of the op code are valid. Therefore, the page 1 map is mostly filled. However, a valid page switch code (18, 1A, CD) forces a double-byte op code resulting in many possible invalid codes in the second byte. Invalid codes lead to large gray areas in the other map pages.

Most-significant digit →

	INH	INH	REL	INH	ACCA	ACCB	IND,X	EXT	IMM	DIR	IND,X	EXT	IMM	DIR	IND,X	EXT	
	0	1	2	3	4	5	6	7	8	9	A	B	C	D	E	F	
0	TEST	SBA	BRA	TSX	NEG				SUB								0
1	NOP	CBA	BRN	INS					CMP								1
2	IDIV	*BRSET*	BHI	PULA					SBC								2
3	FDIV	*BRCLR*	BLS	PULB	COM				SUBD				ADDD				3
4	LSRD	*BSET*	BCC	DES	LSR				AND								4
5	ASLD	*BCLR*	BCS	TXS					BIT								5
6	TAP	TAB	BNE	PSHA	ROR				LDA								6
7	TPA	TBA	BEQ	PSHB	ASR					STA				STA			7
8	INX	PAGE2	BVC	PULX	ASL				EOR								8
9	DEX	DAA	BVS	RTS	ROL				ADC								9
A	CLV	PAGE3	BPL	ABX	DEC				ORA								A
B	SEV	ABA	BMI	RTI					ADD								B
C	CLC	*BSET*	BGE	PSHX	INC				CPX				LDD				C
D	SEC	*BCLR*	BLT	MUL	TST				BSR	JSR			PAGE4	STD			D
E	CLI	*BRSET*	BGT	WAI			JMP		LDS				LDX				E
F	SEI	*BRCLR*	BLE	SWI	CLR				XGDX	STS			STOP	STX			F
	0	1	2	3	4	5	6	7	8	9	A	B	C	D	E	F	

Codes 12 through 15 = DIR
Codes 1C through 1F = IND,X

Op Code Map Page 1 — No prebyte

Most-significant digit →

	INH	IND,Y		INH			IND,Y		IMM	DIR	IND,Y	EXT	IMM	DIR	IND,Y	EXT	
	0	1	2	3	4	5	6	7	8	9	A	B	C	D	E	F	
0				TSY			NEG				SUB				SUB		0
1											CMP				CMP		1
2											SBC				SBC		2
3							COM				SUBD				ADDD		3
4							LSR				AND				AND		4
5				TYS							BIT				BIT		5
6							ROR				LDA				LDA		6
7							ASR				STA				STA		7
8	INY			PULY			ASL				EOR				EOR		8
9	DEY						ROL				ADC				ADC		9
A				ABY			DEC				ORA				ORA		A
B											ADD				ADD		B
C		*BSET*		PSHY			INC		CPY						LDD		C
D		*BCLR*					TST				JSR				STD		D
E		*BRSET*					JMP				LDS				LDY		E
F		*BRCLR*					CLR		XGDY		STS				STY		F
	0	1	2	3	4	5	6	7	8	9	A	B	C	D	E	F	

Codes 1C through 1F = IND,Y

Op Code Map Page 2 — Prebyte = 18

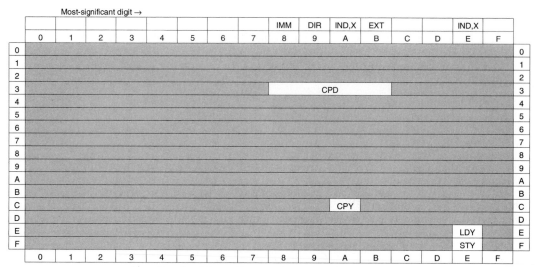

Op Code Map Page 3 — Prebyte = 1A

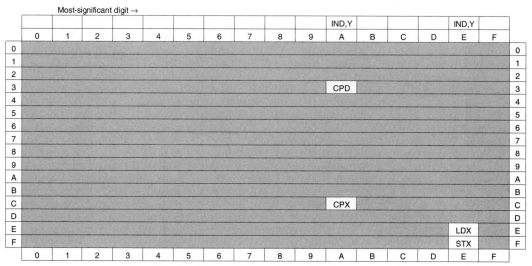

Op Code Map Page 4 — Prebyte = CD

Appendix B

Hardware Control Registers

The 68HC11 microcomputer chip assigns a section of memory for hardware control registers. Most of the registers control input/output hardware. Others control the internal operation of the microcomputer. For example, the highest-priority interrupt is programmable by a control register.

B.1 REGISTER ADDRESSES

The block of memory devoted to hardware control registers extends from address 1000 through address 103F in the 68HC11A8 and 68HC11E9 chips. Other variations may use additional addresses. Because these registers use memory addresses, physical memory cannot also use these addresses.

Software can move the hardware control registers to new locations. Address 1000 is the beginning address of the block of addresses because a reset sets this address. However, the block of control registers can start at any 4K boundary in the memory space. The least-significant four bits of the INIT register at address 103D set the most-significant four bits of the block address. Similarly, the program can move the internal RAM memory to any 4K boundary using the most-significant four bits of INIT. Normally, the RAM begins at address 0000. Therefore, the reset value in INIT is 01.

The INIT register contains only timed-write-once bits that must be set within the first 64_{10} E-clock cycles after a hardware reset. The timed-write-once bits prevent erroneous programs from changing the location of the hardware control registers, which could be dangerous.

B.2 CONTROL REGISTERS AND BITS

The following quick-reference table shows all the control registers and bits. The addresses are those resulting from a hardware reset.

The registers are generally grouped according to the device they are associated with. Some registers are input/output registers while others perform only control functions. Still others contain collections of control bits and flags. Some registers are read only, and others are write only. Reference the appropriate text material to use these bits correctly.

All assigned names and acronyms for both the registers and bits are in the index to this book to facilitate quick reference to the text.

Hardware Control Registers and Bits

Address	Bit 7	Bit 6	Bit 5	Bit 4	Bit 3	Bit 2	Bit 1	Bit 0	Name	Description
$1000	—	—	—	—	—	—	—	—	PORTA	I/O Port A
$1001										Reserved
$1002	STAF	STAI	CWOM	HNDS	OIN	PLS	EGA	INVB	PIOC	Parallel I/O Control Register
$1003	—	—	—	—	—	—	—	—	PORTC	I/O Port C
$1004	—	—	—	—	—	—	—	—	PORTB	Output Port B
$1005	—	—	—	—	—	—	—	—	PORTCL	Alternate Latched Port C
$1006										Reserved
$1007	—	—	—	—	—	—	—	—	DDRC	Data Direction for Port C
$1008			—	—	—	—	—	—	PORTD	I/O Port D
$1009			—	—	—	—	—	—	DDRD	Data Direction for Port D
$100A	—	—	—	—	—	—	—	—	PORTE	Input Port E
$100B	FOC1	FOC2	FOC3	FOC4	FOC5	0	0	0	CFORC	Compare Force Register
$100C	OC1M7	OC1M6	OC1M5	OC1M4	OC1M3	0	0	0	OC1M	OC1 Action Mask Register
$100D	OC1D7	OC1D6	OC1D5	OC1D4	OC1D3	0	0	0	OC1D	OC1 Action Data Register

Hardware Control Registers and Bits

Address	Bit 7	Bit 6	Bit 5	Bit 4	Bit 3	Bit 2	Bit 1	Bit 0	Name	Description
$100E	—	—	—	—	—	—	—	—	TCNT	Timer Counter Register
$100F	—	—	—	—	—	—	—	—		
$1010	—	—	—	—	—	—	—	—	TIC1	Input Capture 1 Register
$1011	—	—	—	—	—	—	—	—		
$1012	—	—	—	—	—	—	—	—	TIC2	Input Capture 2 Register
$1013	—	—	—	—	—	—	—	—		
$1014	—	—	—	—	—	—	—	—	TIC3	Input Capture 3 Register
$1015	—	—	—	—	—	—	—	—		
$1016	—	—	—	—	—	—	—	—	TOC1	Output Compare 1 Register
$1017	—	—	—	—	—	—	—	—		
$1018	—	—	—	—	—	—	—	—	TOC2	Output Compare 2 Register
$1019	—	—	—	—	—	—	—	—		
$101A	—	—	—	—	—	—	—	—	TOC3	Output Compare 3 Register
$101B	—	—	—	—	—	—	—	—		
$101C	—	—	—	—	—	—	—	—	TOC4	Output Compare 4 Register
$101D	—	—	—	—	—	—	—	—		
$101E	—	—	—	—	—	—	—	—	TOC5	Output Compare 5 Register
$101F	—	—	—	—	—	—	—	—		
$1020	OM2	OL2	OM3	OL3	OM4	OL4	OM5	OL5	TCTL1	Timer Control Register 1
$1021	0	0	EDG1B	EDG1A	EDG2B	EDG2A	EDG3B	EDG3A	TCTL2	Timer Control Register 2
$1022	OC1I	OC2I	OC3I	OC4I	OC5I	IC1I	IC2I	IC3I	TMSK1	Timer Interrupt Mask Reg 1
$1023	OC1F	OC2F	OC3F	OC4F	OC5F	IC1F	IC2F	IC3F	TFLG1	Timer Interrupt Flag Reg 1
$1024	TOI	RTII	PAOVI	PAII	0	0	PR1	PR0	TMSK2	Timer Interrupt Mask Reg 2
$1025	TOF	RTIF	PAOVF	PAIF	0	0	0	0	TFLG2	Timer Interrupt Flag Reg 2
$1026	DDRA7	PAEN	PAMOD	PEDGE	0	0	RTR1	RTR0	PACTL	Pulse Accumulator Control Reg
$1027	—	—	—	—	—	—	—	—	PACNT	Pulse Accumulator Count Reg

Hardware Control Registers and Bits

Address	Bit 7	Bit 6	Bit 5	Bit 4	Bit 3	Bit 2	Bit 1	Bit 0	Name	Description
$1028	SPIE	SPE	DWOM	MSTR	CPOL	CPHA	SPR1	SPR0	SPCR	SPI Control Register
$1029	SPIF	WCOL	0	MODF	0	0	0	0	SPSR	SPI Status Register
$102A	—	—	—	—	—	—	—	—	SPDR	SPI Data Register
$102B	TCLR	0	SCP1	SCP0	RCKB	SCR2	SCR1	SCR0	BAUD	SCI Baud Rate Control
$102C	R8	T8	0	M	WAKE	0	0	0	SCCR1	SCI Control Register 1
$102D	TIE	TCIE	RIE	ILIE	TE	RE	RWU	SBK	SCCR2	SCI Control Register 2
$102E	TDRE	TC	RDRF	IDLE	OR	NF	FE	0	SCSR	SCI Status Register
$102F	—	—	—	—	—	—	—	—	SCDR	SCI Data (Rd RDR/ Wr TDR)
$1030	CCF	0	SCAN	MULT	CD	CC	CB	CA	ADCTL	A/D Control Register
$1031	—	—	—	—	—	—	—	—	ADR1	A/D Result Register 1
$1032	—	—	—	—	—	—	—	—	ADR2	A/D Result Register 2
$1033	—	—	—	—	—	—	—	—	ADR3	A/D Result Register 3
$1034	—	—	—	—	—	—	—	—	ADR4	A/D Result Register 4
$1035 to $1038										Reserved
$1039	ADPU	CSEL	IRQE	DLY	CME	0	CR1	CR0	OPTION	Sytem Configuration Options
$103A	—	—	—	—	—	—	—	—	COPRST	Arm/Reset COP Timer Circuitry
$103B	ODD	EVEN	0	BYTE	ROW	ERASE	EELAT	EEPGM	PPROG	EEPROM Programming Control
$103C	RBOOT	SMOD	MDA	IRV	PSEL3	PSEL2	PSEL1	PSEL0	HPRIO	Highest Priority I-Bit Int and Misc
$103D	RAM3	RAM2	RAM1	RAM0	REG3	REG2	REG1	REG0	INIT	RAM and I/O Mapping Register
$103E	TILOP	0	OCCR	CBYP	DISR	FCM	FCOP	TCON	TEST1	Factory Test Control Register
$103F	0	0	0	0	NOSEC	NOCOP	ROMON	EEON	CONFIG	COP, ROM, EEPROM Enable

Appendix C

The Motorola M68HC11EVB Microcomputer Trainer

The Motorola trainer, part number M68HC11EVB, is a single-board computer that requires a DC power supply and a cathode-ray-tube (CRT) terminal. The EVB in the name is an abbreviation for evaluation board. The board contains sufficient read/write RAM memory to hold practical programs. It has a monitor program in permanent ROM memory so that you can control the computer easily. This computer board is a very practical tool for experimenting with and learning the 68HC11 computer. It is called a *trainer* because most people design custom hardware for their applications rather than incorporating this board into the design.

The operation of the Motorola M68HC11EVB2 and S68HC11EVBU trainers is very similar to the operation of the M68HC11EVB trainer, but there are some differences. See the WWW site at WWW.PRENHALL.COM for information on other trainers.

C.1 INTRODUCTION TO THE MOTOROLA TRAINER

Motorola designed the M68HC11EVB microcomputer trainer for making a prototype of a device that will contain a 68HC11 single-chip microcomputer. Learning to use a microcomputer and developing a prototype of a device can both be done with this board.

Many other trainers are available from other companies for Motorola products. All trainers have similar features, but this appendix concentrates on the Motorola 68HC11EVB trainer.

The Trainer Components

A working trainer consists of the microcomputer printed circuit board, a DC power supply, and a CRT terminal—screen display and keyboard. A personal computer can act like a CRT terminal for this purpose if a separate CRT terminal is not available. The following discussion assumes that a CRT terminal is used.

The computer circuit board contains the 68HC11 microcomputer, both read/write and read only memory, and input/output hardware. The read only memory is permanent memory that contains a monitor program named *BUFFALO*. The read/write memory allows you to put your program into memory.

Connect the CRT and trainer board according to the trainer's manual. Further help on setting up this equipment is available on the WWW site at WWW.PRENHALL.COM.

CRT Terminal

A CRT terminal consists of a keyboard and a display device that are independent of each other. The CRT keyboard generates character codes in response to key presses; the codes can be transmitted to a computer. The CRT display receives character codes from a computer and forms character patterns on the screen. Some terminals also can send characters directly from the keyboard to the display.

Incidentally, a CRT terminal display is different from a personal computer monitor. The monitor is essentially just a TV screen that responds to the video signals sent to it. The CRT terminal contains the electronics to form the characters on the screen.

Monitor Program

The BUFFALO monitor program makes the hardware act as a microcomputer trainer. The name is a humorous acronym created by its author who was at a university with a buffalo as its mascot.

Always remember that you are interacting with a program, namely BUFFALO, when you are operating the trainer. The microcomputer responds to the keyboard and controls the screen only if the monitor program is running properly. The monitor program prompts the human operator by placing a right arrow character at the left side of the screen. The prompt means the monitor program is ready to do something for you.

Trainer Startup

The microcomputer must be running in an orderly fashion to make the trainer operational. Turning the power on causes an automatic hardware reset that starts the microcomputer. After the power is on, pushing the reset button on the trainer board forces a hardware reset that acts as if the power was just turned on. Use the reset button only to simulate turning the power on; reset can cause problems if used in other situations.

Pushing and releasing the reset button also starts the monitor program, which prints a message on the screen and leaves the cursor at the end of the message. Following reset, pressing the Enter (Return) key on the keyboard causes the prompt to be printed on the screen. You can feel confident that almost everything in the trainer is working properly if the prompt appears.

Monitor Program Crashes

A *crash* means the program stops operating correctly because something is wrong. Crashing the monitor program is easy. A crash causes no harm to the trainer—pressing the reset button causes it to recover.

A bug in your program is the likely cause of a crash. Your program probably changes something in memory. Your program may destroy itself and the monitor program; therefore, you must check your program carefully before proceeding following a crash.

Don't blame the monitor program for changing your program! After you press the reset button, your program may be gone. The destruction of your program is due most likely to errors in your program. Pressing the reset button does not destroy your program.

C.2   MONITOR PROGRAM FEATURES

The microcomputer trainer will help you write and debug programs. Most microcomputer trainers have similar features regardless of the manufacturer. The monitor program determines most of the characteristics of the trainer. The following lists the principal characteristics of most trainers.

Single-Step

A feature called *single-step* or *trace* executes your program one instruction at a time. After the trainer executes your instruction, the monitor program regains control. You can then single step again, run the program at full speed, or do other things.

Single-step is an excellent tool for learning how instructions work because it shows the effects of the instruction immediately. Seeing the details of how a program works also helps you debug programs easily.

Unfortunately, executing a large group of instructions with single-step requires too many keystrokes, even if you have already checked the instructions. Breakpoints solve this problem.

Breakpoint

A *breakpoint* takes control away from your program and gives it to the monitor program. Your program runs normally to the breakpoint and then breaks out to the monitor program, which displays a message. At a breakpoint, you can use all the monitor commands to check and debug your program. You will frequently want to single-step after a breakpoint is encountered. A command tells the monitor program to return control to your program continuing it from the current point until the next breakpoint.

When using breakpoints, you run segments of your program individually to prove that each segment works correctly. Each segment may contain only a few instructions or hundreds of instructions. Using breakpoints saves you much tedious work.

Download

Downloading is the transmission of binary numbers from a host computer to the memory of the trainer. The word *down* implies that the host computer, probably a personal computer system, is a bigger computer than the trainer. The most common communications medium is a cable connected from the personal computer to the trainer. Programs run on the host computer generate the binary program and data information for the trainer.

This appendix is based on the assumption that you have a personal computer connected to your trainer according to the manufacturer's manual and appropriate software to transmit the download information to the trainer. Further help on setting up this equipment is available on the WWW site at WWW.PRENHALL.COM.

Help Screen

A help screen is one CRT screen of information that describes the program that is in use. Help screens usually document the commands the program recognizes. The operator of the trainer must request the help information through a command. A help screen is a kind of computer manual.

C.3   WRITING PROGRAMS FOR THE MOTOROLA TRAINER

You must follow a few rules when writing programs to run in the Motorola M68HC11EVB trainer. Your program must correctly use the trainer hardware, and it must avoid memory used by the monitor program. The monitor program resides at memory addresses E000 through FFFF inclusively. It also uses read/write memory addresses 0036 through 00FF to hold data that it needs in responding to your commands.

Stopping Your Program

Your program must stop in a way compatible with the monitor program. A *STOP* instruction, op code 3F, at the end of your program will stop your program most easily. The *STOP* instruction is actually an SWI instruction, but is called *STOP* in italics here to emphasize its function in the trainer. The *STOP* instruction transfers control from your program to the monitor program, which then displays the registers. This transfer effectively stops your program but does not stop the computer.

Using Available Memory

Your program may only use the read/write memory locations with addresses 0000 through 0035 and C000 through DFFF. If your trainer has the optional extra memory installed, you also may use locations 6000 through 7FFF.

Don't use memory locations where memory is not installed. If you use nonexistent memory for data, your program may run but the results will be meaningless. Your program will read and use unknown data values from the nonexistent memory locations.

Do not ignore the memory restrictions and assume that you can use memory locations 0036 through 00FF; these locations do exist but the monitor program uses them.

Changing the contents of memory locations 0036 through 00FF probably will crash the monitor program, or, worse yet, cause it to work incorrectly without crashing.

C.4   OPERATING THE MICROCOMPUTER TRAINER

Other than using the reset button to start up the trainer, you use the CRT terminal's keyboard to control all the trainer functions. The trainer responds to commands with messages on the CRT screen. If a message does not appear, your program may have crashed the monitor program, or it may be in an infinite loop.

Entering Monitor Program Commands

Commands entered on the CRT terminal keyboard control the Motorola trainer. The general format of a command is a character string followed by a space followed by one or more numbers separated by spaces. Some commands do not require any numbers. All numbers entered at the keyboard or printed on the screen by the monitor program are hexadecimal numbers. You can omit leading zeros in numbers. Type commands immediately to the right of the prompt and, for most commands, end them with the Enter (Return) key.

Seeing the help screen

Pressing the Enter key twice immediately after the BUFFALO message appears displays the help screen. At all other times, the H command displays the help screen. The help screen lists all possible commands.

Command format

A *command* is a symbolic name followed by symbolic modifiers or numbers. Spaces normally separate the symbols and numbers, but commas work too. The spaces are necessary for all commands, so the discussion below omits mention of the required spaces.

Canceling commands

Pressing the delete key will immediately cancel the function of any command. Some keyboards may have two delete keys, but only one of them may work correctly.

Freezing the screen

Typing Control W—hold the Control key while pressing W—for *wait* will freeze the screen so that it quits printing new characters. Press any key to continue displaying new information.

Repeating commands

Pressing the Enter key without typing anything else will repeat the last command entered. This repetition is very confusing if misunderstood, and very useful otherwise.

The Monitor Program Commands

This section describes some of the BUFFALO monitor commands. The commands covered here are the only commands you should need. The help screen displays more commands than listed here, but the other commands are unnecessary if you have the equipment discussed in this book. The required equipment is an assembler program and a personal computer with a cable connection to the trainer, and software to do downloads.

Assembler/Disassembler   ASM address

This very limited assembler/disassembler command is unnecessary because a complete assembler program should be available on your host computer.

Block Fill   BF address1 address2 data

Block fill puts the *data* value into all memory locations specified by the range from *address1* through and including *address2*. If read/write memory is not at the locations specified, you will get a *rom*-error message.

Breakpoint Insert   BR address

Enters a breakpoint *address* into the breakpoint address table and then lists the current four breakpoint addresses.

Run your program after you have inserted breakpoints at addresses containing instruction op codes. If your program encounters a breakpoint location during the fetch of an instruction, the break occurs. Control goes to the monitor program that displays the microprocessor registers. Then use the monitor commands to check your program. To continue from the breakpoint, use the P command.

When a break occurs, the instruction at the breakpoint location has not yet been executed.

The breakpoint will be transparent to the user. That is, examination of the breakpoint location shows no change when you insert or remove a breakpoint.

Whenever a prompt is displayed, the Breakpoint Insert or Breakpoint Remove commands will modify the breakpoint addresses.

Only put breakpoints at read/write memory locations. Don't use the P command to proceed from a breakpoint set on an SWI or *STOP* instruction—op code 3F.

Pressing the reset button while breakpoints are in place will remove all breakpoints.

Be careful when you use the reset button because the monitor program may alter the numbers at the breakpoint locations.

Reset will be necessary if your program gets into an infinite loop that does not correctly include the breakpoint location.

Breakpoints set at locations that do not contain instruction op codes may cause your program to malfunction.

Breakpoint Remove   BR – or BR –*address*

The breakpoint remove command followed by a minus sign removes all breakpoints. The breakpoint remove command followed by a minus sign and an *address* removes only that address from the breakpoint table.

Bulk Erase EEPROM   BULK

Erases the entire EEPROM memory that is inside the 68HC11 chip. The EEPROM is memory locations B600 through and including B7FF. After erasing, these locations will contain FF. In some versions of the monitor program, this command is of little use because erasing is not necessary to reprogram the EEPROM. In other versions, the EEPROM must be erased using bulk before the move command will correctly program it.

Bulk Erase All   BULKALL

WARNING: DO NOT USE BULKALL, because it is not useful for training purposes. In some versions of the monitor program, BULKALL will switch the trainer to a version of the monitor program in internal memory that is probably obsolete. To check whether this switch has occurred, carefully read the BUFFALO message after pressing the reset button. If (INT) follows the word BUFFALO, reprogram the CONFIG register, address 103F, to contain 0D. See the memory modify command.

Call   CALL *address*

The trainer executes a subroutine only. The *address* is the execution address of the subroutine. A return from subroutine instruction in your subroutine returns control to the monitor program.

Go   G *address*

Run the program starting at the *address* specified. If you omit the *address*, the monitor program uses the current value of the program counter it displays with a register modify command.

The program will not proceed if the location specified by the address has a breakpoint or contains a *STOP* instruction—op code 3F.

Starting the computer at an address of nonexistent memory probably will crash the computer.

Help   H

Display the help screen that lists all monitor commands.

Load   LOAD FILE

Download a binary program into the memory of the trainer from the host computer system. The monitor program prints the word FILE on the next line after the command and puts the cursor on the word FILE. The trainer will wait for the host computer to start the download. After the download is complete and correct, the trainer prints the word *done* on the following line.

If the download does not occur for some reason, press the reset button to regain control of the trainer. Otherwise, the trainer continues to wait for a download.

The trainer cannot download directly to the EEPROM. See the move command.

The information downloaded must be in the Motorola S19 format. The 2500AD assembler (and linker) can generate a file in the correct format.

If the file contains load information that specifies an invalid memory location, the monitor program displays an error message.

Memory Display   MD *address1 address2*

Show the contents of a block of memory from *address1* through and including *address2*. The display format is lines of three regions that contain addresses, data numbers, and characters. First, each line starts with a memory address. Second, the data is 16_{10} bytes that are the numbers in 16_{10} memory registers beginning with the address at the left. Third, 16_{10} ASCII characters that correspond to the memory bytes. The monitor program substitutes a space if there is no corresponding ASCII character.

If you omit *address2*, the screen displays nine lines beginning at *address1*.

If you omit both addresses, the screen displays nine lines beginning at the last memory location accessed.

Memory Modify   MM *address*

Display the contents of the memory location specified by *address* and leave the cursor at the right of the number. Whether hardware for that memory location exists is not considered, so the number displayed could be garbage.

Three subcommands can follow memory modify. First, the Enter key will stop the memory modify. Second, Control J, or cursor down, will examine the next memory location. Third, Control H, or backspace, will examine the previous memory location. Typing a number following *address* and before a subcommand puts that number into the memory location specified.

Always check your entries before proceeding. Since the Enter key repeats the last command, it is easy to restart the memory modify from the beginning by pressing Enter. After restarting the command, use Control J to inspect and correct your entries.

If you try to enter numbers into nonexistent or permanent read only memory, you will get a *rom*-error message.

Memory modify will erase and reprogram individual memory locations in the EEPROM.

Move Memory   MOVE *address1 address2 address3*

Copies the block of memory locations *address1* through and including *address2* to the locations beginning at *address3*. Omitting *address3* moves the block one location higher.

Move memory programs the EEPROM by copying read/write memory to it. In some versions of the monitor program, erasing the EEPROM using the bulk command will be necessary before move will work correctly.

Moving to nonexistent or permanent read only memory does not cause an error message.

Do not modify the CONFIG register (address 103F). See the bulkall command.

Proceed   P

Proceed or continue program execution after encountering a breakpoint. If P is used after a trace command, it works the same as the G command.

Register Display   RD

Register display is a second name for the register modify command.

Register Modify   RM

Modify the contents of the internal 68HC11 registers P, X, Y, A, B, C, and S. The command displays two lines. The first lists the contents of the 68HC11 internal registers. The second line is P-*number* where *number* is the contents of the program counter.

After register modify displays the registers, two subcommand keys are effective. First, the Enter key ends the command. Second, each press of the space bar shows the next register on the next line. Typing a number before the subcommand changes the contents of the register most recently displayed to that value.

A variation of the register modify command is RM *register*, which displays *register* initially. *Register* is the single character designation of a register including P, X, Y, A, B, C, and S.

Trace   T number

Single-step or trace the *number* of instructions. Execution begins at the location specified by the program counter; use register modify to see or change the program counter. After each instruction execution, trace displays two lines. The first line displays Op-*number* where *number* is the first byte of the op code of the instruction that was executed. The second line lists the contents of the 68HC11 internal registers P, X, Y, A, B, C, and S.

Omitting *number* causes the number to default to one and T single-steps only one instruction. You can then repeat trace by pressing only the Enter key.

After single-stepping, you can continue the program at full speed by using the G or P commands.

Messages from the Monitor Program

When error conditions occur, the monitor program will print a message on the next line following the command. The message often requires considerable interpretation to understand how it applies to your circumstances.

What?

You typed an invalid command. Type a new and correct command.

rom-

You have tried to change read only memory that cannot change. The message implies that the monitor program did not store your number, which is not what happens. Actually, the monitor program stores your number and then reads it back. The *rom*-message is displayed if the number read back is different from the number stored. The monitor program stores all the bits of your number that the hardware can accept.

Many different situations cause this message including an actual read only memory register and a register that has some read/write bits and some permanent bits. A defective bit in a read/write memory register that reads incorrectly for the desired number also causes the error message. If the defective bit reads correctly by accident, no message would result.

You can confirm this method of error checking. Examine an actual read only memory location, and then store the same value into it.

Command?

Incorrectly entered hexadecimal digits are the usual cause for this message. For example, entering FZ as a hexadecimal number causes the monitor program to ask if you are trying to input a command.

Bad argument

Usually the error will be apparent. You probably entered a command with incorrect arguments or parameters. Be careful because many commands substitute default values for omitted arguments. Such substitution is not an error to the monitor program.

Too Long

You probably entered a command name with more than eight characters. This message may occur for other reasons also, but the problem is usually obvious.

done

The download from the host computer system finished successfully. You may now use your program, which is in the trainer memory.

error address

An error occurred at the *address* during downloading from the personal computer. The error was probably due to a download into nonexistent memory, read only memory, or to the EEPROM. If so, rewrite your program and download the corrected version.

checksum error

An error occurred during the downloading of information from the personal computer system. Check the cable connections between the computers and try again.

No host port available

A hardware failure has occurred.

C.5   SUGGESTIONS ON USING THE TRAINER

Debugging programs is easy if you use the features available in the trainer. You must be patient and use the trainer effectively. Testing begins with the assumption that your program has errors; then you must check its performance in an orderly manner. Running a faulty program over and over with poorly thought out changes inserted haphazardly will waste time.

The single-step and breakpoint features make checking the operation of a program and finding problems easy. Check every new program with these tools.

The fastest approach to get a program working is to begin by single-stepping it. You must examine the memory and microprocessor registers that are important to the program after each step.

If the program has a loop, you may have some confidence in the loop after single-stepping around it once. If so, put a breakpoint at the beginning of the loop. Then use the P command to run the loop once at full speed. If the results are good, continue to run the loop with the P command. When the loop exits, you can single-step through the next section. The loop may require too many iterations for this approach to be practical. If so, remove the breakpoint and put a new one just beyond the loop. Then use P again to run the rest of the loop at full speed.

By using single-step and breakpoint, you can check the entire program in a few minutes and find the mistakes. You will waste no time at all. On the contrary, just starting up a new program and finding out that it doesn't work provides almost no information—you knew it wouldn't work, but you still don't know the problem. Running it probably will destroy the information that you entered and crash the monitor program in the trainer. So running it is a waste of time. Slow, careful, and orderly techniques to programming and testing programs is the quickest way to success.

Using Interrupts

If you have not yet learned about the hardware interrupt system, skip this section and return to it when you want to use interrupts.

Running a program using interrupts and using the monitor program at the same time are very difficult. For example, tracing or single-stepping a program while interrupts are occurring is troublesome. However, you can avoid any problems by executing parts of your program with the interrupt system disabled and by using breakpoints.

For example, you can run the initialization part of your program with a breakpoint set at the CLI instruction. The breakpoint prevents the CLI from enabling the interrupt system, but the initialization will be correct. Do not proceed with the P command because that will execute the CLI instruction. Instead, you can run the main program loop using the G command with breakpoints set in the loop. Next, you can run an interrupt service routine with a breakpoint at the RTI instruction. The breakpoint prevents the RTI from executing, but the entire interrupt service routine can be executed and tested. The goal here is to avoid enabling the interrupt system. Therefore, if your interrupt service routine contains a CLI instruction, replace it with a NOP during this test.

The recommendations here avoid using the trace feature because it uses interrupt and the programmable timer device. If your program uses the timer, interactions will occur that can be very confusing. The timer is also used by the proceed command, so it too must be avoided.

The section on trainer hardware provides further information on setting up the interrupt vectors for your program. Read it carefully before using interrupts.

C.6   A PRACTICE SESSION WITH THE TRAINER

The following tutorial teaches you to use the Motorola M68HC11EVB trainer. The procedures are for people who have very little knowledge of the microcomputer instruction set; therefore, you can learn to use the trainer before learning much about the 68HC11 computer.

The practice session that follows only covers using the trainer with a CRT terminal. If you are using a personal computer (with terminal-emulation software) as a terminal, when the tutorial refers to the terminal, interpret *terminal* to mean the personal computer used as a terminal.

Turn on the Trainer

Begin by turning on the power to the CRT terminal; then turn on the power to the trainer. The DC power supply for the trainer may have an indicator light to show that it is working.

BUFFALO Message

When the trainer powers up, it will print a message on the CRT screen identifying the BUFFALO monitor program. It then positions the cursor at the right end of this character string.

You next need the carriage return key that is usually labeled Enter. On some keyboards the label on the Enter key is the word *Return* or a bent arrow. Others have both a Return and an Enter key. On most keyboards, these two keys will have the same effect on the monitor program.

Press the Enter key on the keyboard to cause the monitor program to begin normal operation. The monitor program shows that it is responding by printing a right arrow character at the left side of the screen. The right arrow character, called a *prompt*, alerts or prompts the human operator. The prompt means that the trainer is ready for a command. The appearance of the prompt also means that most of the trainer works properly.

Reset the Trainer

Some mistakes crash the monitor program, and the trainer will apparently stop working. You recover from these mistakes by pressing the reset button on the trainer circuit board. A reset acts like you just turned on the power. Press the reset button now, and then press the Enter key again.

Next, press the Enter key a second time. This time, and in this circumstance only, pressing the Enter key causes the monitor program to display the help screen. The help screen is a command summary. It shows all the commands that the monitor program understands; you won't need all of them.

The prompt should have returned at the left side of the screen. You may now type in other commands.

Displaying and Changing Memory Contents

You need to access memory to examine and change the stored numbers. Consider read/write memory—called *RAM*—first. Initially, unknown numbers are in the RAM memory. During power up, the flip flops in the memory ICs will flip and flop however they like. This flip flopping leaves a strange pattern of ones and zeros in the memory accurately called *garbage*. The pattern is not random, since it depends on the characteristics of the transistors in the memory ICs. If the trainer is turned off and then back on, most of the bits will come back to the same pattern. Some bits will change because the heat generated due to the operation of the memory chip changes the transistor characteristics.

Display memory

The trainer displays the contents of memory in several different ways. Let's begin with the simplest. Type the memory display command MD C000 and press Enter—always press Enter to delimit the commands discussed here. The MD command prints nine lines of information with three regions to each line. First, the number at the left of each line is a beginning address for the next region. The second region is 16_{10} bytes of memory contents in hexadecimal format. The third region is 16_{10} ASCII characters corresponding to the memory bytes. If there is no character for a byte value, the monitor program substitutes a space.

Remember, pressing only the Enter key repeats this command. Try it now.

You select a different size block of memory by specifying a beginning and ending address. Type MD C005 C103. The display now contains more locations than you requested, but it includes all the specified locations.

Now let's try a different way to display memory. Type MM C100. Next press the cursor down key if you have one, or hold the Control key down and press J to make a Control J. Moving the cursor down repeatedly shows successive locations in a vertical format. End this command by pressing the Enter key. If you forget to use cursor down and press Enter too soon, remember to repeat your original command by pressing the Enter key.

The MM command also allows you to go backward through locations; that is, show the contents of successively lower addresses. Type MM C100 again. Next, repeatedly press the backspace key or Control H and end with the Enter key.

You can display memory that doesn't exist, but be careful. The values you get may be garbage because the memory is not there. However, displaying nonexistent memory is even more complicated than that. The hardware in the trainer cannot distinguish all possible addresses; therefore, several different addresses actually access the same memory location. If you type MD 200, you will get values that include ASCII characters for the alphabet. Now type MD 300, and then MD 400. You get the same values because the hardware cannot distinguish these addresses; they all refer to the same memory hardware.

Change memory

The MM command will also change the contents of memory registers. Type MM C100 to display the contents of memory location C100. The cursor will be at the right end of the contents. Now type 23 and press the Enter key—the monitor program stores the number 23 at location C100. Demonstrate this by typing MM C100 once again to look at the contents.

Displaying and Changing Microprocessor Registers

The trainer displays the microprocessor registers together, and then modifies them individually. The values displayed are the values put into the registers when the monitor program transfers control to your program. So, if you run your program, it will start with the displayed values in the microprocessor registers. The number in the program counter register determines where your program must be in memory.

If you have already run your program and it stopped, the register values that are displayed were the numbers in the registers when your program stopped and returned control to the monitor program. They are not the values in the registers at the instant that you type the command. The monitor program is running and using the registers, so it saves a copy of your program registers.

Enter a Register Modify command by typing RM and look at the message on the display. The first line shows seven registers, each identified by a single letter name, and the numbers in those registers. Since some registers hold 8-bit numbers and others hold 16-bit numbers, the numbers have either two or four hexadecimal digits.

The next line shows the program counter register and its contents. If you now press the Enter key, the RM command is finished. If instead you type a new number and then press Enter, the RM command puts that number into the program counter. Change the program counter to C100, and then type RM again to see the new value.

Now let's put a number into the X register. Type RM X. The first line will look the same as before, but the second line shows the X register and its contents. Type 321 and press Enter to put this number into the X register. Type RM to see the results.

Entering First Practice Program

Now let's enter a simple program to demonstrate some features of the trainer. The example was selected for people who have just started learning the microcomputer instructions and addressing modes. You do not need to understand the instructions to follow this example.

The first example program has four instructions. The program loads the A accumulator with 22, loads the A accumulator with 33, loads the A accumulator with 44, and then jumps back to the beginning. So it is an infinite loop that never stops.

Always carefully check entered codes, since errors are likely and incorrect codes might destroy your program when you run it.

Enter the program in Figure C-1 into the memory using the memory modify command. Type MM C100, and then each number followed by cursor down or Control J to access the next sequential memory location. When you have reached the end of the required values, press the Enter key to end the memory modify command.

After entering the program, use the MD command to look at all your numbers to be sure that they are perfectly correct. Correct any errors with the memory modify command.

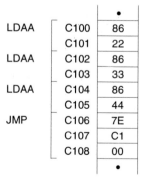

		•
LDAA	C100	86
	C101	22
LDAA	C102	86
	C103	33
LDAA	C104	86
	C105	44
JMP	C106	7E
	C107	C1
	C108	00
		•

Figure C-1   First practice program.

Running the First Practice Program

Running a program means placing the address of the first instruction into the program counter and then starting the computer. The GO or G command runs a program. It puts a number into the program counter and transfers control from the monitor program to your program, effectively starting the computer.

When using the G command to run a program, carefully check what you have typed before you press Enter. If you have typed the command incorrectly, your program may be destroyed.

Type G C100 to run your program starting at address C100. Correct typing errors by backspacing and reentering characters. Backspacing does not erase the original character, but typing a new character will overwrite the old one.

While the example program is running, the keyboard and display will not respond, because your program is an infinite loop and it has complete control of the computer. The monitor program is not running, so the terminal does not respond. To regain control of the trainer by making the monitor program run again, press the reset button. You should get the BUFFALO message.

You should learn an important lesson from running this infinite loop program. The infinite loop prevents the CRT terminal from responding while the program is working properly. So, if you ever have the trainer stop working, you should investigate the chance that you have a program with an infinite loop. Don't immediately assume that the trainer has failed.

Resetting the trainer will not change your program numbers. Use the memory display command to see that your program still is in memory; type MD C100. If your program is gone, you made a mistake. Don't blame the trainer or the monitor program.

Here is a second way to run your program. First, manually enter the address of the first instruction into the program counter register using the register modify command; type RM, press Enter, and then type C100. Now type G alone (you can also use GO) to run your program. The G command without a number starts the computer at the address that is already in the program counter.

Running the Second Practice Program

Enter the program in Figure C-2 into the trainer memory. This program has a *STOP* instruction at the end instead of the jump in the previous example. The *STOP* instruction transfers control from your program back to the monitor program. It does not stop the computer from running—it stops your program from running.

When the *STOP* instruction sends control to the monitor program, the program automatically displays the microprocessor registers as the register modify command does. The display after a *STOP* shows the program counter containing the address of the *STOP* instruction. Normally the address in the program counter advances to the next instruction.

Run your program by typing G C200 and read the register contents. The first instruction in the program put the number 22 into the A accumulator, the second put 33 into the A accumulator, and the third put 1234 into the X index register. The program counter P should

		•
LDAA	C200	86
	C201	22
LDAA	C202	86
	C203	33
LDX	C204	CE
	C205	12
	C206	34
STOP	C207	3F
INVALID	C208	00
		•

Figure C-2   Second practice program.

have C207, which is the address of the *STOP* instruction. You should look carefully at the numbers listed for each register.

Experiment with this program by changing the data numbers in the instruction codes. For example, change the program so that it ends with 44 in A and 6789 in X.

Using Single-Step

The single-step or trace command requires that the program counter be set to the address of the instruction to be executed. Let's use the program in Figure C-2. Enter the program into memory or check it if you already have it in memory.

Use the register modify command to put C200 into the program counter to set up the single-step. Then type the T command and press the Enter key. The T command executes only one instruction and then it displays the instruction op code and the 68HC11 internal registers. After using T once, you should have an op code of 86, the number C202 in the program counter, and 22 in the A register. The program counter is ready for the next instruction at address C202.

Continue by executing two more instructions using the trace command twice more. You could do this by typing T, Enter, T, and Enter again. However, remember that pressing Enter by itself repeats the last command. So type T, Enter, and then just Enter again. Now the program counter should contain C207.

Next, single-step the *STOP* instruction by typing T again. Single-stepping the *STOP* does nothing but advance to the next instruction in order, so be careful.

You can easily single-step beyond the end of your program, which can cause problems. For example, the number in the next register after the *STOP* instruction is an invalid instruction code. Use T to single-step once more. The trainer will go dead because the invalid instruction physically stops the trainer from running. You must press the reset button to restart the trainer. However, this error will not cause anything in memory to be lost.

Let's do a multiple trace from the beginning of the program. First, load the program counter with C200 using the register modify command. Then type T 3 to single-step three instructions at once.

Finally, consider making an error. Set your P register to C201, and then single-step the trainer. Of course, location C201 does not contain the op code of your instruction.

Using Breakpoints

Breakpoints make your program transfer control to the monitor program at selected places. By breaking out of your program, you can check whether it is working correctly up to the breakpoint.

Let's use the program in Figure C-1 again. Set a breakpoint on the second instruction by typing BR C102. The monitor program will display the breakpoint table, which contains four addresses. The first will be C102, the address of your breakpoint, and then three zeros, which mean there are no more breakpoints. If this is not true, press reset and try again. The reset will remove any breakpoints that have been left in place.

Now run the program from the beginning by typing G C100. When the program gets to the breakpoint, the monitor program gets control and displays the register modify message. The program counter should contain C102, which is the location of the breakpoint. Notice that the instruction at location C102 has not yet been executed, because the program counter contains C102.

Now continue running your program by typing the P command for proceed from breakpoint. The program will run through the loop and come back to the breakpoint; the monitor program displays the same message again. It will appear that nothing has happened, but the loop did run.

Let's add a second breakpoint to make the operation easier to observe. Put a breakpoint at address C106 by typing BR C106. Now type P and see that a break occurs at address C106. Type P again and a break will occur at address C102. Look at the contents of all the registers as you continue to use the P command several times.

To prove that it can be done, single-step a few instructions with the T command and then type P to proceed to the next breakpoint.

Next let's remove the breakpoints. First, remove the breakpoint at address C102 by typing BR –C102. The minus sign means remove that breakpoint. To see the effect, type BR alone to display the breakpoint table. Now type BR – without a number to remove all breakpoints. Again type BR alone to see the breakpoint table.

Breakpoint problems

All trainers that have breakpoint capability can cause problems and confusion if you make certain errors. So let's look more carefully at breakpoints.

Use the program in Figure C-2 and put breakpoints at addresses C200 and C204. To be sure the breakpoints are correctly in place, type BR alone and check the breakpoint table. Now type MD C200 and look at memory locations C200 and C204. Your instructions should be shown as if the breakpoints did nothing.

The monitor program makes a breakpoint by putting a 1-byte instruction at the breakpoint address. The breakpoint instruction replaces your instruction, but the monitor program saves a copy of your instruction. However, the monitor program puts the breakpoint instruction into the program only when you run the program, and it puts your instruction back when the program stops due to a breakpoint or *STOP* instruction. This replacement of your instruction is why you don't see any changes when using the memory display command while breakpoints are set. The breakpoints are transparent to the user.

Incorrectly located breakpoint

Unfortunately, transparent features, such as breakpoint, can be very confusing when something goes wrong. To demonstrate this, remove any breakpoints by typing BR – and then BR alone to be sure. Now put a breakpoint at address C203. This breakpoint is incorrect because location C203 does not contain the op code of an instruction.

Now run the program by typing G C200 and observe that the program stops at address C207, and the breakpoint appears to do nothing. However, the breakpoint instruction was put into place as you can see by looking at the contents of the A accumulator register. The LDAA instruction loaded 3F into the A register instead of the correct data 33.

To be sure that the problem is clear, type MD C200 and look at the second LDAA instruction to see that the instruction appears to be correct even though it loaded the number 3F into the A register.

If you ever have a program that seems to work correctly until you put in the break-points, an incorrectly located breakpoint is likely to be the problem.

An incorrectly located breakpoint is a serious problem because it prevents your program from working correctly. The problem is also very difficult to find because the instruction appears to be correct, but it works incorrectly when the program runs. Of course, if the breakpoint is erroneously located outside your program and data area, then it will not affect your program and a break will never occur.

Reset while breakpoint is in place

Use the program in Figure C-3 for this example, but press reset before entering it into memory to be sure that the monitor program is initialized. The program is the same as that in Figure C-1 except that the jump instruction jumps to itself making a tight infinite loop.

Put in a correct breakpoint by typing BR C102. Now run the program by typing G C100 and then type the P command to continue from the breakpoint. The program will go into the infinite loop. The only way out of the loop is to press the reset button, so press it. Now display the program by typing MD C100 and then look carefully at the codes. You should see that your instruction at the breakpoint location was not put back and that the breakpoint instruction 3F is still at address C102. Your instruction is permanently lost. You must reenter your instruction using the memory modify command.

The problem is caused by a reset while a breakpoint instruction is in place. The trainer could not put your instruction back. The breakpoint instruction was left in place because the infinite loop did not encounter a *STOP* or a breakpoint.

If you must use the reset button while also using breakpoints, always check to see that your program has not been modified.

When this problem is combined with an incorrectly located breakpoint in a long program, you may have difficulty recognizing it. Always be very careful when entering break-points so that you enter them at the correct addresses. Trying to find a problem in your program when the problem is caused by how you used the trainer can be very frustrating.

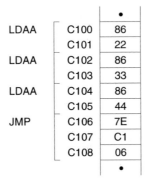

		•
LDAA	C100	86
	C101	22
LDAA	C102	86
	C103	33
LDAA	C104	86
	C105	44
JMP	C106	7E
	C107	C1
	C108	06
		•

Figure C-3   Practice program.

C.7   TRAINER HARDWARE

In using the microcomputer trainer, you always need to know which memory addresses are occupied and what kind of memory is present at each address. In addition, you may need to know the interrupt vector addresses and how the trainer uses these interrupt vectors. The following sections tabulate this information. It also points out details about the input/output ports and the connections to these ports that may cause problems if you ignore certain hardware on the trainer board.

Trainer Memory Map

Table C-1 shows the memory addresses used by the hardware in the Motorola trainer. Each region of memory is designated by a range of addresses. The table lists the type of memory or I/O hardware at those addresses. No memory hardware exists in those regions listed as *not used.*

Trainer Interrupts

Table C-2 shows the mapping of the 68HC11 interrupt vectors. Since the interrupt vector locations in the trainer are in permanent ROM memory, the vector addresses are mapped into a region of read/write RAM memory. That is, an interrupt sends control to a location in RAM. Type **MD FFD6** to do a memory display that shows the addresses of the interrupt vectors. Also try changing one of the interrupt vectors to confirm that they are in permanent ROM memory.

Each vector in read/write memory has three bytes of room allocated. You may put jump instructions into these locations to redirect the interrupts to your interrupt service routines. Unless you are using interrupts in the trainer, you need not be concerned with these jump instructions.

Reset state of jump table

If you have put a jump instruction in the jump table so that you can use an interrupt, the trainer reset will not affect your jump instruction. However, when the trainer is reset and the monitor program finds something other than a jump instruction in the jump table, the monitor program replaces it with a new jump instruction. This replacement occurs when the

TABLE C-1   TRAINER MEMORY MAP

Address Range	Function
0000 0035	54_{10} bytes user RAM in 68HC11
0036 00FF	Monitor RAM in 68HC11
0100 0FFF	Not used
1000 17FF	Port replacement and register decode
1800 3FFF	Not used
4000 5FFF	Flip flop decode—switches SCI input port
6000 7FFF	Optional 8K RAM
8000 B5FF	Not used
B600 B7FF	EEPROM in 68HC11
B800 BFFF	Not used
C000 DFFF	8K bytes user RAM
E000 FFFF	Monitor program EPROM

TABLE C-2   TRAINER INTERRUPT JUMP ADDRESSES

Vector Address	Jump Address	Interrupt Device
FFD6	00C4	SCI Receive Data Register Full
		SCI Receiver Overrun
		SCI Idle Line Detect
		SCI Transmit Data Register Empty
		SCI Transmit Complete
FFD8	00C7	SPI Serial Transfer Complete
FFDA	00CA	Pulse-Accumulator Input Edge
FFDC	00CD	Pulse-Accumulator Overflow
FFDE	00D0	Timer Overflow
FFE0	00D3	Timer Output Compare 5
FFE2	00D6	Timer Output Compare 4
FFE4	00D9	Timer Output Compare 3
FFE6	00DC	Timer Output Compare 2
FFE8	00DF	Timer Output Compare 1
FFEA	00E2	Timer Input Capture 3
FFEC	00E5	Timer Input Capture 2
FFEE	00E8	Timer Input Capture 1
FFF0	00EB	Real Time Interrupt
FFF2	00EE	IRQ (External Pin or Parallel I/O)
FFF4	00F1	XIRQ Pin
FFF6	00F4	SWI
FFF8	00F7	Illegal Opcode Trap
FFFA	00FA	COP Timeout
FFFC	00FD	Clock Monitor Timeout
FFFE	None	Hardware Reset

trainer is powered up. Any new jumps provided by the monitor program all send control to a hardware STOP instruction. So, if you don't change the jump instruction and a corresponding interrupt occurs, the STOP instruction will be executed. The STOP instruction physically stops the microprocessor from running, so the trainer will no longer respond to the keyboard. A reset or an interrupt can restart the computer.

Don't confuse a hardware STOP instruction with the trainer *STOP* function that stops a program.

Breakpoints and STOP

The trainer implements the breakpoint feature by replacing your instruction op code with a software interrupt instruction or SWI. The monitor program saves a copy of the op code and replaces it after the break occurs. The *STOP* instruction used to stop your programs is the SWI instruction, so the *STOP* is effectively a breakpoint. Since the SWI instruction causes the microprocessor registers to be saved into the stack, the monitor program knows all the register contents from your program whenever a break occurs.

Illegal instruction execution

If you have not modified the interrupt jump table and then execute an illegal instruction in a program, the illegal op code interrupt will lead to a hardware STOP instruction. The trainer will go dead. Pressing the reset button will restart it.

User Reset Vector

You may want to use the reset button on the trainer to perform a reset to your program instead of the monitor program. The monitor program can simulate the reset by jumping to your program upon reset.

At the very beginning of the monitor program, the PORTE bit PE0 is examined as an input bit. If the input is logic 0, the monitor program continues normally. If the input is logic 1, control is sent to the beginning of the EEPROM. So you can use the reset button to send control to your program in the EEPROM.

Jumper J4 on the trainer board determines which program will operate upon reset. See the next section on the A/D converter.

Analog-to-Digital Converter

The A/D converter input pins connect directly to the ribbon cable connector with one exception. PORTE bit PE0 is used by the monitor program to decide whether to run the monitor program or a user program following a reset. Jumper J4 connects a 10K-ohm resistor to either ground or +5 VDC to indicate which program to run. When using channel 0 of the A/D converter, you must account for this resistor. The jumper can be removed to disconnect the resistor if you properly take care of selecting the correct program following reset.

SCI Receiver Connection

The input signal to the SCI serial receiver connects to PORTD pin PD0 of the 68HC11. In the Motorola trainer, the signal to pin PD0 passes through an electronic switch. The switch allows two different sources of serial input; namely, the host port serial connector or the 60-pin I/O connector. The switch is controlled by a flip-flop that implements bit 0 of memory location 4000.

Address 4000 consists of only a single bit that can be written but not read. When 1 is stored in this bit, the SCI serial input comes from the host serial port. When 0 is stored in the bit, the SCI serial input comes from the 60-pin I/O connector.

The monitor program sets this switch-controlling bit to 0 when reset is pressed and at other times. Therefore, if you want to use the host port connector with the SCI receiver, your program must put a 1 into bit 0 of address 4000. Using the monitor program to control the bit before running your program won't work.

Furthermore, setting the electronic switch incorrectly can cause confusing problems. Suppose that you connect a serial input device to the host port and nothing to the 60-pin connector. The incorrect switch position directs input from the 60-pin connector. However, crosstalk from the host port may cause noise and framing errors in the receiver. This problem can lead you to think your input device is correctly connected but defective!

Trainer Ribbon Cable Connector

The 60-pin ribbon cable connector on the trainer board makes direct connection of external devices to all 68HC11 input/output port pins possible. Some other control and power pins on the chip are available at the cable connector.

Figure C-4 shows the pin locations for the various signals. Pin 1 is the ground or common for all other signals.

Power supply

One connector pin, pin 26, connects directly to the +5 VDC power supply for the trainer. Therefore, power for the external hardware can be obtained through the ribbon cable connector. Be careful with this power pin—incorrect connections could easily damage your hardware.

GND	1	• •	2	MODB
MODA	3	• •	4	STRA
E	5	• •	6	STRB
EXTAL	7	• •	8	XTAL
PC0	9	• •	10	PC1
PC2	11	• •	12	PC3
PC4	13	• •	14	PC5
PC6	15	• •	16	PC7
RESET	17	• •	18	XIRQ
IRQ	19	• •	20	PD0
PD1	21	• •	22	PD2
PD3	23	• •	24	PD4
PD5	25	• •	26	V_{DD}
PA7	27	• •	28	PA6
PA5	29	• •	30	PA4
PA3	31	• •	32	PA2
PA1	33	• •	34	PA0
PB7	35	• •	36	PB6
PB5	37	• •	38	PB4
PB3	39	• •	40	PB2
PB1	41	• •	42	PB0
PE0	43	• •	44	PE4
PE1	45	• •	46	PE5
PE2	47	• •	48	PE6
PE3	49	• •	50	PE7
V_{RL}	51	• •	52	V_{RH}
NC	53	• •	54	NC
NC	55	• •	56	NC
NC	57	• •	58	NC
NC	59	• •	60	NC

Figure C-4   Ribbon cable connector pin assignments.

MODA and MODB

The MODA and MODB pins have pull-up resistors to select expanded mode. Usually jumper J2 will select the on-board crystal clock, and the EXTAL pin on the cable connector will not be connected.

A/D converter

When using the A/D converter, the external circuits must supply its reference voltage because the V_{RH} and V_{RL} pins are not connected on the trainer board.

IRQ and XIRQ

The IRQ pin has an on-board pull-up resistor, but the XIRQ pin only connects directly to the cable connector.

Hardware reset

The reset pin can reset external circuits as the computer hardware is reset, but jumper J1 can disconnect this signal.

Appendix D

Answers to Selected Exercises

D.1 CHAPTER 1 ANSWERS

1-1. (a) 00010111 + 10001000 = 10011111   No overflow
         (b) 11101100 + 11100000 = 11001100   No overflow
         (e) 00101111 + 01011100 = 10001011   Overflow
         (i) 11001000 + 10111000 = 10000000   No overflow

1-2. (a) FD   (e) 153
         (b) 00110111   (f) 203
         (c) 00110100   (g) 47

1-3. (b) 01000100
         (e) 10000000   Overflow
         (f) 1000001111111111
         (j) 01010100

1-4. (b) −8   (g) −128
         (c) +63   (i) +64
         (e) −32768   (j) −1
         (f) +127

1-5. (b) B5   (f) E7
         (c) 3D   (g) 9E
         (e) AB6C   (j) 6EEF

1-9. No conclusions possible. Borrows mean nothing to two's complement numbers.

1-10. The number C is higher than or larger than number D. Don't use the words *greater than* because the numbers are unsigned.

1-11. 7FFF, 32,767

1-13. Never.

1-15. 0000000001011001, 1111111111000110

1-22. You must convert the 8-bit number to a 16-bit number before adding to get 1101100101101100.

1-23. KB is kilobyte, Kb is kilobit.

1-30. Low is −12 volts and high is 0 volts.

1-32. START = 1 if A = 1 and B = 0.

1-41. A small change in the circuit, such as that caused by temperature variations, may push the system into failure.

1-45. Yes. Be careful when making comparisons.

1-46. In this book, *fetch* always refers to instruction numbers, but some people use the word with both instruction numbers and data numbers.

1-47. For most computers, the answer is no. However, some computers allow the instruction set to be altered.

1-49. The top connection is the electrically safer connection. Furthermore, an inadvertent connection to the common should not cause the circuit to do dangerous things.

D.2   CHAPTER 2 ANSWERS

2-6. 5F, which is CLRB, or C6 00, which is LDAB with zero; 7F C111, which is the CLR instruction with extended addressing.

2-7. Unsigned numbers.

2-8. V will never be 1.

2-10. 89 00, which is ADCA with immediate addressing.

2-12. BHI

2-13. SEC, none that directly affects N.

2-15. 20 FE, which causes an infinite loop that locks up the program.

2-18. The branch instructions do not affect the condition code bits, so multiple branches can be done without resetting the condition codes.

2-19. After an addition of unsigned numbers, the C bit is meaningful and the V bit isn't. After an addition of signed numbers, the V bit is meaningful and the C bit isn't.

2-21. In general, no. The C bit may be different.

2-22. Unless the number is needed in the A accumulator, use the TST instruction.

2-34. Use the ADDD instruction with immediate addressing and data of 0001, namely C3 0001.

2-35. 4, A

2-40. There is neither an advantage nor a disadvantage to this order.

2-42. The first instruction of a program can never have indexed addressing because the index register will contain an unknown number.

2-47. Yes. To prove this, use the circular representation of numbers, but interpret the numbers as unsigned numbers.

D.3   CHAPTER 3 ANSWERS

3-1. LDD #3041

3-2. END

3-3. In most practical programs, the FCB could be used instead of the RMB. However, using RMB conveys the intent that the number in memory after the program is loaded is of no significance. FCB implies that a useful number is to be loaded into memory at the specified location. Be careful to distinguish between load time and run time.

3-4. The same binary instruction codes would be loaded at the new address. The program would function the same as it did. It would work on the same data numbers.

3-5.
(a)	000C	(b)	TEMP
(c)	D300	(d)	0103
(e)	CE D301	(f)	F6 D300
(g)	17	(h)	TEMP
(i)	NEXT	(j)	6A 01
(k)	20 F9	(l)	AGAIN
(m)	C009	(n)	C012
(o)	3	(p)	11
(q)	TYPE	(r)	D300

3-6.
(a)	0C	(b)	Did
(c)	00	(d)	D301
(e)	0C	(f)	Unknown
(g)	No	(h)	14, 19

3-7.
(a)	RMB 2	(b)	D002
(c)	0A	(d)	3C
(e)	FC D002	(f)	THERE
(g)	CE D001	(h)	CLR 0,X
(i)	DOWN	(j)	20 F7
(k)	FD D005	(l)	21
(m)	19	(n)	D002
(o)	D005		

D.4   CHAPTER 4 ANSWERS

4-2. DO-WHILE.

4-4. The second comment line is at the beginning of a lower-level module that is contained within the higher-level module identified by the first comment line.

4-5. Unless the module is a loop, the program likely is incorrectly structured because some part of the program is probably branching to the line with the address label which is not the beginning of this module.

4-16. DO-WHILE.

4-17. SEQUENTIAL.

4-18. DO-WHILE.

4-19. IF-THEN-ELSE.

4-20. SEQUENTIAL.

D.5 CHAPTER 5 ANSWERS

5-1. LSLD, ADCB #0

5-3. No.

5-4. True if no overflow occurs.

5-5. No, the condition code bits are incorrect.

5-7. 00

5-8. C31F, C41E

5-9. 04, A accumulator

5-11. C204

5-15. (a) 06, 04   (b) DFFA
        (c) 60   (d) 08
        (e) 22   (f) None
        (g) 09

5-16. (a) 01   (b) 10000000
        (c) MASK1   (d) BEQ
        (e) RESULT   (f) ORAA
        (g) 11111110

5-18. 12 F4 50 1F, BRSET

5-21. nothing

5-22. true

5-23. PSHA, TBA, PULB, RTS

5-24. (a) 2,X   (b) BGE
        (c) LDAA 0,X   (d) #1
        (e) 4,X   (f) #2
        (g) D000   (h) D10B
        (i) D1FA

D.6 CHAPTER 6 ANSWERS

6-2. Read only memory or ROM.

6-3. Nonvolatile memory is needed for the RESET vector at the highest addresses, and read/write memory is needed in the direct addressing range.

6-4. To be sure that the EPROM will not get erased by stray light.

6-5. No, EEPROMs wear out in relatively few erase and program cycles.

6-7. During execute phase of LDAA, the data bus carries a number into the microprocessor. During the fetch phase of any instruction, the data bus carries numbers into the microprocessor.

6-9.   Execute phase.

6-10.   The 68HC11 can directly address 64 KB of memory without using any switching or paging techniques.

6-12.   Yes, this instruction is a read-modify-write instruction.

6-16.   1FFF

6-17.   4FFF, 4003

6-20.   Yes, because the R/W signal will distinguish them.

6-23.   Yes.

6-24.   All eight bits of the output will be cleared instead of only a single bit. The effects depend on the hardware, if any, controlled by those bits, but good practice is to control only the bits that are in use.

6-27.   Using BCLR to clear STAF has no practical purpose.

6-31.   The computer responds only to the change in the push button because the flag input responds only to transitions. Thereafter, holding the push button does nothing so that the program will not read the thumbwheel switch again.

6-32.   The main program never again executes even a single instruction. The ISR runs repeatedly as fast as it can continuously reading the thumbwheel switch and updating the display. The push button switch has no effect.

6-33.   The main program runs forever. The ISR never runs, so the display is never updated. The push button switch has no effect.

6-34.   The CLI at line 69 causes continuous interrupts after the flag is set because the flag is never cleared. The main program never again runs. The ISR never runs beyond the CLI, so the display is never updated. The push button switch has no effect.

6-35.   Here is a clue. The program reads the signals from the PORTC pins every time ROR is executed.

D.7   CHAPTER 7 ANSWERS

7-1.   Changing modes with software is impossible because such a change could cause a dangerous situation.

7-3.   None.

7-4. 32_{10} microseconds.

7-5.   If an external memory chip was connected to implement addresses beginning at 0000, the internal RAM could be moved so that it is useful.

7-7.   The reset pin is a bidirectional signal pin. External devices can reset the 68HC11 chip, and I/O devices inside the 68HC11 chip can reset external devices.

7-9.   The I bit cannot prevent the SWI instruction from causing an interrupt. The SWI instruction affects the I bit the same as hardware interrupts.

7-10.   The flag will not be cleared, so the main program loop never again executes even a single instruction. The port C bits will toggle as fast as the computer can repeatedly execute the interrupt service routine.

7-17. Yes, this waveform makes an efficient digital-to-analog converter that is easy to make if the device using the waveform provides sufficient filtering. For example, dimming incandescent light bulbs works well.

7-18. Output compare 1 can control the five port A pins PA3 through PA7. The PORTA register has no effect while the output compare hardware is enabled to control the output pins.

D.8   CHAPTER 8 ANSWERS

8-1. All!

8-2. Yes. Real-time does not precisely specify the required response time.

8-3. There is no relationship.

8-4. Yes, with appropriate other changes. Stopping the computer may reduce electromagnetic interference from a lightly loaded system.

8-5. Yes, unless the subroutine is called by only a single task.

8-6. The tasks do not run in a predictable order, so a shared last-in-first-out stack is not useful.

8-7. No.

8-8. Yes.

8-10. Yes, but it introduces new concurrency problems.

D.9   CHAPTER 9 ANSWERS

9-2. Only in the startup initialization.

9-3. Yes, those scheduled periodically have periods chosen so that they sometimes run on the same tick. Task0 that is scheduled by external interrupts likely will hit the same tick as some other task.

9-5. System service functions eliminate concurrency problems.

9-6. No effect on higher-priority tasks. Lower-priority tasks don't run. Some operating systems increase task priorities according to some algorithm so that all tasks get a chance to run sooner or later.

9-9. The task runs at the current time-of-day tomorrow!

9-13. As written in this chapter, the operating system does run the dispatcher after each character.

9-14. Probably not much of an effect, but the system overhead would be reduced a little.

9-19. If instructions that operate on double-byte data numbers were used to access the double-byte wind speed, there would not be concurrency problems because these instructions do not allow interrupts between memory accesses. More than one instruction will be necessary to access triple-byte numbers, so there will certainly be concurrency problems.

9-26. TCBs and DCBs must be in read/write (RAM) memory because the numbers stored in them must change as the system runs. SCBs can be in either RAM or ROM memory depending on the need for the program to alter their contents as the system runs.

9-28. All those tasks currently scheduled to run at times between the actual current time and the new time in the system clock will not run until tomorrow.

General Index